Chinese Medicine for Upper Body Pain

Pain medication is widely used to treat patients who suffer from acute and chronic pain. However, it is not the only treatment option available for managing traumatic and chronic upper body pain. Chinese medicine is a popular option without the risk of adverse side effects that may come with opioid use.

Chinese Medicine for Upper Body Pain further explores other pain management options, introducing treatments such as acupuncture, herbal medicine, and Tui Na (Chinese bodywork). This book provides concise explanations in terms of anatomy, pathology, and etiology for both Western and Chinese medicine, and also provides self-care instructions that are effective and easy to follow.

Key Features:

- Introduces treatments in Chinese medicine including acupuncture with traditional methods such as acupuncture point selection and its modifications, electroacupuncture, point bleeding, auricular medicine, cupping, and moxibustion; herbal formulas with their sources and modifications and Tui Na therapy.
- Recommends dietary, exercise, and self-care protocols for patients.
- Includes photos for reference and ease of understanding.

This book is suitable for self-care practitioners and students of Chinese medicine practices as well as patients who want to better understand and mitigate their pain.

Along with *Chinese Medicine for Lower Body Pain* (ISBN: 9780367235857), the author covers the use of Chinese medicine for pain management for the complete body.

Chinese Medicine for Upper Body Pain

Benjamin Apichai

CRC Press
Taylor & Francis Group
Boca Raton London New York

CRC Press is an imprint of the
Taylor & Francis Group, an **informa** business

Cover image credit: Benjamin Apichai

First edition published 2023
by CRC Press
6000 Broken Sound Parkway NW, Suite 300, Boca Raton, FL 33487-2742

and by CRC Press
4 Park Square, Milton Park, Abingdon, Oxon, OX14 4RN

CRC Press is an imprint of Taylor & Francis Group, LLC

Library of Congress Cataloging-in-Publication Data
Names: Apichai, Benjamin, author.
Title: Chinese medicine for upper body pain / Benjamin Apichai.
Description: First edition. | Boca Raton : CRC Press, 2023. | Includes
bibliographical references and index.
Identifiers: LCCN 2022045397 (print) | LCCN 2022045398 (ebook) | ISBN
9781032066004 (paperback) | ISBN 9781032066011 (hardback) | ISBN
9781003203018 (ebook)
Subjects: LCSH: Pain--Alternative treatment. | Medicine, Chinese. |
Medicine, Chinese--Formulae, receipts, prescriptions. | Acupuncture.
Classification: LCC RB127 .A65 2023 (print) | LCC RB127 (ebook) | DDC
615.8/92--dc23/eng/20221207
LC record available at https://lccn.loc.gov/2022045397
LC ebook record available at https://lccn.loc.gov/2022045398

ISBN: 978-1-032-06601-1 (hbk)
ISBN: 978-1-032-06600-4 (pbk)
ISBN: 978-1-003-20301-8 (ebk)

DOI: 10.1201/9781003203018

Typeset in Times
by Deanta Global Publishing Services, Chennai, India

To my father, Mr. Prayong Ratimart, the person I admire the most. In my teenage time, with your encouragement, your words still stay deeply in my mind. You wanted me to pass on the family tradition. Medicine has been practiced in our family for several generations, I went to the medical school.

To my dearest mother and brothers, with your love and support both mentally and financially, I write this book today.

To my most important mentor Professor Dr. Chongyun Liu of Bastyr University who taught me and has been my best consultant.

To my wife Agnes Wong and my son Dominic Apichai, who edited and helped me with the photography and making numerous meals.

Contents

Preface...lxiii
About the Author lxv

Chapter 1   Conditions with Multiple Pain Locations..............................1

Bi Syndrome 痹證..1
Introduction 1
Chinese Medicine Etiology and Pathology 1
Internal Causes 1
External Causes 4
Classification of Bi Syndrome...4
Manifestations, Diagnosis and Treatments 6
Wind Bi行痹型 6
Herbal Formula 6
Herbal Ingredients 6
Ingredient Explanations...6
Modifications 7
Formula Indications 7
Pattern Herbs 7
Cold Bi痛痹型 7
Herbal Formulas 7
Herbal Ingredients 8
Ingredient Explanations...8
Formula Indications 8
Herbal Ingrcdients 8
Ingredient Explanations...8
Modifications 9
Formula Indications 9
Cooking Instructions 9
Contraindications 9
Herbal Ingredients 9
Ingredient Explanations...9
Modifications 9
Formula Indications 10
Cautions 10
Patent Formulas 10
Damp Bi着痹型 11
Herbal Formulas 11
Herbal Ingredients 11
Ingredient Explanations..11
Modifications 12

Formula Indications 12
Herbal Ingredients 12
Ingredient Explanations 12
Modifications 12
Formula Indications 13
Contraindications.. 13
Patents.. 13
Heat Bi熱痹型 13
Herbal formula.. 13
Herbal Ingredients 13
Ingredient Explanations 13
Modifications 13
Formula Actions 14
Formula Indication 14
Cooking Instructions 14
Dose 14
Bone Bi骨痹型.. 14
Wind-Cold-Damp Type風寒濕痹症 14
Herbal Formula... 14
Herbal Ingredients 14
Ingredient Explanations 14
Modifications 15
Formula Indications 15
Damp Heat Type濕熱蘊結證 15
Herbal Formula... 15
Herbal Ingredients 15
Ingredient Explanations 15
Modifications 16
Formula Actions 16
Formula Indication 16
Cooking Instructions 16
Dose 16
Liver Kidney Deficiency Type肝腎虧損證 16
Herbal Formula... 17
Herbal Ingredients 17
Ingredient Explanations 17
Modifications 17
Formula actions 17
Formula Indication 17
Contraindications.. 17
Cooking Instructions 18
Dose 18
Accumulation of Phlegm and Blood Type 痰瘀互結證 18
Phlegm Obstruction in the Joints.. 18

Blood Stagnation in Joints 19
Acupuncture Treatments 20
Modifications 20
Patient Advisory 21
Thoracic Outlet Syndrome 胸廓出口症候群 21
Introduction 21
Anatomy 21
Western Medicine Etiology and Pathology 23
Chinese Medicine Etiology and Pathology 24
Acute State 24
Chronic State 25
Manifestations 25
In Western Medicine 25
In Chinese Medicine 25
Wind Invading Tendons 26
Physical Examination 26
Differential Diagnosis 27
Carpal Tunnel Syndrome 27
Lateral Epicondylitis 27
Medial Epicondylitis 27
Raynaud's Disease 27
Treatments 27
Herbal Formula 27
Formula Ingredients 27
Ingredient Explanations 27
Modifications 28
Formula Indications 28
Contraindications 28
Liver Blood Deficiency 28
Modifications 29
Formula Indications 29
Wind Invading Tendons 29
Herbal Ingredients 29
Ingredient Explanations 29
Modifications 30
Formula Indications 30
Acupuncture Point Formula 30
Tui Na Techniques 30
Fibromyalgia Syndrome 纖維肌痛綜合征 31
Introduction 31
Anatomy 32
Western Medicine Etiology and Pathology 32
Chinese Medicine Etiology and Pathology 32
Liver Qi Stagnation 32

Manifestations 32
   In Western Medicine.. 32
   In Chinese Medicine.. 33
Physical Examination 33
Differential Diagnosis 34
   Rheumatoid Arthritis.. 34
   Myofascial Pain 35
   Heart Disease... 35
Treatments 35
   Herbal Treatments 35
   Treatment Principle 35
   Herbal Formula... 35
   Formula Ingredients... 35
   Ingredient Explanations... 35
   Modifications 36
   Formula Indications.. 36
   Cooking Instructions 36
   Dose 36
   Treatment Principle 36
   Herbal Formula... 36
   Formula Ingredients... 36
   Ingredient Explanations... 36
   Modifications 37
   Formula Indications.. 37
   Contraindications.. 37
   Treatment Principle 37
   Herbal Treatments 37
   Formula Ingredients... 37
   Ingredient Explanations... 37
   Modifications 38
   Formula Indications.. 38
   Contraindications.. 38
   Treatment Principles.. 38
   Herbal Formula... 38
   Herbal Ingredients 38
   Ingredient Explanations... 38
   Modifications 38
   Formula Indications.. 39
   Contraindications.. 39
   Acupuncture Treatments.. 39
   Acupuncture Points... 39
   Auricular Therapy 39
   Tui Na Treatments... 40
Notes.. 43

Chapter 2 Headache and Facial Pain 45

    Headache in Chinese Medicine 頭痛（中醫） 45
      Introduction 45
      Chinese Medicine Etiology and Pathology 45
      Mechanism According to the Type of Exterior Factors 46
        Six Qi 六氣 46
        Six Yin六淫 46
      Six Yin 六淫, Six Meridians, and the Headaches 48
        The Location of Tai Yang太陽 49
        The Character of Tai Yang 太陽 49
        The Source of the Causes 49
        The Location of Yang Ming陽明 50
        The Character of Yang Ming陽明 50
        The Source of Causes 50
        The Location of Shao Yang少陽 51
        The Character of Shao Yang少陽: 51
        The Source of Causes 51
        The Location of Jue Yin厥阴 52
        The Character of Jue Yin厥陰 52
        The Source of the Causes 53
      Manifestations 55
      The Occipital Headache or Tai Yang Headache太陽頭痛 56
        Characteristics of Tai Yang Headache 57
      Treatments 57
        Treatment Principle 57
        Herbal Treatments 57
        Herbal Ingredients 57
        Ingredient Explanations.. 58
        Modifications 58
        Formula Actions 58
        Formula Indication 58
        Cooking Instructions 58
        Dose 58
      The Frontal Headache or Yang Ming Headache陽明頭痛........... 59
        Characteristics of Yang Ming Headache 60
      Treatments 60
        Treatment Principle 60
        Herbal Treatments 60
        Herbal Ingredients 60
        Ingredient Explanations.. 60
        Modifications 61
        Formula Actions 61
        Formula Indications 61
        Contraindications.. 61

Cooking Instructions 61
Dose 61
The Temporal Headache or Shao Yang Headache少陽頭痛 61
Characteristics of Shao Yang Headache 62
Treatments 63
Treatment Principle 63
Herbal Treatments 63
Herbal Ingredients 63
Ingredient Explanations 63
Modifications 63
Formula Actions 63
Formula Indications 63
Contraindications 63
Cooking Instructions 64
Dose 64
The Vertex Headache or Jue Yin Headache厥陰頭痛 64
Characteristics of Jue Yin Headache 65
Treatments 65
Treatment Principle 65
Herbal Treatments 65
Herbal Ingredients 65
Ingredient Explanations 65
Modifications 65
Formula Actions 66
Formula Indication 66
Contraindications 66
Cooking Instructions 66
Dose 66
The Headache Is Behind the Eyes眼後頭痛 66
Treatment Principle 67
Herbal Treatments 67
Herbal Ingredients 67
Ingredient Explanations 67
Modifications 68
Formula Actions 68
Formula Indication 68
Contraindications 68
Cooking Instructions 68
Dose 68
Treatments 69
Treatment Principle 69
Herbal Treatments 69
Herbal Ingredients 69
Ingredient Explanations 69
Modifications 69

Formula Actions 69
Formula Indication 69
Contraindications...70
Cooking Instructions 70
Dose 70
Treatments 70
Treatment Principle 70
Herbal Treatments 70
Herbal Ingredients 71
Ingredient Explanations...71
Modifications..71
Formula Actions 71
Formula Indication 71
Contraindications...71
Cooking Instructions 72
Dose 72
Treatments 72
Treatment Principle 72
Herbal Treatments 72
Herbal Ingredients 73
Ingredient Explanations...73
Modifications..73
Formula Actions 73
Formula Indication 73
Contraindications...73
Cooking Instructions 73
Dose 74
The Headache Is in the Whole Head全頭痛 74
Treatments 74
Treatment Principle 74
Herbal Treatments 75
Herbal Ingredients 75
Ingredient Explanations...75
Modifications..75
Formula Actions 75
Formula Indication 75
Contraindications...75
Cooking Instructions 76
Dose 76
Treatments 77
Treatment Principle 77
Herbal Treatments 77
Herbal Ingredients 77
Ingredient Explanations...77
Modifications..77

Formula Actions 78
Formula Indication 78
Contraindications...78
Cooking Instructions 78
Dose 78
Treatments 78
Treatment Principle 78
Herbal Treatments 78
Herbal Ingredients 79
Ingredient Explanations...79
Modifications 79
Formula Actions 79
Formula Indication 79
Contraindications...79
Cooking Instructions 79
Dose 80
Headache Due to Wind Heat風熱頭痛 80
Treatments 80
Treatment Principle 80
Herbal Treatments 80
Herbal Ingredients 80
Ingredient Explanations...80
Modifications 81
Formula Indication 81
Cooking Instructions 81
Dose 81
Headache Due to Wind Damp風濕頭痛 81
Treatments 82
Treatment Principle 82
Herbal Treatments 82
Herbal Ingredients 82
Ingredient Explanations...82
Modifications 82
Formula Actions 83
Formula Indication 83
Contraindications...83
Cooking Instructions 83
Dose 83
Headache Due to Liver Yang Rising肝陽頭痛 83
Treatments 83
Treatment Principle 83
Herbal Treatments 83
Herbal Ingredients 84
Ingredient Explanations...84
Modifications 84

Headache Due to Blood Stasis瘀血頭痛 84
Treatments 85
   Treatment Principle 85
   Herbal Treatments 85
   Herbal Ingredients 85
   Ingredient Explanations..85
   Modifications 86
   Formula Actions 86
   Formula Indication 86
   Contraindications..86
   Cooking Instructions 86
   Dose 86
Headache Due to Phlegm-Dampness Obstruction
痰濁阻絡頭痛..86
Treatments 87
   Treatment Principle 87
   Herbal Treatments 87
   Herbal Ingredients 87
   Ingredient Explanations..87
   Modifications 87
   Formula Actions 87
   Formula Indication 87
   Contraindications..87
   Cooking Instructions 88
   Dose 88
The Headache During a Menstrual Cycle月經期頭痛 88
   Herbal Formula...88
   Herbal Ingredients 89
   Ingredient Explanations..89
   Modifications 89
   Treatment Principle 90
   Herbal Treatments 90
   Herbal Ingredients 90
   Ingredient Explanations..90
   Modifications 90
   Cooking Instructions 91
   Dose 91
   Formula action 91
   Formula Indication 91
   Contraindications..91
   Treatment Principle 91
   Herbal Formula...91
   Herbal Ingredients 91
   Ingredient Explanations..92
   Acupuncture Treatments...92

Acupuncture Points...92
Modifications 92
Bleeding Therapy..93
Procedures 93
Electro-Acupuncture and Moxibustion Therapy 94
Treatment Frequency 94
Tui Na Treatments...94
Tui Na Procedures 94
Patient Advisory 95
The Types of Pain 96
Throbbing Headache跳痛性頭痛...96
Heaviness Headache 重痛性頭痛 97
Stabbing Headache刺痛性頭痛..97
Emptiness Headache虛型性頭痛..97
Electric Shock-Like Headache電擊樣頭痛................................97
Thunderclap Headache炸裂样痛頭痛.......................................98
Tension Headache紧箍样頭痛..98
The Headache under Certain Circumstances 98
Morning Headache晨起頭痛..98
Nighttime Headache 入夜頭痛 99
Fatigue Headache 勞累性頭痛 99
Hangover Headache 酒精性頭痛 100
Caffeine Headache 咖啡依賴性頭痛 100
Orgasm Headache 房事頭痛 100
Weekend Headache 週末頭痛 100
Food Triggered Headaches 食物引起疼痛 101
Weather Triggered Headaches 天氣變化頭痛 101
Migraine Headache偏頭痛 101
Introduction 101
Anatomy 102
Western Medicine Etiology and Pathology 103
Chinese Medicine Etiology and Pathology 103
Liver Yang Rising..103
Blood Stasis 104
Wind Cold Invasion 104
Liver Qi Stagnation...105
Liver Blood Deficiency 105
Phlegm Blocking Orifice 105
Liver and Kidney Yin Deficiency 105
Manifestations 105
In Western Medicine..105
In Chinese Medicine..106
Physical Examination 108
Differential Diagnosis 109
Subarachnoid Hemorrhage 109

Meningitis 109
Cerebral Hemorrhage 109
Treatments 110
Treatment Principle 110
Herbal Treatments 110
Herbal Ingredients 110
Ingredient Explanations 110
Modifications 110
Formula Actions 110
Formula Indication 111
Contraindications 111
Cooking Instructions 111
Dose: 111
Treatment Principle 111
Herbal Treatments 111
Herbal Ingredients 111
Ingredient Explanations 111
Modifications 112
Formula Actions 112
Formula Indication 112
Contraindications 112
Cooking Instructions 112
Dose 113
Treatment Principle 113
Herbal Treatments 113
Herbal Ingredients 113
Ingredient Explanations 113
Modifications 113
Formula Actions 113
Formula Indication 113
Contraindications 114
Cooking Instructions 114
Dose: 114
Treatment Principle 114
Herbal Treatments 114
Herbal Ingredients 114
Ingredient Explanations 114
Modifications 115
Formula Actions 115
Formula Indication 115
Contraindications 115
Cooking Instructions 115
Dose: 116
Treatment Principle 116
Herbal Treatments 116

Herbal Ingredients 116
Ingredient Explanations 116
Modifications 116
Formula Actions 116
Formula Indication 117
Contraindications 117
Cooking Instructions 117
Dose 117
Treatment Principle 117
Herbal Formula 117
Herbal Ingredients 117
Ingredient Explanations 117
Modifications 118
Formula Actions 118
Formula Indication 118
Contraindications 118
Cooking Instructions 118
Dose 118
Treatment Principle 118
Herbal Formula 119
Herbal Ingredients 119
Ingredient Explanations 119
Modifications 119
Formula Actions 119
Formula Indication 119
Contraindications 119
Cooking Instructions 119
Dose: 120
Acupuncture Treatments 120
Modifications 120
Bleeding Therapy 121
Procedures 121
Auricular Therapy 121
Aquapuncture Therapy 121
Tui Na Treatments 121
Tui Na procedures 122
Patient Advisory 123
Dietary Recommendations 123
Lifestyle 123
Post-Concussion Syndrome 脑震荡症候群 124
Introduction 124
Anatomy 124
Western Medicine Etiology and Pathology 125
Chinese Medicine Etiology and Pathology 125
Manifestations 127

Physical Examination 128
Differential Diagnosis 130
    An epidural hematoma (EDH) occurs when blood
    accumulates between the skull and the dura mater. 131
Treatments 132
    Treatment Principle 132
    Herbal Formula.. 132
    Herbal Ingredients 133
    Ingredient Explanations... 133
    Formula Actions 133
    Contraindications... 133
    Modifications 133
    Cooking Instructions 133
    Treatment Principle 134
    Herbal Formula.. 134
    Herbal Ingredients 134
    Ingredient Explanations... 134
    Formula Actions 134
    Contraindications... 134
    Modifications 134
    Cooking Instructions 135
    Dose 135
    Treatment Principle 135
    Herbal Formula.. 135
    Herbal Ingredients 135
    Ingredient Explanations... 135
    Formula Actions 136
    Caution and Contraindications 136
    Modifications 136
    Cooking Instructions 136
    Dose 137
    Treatment Principle 137
    Herbal Formula.. 138
    Herbal Ingredients 138
    Ingredient Explanations... 138
    Modifications 138
    Formula Actions 138
    Contraindications... 139
    Cooking Instructions 139
    Dose 139
    Treatment Principle 139
    Herbal Treatments 139
    Herbal Ingredients 139
    Ingredient Explanations... 139
    Modifications 140

Formula Actions 140
Formula Indication 140
Contraindications.. 140
Cooking Instructions 140
Dose 141
Acupuncture Treatments.. 141
Acupuncture point formula... 141
Modifications 141
Needling Technique 141
Treatment Frequency 141
Tui Na Techniques 141
Tui Na Procedures 141
Patient Advisory 143
Temporomandibular Joint Disorders 顳頜關節功能紊亂癥 143
Introduction 143
Anatomy 143
Western Medicine Etiology and Pathology 145
Stress Factors 145
Traumatic Factors 145
Chinese Medicine Etiology and Pathology 145
Manifestations 147
Physical Examination 149
Differential Diagnosis 150
Treatments 151
Treatment Principle 151
Herbal Formula... 151
Herbal Ingredients 151
Ingredient Explanations.. 151
Modifications 152
Formula Actions 152
Cooking Instructions 152
Dose 152
Treatment Principle 152
Herbal Formula... 152
Herbal Ingredients 152
Ingredient Explanations.. 152
Formula Action 153
Modifications 153
Formula Indication 153
Contraindications.. 153
Cooking Instructions 153
Dose 153
Treatment Principle 154
Herbal Treatments 154
Herbal Ingredients 154

Ingredient Explanations 154
Modifications 154
Formula Actions 154
Formula Indication 154
Cooking Instructions 155
Dose 155
Treatment Principle 155
Herbal Treatments 155
Herbal Ingredients 155
Ingredient Explanations 155
Modifications 156
Formula Actions 156
Formula Indication 156
Cooking Instructions 156
Dose 156
Treatment Principle 156
Herbal Ingredients 156
Ingredient Explanations 156
Modifications 157
Formula Actions 157
Formula Indication 157
Contraindications 157
Cooking Instructions 157
Dose 157
Acupuncture Treatment 158
Modifications 158
Tui Na Treatments.. 158
Tui Na Procedures 158
Patient Advisory 159
Temporomandibular Joint Dislocation顳下頜關節脫位 159
Introduction 159
Anatomy 159
Western Medicine Etiology and Pathology 159
Chinese Medicine Etiology and Pathology 161
Manifestations 161
Physical Examination 161
Differential Diagnosis 163
Treatments 163
Treatment Principle 163
Herbal Treatments 163
Herbal Formula.. 163
Herbal Ingredients 163
Ingredient Explanations 163
Cooking Instructions 164
Dose 164

Formula Action 164
Formula Indication 164
Intermediate and Later Dislocation Stage 164
Herbal Formula 164
Herbal Ingredients 164
Ingredient Explanations 164
Modifications 165
Habitual Dislocation Stage 165
Herbal Formula 165
Herbal Ingredients 165
Ingredient Explanations 165
Modifications 166
Formula Actions 166
Acupuncture Treatments 166
Tui Na Treatments 166
Reduction Procedures 166
Tui Na Techniques 167
Tui Na Procedures 167
Patient Advisory 167
Sinusitis 鼻竇炎 167
Introduction 167
Anatomy 168
Western Medicine Etiology and Pathology 168
Chinese Medicine Etiology and Pathology 168
Manifestations 169
In Western Medicine 169
In Chinese Medicine 170
Physical Examination 171
Differential Diagnosis 171
Treatments 172
Wind Cold Type 風寒型 172
Herbal Treatments 172
Herbal Ingredients 172
Ingredient Explanations 172
Modifications 173
Formula Actions 173
Instructions 173
Dose 173
Wind Heat type 風熱型 173
Herbal Treatments 173
Herbal Ingredients 173
Ingredient Explanation 173
Modifications 174
Formula Actions 174
Cooking Instructions 174

Dose 174
Treatment Principle 174
Herbal Treatments 174
Herbal Ingredients 174
Ingredient Explanations 175
Modifications 175
Formula Actions 175
Cooking Instructions 175
Dose 176
Treatment Principle 176
Herbal Treatments 176
Herbal Ingredients 176
Ingredient Explanations 176
Modifications 176
Formula Actions 176
Cooking Instructions 176
Dose 177
Treatment Principle 177
Herbal Treatments 177
Herbal Ingredients 177
Ingredient Explanations 177
Modifications 178
Formula Actions 178
Formula Indication 178
Contraindications.. 178
Cooking Instructions 178
Dose 178
Acupuncture Treatments.. 178
Modifications 178
Auricular Therapy 179
Tui Na Treatments... 179
Tui Na Procedures 179
Tui Na Techniques 179
Tui Na on the Face 180
Patient Advisory 180
Trigeminal Neuralgia 三叉神經痛 181
Introduction 181
Anatomy 181
Western Medicine Etiology and Pathology 182
Chinese Medicine Etiology and Pathology 182
Manifestations 183
Physical Examination 184
Differential Diagnosis 185
Treatments 185
Treatment Principle 185

Herbal Treatments 185
Herbal Ingredients 186
Ingredient Explanations 186
Modifications 186
Formula Actions 186
Cooking Instructions 186
Dose 186
Treatment Principle 186
Herbal Treatments 186
Herbal Ingredients 187
Ingredient Explanation 187
Modifications 187
Formula Actions 187
Cooking Instructions 187
Dose 187
Treatment Principle 187
Herbal Treatments 187
Herbal Ingredients 188
Ingredient Explanations 188
Modifications 188
Formula Actions 188
Formula Indication 188
Contraindications 188
Cooking Instructions 188
Dose 189
Treatment Principle 189
Herbal Treatments 189
Herbal Ingredients 189
Ingredient Explanations 189
Modifications 190
Formula Actions 190
Formula Indication 190
Contraindications 190
Cooking Instructions 190
Dose 190
Acupuncture Treatments.. 190
Modifications 190
Extra Points... 191
Auricular Therapy 191
Tui Na Treatments... 191
Tui Na Procedures 191
Patient Advisory 192
Bell's Palsy 面瘫 192
Introduction 192
Anatomy 193

Intracranial 193
   Extracranial 194
Western Medicine Etiology and Pathology 194
Chinese Medicine Etiology and Pathology 195
Manifestations 196
   In Western Medicine... 196
   In Chinese Medicine... 196
Physical Examination 197
Differential Diagnosis 199
Treatments 199
   Treatment Principle 199
   Herbal Treatments 199
   Herbal Ingredients 199
   Ingredient Explanation 199
   Modifications 199
   Formula Actions 200
   Cooking Instructions 200
   Dose 200
   Treatment Principle 200
   Herbal Formula.. 200
   Herbal Ingredients 200
   Ingredient Explanation 200
   Formula Actions 201
   Contraindications... 201
   Modifications 201
   Cooking Instructions 201
   Dose 201
   Treatment Principle 201
   Herbal Formula.. 201
   Herbal Ingredients 201
   Ingredient Explanation 202
   Formula Actions 202
   Contraindications... 202
   Modifications 202
   Cooking Instructions 202
   Dose 203
   Herbal Formula.. 203
   Herbal Ingredients 203
   Ingredient Explanations... 203
   Modifications 204
   Formula Actions 204
   Cooking Instructions 204
   Dose 204
   Treatment Principle 204
   Herbal Formula.. 204

Herbal Ingredients 204
Ingredient Explanations 204
Modifications 205
Formula Actions 205
Formula Indication 205
Contraindications 205
Cooking Instructions 205
Dose 206
External Formula 206
Herbal Ingredients 206
Ingredient Explanations 206
Direction 206
Acupuncture Treatments...206
Tui Na Treatments..207
Tui Na Procedures 207
Patient Advisory 207
Notes..207

Chapter 3   Neck Pain 209

Acute Cervical Fibrositis or Stiff Neck 落枕 209
Introduction 209
Anatomy 209
Western Medicine Etiology and Pathology 209
Chinese Medicine Etiology and Pathology 209
Local Qi Stagnation 頸筋受挫型 209
Wind Cold Invasion風寒侵淫型 210
Liver and Kidney Deficiency肝腎虧虛型 210
Manifestations 210
In Western Medicine.. 210
Local Qi Stagnation頸筋受挫型 210
Wind Cold Invasion風寒侵淫型 210
Liver and Kidney Deficiency肝腎虧虛型 210
Physical Examination 211
Differential Diagnosis 211
Treatments 211
Local Qi Stagnation頸筋受挫型 211
Herbal Treatments 212
Ingredient Explanations 212
Modifications 212
Formula Actions 212
Wind Cold Invasion風寒侵淫型 212
Ingredient Explanations 213
Modifications 213
Formula Actions 213

Cautions 213
Pattern Herbs 213
Treatment Principle 213
Herbal Treatments 213
Herbal Ingredients 213
Ingredient Explanations 214
Modifications 214
Formula Actions 214
Liver and Kidney Deficiency 肝腎虧虛型 214
Herbal Ingredients 214
Ingredient Explanations 214
Formula Actions 215
Modifications 215
Contraindications 215
Acupuncture Treatments 215
Moxibustion Therapy 216
Bleeding Therapy 216
Electro-Acupuncture Therapy 217
Tui Na Treatments 217
Tui Na Procedures 217
Cervical Spondylosis 頸椎病 220
Introduction 220
Anatomy 220
Intervertebral Discs 221
Facet Joints 221
Neuroforamen 221
Transverse Foramina 221
Ligaments 221
Nerves 222
Etiology and Pathology in Western Medicine 222
Four Distinct Types of Cervical Spondylosis 223
Manifestations in Western Medicine 224
Type I. Cervical Radiculopathy 224
Type II. Cervical Myelopathy 224
Type III Axial Joint Pain 225
Type IV. Cervical Spondylotic Arteriopathy 225
Type V. Sympathetic Cervical Spondylosis 225
Physical Examination 225
Differential Diagnosis 227
Etiology and Pathology in Chinese Medicine 228
Manifestations in Chinese Medicine and Treatments 229
Treatment Principle 229
Herbal Treatments 229
Herbal Ingredients 230
Ingredient Explanations 230

Modifications 230
Formula Actions 230
Treatment Principle 231
Herbal Treatments 231
Herbal Ingredients 231
Ingredient Explanations 231
Formula Actions 231
Contraindications..231
Cautions 231
Treatment Principles 232
Herbal Formula: 232
Herbal Ingredients 232
Ingredient Explanations 232
Modifications 232
Formula Actions 233
Formula Indication 233
Contraindications..233
Treatment Principles 233
Herbal Formula 233
Herbal Ingredients 233
Ingredient Explanations 233
Modifications 234
Pattern Herbs 234
Treatment Principles 235
Move Blood and Qi, stop pain. 235
Herbal Formula 235
Shen Tong Zhu Yu Tang身痛逐瘀湯 235
Herbal Ingredients 235
Ingredient Explanations 235
Modifications 235
Formula Indications 236
Contraindications..236
Treatment Principle 236
Herbal Formula 236
Herbal Ingredients 236
Ingredient Explanations 236
Modifications 237
Formula Indications 237
Contraindications..237
Acupuncture Treatments..237
Modifications 237
Tui Na Treatments...237
Tui Na Procedures 238
Upper Cervical Subluxation 寰樞關節半脫位 240
Introduction 240

Anatomy 240
    Atlas (C1) 240
    Axis (C2) 240
    Intervertebral Discs 241
    Atlantooccipital Joint 241
    Atlantoaxial Joint 242
Western Medicine Etiology and Pathology 243
Chinese Medicine Etiology and Pathology 245
Manifestations 246
    In Western Medicine 246
    In Chinese Medicine 246
Physical Examination 247
    Lateral cervical spine radiograph 248
    Open mouth (odontoid view) radiograph 248
Differential Diagnosis 249
Treatments 249
    Treatment Principle 249
    Herbal Treatments 249
    Herbal Formula 249
    Herbal Ingredients 250
    Ingredient Explanations 250
    Modifications 250
    Formula Actions 250
    Direction 250
    Treatment Principle 250
    Herbal Treatments 250
    Herbal Ingredients 251
    Ingredient Explanations 251
    Modifications 251
    Treatment Principles 251
    Herbal Formula 251
    Herbal Ingredients 252
    Ingredient Explanations 252
    Formula Actions 252
    Contraindications 252
    Modifications 252
    Treatment Principles 252
    Herbal Formula: 252
    Herbal Ingredients 253
    Ingredient Explanations 253
    Formula Actions 253
    Contraindications 253
    Modifications 253
    Directions 253
    Acupuncture Treatments 254

           Modifications 254
           Tui Na Treatments...254
           Tui Na Procedures 254
      Notes...256

Chapter 4   Chest Pain...257

      Intercostal Neuralgia 肋間神經痛.....................................257
      Introduction 257
           Anatomy 257
                Typical Ribs 257
                Atypical Ribs 257
                Anterior Articulation 257
                Posterior Articulation 257
                Intercostal Nerves 258
           Western Medicine Etiology and Pathology 258
           Chinese Medicine Etiology and Pathology 258
           Manifestations 259
                In Western Medicine..259
                In Chinese Medicine 259
           Physical Examination 260
           Differential Diagnosis 261
           Treatments 261
                Herbal Formula...261
                Herbal Ingredients 262
                Ingredient Explanations...262
                Cooking Instructions 262
                Dose 262
                Treatment Principle 263
                Herbal Formula...263
                Herbal Ingredients 263
                Ingredient Explanations...263
                Modifications 263
                Formula Actions 263
                Formula Indication 263
                Contraindications..264
                Cooking instructions..264
                Dose 264
                Treatment Principle 264
                Herbal Formula...264
                Herbal Ingredients 264
                Ingredient Explanations...264
                Modifications 265
                Formula Actions 265
                Formula Indication 265

Contraindications...265
Dose...265
Acupuncture Treatments..265
Modifications..266
Extra Points..266
Auricular Therapy 266
Cupping Therapy 266
Cutaneous Acupuncture Therapy 266
Procedures 266
Tui Na Treatments...267
Tui Na Procedures 267
Patient Advisory 268
Sprain of the Chest and Hypochondrium胸肋屏傷 268
Introduction 268
Anatomy 268
Pectoralis Major..268
Origin..269
Insertion..269
Action...269
Nerve Innervation 269
Blood Supply...269
Intercostal Muscle...269
Origin..269
Insertion..269
Action..270
Nerve Innervation 270
Blood Supply...270
Western Medicine Etiology and Pathology 270
Chinese Medicine Etiology and Pathology 271
Manifestations 271
In Western Medicine..271
In Chinese Medicine..272
Physical Examination 272
Differential Diagnosis 273
Treatments 274
Treatment Principle 274
Herbal Treatments 274
Herbal Ingredients 274
Ingredient Explanations 274
Modifications...274
Formula Indications 274
Contraindications...274
Treatment Principle 275
Herbal Formula..275
Herbal Ingredients 275

Ingredient Explanations 275
Formula Action 275
Formula Indication 275
External Herbal Applications 275
Acupuncture Treatments 275
Modifications 276
Bleeding and Cupping with Cutaneous Acupuncture
Treatments 276
Acupuncture Points 276
Procedures 276
Tui Na Treatments 276
Tui Na Procedures 276
Patient Advisory 277
Note 278

Chapter 5   Abdominal Pain 279

General Concepts of Abdominal Pain in Chinese Medicine 中
醫腹痛 279
Introduction 279
Anatomy 279
Chinese Medicine Etiology and Pathology 280
Manifestations 281
Quality of Pain 281
Location of Pain 281
Duration of Pain 281
Condition of Pain 282
Classifications and Treatments 282
Treatment Principle 282
Herbal Treatments 282
Herbal Ingredients 282
Ingredient Explanations 283
Modifications 283
Formula Indications 283
Contraindications 283
Treatment Principle 283
Herbal Treatments 283
Herbal Ingredients 283
Ingredient Explanations 284
Modifications 284
Formula Indications 284
Contraindications 284
Treatment Principle 284
Herbal Treatments 284
Herbal Ingredients 284

Ingredient Explanations 285
Modifications 285
Formula Actions 285
Formula Indication 285
Contraindications 285
Herbal Treatments 285
Herbal Ingredients 285
Ingredient Explanations 286
Modifications 286
Formula Indications 286
Contraindications 286
Treatment Principle 286
Herbal Treatments 286
Herbal Ingredients 287
Ingredient Explanations 287
Modifications 287
Formula Indications 287
Contraindications 287
Treatment Principle 287
Herbal Treatments 287
Ingredients 287
Ingredient Explanations 288
Modifications 288
Formula Indications 288
Contraindications 288
Acupuncture Treatments 288
Modifications 288
Auricular Therapy 288
Tui Na Treatments 289
Tui Na Procedures 289
Patient Advisory 289
Epigastric Pain 胃脘痛 290
Introduction 290
Anatomy 290
Western Medicine Etiology and Pathology 290
Pain in Local Organs 291
Radiating Pain to the Epigastrium 291
Chinese Medicine Etiology and Pathology 291
Excessive Conditions 291
Deficient Conditions 293
Manifestations 294
Excessive Conditions 294
Deficient Conditions 296
Physical Examination 296
Differential Diagnosis 297

Herbal Treatments 297
  Excessive Conditions 297
  Treatment Principle 298
  Herbal Treatments 298
  Herbal Ingredients 298
  Ingredient Explanations 298
  Modifications 298
  Formula Indications 298
  Contraindications 298
  Treatment Principle 298
  Herbal Treatments 298
  Ingredients 298
  Ingredient Explanations 299
  Modifications 299
  Formula Indications 299
  Contraindications 299
  Treatment Principle 299
  Herbal Treatments 299
  Herbal Formula 299
  Herbal Ingredients 299
  Ingredient Explanations 299
  Modifications 300
  Formula Indications 300
  Contraindications 300
  Treatment Principle 300
  Herbal Treatments 300
  Herbal Ingredients 300
  Ingredient Explanations 300
  Zhi Shi枳實 breaks up stagnation and transforms Phlegm. 300
  Modifications 300
  Formula Indications 300
  Contraindications 301
  Treatment Principle 301
  Herbal Treatments 301
  Herbal Ingredients 301
  Ingredient Explanations 301
  Modifications 301
  Formula Actions 301
  Formula Indication 301
  Contraindications 302
  Treatment Principle 302
  Herbal Treatments 302
  Ingredients 302
  Ingredient Explanations 302
  Formula Indications 302

Cautions 302
Treatment Principle 303
Herbal Treatments 303
Herbal Ingredients 303
Ingredient Explanations 303
Modifications 303
Formula Indications 303
Contraindications 303
Treatment Principle 304
Herbal Treatments 304
Herbal Ingredients 304
Ingredient Explanations 304
Modifications 304
Formula Indications 304
Contraindications 305
Deficient Conditions 305
Herbal Treatments 305
Herbal Ingredients 305
Ingredient Explanations 305
Modifications 305
Formula Indications 305
Contraindications 306
Treatment Principle 306
Herbal Formula 306
Herbal Ingredients 306
Ingredient Explanations 306
Modifications 306
For insomnia, add Bai Zi Ren柏子仁, Suan Zao Ren酸棗
仁 Wu Wei Zi五味子...306
Formula Actions 306
Formula Indications 306
Contraindications 307
The formula contains many herbs that are sweet; patients
who are weak in digestion should be cautious.....................307
Acupuncture Treatments 307
Acupuncture Point Formula....................................307
Basic Points 307
Excessive Conditions 307
Deficient Conditions 308
Cupping Therapy 308
Auricular Therapy 308
Tui Na Treatments 308
Tui Na Techniques 308
Tui Na Procedures 308
Patient Advisory 311

Hypochondriac pain in Chinese medicine中醫脅痛...................... 311
    Introduction 311
    Anatomy 311
        Right Upper Quadrant 311
        Left Upper Quadrant 312
    Western Medicine Etiology and Pathology 312
    Right Upper Quadrant 312
        Left Upper Quadrant 313
    Chinese Medicine Etiology and Pathology 313
        Excessive Condition.. 313
        Deficiency Condition: 315
    Manifestations: 316
        Excessive Condition.. 316
        Deficiency Condition: 316
    Physical Examination: 317
    Differential Diagnosis: 317
    Treatments 317
    Treatment Principle 317
        Herbal Treatments 317
        Herbal Ingredients 318
        Ingredient Explanations.. 318
        Modifications 318
        Formula Actions 318
        Formula Indications 318
        Contraindications.. 318
        Herbal Formula... 318
        Formula Ingredients... 318
        Ingredient Explanations.. 319
        Treatment Principle 319
        Herbal Formula... 319
        Herbal Ingredients 319
        Ingredient Explanations.. 319
        Modifications 320
        Formula Actions 320
        Formula Indication 320
        Contraindications.. 320
        Deficiency Condition 320
        Treatment Principle 320
        Herbal Formula... 320
        Herbal Ingredients 320
        Ingredient Explanations.. 320
        Modifications 321
        Formula Actions 321
        Formula Indication 321

Contraindications.. 321
Treatment Principle 321
Herbal Treatments 321
Herbal Ingredients 321
Ingredient Explanations.. 321
Modifications 322
Formula Actions 322
Treatment Principle 322
Herbal Treatments 322
Herbal Ingredients 322
Ingredient Explanations.. 322
Modifications 322
Formula Actions 322
Acupuncture Treatments... 323
Local Points 323
Auricular Therapy 323
Electro-Acupuncture Therapy 323
Tui Na Treatments.. 323
Patient Advisory 324
Irritable Bowel Syndrome 腸易激綜合症 324
Introduction 324
Anatomy 325
Western Medicine Etiology and Pathology 325
Chinese Medicine Etiology and Pathology 326
Manifestations 327
In Western Medicine... 327
In Chinese Medicine... 327
Physical Examination 328
Differential Diagnosis 329
Inflammatory Bowel Disease 329
Colorectal Cancer 329
Treatments 330
Treatment Principle 330
Herbal Treatments 330
Herbal Ingredients 330
Ingredient Explanations.. 330
Modifications 330
Formula Indications 330
Contraindications.. 330
Treatment Principle 330
Herbal Treatments 330
Herbal Ingredients 331
Ingredient Explanations.. 331
Modifications 331
Formula Indications 331

Contraindications 331
Treatment Principle 331
Herbal Treatments 332
Herbal Ingredients 332
Ingredient Explanations 332
Modifications 332
Formula Indications 332
Contraindications 332
Treatment Principle 332
Herbal Treatments 332
Herbal Ingredients 333
Ingredient Explanations 333
Modifications 333
Formula Indications 333
Contraindications 333
Constipation Predominant IBS 333
Treatment Principle 333
Herbal Treatments 333
Herbal Ingredients 333
Ingredient Explanations 333
Modifications 334
Formula Indications 334
Contraindications 334
Treatment Principle 334
Herbal Treatments 334
Herbal Ingredients 334
Ingredient Explanations 334
Modifications 335
Formula Indications 335
Contraindications 335
Treatment Principle 335
Herbal Treatments 335
Herbal Ingredients 335
Ingredient Explanations 335
Modifications 336
Formula Indications 336
Contraindications 336
Acupuncture Treatments 336
Basic Formula 336
Moxibustion Therapy 337
Tui Na Treatments 337
Tui Na Procedures 337
Patient Advisory 339
Lower Abdominal Pain 下腹痛 339
Introduction 339

Anatomy 340
Right Iliac Region...340
  Hypogastric Region 340
  Left Iliac Region 340
Western Medicine Etiology and Pathology 341
Multiple Regions...341
  Left Iliac Region...341
  Right Iliac Region...341
  Suprapubic Region 341
  Referred Pain 341
  In Males 342
Manifestations 342
  In Western Medicine...342
  In Chinese Medicine..343
  Both Left and Right Iliac Region Disorders............344
Physical Examination 344
Differential Diagnosis 348
  Abdominal Aortic Aneurysm 348
  Ectopic Pregnancy 349
  Ovarian Torsion 349
  Testicular Torsion...349
  Acute Appendicitis 349
Treatments 350
  Suprapubic Region Disorders 350
  Treatment Principle 350
  Herbal Treatments 350
  Herbal Ingredients 350
  Ingredient Explanations.....................................350
  Modifications 350
  Formula Indications 350
  Contraindications...351
  Treatment Principle 351
  Herbal Treatments 351
  Herbal Ingredients 351
  Administrations 351
  Ingredient Explanations.....................................351
  Modifications 352
  Formula Indications 352
  Blood Stasis瘀血阻滞..352
  Treatment Principle 352
  Herbal Treatments 352
  Formula Ingredients..352
  Ingredient Explanations.....................................352
  Modifications 352
  Formula Indications 352

Contraindications..352
Treatment Principle 353
Herbal Treatments 353
Formula Ingredients...353
Ingredient Explanations..353
Modifications 353
Formula Indications..353
Contraindications..354
Treatment principle..354
Herbal Formula for Kidney Yang deficiency.......................354
Herbal Ingredients 354
Ingredient Explanations..354
Modifications 354
Formula Indications 355
Contraindications..355
Both Left and Right Iliac Region Disorders.........................355
Treatment Principle 355
Herbal Treatments 355
Herbal Ingredients 355
Ingredient Explanations..355
Modifications 355
Formula Indications 355
Contraindications..355
Treatment Principle 355
Herbal Treatments 356
Herbal Ingredients 356
Ingredient Explanations..356
Modifications 356
Formula Indication 356
Contraindications..356
Treatment Principle 356
Herbal Treatments 356
Herbal Ingredients 357
Ingredient Explanations..357
Modifications 357
Formula Indications 357
Contraindications..357
Treatment Principle 357
Herbal Treatments 357
Herbal Ingredients 357
Ingredient Explanations..357
Modifications 358
Formula Indications 358
Contraindications..358
Acupuncture Treatments...358

Modifications 358
Both Left and Right Iliac Region Disorders 358
Modifications 359
Electro-Acupuncture Therapy 359
Tui Na Treatments 359
Tui Na Procedures 359
Auricular Therapy 360
Patient Advisory 360
Dysmenorrhea 痛經 361
Introduction 361
Anatomy 361
The Uterus 361
The Fallopian (Uterine) Tubes 362
The Cervix 362
Western Medicine Etiology and Pathology 363
Primary Dysmenorrhea 363
Secondary Dysmenorrhea 364
Chinese Medicine Etiology and Pathology 364
Excessive Types 364
Deficient Types 365
Manifestations 365
In Western Medicine 365
Secondary Dysmenorrhea 366
Ectopic Pregnancy 366
In Chinese Medicine 366
Deficient Types 367
Physical Examination 367
Differential Diagnosis 369
Endometriosis 369
Treatments 369
Excessive Types 370
Treatment Principle 370
Herbal Treatments 370
Herbal Ingredients 370
Ingredient Explanations 370
Modifications 370
Formula Indications 371
Contraindications 371
Treatment Principle 371
Herbal Treatments 371
Herbal Ingredients 371
Ingredient Explanations 371
Modifications 371
Formula Indications 371
Contraindications 371

Treatment Principle 372
Herbal Treatments 372
Herbal Ingredients 372
Ingredient Explanations 372
Modifications 372
Formula Indications 372
Contraindications..372
Deficient Types 373
Treatment Principle 373
Herbal Treatments 373
Herbal Ingredients 373
Ingredient Explanations 373
Modifications 373
Formula Indications 373
Treatment Principle 373
Herbal Formula..373
Herbal Ingredients 374
Ingredient Explanations 374
Modifications 374
Formula Actions 374
Formula Indication 374
Contraindications..374
Acupuncture Treatments...374
Modifications 375
Deficient Types 375
Auricular Therapy 375
Moxibustion Therapy...375
Tui Na Treatments..375
Tui Na Procedures 375
Patient Advisory 376
Always keep the abdomen, lower back, and both feet warm. 376
Notes...377

Chapter 6   Upper Back Pain...379

Inflammation of the Trapezius Muscle 斜方肌筋膜炎 379
Introduction 379
Anatomy 379
Western Medicine Etiology and Pathology 379
Chinese Medicine Etiology and Pathology 380
Manifestations 380
Physical Examination 380
Differential Diagnosis 381
Treatment...382
Treatment Principle 382

Herbal Medicine 382
Herbal Formula 382
Herbal Ingredients 382
Ingredient Explanations 382
Formula Ingredients 383
Ingredient Explanations 383
Herbal Formula 383
Formula Ingredients 383
Ingredient Explanations 383
Acupuncture Treatments 384
Distal Acupuncture Point Formula 384
Cupping Therapy 384
Moxibustion Therapy 384
Bleeding Therapy 384
Procedures 384
Electro-Acupuncture Therapy 385
Tui Na Treatments 385
Tui Na Procedures 385
Patient Advisory 386
Rhomboid Muscle Pain 菱形肌勞損 387
Introduction 387
Anatomy 387
Western Medicine Etiology and Pathology 387
Chinese Medicine Etiology and Pathology 387
Manifestations 388
In Western Medicine 388
In Chinese Medicine 389
Physical Examination 389
Differential Diagnosis 390
Treatments 391
Treatment Principle 391
Herbal Formula 391
Herbal Ingredients 391
Ingredient Explanations 391
Modifications 391
Formula Indication 392
Formula Contraindication 392
Treatment Principle 392
Herbal Treatments 392
Herbal Ingredients 392
Administrations 392
Ingredient Explanation 392
Formula Indication 392
Formula Contraindication 392
Treatment Principle 393

Herbal Treatments 393
Herbal Ingredients 393
Ingredient Explanations... 393
Modifications 393
Formula Actions 393
Formula Indication 393
Contraindications... 393
Acupuncture Treatments.. 393
Administrations 394
Auricular Therapy 394
Cupping Therapy 394
Moxibustion Therapy... 394
Bleeding Therapy... 394
Tui Na Treatments.. 394
Tui Na Procedures 394
Patient Advisory 395
Costovertebral Joint Sprain 胸壁扭挫傷.............................. 395
Introduction 395
Anatomy 395
Western Medicine Etiology and Pathology 396
Chinese Medicine Etiology and Pathology 396
Qi and Blood stagnation 396
Manifestations 396
Physical Examination 396
Differential Diagnosis 397
Treatment.. 397
Treatment Principle 397
Herbal Treatments 397
Herbal Formula.. 397
Formula Ingredients... 397
Ingredient Explanations... 398
Blood Stasis Sprain.. 398
Herbal Formula.. 398
Formula Ingredients... 398
Ingredient Explanations... 398
Procedures 399
Cupping Therapy 399
Moxibustion Therapy... 399
Bleeding Therapy... 399
Procedures 399
Electro-Acupuncture Therapy 400
Tui Na Treatments.. 400
Tui Na Procedures 400
Patient Advisory 402
Notes.. 402

Chapter 7   Shoulder Pain 403

    General Concept of Shoulder Pain in Chinese Medicine................ 403
      Introduction 403
      Chinese Medicine Mechanism of Disease 403
      Chinese Medicine Etiology and Pathology 405
        Wind Cold Invasion Shoulder Pain风寒肩痛 405
        Phlegm Damp Obstruction Shoulder Pain痰濕肩痛 405
        Blood Stasis Shoulder Pain 淤血肩痛 406
        Blood Deficiency Shoulder Pain血虚肩痛 406
      Physical Examination 406
        Normal Range...408
        Abduction...408
        Adduction...408
        Flexion 409
        Hyperextension 409
        Internal Rotation 409
        External Rotation 409
        Pathogenic Range 409
        Empty Can Test (Jobe's Test)..409
        Patte Test (Hornblower's Test) 410
        Bear Hug Test 410
        Apley's Scratch Test... 410
        Hawkins Test.. 410
        Speed's Test 410
        Yergason Test... 411
        Cross-Arm Adduction Test (Scarf Test)..................................... 411
      Manifestations 411
        Wind Cold Invasion Shoulder Pain风寒肩痛 411
        Phlegm Damp Obstruction Shoulder Pain痰濕肩痛 411
        Blood Stasis Shoulder Pain 淤血肩痛 412
        Blood Deficiency Shoulder Pain血虚肩痛 412
      Treatments 412
        Wind Cold Invasion Shoulder Pain风寒肩痛 412
        Herbal Treatments 412
        Herbal Ingredients 412
        Ingredient Explanations..412
        Modifications 413
        Formula Indication 413
        Formula Contraindication...413
        Phlegm Damp Obstruction Shoulder Pain痰濕肩痛 413
        Herbal Treatments 413
        Herbal Ingredients 413
        Ingredient Explanations..413
        Modifications 414

Formula Indication 414
Contraindications... 414
Blood Stasis Shoulder Pain 淤血肩痛 414
Herbal Formula.. 414
Herbal Ingredients 414
Ingredient Explanations.. 414
Modifications 415
Formula Indication 415
Formula Contraindication... 415
Blood Deficiency Shoulder Pain血虚肩痛 415
Herbal Treatments 415
Formula Ingredients... 415
Ingredient Explanation 415
Modifications 416
Formula Indication 416
Acupuncture Treatments... 416
Acupuncture Point Formula.. 416
Modifications 416
Electro-Acupuncture Therapy 416
Moxibustion Therapy.. 416
Cupping Therapy 416
Bleeding Therapy.. 417
Auricular Therapy 417
Teding Diancibo Pu Therapy... 417
Tui Na Treatments... 417
Tui Na Procedures 417
Patient Advisory 420
Frozen Shoulder五十肩 420
Introduction 420
Anatomy 421
The Glenohumeral Joint 421
Acromioclavicular Joint.. 421
Sternoclavicular Joint 421
The Scapulothoracic Joint.. 422
The Synovial Bursae.. 422
Western Medicine Etiology and Pathology 422
Chinese Medicine Etiology & Pathology 423
Manifestations 423
In Western Medicine.. 423
In Chinese Medicine.. 424
Physical Examination 424
Normal Range... 426
Abduction.. 426
Adduction.. 426
Flexion 426

Hyperextension 426
Internal Rotation 426
External Rotation 427
Pathogenic Range 427
Empty Can Test (Jobe's Test) 427
Patte Test (Hornblower's Test) 427
Bear Hug Test 427
Apley's Scratch Test 428
Hawkins Test 428
Neer's Test 428
Yergason Test 428
Speed's Test 428
Cross-arm Adduction Test (Scarf Test) 429
Apprehension Test 429
Differential Diagnosis 429
Biceps Tendinopathy 429
Cervical Disk Degeneration 429
Rotator Cuff Tendinopathy 429
Subdeltoid Bursitis 429
GH Joint Dislocation 429
Treatments 429
Treatment Principle 430
Herbal Treatments 430
Herbal Ingredients 430
Ingredient Explanations 430
Modifications 430
Formula Indication 430
Treatment Principle 430
Herbal Formula 430
Herbal Ingredients 431
Ingredient Explanations 431
Modifications 431
Formula Indication 431
Formula Contraindication 431
Treatment Principle 431
Herbal Treatments 432
Herbal Ingredients 432
Ingredient Explanations 432
Modifications 432
Formula Indication 432
Acupuncture Treatments 432
Special Points 433
Modifications 433
Yujian魚肩穴 433
Auricular Therapy 433

Cupping Therapy 433
Moxibustion Therapy 434
Bleeding Therapy 434
Electro-Acupuncture Therapy 434
Tui Na Treatments 434
Tui Na Procedures 434
Patient Advisory 435
Supraspinatus Tendonitis岡上肌肌腱炎 436
Introduction 436
Anatomy 437
Four Rotator Cuff Muscles 437
Nerve Supply 437
Blood Supply 437
Function 437
Western Medicine Etiology and Pathology 437
Chinese Medicine Etiology and Pathology 438
Jin Bi Syndrome 筋痹 438
Manifestations 438
In Western Medicine 438
In Chinese Medicine 439
Wind-Cold-Damp Bi Syndrome 虛寒痹型 439
Physical Examination 439
X-Ray 440
Differential Diagnosis 441
Frozen Shoulder 441
Acromioclavicular Arthritis 441
Suprascapular neuropathy and cervical radiculopathy 441
Treatments 441
Blood Stasis瘀滯型 441
Herbal Treatments 441
Herbal Ingredients 442
Ingredient Explanations 442
Formula action 442
Formula Indication 442
Wind-Cold-Damp Bi Syndrome 虛寒痹型 442
Herbal Treatments 442
Herbal Ingredients 442
Ingredient Explanations 443
Modifications 443
Formula Actions 443
Acupuncture Treatments 443
Modifications 443
Auricular Therapy 443
Cupping Therapy 443
Moxibustion Therapy 444

Bleeding Therapy...444
Electro-Acupuncture Therapy 444
Tui Na Treatments..444
Tui Na Procedures 445
Patient Advisory 445
Short Head Biceps Injury 肱二頭肌短頭腱損傷...........................446
Introduction 446
Anatomy 446
Western Medicine Etiology and Pathology 446
Chinese Medicine Etiology and Pathology 447
Manifestations 447
Physical Examination 447
Differential Diagnosis 448
Treatments 448
Treatment Principle 448
Herbal Treatments 448
Herbal Ingredients 448
Ingredient Explanations 449
Modifications 449
Formula Indications 449
Contraindications 449
Chronic Stage 449
Herbal Formula..449
Herbal Ingredients 449
Ingredient Explanations 450
Modifications 450
Formula Indication 450
Formula Contraindication...450
Acupuncture Treatments...450
Modifications 451
Auricular Therapy 451
Cupping Therapy 451
Moxibustion Therapy...451
Bleeding Therapy...451
Electro-acupuncture Therapy 451
Tui Na Treatments..451
Tui Na Procedures 451
Patient Advisory 452
Long Head Biceps Tendonitis 肱二頭肌長頭肌腱炎 452
Introduction 452
Anatomy 452
Western Medicine Etiology and Pathology 453
Chinese Medicine Etiology and Pathology 453
Manifestations 453
Physical Examination 454

Speed's Test 454
Yergason Test 454
Differential Diagnosis 454
Treatments 455
Treatment Principle 455
Herbal Treatments 455
Herbal Ingredients 455
Ingredient Explanations 455
Modifications 456
Topical Application 456
Herbal Ingredients 456
Ingredient Explanations 456
Application Procedures 456
Chronic Stage 457
Herbal Formula 457
Herbal Ingredients 457
Ingredient Explanations 457
Modifications 458
Formula Indication 458
Formula Contraindication 458
Acupuncture Treatments 458
Modifications 458
Auricular Therapy 458
Cupping Therapy 458
Moxibustion Therapy 458
Bleeding Therapy 458
Electro-Acupuncture Therapy 458
Tui Na Treatments 459
Tui Na Procedures 459
Patient Advisory 460
Subluxation of the Long Head Biceps Tendon 肱二頭肌長頭腱
滑脫 461
Introduction 461
Anatomy 461
Western Medicine Etiology and Pathology 461
Chinese Medicine Etiology and Pathology 462
Liver and Kidney Deficiency 462
Manifestations 462
Physical Examination 463
Speed's Test 463
Shoulder Abduction 90° and External Rotation 90° Test 463
Differential Diagnosis 464
Treatments 464
Treatment Principle 464
Herbal Treatments 464

Herbal Ingredients 464
Ingredient Explanations 465
Modifications 465
Formula Indications 465
Contraindications 465
Chronic Stage 465
Herbal Formula 465
Herbal Ingredients 465
Ingredient Explanations 466
Modifications 466
Formula Indication 466
Formula Contraindication 466
Acupuncture Treatments 466
Modifications 467
Auricular Therapy 467
Cupping Therapy 467
Moxibustion Therapy 467
Bleeding Therapy 467
Electro-Acupuncture Therapy 467
Tui Na Treatments 467
Tui Na Procedures 467
Patient Advisory 468
Subacromial Bursitis 肩峰下滑囊炎 469
Introduction 469
Anatomy 469
Western Medicine Etiology and Pathology 469
Chinese Medicine Etiology and Pathology 470
Manifestations 471
Physical Examination 471
Hawkins Test 472
Neer's Test 472
Speed's Test 472
Differential Diagnosis 472
Treatments 472
Treatment Principle 472
Herbal Treatments 473
Herbal Ingredients 473
Ingredient Explanations 473
Modifications 473
Formula Indications 473
Contraindications 473
Administrations 473
Acupuncture Treatments 474
Modifications 474
Auricular Therapy 474

Cupping Therapy 474

Moxibustion Therapy 474

Bleeding Therapy 474

Electro-Acupuncture Therapy 474

Tui Na Treatments 474

Tui Na Procedures 474

Patient Advisory 476

Notes 476

Chapter 8   Elbow Pain 477

Lateral Epicondylitis 肱骨外上髁炎 477

Introduction 477

Anatomy 477

Western Medicine Etiology and Pathology 479

Chinese Medicine Etiology and Pathology 479

Manifestations 479

In Western Medicine 479

In Chinese Medicine 480

Physical Examination 480

Differential Diagnosis 481

Cervical radiculopathy 481

Treatments 482

Herbal Treatments 482

Treatment Principle 482

Herbal Treatments 482

Herbal Ingredients 482

Ingredient Explanations 482

Formula Action 482

Formula Indication 482

Treatment Principle 483

Herbal Formula 483

Herbal Ingredients 483

Ingredient Explanations 483

Modifications 483

Formula Indication 483

Formula Contraindication 483

Treatment Principle 484

Herbal Formula 484

Herbal Ingredients 484

Ingredient Explanations 484

Modifications 485

Formula Indication 485

Formula Contraindication 485

Acupuncture Treatments 485

Modifications 485
Auricular Therapy 485
Moxibustion Therapy..485
Bleeding Therapy..485
Cupping Therapy 485
Tui Na Treatments..486
Tui Na Procedures 486
Patient Advisory 486
Medial Epicondylitis 肱骨內上髁炎 487
Introduction 487
Anatomy 487
Western Medicine Etiology and Pathology 489
Manifestations 489
Physical Examination 489
Differential Diagnosis 490
Cervical Radiculopathy 490
Elbow Osteoarthritis..490
Anterior Interosseous Nerve Entrapment 490
Treatments 490
Herbal Treatments 490
Treatment Principle 490
Herbal Treatments 490
Herbal Ingredients 490
Ingredient Explanations 490
Formula Action 491
Formula Indication 491
Treatment Principle 491
Herbal Formula 491
Herbal Ingredients 491
Ingredient Explanations 491
Modifications 492
Formula Indication 492
Formula Contraindication..492
Treatment Principle 492
Herbal Formula 492
Herbal Ingredients 492
Ingredient Explanations 492
Modifications 493
Formula Indication 493
Formula Contraindication..493
Acupuncture Treatments..493
Modifications 493
Auricular Therapy 494
Moxibustion Therapy..494
Bleeding Therapy..494

Cupping Therapy 494
Tui Na Treatments...494
Tui Na Procedures 494
Patient Advisory 495
Olecranon Bursitis 尺骨鷹嘴滑囊炎..495
Introduction 495
Anatomy 495
Olecranon Bursae 495
Cubital Bursae 496
Western Medicine Etiology and Pathology 496
Chinese Medicine Etiology and Pathology 496
Qi and Blood Stagnation 氣滯血瘀型 496
Manifestations 497
Physical Examination 497
Differential Diagnosis 498
Fracture of the Olecranon Process 498
Distal Partial Ruptures of Triceps Brachii Tendon 498
Treatments 498
Treatment Principle 498
Herbal Treatments 498
Herbal Ingredients 498
Ingredient Explanations..498
Modifications 499
Formula Indications 499
Contraindications...499
Administrations 499
Topical Herbal Formula..499
Formula Ingredients...499
Dose...499
Contraindication 499
Suggested Usage 499
Red-Colored Spray – Yunnan Baiyao Aerosol Baoxianye
雲南白藥氣霧劑保險液..500
Cream-Colored Spray – Yunnan Baiyao Aerosol雲南白藥
氣霧劑..500
Acupuncture Treatments..500
Modifications 500
Auricular Therapy 500
Bleeding Therapy...500
Cupping Therapy 501
Tui Na Treatments...501
Tui Na Procedures 501
Patient Advisory 501
Notes...501

Chapter 9   Wrist Pain 503

        Carpal Tunnel Syndrome 腕管綜合症 503
           Introduction 503
           Anatomy 503
               The Nerves 503
               The Tendons 504
               The Boundary Forming the Tunnel 504
           Western Medicine Etiology and Pathology 504
           Chinese Medicine Etiology and Pathology 505
           Manifestations 505
           Physical Examination 505
           Differential Diagnosis 507
               Cervical Radiculopathy 507
               Median Neuropathy in the Forearm 507
           Treatments 508
               Treatment Principle 508
               Herbal Treatments 508
               Herbal Ingredients 508
               Ingredient Explanations 508
               Modifications 508
               Formula Indications 508
               Contraindications 508
               Administrations 508
               Acupuncture Treatments 509
               Moxibustion Therapy 509
               Tui Na Treatments 509
               Tui Na Procedures 509
           Patient Advisory 510
        Wrist Sprain 腕關節扭傷 510
           Introduction 510
           Anatomy 511
               Ligaments Associated with the Radiocarpal Joint 511
           Western Medicine Etiology and Pathology 512
           Chinese Medicine Etiology and Pathology 512
               Qi and Blood Stagnation 512
           Manifestations 513
           Physical Examination 513
           Differential Diagnosis 514
               De Quervain's Tendinosis 514
               Ganglion Cyst 514
               Scaphoid Fracture of the Wrist 514
               Distal Radius Fracture 514
           Treatments 514

Treatment Principle 514
Herbal Treatments 515
Herbal Formula.. 515
Herbal Ingredients 515
Ingredient Explanations 515
Modifications 515
Formula Indications 515
Contraindications.. 515
Administrations 515
External Wash Formula 516
Herbal Ingredients 516
Ingredient Explanations 516
Modification.. 516
Direction 516
Topical Herbal Formula... 516
Formula Ingredients.. 516
Dose.. 516
Contraindication 517
Suggested Usage 517
Red-Colored Spray – Yunnan Baiyao Aerosol Baoxianye
雲南白藥氣霧劑保險液... 517
Cream-Colored Spray – Yunnan Baiyao Aerosol雲南白藥
氣霧劑.. 517
Acupuncture Treatments.. 517
Modification.. 517
Auricular Therapy 517
Moxibustion Therapy.. 517
Bleeding Therapy.. 517
Tui Na Treatments... 517
Tui Na Procedures 517
Patient Advisory 518
Dislocation of Distal Radioulnar Joint 遠端橈尺關節脫位 519
Introduction 519
Anatomy 519
Articulation... 519
Ligaments 519
Innervation... 519
Blood Supply.. 519
Movements... 519
Western Medicine Etiology and Pathology 520
Volar Dislocation 520
Dorsal Dislocation 520
Chinese Medicine Etiology and Pathology 520
Qi and Blood Stagnation... 520
Manifestations 520

Physical Examination 521
Differential Diagnosis 521
 Distal Radius Fracture 521
Treatments 522
 Treatment Principle 522
 Herbal Treatments 522
 Herbal Ingredients 522
 Ingredient Explanations 522
 Modifications 522
 Formula Indications 522
 Contraindications 522
 Administrations 522
 External Wash Formula 523
 Herbal Ingredients 523
 Ingredient Explanations 523
 Modification 523
 Procedure 523
 Topical Herbal Formula 523
 Formula Ingredients 523
Dose 523
 Contraindication 523
 Suggested Usage 524
 Red-Colored Spray-Yunnan Baiyao Aerosol Baoxianye雲
 南白藥氣霧劑保險液 524
 Cream-Colored Spray-Yunnan Baiyao Aerosol雲南白藥氣
 霧劑 524
 Acupuncture Treatments 524
 Modifications 524
 Moxibustion Therapy 524
 Bleeding Therapy 524
 Auricular Therapy 524
 Tui Na Treatments 524
 Tui Na Procedures 524
 Dorsal Ulna Dislocation 525
Patient Advisory 525
Dislocation of the Wrist Joint 腕關節脫位 526
Introduction 526
Anatomy 527
Western Medicine Etiology and Pathology 528
 Perilunate Wrist Dislocation 528
Chinese Medicine Etiology and Pathology 528
 Qi and Blood Stagnation 528
Manifestations 528
Physical Examination 529
Differential Diagnosis 530

Distal Radius Fracture 530
Treatments 530
Treatment Principle 530
Herbal Treatments 530
Herbal Ingredients 530
Ingredient Explanations 530
Modifications 531
Formula Indications 531
Contraindications 531
Administrations 531
External Wash Formula 531
Herbal Ingredients 531
Ingredient Explanations 531
Modification 532
Direction 532
Topical Herbal Formula 532
Formula Ingredients 532
Dose 532
Contraindication 532
Suggested Usage 532
Red-Colored Spray – Yunnan Baiyao Aerosol Baoxianye
雲南白藥氣霧劑保險液...532
Cream-Colored Spray – Yunnan Baiyao Aerosol雲南白藥
氣霧劑...532
Acupuncture Treatments 533
Acupuncture Point Formula 533
Modifications 533
Cupping Therapy 533
Moxibustion Therapy 533
Bleeding Therapy 533
Auricular Therapy 533
Tui Na Treatments 533
Tui Na Procedures 533
Patient Advisory 534
Notes..534

Chapter 10 Hand and Finger Pain..535

Trigger Finger 板機指 535
Introduction 535
Anatomy 535
Flexor Digitorum Superficialis Tendons..................................535
A1 Pulley 535
Western Medicine Etiology and Pathology 536

Chinese Medicine Etiology & Pathology 537
   Qi and Blood Stagnation 537
Manifestations 537
Physical Examination 537
Differential Diagnosis 538
   Metacarpophalangeal Joint Sprain 538
   Metacarpophalangeal Joint Arthritis 538
Treatments 538
   Treatment Principle 538
   Herbal Treatments 538
   Herbal Ingredients 539
   Ingredient Explanations 539
   Modifications 539
   Formula Indications 539
   Contraindications 539
   Administrations 539
   Acupuncture Treatments 539
   Modifications 539
   Extra Points 540
   Auricular Therapy 540
   Moxibustion Therapy 540
   Tui Na Treatments 540
   Tui Na Procedures 540
Patient Advisory 541
Ganglion Cyst in Wrist 腱鞘囊腫 541
Introduction 541
Anatomy 542
Western Medicine Etiology and Pathology 542
Chinese Medicine Etiology & Pathology 542
Manifestations 542
Physical Examination 543
Differential Diagnosis 543
   Lipoma 543
Treatments 543
   Treatment Principle 543
   Herbal Treatments 543
   Herbal Ingredients 543
   Ingredient Explanations 544
   Modifications 544
   Formula indication 544
   Contraindications 544
   Acupuncture Treatments 544
   Tui Na Treatments 545
   Tui Na Procedures 545

Patient Advisory 546
Finger Sprain 手指扭傷 546
Introduction 546
   Anatomy 546
      Bones 546
      Joints 546
      Ligaments 546
      Muscles 547
      Nerves 547
      Blood Supplies 547
   Western Medicine Etiology and Pathology 547
   Chinese Medicine Etiology and Pathology 547
      Qi and Blood Stagnation... 547
   Manifestations 547
   Physical Examination 548
   Differential Diagnosis 548
      Phalangeal Fracture 548
   Treatments 549
      Treatment Principle 549
      Herbal Treatments 549
      Herbal Ingredients 549
      Ingredient Explanations 549
      Modifications 549
      Formula Indications 549
      Contraindications 549
      Administrations 550
      External Wash Formula: 550
      Herbal Ingredients 550
      Ingredient Explanations 550
      Modification.. 550
      Procedures 550
      Topical Herbal Formula... 550
      Formula Ingredients.. 550
   Dose 551
      Contraindication 551
      Suggested Usage 551
      Red Colored Spray-Yunnan Baiyao Aerosol Baoxianye雲
      南白藥氣霧劑保險液 551
      Cream Colored Spray-Yunnan Baiyao Aerosol雲南白藥氣
      霧劑 551
      Acupuncture Treatments.. 551
      Modifications 551
      Auricular Therapy 551
      Bleeding Therapy... 551

Contents

       Tui Na Treatments.. 552
       Tui Na Procedures 552
     Patient Advisory 552
   Notes.. 553
Bibliography 555
Index... 561

Preface

In the present day, we live with the technology, computers and cellphones that have had a major impact on our lives and the way that we perform everyday tasks; it seems that we cannot live and work without them. Since I came to the US in 1994, I have noticed that the number of patients who suffer from pain in the upper body from using these electronics is increasing dramatically each year, and the patient's age is getting younger as they have fewer barriers to technology use and more engagement in digital platforms.

Overusing electronics is going to cause so many health issues especially pain in the joints and muscles of the upper body. I've noticed that patients who suffer from both acute and chronic pain, including postoperative pain, and the pain caused by the degenerative change in the joints are prescribed hydrocodone, methadone, oxycodone, fentanyl, and many other narcotic analgesic medications. These pain medications attach to receptors on nerves in the brain that increases the threshold to pain, the level of pain stimulus at which we start to notice it. Then these medicines reduce the perception of pain. When taken as directed, these medications can relieve both acute and chronic pain. However, narcotic pain medications have a high potential for misuse, abuse and, worse, it becomes addictive. They can have serious side effects including constipation, headache, trouble breathing, abnormal heart beat, cardiac arrest, and death.

With the opioid crisis in America still at record levels, it is time we educate the communities that pain medications, especially narcotics, are not the only options for pain management. Treatments with Chinese medicine, such as acupuncture, Chinese herbal medicine, Tui Na, and posture correction work effectively to ease pain, help patients with acute and chronic conditions feel more comfortable, and improve their daily function. Acupuncture has no side effects, it activates the body's own self-healing mechanism, and it works through repairing the source of pain and stimulates the body to self-recovery.

Even more commendable, acupuncture treatments also work for opioid dependence. The acupuncture needle is just a piece of metal, there is no opioid medication on it nor any other chemical pain relievers on it. Thanks to the US FDA, on December 6, 1996 and revised as of April 1st, 2019, the FDA Title 21, Volume 8 classified acupuncture needles as class II medical therapeutic devices.

I am a clinical practitioner, I am also an educator. My intention in writing this book is not only to educate the patients who are seeking a solution for managing their pain but also to address several treatment modalities in Chinese medicine for Chinese medicine professionals.

For patients, this book helps you to easily understand your pain, as well as the simple but effective self-care in the section of Patient Advisory at the end of each topic. It advises you on the self-management of your pain and gives you suggestions for prevention and diet.

For medical professionals, this book explains the mechanism of each disease including the concise regional anatomy that relates to the pain, Western and Chinese medicine pathology and etiology, and the best treatment plans.

About the Author

Dr. Apichai is a clinical supervisor at Bastyr Center for Natural Health and a core faculty member in the Department of Acupuncture and East Asian Medicine at Bastyr University, Seattle, Washington. He also has a private practice in Seattle. Dr. Apichai serves as an advisor for the Traditional Chinese Medicine (TCM) Sports Medicine Club. Dr. Apichai specializes in musculoskeletal concerns, such as pain associated with the neck, back, and legs, and in headaches. He earned his Doctor of Medicine (MD) degree from Jinan University, Guangzhou, The People's Republic of China in 1992, where he also completed several medical internships and research projects. He joined the Tui Na Rehab Department at First Branch of Guangzhou TCM Hospital in 1993.

1 Conditions with Multiple Pain Locations

BI Syndrome 痹 證

INTRODUCTION

The word "Bi" (痹) is a Chinese Mandarin term that means an obstruction in the meridians.

Bi Syndrome is a disorder manifested as pain, numbness and heaviness of muscles, tendons, and joints or swelling, hotness, and limitation of movement of joints, resulting from the obstruction of the meridians and sluggishness of Qi and Blood circulation after the invasion of pathogenic Wind, Cold, Dampness, or Heat.

CHINESE MEDICINE ETIOLOGY AND PATHOLOGY

Bi Syndrome in Chinese medicine is caused by either internal factors or external factors.

Internal Causes

a) Weak natural endowment

General weakness of the body inhibits resistance to pathogens; as a result, Wind, Damp, Cold, and Heat can easily invade the body causing Bi Syndrome. For example, a patient with Yang deficiency may experience slow or sluggish blood flow which is susceptible to the effect of Wind, Cold, and Damp; thus, a Bi Syndrome of the Wind-Damp-Cold type may develop (Apichai 2020). A patient with a constitution of Yin deficiency is susceptible to the effect of Wind, Damp, and Heat and is liable to get the Bi Syndrome of the Wind-Damp-Heat type.

b) Indulgence

- Overstrain refers to protracted physical work. Too much physical hard work consumes great amounts of Qi and Blood and can cause disease. Overstrain includes protracted standing, prolonged walking, too much sports activity, and even includes speaking for a long time.
- Over-mental work consumes Heart Blood and can also impair the Spleen. Heart is the mother of the Spleen. Insufficiency of Heart Blood causes the inability to nourish the Spleen.
- Over-sexual activity consumes Kidney Essence, resulting in malnourishment in the five Zang and six Fu organs.

DOI: 10.1201/9781003203018-1

- Over-rest leads to the slow flow of Qi and Blood, resulting in insufficient nourishment in the Zang-fu organs.

Chinese medicine holds that moderation is helpful for the flow of Qi and Blood and the regulation of Yin and Yang. Be active and use your body, but not too much, even in the most common activities – using your eyes, lying down, sitting, standing, and walking.

Excessive activity and overloading ourselves can also diminish our life Essence. "Five Strains" refer to five overload activities that can consume and harm the body.

The Five Strains are:

- Excessive use of the eyes
- Excessive lying down or hunching over
- Excessive sitting
- Excessive standing
- Excessive exercise or overwork
  1) Excessive use of the eyes consumes Liver Blood 久 視 傷 血.

     "Zoom Fatigue" describes the tiredness, worry, or burnout associated with overusing virtual platforms of communication. It is hip terminology that originated during the coronavirus (COVID-19) pandemic. People work, study, and communicate with each other using a virtual platform called Zoom. The eyes stare at the cellphone screen, tablet screen, or computer monitor for hours each day, resulting in Zoom Fatigue syndrome.

     Using the eyes too much causes eye strain, fatigue, poor concentration, poor memory, poor appetite, poor sleep, and vision deterioration.

     "Liver opens into the eyes", as the Yellow Emperor says. The Liver stores Blood, dominates Blood production, and keeps Blood clean. Excessive use of the eyes consumes Liver Blood. The symptoms include dry eyes, blurred vision, poor night vision, insomnia, poor concentration, fatigue, dizziness, thin hair, dry hair, graying of the hair, muscle weakness, muscle cramps, and numbness. In women, it may manifest in scanty and painful menstruation.

  2) Excessive lying down or hunching over consumes Qi久 臥 傷 氣.

     Although sleeping or lying down on the bed helps to restore energy, sleeping too long isn't good for health, as it doesn't restore energy – it drains Qi. Smooth circulation of Qi relies on physical motion; staying in one position for too long leads to slow circulation.

  3) Excessive sitting gains weight久 坐 傷 肉.

     After activities, sitting helps to regain energy, and rest helps muscles to relax. There is less Blood- and energy-consuming when at rest. Too much sitting, however, impairs the Spleen, causing poor appetite and indigestion. Spleen refers to the digestive system in Chinese medicine. Spleen deficiency from sitting too long causes weight gain, fat gain, and excessive fluid accumulation.

4) Excessive standing damages Bones 久 立 傷 骨.

Standing requires strong bones and stabilized joints. Kidney in Chinese medicine provides energy to keep the bones and joints healthy. The back, spine, pelvis, and legs get tired when we stand for too long, resulting in the depletion of consuming Kidney energy. Kidney deficiency leads to weak and unhealthy bones and joints.

5) Excessive walking damages Tendons 久 行 傷 筋.

Walking reduces the risk of heart disease and stroke and improves the flexibility of joints and muscles. Over-walking, however, can strain the muscles and tendons. It deprives the body of Qi and Blood, leading to brittle Tendons.

c) Prolonged illness, a severe disease, or after pregnancy

These conditions lead to the consumption of Qi and Blood and deficiency of the Liver and Kidney, resulting in a deficit of Yin and Yang. Corresponding manifestations, such as fragile bones and joints lack of support from healthy tendons, may occur.

d) Irregular diet

Eating the right food helps to maintain the supply of Qi and Blood. Eating the wrong foods causes illness.

All foods in Chinese medicine are assigned according to the five colors, five flavors, and five temperatures.

The five colors of food nourish the five different Zang organs in our body:

- Green foods cleanse the Liver.
- Red foods nourish the Heart.
- Yellow foods nourish the Spleen.
- White foods nourish and moisten the Lungs.
- Black foods nourish the Kidneys.

The five flavors of food that nourish the five different Zang organs in our body are:

- Bitter taste clears Heat, especially in the Heart.
- Pungent taste helps the Lung to expel the Wind and Cold pathogenic factors that attack the body.
- Salty taste helps to strengthen the Kidney.
- Sweet taste tonifies the Spleen Qi.
- Sour taste can calm the emotions.

The five natural temperatures of food nourish the five different Zang organs in our body:

- Cold temperature cools Fire for those who have too much Heat in the body.
- Cool temperature clears Heat condition.
- Eating neutral temperature foods, such as rice, is recommended.
- Warm temperature foods warm the body to prevent Cold invasion.
- Hot temperature eliminates Coldness and warms the deficiency of Yang.

The natural temperature of foods is not determined by their cooking temperature but rather by what effects they have on a person's body after consumption. When a person continually eats one temperature type of food, it creates an imbalance in their body and affects their immune system. In Chinese medicine, the goal is to keep the body "neutral". Thus, it is important to apply the food temperature types accordingly.

e) Injury

Physical trauma can lead to body deficiency. Soft tissue injury or fractures can stop the flow of Blood leading to the stagnation of Qi and the Blood. Endurance conditions impair the Liver and Kidney, leading to the deficiency of Essence and resulting in the undernourishment of the bones.

External Causes

a) Seasonal effects

Wind, Damp, Heat, and Cold are the main causes of seasonal effects. Chapter 43, the *Bi Syndrome* 痹 論, of *Su Wen* (素 問 *Plain Questions*), a part of The Yellow Emperor's Canon of Internal Medicine (黃 帝 內 經, BC475–221), contains a discussion between the Yellow Emperor and his physician Qi Bo岐 伯: "If pathogenic Wind, Cold, and Dampness invade the body together, it will lead to obstruction in the meridians, and *Bi* Syndrome may take place". 黃 帝 問 曰 ： 痹 之 安 生 。 歧 伯 對 曰 ： 風 寒 濕 三 氣 雜 至 ， 合 而 為 痹 也 。

Usually, these four factors will not be the cause simultaneously – commonly, only two or three join together to invade the meridians. They do not equally affect the disease, as one of them often is emphasized. For example, if a patient presents a chief complaint of right shoulder pain, even when Wind, Damp, and Cold are the causes of the pain, Cold is the primary.

Abnormal weather can be the main cause of the invasion. For instance, the weather is supposed to be warm in Spring, but it is cold; or the climate is hot in Winter. On the other hand, the change of climate might be too sudden, or the temperature is extreme, exceeding the human body's ability to adapt and regulate. Clinically, the pain intensity in patients with rheumatoid arthritis often is aggravated or alleviated with climate change.

b) Living environment effects

Living in a place where it is cold or very humid may lead to body pain, while working in a windy environment leads to convulsions and pain.

c) Equipment effects

The following factors are examples of common equipment effects that cause Bi Syndrome: improper clothing, such as wearing shorts in windy, cold, and rainy weather; sleeping with an electric fan blowing; and sitting or sleeping on a wet lawn. These effects lead to the invasion of Wind, Damp, and Cold that cause blockages resulting in the stagnation of Qi and Blood.

CLASSIFICATION OF BI SYNDROME

1. Wind Bi Syndrome 風 痹
   - It is also called Wandering Bi行 痹.

- There are two sources of Wind in Chinese medicine: Internal Wind (from the Liver) and External Wind (from the environment). Wind Bi Syndrome comprises the symptoms caused mainly by external Wind.
- External Wind often penetrates the body through the back of the neck.
- External Wind is Yang energy, characterized by rapid onset and quickly changing symptoms that affect the upper body and the surface of the body.
- Wind usually carries Damp, Cold, and Heat to cause blockages in the meridians, resulting in pain. Wind dominates among the symptoms.
- The outcome of blockages caused by Wind invasion is Qi and Blood stagnation.

2. Cold Bi Syndrome寒 痹
   - It is also called Painful Bi Syndrome 痛 痹.
   - Cold is extremely Yin and, therefore, can easily injure the Yang Qi of the body.
   - Cold can enter through the skin and pores as well as the mouth, nose, and ears.
   - Cold constricts the sweat pores and causes contraction, tension, and pain in the muscles and joints.
   - Yang Qi's functions are movement and transformation. Cold can impair these functions, disrupt digestion, and cause fluid accumulation.

3. Dampness Bi Syndrome 濕 痹
   - It is also called Fixed Bi Syndrome 着 痹.
   - Damp is thick and hard to move. Even when Qi is the commander, Damp is not easily regulated by Qi; thus, it can be difficult to drain or dissolve.
   - The Spleen is a Zang organ that transforms Damp, but Damp often disrupts the Spleen's transformation and transportation functions.
   - The thickness of Damp also blocks the movement and transformation of Yang Qi causing the sensation of heaviness, distention, and fatigue that may manifest in swelling.

4. Heat Bi Syndrome 熱 痹
   - Onset is usually rapid.
   - It can be either an excessive condition or a Heat condition as well as a mixed condition of both deficiency and excess.
   - Heat is Yang energy. It attacks Yin energy, including Blood and Body Fluids, resulting in dryness of the skin, lips, mouth, and tongue.
   - Heat can stir up Wind and block the circulation of Qi.
   - Heat Bi Syndrome can cause Qi and Blood stagnation due to the direct invasion of Heat or the latent condition of Wind-Cold that is transformed to Heat. It can also be due to the Yang-type constitution and Yin deficiency constitution.
   - Damp gives rise to Heat, but Heat predominates.

5. Blood stasis Bi Syndrome血 痹
   - Pathogenic factors enter the body at the Blood stage, causing deficiency of Spleen and Kidney.

- Underlying Qi and Blood deficiency, sleeping in blowing wind, or sweating while working give chance to Wind invasion, resulting in obstruction of the flow of Qi and Blood.

6. Bone Bi Syndrome骨 痹

   A more advanced and chronic form, Bone Bi Syndrome develops from any of the previous Bi Syndrome that obstruct body fluids, which then congeal into Phlegm. This further obstructs the flow of Qi, resulting in muscle atrophy and deformity of bones and joints. Internal organs, especially the Liver and Kidney, may be more highly affected.

7. Muscle Bi Syndrome 肌 痹
   - Muscle Bi Syndrome develops from the invasion of the muscles by Wind, Damp, Cold, or Heat.

8. Vessel Bi Syndrome 脈 痹
   - Vessel Bi Syndrome is the invasion of Wind, Damp, and Cold, which obstruct the vessel.

MANIFESTATIONS, DIAGNOSIS AND TREATMENTS

1. Wind Bi行 痹 型

Main symptoms: arthralgia-pain and aching of muscles and joints, limited range of motion, and pain moving between different areas; fever and aversion to wind.
Tongue: pale body, thin white coating.
Pulse: superficial.
Treatment principles: disperse Wind and Cold, dredge the meridians, and eliminate Dampness.

Herbal Formula

*Fang Feng Tang*防 風 湯
This formula is from "Elucidated Prescriptions and Expositions" of Huangdi's *Plain Questions* 黃 帝 素 問 宣 明 論 方", authored by Liu Wansu 劉 完 素 (1100–1180), in Jin dynasty 金 朝 (1115–1234) and published in 1172.

Herbal Ingredients

Fang Feng 防 風6g, Ma Huang 麻 黃3g, Qin Jiao秦 艽 3g, Xing Ren杏 仁 2g, Ge Gen葛 根 3g, Rou Gui 肉 桂1.5g, Fu Ling 茯 苓3 g, Dang Gui 當 歸3g, Huang Qin 黃 芩2g, Gan Cao 3g, Sheng Jiang 生 姜3 slices, Da Zao 大 棗3 pieces

Ingredient Explanations

- Fang Feng防 風 expels Wind and drains Damp.
- Ma Huang 麻 黃 induces sweating, releases the Exterior, and warms and disperses Cold pathogens.

- Qin Jiao秦 芄 expels Wind-Dampness, opens the channels, and soothes the Sinews and collaterals.
- Xing Ren杏 仁 stops cough, calms wheezing, and moistens the Intestines.
- Ge Gen葛 根 relieves Heat and releases muscles, especially of the neck and upper back.
- Rou Gui 肉 桂 warms the meridians, expels Cold, and opens the meridians to promote Blood circulation.
- Fu Ling茯 苓 drains Damp and tonifies the Spleen.
- Dang Gui當 歸 moves the Blood, disperses Bruising, and stops pain due to Blood stasis.
- Huang Qin黃 芩 clears Heat, dries Damp, cools the Blood, stops bleeding, and calms the Liver Yang.
- Zhi Gan Cao炙 甘 草 reduces swelling and stops pain.
- Sheng Jiang 生 姜 unblocks the pure Yang pathway and harmonizes rebellious Qi.
- Da Zao大 枣 tonifies the Spleen Qi, nourishes the Blood, and moderates and harmonizes the harsh properties of the fragrant herbs.

Modifications

To make the formula stronger, add Gui Zhi 桂 枝, Qiang Huo 羌 活, Du Huo 獨 活, Wei Ling Xian威 靈 仙, and Fang Ji防 己.

Formula Indications

- Wind-Damp *Bi* is Wind predominant.

Pattern Herbs

- Feng Shi Pian 風 濕 馬 錢 片.
- Zhui Feng Huo Xue Pian 追 風 活 血 片.

2. Cold Bi痛 痹 型

Main symptoms: extremely severe pain in muscle or joint, and the location is fixed with limited range of motion; symptoms are made worse by invasion of Cold weather, or exposure to Cold, and alleviated by warmth; skin color is not red, palpation is not hot.
Tongue: thin white coating.
Pulse: tight, wiry.
Treatment principles: warm the meridians, disperse Cold and Wind, and eliminate Dampness.

Herbal Formulas

1) **Wu Tou Tang** 烏 頭 湯

This formula is from "Shang Han Lun 傷 寒 雜 病 論", authored by Zhang Zhong-Jing 張 仲 景 (150–219), in eastern Han dynasty 東 漢 (25–220), and published in 200–210.

Herbal Ingredients

Zhi Fu Zi制 附 子6 g, Ma Huang, 麻 黄 9 g, Gui Zhi桂 枝6-10 g, Chi Shao赤 芍9 g, Zhi Gan Cao炙 甘 草6 g

Ingredient Explanations

- Zhi Fu Zi制 附 子 warms Ming Men Fire.
- Ma Huang 麻 黄 induces sweating, releases the Exterior, and warms and disperses Cold pathogens.
- Gui Zhi桂 枝 warms the meridians and relieves pain.
- Chi Shao赤 芍 moves the Blood, disperses Bruising, and stops pain.
- Zhi Gan Cao炙 甘 草 tonifies the Spleen, harmonizes the other herbs, and guides the herbs to all 12 meridians.
- Huang Qi黄 芪 tonifies Qi, strengthens the Spleen, raises the Yang Qi of the Spleen and Stomach, tonifies Wei Qi, stabilizes the exterior, and tonifies the Blood.
- Sheng Jiang生 姜 unblocks the pure Yang pathway and harmonizes rebellious Qi.
- Da Zao大 枣 tonifies the Spleen Qi, nourishes the Blood, and moderates and harmonizes the harsh properties of the fragrant herbs.

Formula Indications

- Wind-Cold-Damp Bi (Liu, Tseng, and Yang 2005) with Cold predominant.

2) **Wu Fu Ma Xin Gui Jiang Tang**烏 附 麻 辛 桂 姜 湯

This formula is from "Chinese Medicine Treatment and Formula中 醫 治 法 與 方 劑", authored by Chen Chaozu 陳 潮 祖 (1929–2018), in the People's Republic of China中 華 人 民 共 和 國, and published on December 12, 2009.

Herbal Ingredients

- Zhi Wu Tou 制 烏 頭3g, Zhi Fu Zi 制 附 子1.5g, Ma Huang 麻 黄3g, Xi Xin 細 辛1.5g, Gui Zhi桂 枝 3g, Gan Jiang 乾 薑1.5g, Gan Cao甘 草 2g

Ingredient Explanations

- Zhi Chuan Wu 制 烏 頭 expels Wind-Damp, warms the meridians, and stops pain.
- Zhi Fu Zi 制 附 子 warms Ming Men Fire.
- Ma Huang 麻 黄 induces sweating, releases the exterior, and warms and disperses Cold pathogens.
- Xi Xin 細 辛 enters the Shao Yin meridian to expel Wind, Damp, and Cold.
- Gui Zhi桂 枝 warms the meridians and relieves pain.
- Gan Jiang 乾 薑 warms Middle Jiao, expels interior Cold and dispels Wind-Dampness.
- Gan Cao甘 草 harmonizes the herbs, tonifies Middle Jiao Qi, clears Heat, expectorates Phlegm, and stops cough.

Modifications

- For pain on upper limb, add Qiang Huo羌活, Wei Ling Xian威靈仙, and Qian Nian Jian千年健.
- For pain on lower limb, add Du Huo獨活, Niu Xi牛膝, and Fang Ji防己.
- For pain on the lumbar, add Sang Ji Sheng桑寄生, Du Zhong杜仲, Xu Duan續斷, and Yin Yang Huo淫羊藿.

Formula Indications

Wind-Cold-Damp with Cold Predominant.

Cooking Instructions

Precook Zhi Chuan Wu 制烏頭 and Zhi Fu Zi 制附子.

Contraindications

Notice that this is a very hot formula. Use appropriate precautions. May not be appropriate on patients with hypertension or conditions with Heat signs. Monitor patient closely, especially when starting the formula.

3) **Gui Zhi Jia Fu Zi Tang** 桂枝加附子湯

This formula is from "Shang Han Lun 傷寒雜病論", authored by Zhang Zhong-Jing 張仲景 (150–219), in eastern Han dynasty 東漢 (25–220), and published in 200–210.

Herbal Ingredients

Fu Zi 附子3g, Gui Zhi 桂枝6g, Gan Cao 甘草3g, Sheng Jiang 生薑3 slices, Da Zao 大棗3 dates

Ingredient Explanations

- Zhi Fu Zi 制附子disperses Cold and Dampness, warms the meridians, and stops pain.
- Gui Zhi桂枝 warms the meridians and relieves pain.
- Gan Cao甘草 harmonizes the herbs, tonifies Middle Jiao Qi, clears Heat, and expectorates Phlegm, Sheng Jiang 生薑. Unblocks the pure Yang pathway and harmonizes rebellious Qi.
- Da Zao 大棗 tonifies the Spleen Qi, nourishes the Blood, and moderates and harmonizes the harsh properties of the fragrant herbs.

Modifications

- To make the formula stronger, add Zhi Chuan Wu制川烏.
- For pain on upper limb, add Qiang Huo羌活, Chuan Xiong川芎, and Jiang Huang薑黃.
- For pain on lower limb, add Du Huo獨活, Niu Xi牛膝, Fang Ji防己.
- For pain on lumbar, add Du Zhong杜仲, Xu Duan續斷 and Sang Ji Sheng桑寄生.

Formula Indications
- Wind-Cold Invasion with underlying Yang Deficiency.

Cautions
Even though this formula is for conditions with profuse sweating, it may be too weak for the condition of Yang Qi collapse. In that case, Si Ni Tang 四逆湯should be used.

Patent Formulas
- Xiao Huo luo Dan小活絡丹.
- Feng Shi Ling Pian 風濕靈片.

4) **Warming soak** – external use only
- Chuan Wu川烏, Cao Wu草烏, Chuan Jiao川椒, Tou Gu Cao透骨草, Ai Ye艾葉, Cang Zhu蒼朮, Du Huo獨活, Gui Zhi桂枝, Fang Feng防風, Hong Hua紅花, Shen Jin Cao伸筋草, Liu Ji Nu劉寄奴 – all 9 grams.

Directions for soaking:
- Add the herbs to 2 gallons of water.
- Bring to a boil.
- Reduce the flame and cook for 20–25 minutes.
- Remove the pot from the stove.
- When cool enough, put the injured part in enough water to cover the area, or soak a towel and wrap the area.
- The liquid needs to stay warm to penetrate. Use warm towels or a heat lamp. The soak should last 20–30 minutes.
- The remaining liquid can be reused. Simply reheat.

Adding alcohol or vinegar:
- These can add to the effectiveness of the herbal soak.
- Add after you have removed the herbs from the stove.
- Vinegar – reduces spasms in muscles and tendons and soothes flow of Qi.
- Alcohol – warms tissues and improves local circulation of Qi and Blood.

5) **Tendon lotion (linament)**
- Cao Wu草烏, Chuan Wu, Tao Ren桃仁, Ma Huang麻黃, Zi Ran Tong自然銅, Mo Yao沒藥, Ru Xiang乳香, Da Huang大黃, Lu Lu Tong路路通, Zhang Mu樟木.
- Put the herbs in 1 gallon of 80–100 proof vodka or rice wine. Soak for a month or longer.
- Cautions: Do not use liniment if patient is hypertensive or uses blood thinners.
- Use for chronic injuries to tendons and ligaments.
- The emphasis is on warming herbs (although it contains both cooling and warming herbs) in order to increase circulation locally.

- Do not use if inflammation is present.
- This increases local circulation and keeps Cold and Damp from settling in this area.
- Put liniment on fingers and thumb and massage into the area twice each day.
- If the liniment makes this area feel worse, there is trauma present. Discontinue use until inflammation subsides.

3. **Damp Bi**着 痹 型

Main symptoms: sore pain, achiness and swollen/distended feeling, muscles and joints may feel heavy and numb; exposure to Damp weather can worsen the outbreak.
Tongue: white greasy coating.
Pulse: slow pulse.
Treatment principles: eliminate Dampness, dredge the meridians, and expel Wind and Cold.
Herbal Formulas

1) **Yi Yi Ren Tang** 薏 苡 仁 湯

This formula is from "Wonderful Well-Tried Recipes奇 效 良 方", authored by Dong Su董 宿, Fang Xian 方 賢 (unknown year of birth and death), in Ming dynasty 明 朝 (1368–1644), and published in 1470.

Herbal Ingredients

Yi Yi Ren薏 苡 仁 6g, Dang Gui 當 歸6g, Bai Shao 白 芍 10g, Cang Zhu 苍 术 6g, Ma Huang 麻 黄3g, Gui Zhi 桂 枝3g, Gan Cao 甘 草 3g, Sheng Jiang 生 姜3 slices

Ingredient Explanations

- Yi Yi Ren薏 苡 仁 promotes urination, leaches out Dampness, strengthens the Spleen, expels Wind-Dampness, and clears Damp-Heat.
- Dang Gui 當 歸 moves the Blood, disperses Bruising, and stops pain due to Blood stasis.
- Bai Shao 白 芍 nourishes the Blood, soothes the Liver, and stops pain.
- Cang Zhu 苍 术 strongly dries Dampness, tonifies the Spleen, and clears Dampness from the Lower Jiao.
- Ma Huang 麻 黄 induces sweating, releases the Exterior, and warms and disperses Cold pathogens.
- Gui Zhi 桂 枝 warms the meridians and relieves pain.
- Gan Cao 甘 草 harmonizes the herbs, tonifies Middle Jiao Qi, clears Heat, expectorates Phlegm, and stops cough.
- Sheng Jiang 生 姜(slices) unblocks the pure Yang pathway and harmonizes rebellious Qi.

Modifications
- For severe pain and aversion to Cold, add Du Huo獨活, Fang Feng防風, Wu Tou 川 鳥, Chuan Xiong 川 芎, Fu Zi 附 子, and Qiang Huo羌活.

Formula Indications
- Wind-Cold Damp Bi with Dampness predominant.

2) **Juan Bi Tang**蠲痹湯

This formula is from "A Book of Formulas to Promote Well-Being 嚴氏濟生方", authored by Yan Yonghe 嚴用和 (1200~1268), in Song dynasty南宋 (1127–1239), and published in 1253.

Herbal Ingredients
Dang Gui當歸9g, Chi Shao赤芍9g, Jiang Huang薑黃9g, Huang Qi黃耆9g, Qiang Huo羌活9g, Gan Cao甘草3g, Sheng Jiang生薑15g, Da Zao大棗3 pieces

Ingredient Explanations
- Dang Gui當歸 nourishes the Blood, benefits the Liver, and regulates menstruation.
- Chi Shao赤芍 moves the Blood, disperses Bruising, and stops pain.
- Jiang Huang薑黃 moves the Blood, opens the meridians, expels Wind, and reduces swelling.
- Huang Qi黃耆 tonifies Qi, strengthens the Spleen, raises the Yang Qi of the Spleen and Stomach, tonifies Wei Qi, stabilizes the Exterior, and tonifies the Blood.
- Qiang Huo羌活 expels Wind-Cold-Dampness, unblocks painful obstructions, and alleviates pain.
- Gan Cao甘草 harmonizes the herbs, tonifies Middle Jiao Qi, clears Heat, expectorates Phlegm, and stops cough.
- Sheng Jiang生薑 unblocks the pure Yang pathway and harmonizes rebellious Qi.
- Da Zao大棗 tonifies the Spleen Qi, nourishes the Blood, and moderates and harmonizes the harsh properties of the fragrant herbs.

Modifications
- For bringing the formula to the arm, add Gui Zhi桂枝.
- For pronounced swelling of joints, add Bi Xie萆薢, Mu Tong木通, and Jiang Huang薑黃.
- For Cold symptoms, add Fu Zi附子.
- For Damp symptoms, add Cang Zhu蒼朮, Fang Ji防己, Yi Yi Ren薏苡仁, Hai Tong Pi海桐皮, or Xi Xian Cao豨薟草.
- For Wind symptoms, add Fang Feng防風.
- For Blood stasis, add Tao Ren桃仁, Hong Hua紅花, and Di Long 地龍.

Formula Indications
- Wind-Cold-Damp Bi Syndrome.

Contraindications
- Contraindicated for patients with Damp-Heat in the meridians.

3. **Patents**
- Guan Jie Yan Wan 關 節 炎 丸

4. **Heat Bi**熱 痹 型

Main symptoms: severe pain and hot sensation in the muscles and joints; warm to the touch, the skin appears red and swollen.
Tongue: yellow tongue coating.
Pulse: smooth and rapid pulse.
Treatment principles: clear Heat, dredge the meridians, expel Wind, and eliminate Dampness.

Herbal formula

Xuan Bi Tang宣 痹 湯

Herbal Ingredients
Han Fang Ji防 己15g, Xing Ren杏 仁15g, Yi Yi Ren薏 苡 仁15g, Can Sha蚕 砂9g, Ban Xia半 夏9g, Lian Qiao連 翹9g, Zhi Zi山 栀 子9g, Hua Shi滑 石15g, Chi Xiao Dou赤 小 豆9g

Ingredient Explanations
- Fang Ji防 己 promotes urination and reduces edema, especially in the lower body, expels Wind-Dampness, and alleviates pain.
- Xing Ren杏 仁 stops cough, calms wheezing, and moistens the Intestines.
- Yi Yi Ren薏 苡 仁 promotes urination, leaches out Dampness, strengthens the Spleen, expels Wind-Dampness, and clears Damp-Heat.
- Can Sha蠶 沙 expels Wind, drains Damp, and harmonizes the Stomach.
- Ban Xia半 夏 transforms Phlegm.
- Lian Qiao連 翹 clears Upper Jiao Heat.
- Zhi Zi栀 子 clears Heat, cools Blood, resolves Damp, reduces swelling, and opens the meridians.
- Hua Shi滑 石 clears Phlegm Heat and Heat from Urinary Bladder, promotes urination, and drains Damp.
- Chi Xiao Dou赤 小 豆 clears Damp-Heat, reduces swelling, and promotes urination.

Modifications
- For severe pain, add Jiang Huang薑 黃 and Hai Tong Pi海 桐 皮.
- For Blood deficiency, add Dang Gui當 歸and/or Bai Shao白 芍.
- For Yin deficiency, add Sheng Di Huang生 地 黃and Zhi Mu知 母.
- For Damp-Heat, add Er Miao San二 妙 散.

Formula Actions
- Clears and transforms Heat, opens the meridians, and stops pain.

Formula Indication
- Wind-Damp-Heat Bi Syndrome.

Cooking Instructions
1) Place the herbs into an herb pot. Traditionally, a clay cooker is used. Gently rinse the herbs once with a strainer.
2) Add 1.6 liters water to the pot.
3) Presoak the herbs for 15 minutes.
4) Turn on the stove to high heat and bring the herbs to a boil. Then reduce the heat to medium low to simmer with light bubbling. Turn off the heat when the decoction is reduced to 600 milliliters.
5) Drain the decoction into a container.
6) Divide the decoction into three separate cups.

Dose
Drink one cup of the decoction after meals three times daily.

5. Bone Bi骨痹型

Wind-Cold-Damp Type風寒濕痹症
Main symptoms: extremely severe fixed pain in joints and muscles along the back and limbs; sore pain, distended pain, heavy sensation in limbs; edema in joints or dysarthrosis; stiffness of limbs and movement difficulty; profuse sweating, irritability, fatigue, low grade fever, aversion to wind and cold.
Tongue: pale red tongue body, thin white (greasy) coating.
Pulse: thready, weak.
Treatment principles: drain the Damp, warm the channels, expel Wind, and open the meridians.

Herbal Formula
Yi Yi Ren Tang 薏苡仁湯

Herbal Ingredients
Yi Yi Ren薏苡仁 30g, Chuan Xiong川芎12g, Dang Gui當歸6g, Qiang Huo羌活 15g, Du Huo獨活20g, Ma Huang麻黃3g, Gui Zhi桂枝3g, Niu Xi牛膝 20g, Fang Feng防風12g

Ingredient Explanations
- Yi Yi Ren薏苡仁 promotes urination, leaches out Dampness, strengthens the Spleen, expels Wind-Dampness, and clears Damp-Heat.
- Chuan Xiong川芎 moves the Blood, disperses Bruising, and stops pain due to Blood stasis.

- Dang Gui 當歸 moves the Blood, disperses Bruising, and stops pain due to Blood stasis.
- Qiang Huo羌活 expels Wind-Cold-Dampness, unblocks painful obstructions, and alleviates pain.
- Du Huo獨活 expels Wind, Damp, and Cold in Lower Jiao and joints; and removes chronic Bi stagnation in lower limbs.
- Ma Huang 麻黃 induces sweating, releases the Exterior, and warms and disperses Cold pathogens.
- Gui Zhi 桂枝 warms the meridians and relieves pain.
- Niu Xi牛膝 moves Blood and releases Bruising.
- Fang Feng防風 expels Wind and drains Damp.

Modifications
- For severe Coldness, add Fu Zi 附子 and Wu Tou 川烏.
- For swollen joints, add Fu Ling茯苓, Ze Xie澤瀉, and Che Qian Cao車前草.
- For pain in the upper limb, add Xi Xin细辛 and Jiang Huang姜黃.
- For pain in the lower limb, add Song Jie松節 and Zuan Di Feng鑽地風.
- For dry or sore throat after drinking the formula, add Mai Men Dong麦冬, Sheng Di Huang生地, and Xuan Shen玄參.

Formula Indications
- Wind-Cold Damp Bi with Dampness predominant.

Damp Heat Type濕熱蘊結證
Main symptoms:
red and swollen joints, hot sensation and warm to touch, edema in the joints.

Difficulty moving the joints, fever, sweating, irritability, bitter taste and dry sticky feeling in the mouth. Poor appetite, dark yellow urine.
Tongue: red tongue, yellow greasy coating.
Pulse: rapid, slippery.
Treatment principles: drain Damp, clear Heat, and expel Wind.

Herbal Formula
Xuan Bi Tang宣痺湯

Herbal Ingredients
Han Fang Ji防己15g, Xing Ren杏仁15g, Yi Yi Ren薏苡仁15g, Can Sha蚕砂9g, Ban Xia半夏9g, Lian Qiao連翹9g, Zhi Zi山栀子9g, Hua Shi滑石15g, Chi Xiao Dou赤小豆9g

Ingredient Explanations
- Fang Ji防己 promotes urination and reduces edema, especially in the lower body; expels Wind-Dampness and alleviates pain.

- Xing Ren杏 仁 stops cough, calms wheezing, and moistens the Intestines.
- Hua Shi滑 石 clears Phlegm Heat and Urinary Bladder Heat, promotes urination, and drains Damp.
- Lian Qiao連 翹 clears Upper Jiao Heat.
- Zhi Zi梔 子 clears Heat, cools the Blood, resolves Damp, reduces swelling, and opens the meridians.
- Yi Yi Ren薏 苡 仁 promotes urination, leaches out Dampness, strengthens the spleen, expels Wind-Dampness, and clears Damp-Heat.
- Ban Xia半 夏 transforms Phlegm.
- Can Sha蠶 沙 expels Wind, drains Damp, and harmonizes the Stomach.
- Chi Xiao Dou赤 小 豆 clears Damp-Heat, reduces swelling, and promotes urination.

Modifications

- For severe pain, add Jiang Huang 薑 黃, Hai Tong Pi海 桐 皮.
- For Blood deficiency, add Dang Gui 當 歸, and/or Bai Shao 白 芍.
- For Yin deficiency, add Sheng Di Huang 生 地 黃and Zhi Mu知 母.
- For Damp-Heat, add Er Miao San二 妙 散.

Formula Actions

- Clears and transforms Heat, opens the meridians, and stops pain.

Formula Indication

- TMD due to Wind-Damp-Heat Bi Syndrome.

Cooking Instructions

1) Place the herbs into an herb pot. Traditionally, a clay cooker is used. Gently rinse the herbs once with a strainer.
2) Add 1.6 liters water to the pot.
3) Presoak the herbs for 15 minutes.
4) Turn on the stove to high heat and bring the herbs to a boil. Then reduce the heat to medium low to simmer with light bubbling. Turn off the heat when the decoction is reduced to 600 milliliters.
5) Drain the decoction into a container.
6) Divide the decoction into three separate cups.

Dose

- Drink one cup of the decoction after meals three times daily.

Liver Kidney Deficiency Type肝 腎 虧 損 證

Main symptoms:
Pain is located mainly in the low back but can radiate to upper back and neck or to hip and knees; stiffness sensation, difficulty with movement, especially leaning forward; pain at inguinal region; afternoon fever, spontaneous sweating, and night sweats.

Tongue: red tip, peeled coating or thin white.

Pulse: deep, thin, or rapid thin.

Treatment principles: tonify the Kidney and the Liver, move the Blood, and open the channels.

Herbal Formula
Da Bu Yuan Jian 大 補 元 煎

This formula is from "The Complete Compendium of Jingyue景 岳 全 書", authored by Zhang Jiebin張 介 賓 (1111–1117 AD), in Ming dynasty明 代 (1624 AD).

Herbal Ingredients
Ren Shen人 參6g, Shan Yao山 藥 炒6g, Shu Di Huang熟 地6g, Du Zhong杜 仲6g, Dang Gui當 歸6g, Shan Zhu Yu山 茱 萸3g, Gou Qi Zi枸 杞 子6g, Zhi Gan Cao甘 草 炙3g

Ingredient Explanations
- Ren Shen人 參 tonifies Qi and Yang.
- Shan Yao山 藥 炒 tonifies Kidneys and benefits the Yin and the Spleen Qi.
- Shu Di Huang熟 地 黃 tonifies the Liver and Kidney Yin and benefits the Essence and the Blood.

The characteristics of Damp濕:

- Du Zhong杜 仲 tonifies the Liver and Kidney Yang, strengthens the Bones, and benefits the Essence and the Blood.
- Dang Gui當 歸 moves the Blood, disperses Bruising, and stops pain due to Blood stasis.
- Shan Zhu Yu山 茱 萸 tonifies the Liver and Kidney, strengthens the Bones, and benefits the Essence and the Blood.
- Gou Qi Zi枸 杞 子 tonifies the Liver and Kidney, strengthens the Bones, and benefits the Essence and the Blood.
- Zhi Gan Cao甘 草 炙 reduces swelling and stops pain.

Modifications
- For Cold symptoms due to Yang deficiency, add Fu Zi附 子, Rou Gui肉 桂, Pao Jiang炮 薑.
- For Blood stasis headache, remove Shan Zhu Yu山 茱 萸 and add Chuan Xiong川 芎.

Formula actions
- Severe depletion of Qi and Blood and Kidney Essence deficiency.

Formula Indication
- Headache due to Kidney Essence deficiency and severe loss of Yuan Qi and Blood.

Contraindications
- Exercise caution for patients who suffer from Yin deficiency with deficient Fire, Heat in Blood stage, Stomach Fire, Lung Phlegm Heat, Wind invasion, common cold, or Wind Heat.

- While taking this formula, avoid consuming spicy and hot foods, cold and raw foods, or fatty foods.

Cooking Instructions

1. Place the herbs into an herb pot. Traditionally, a clay cooker is used. Gently rinse the herbs once with a strainer.
2. Add 400 ml of water to the pot.
3. Turn on the stove to high heat and bring the herbs to a boil. Turn off the heat when the decoction is reduced to 280 ml.
4. Drain the decoction into a container.

Dose

- Drink one cup of the decoction before meals twice daily.
- Stop taking the formula if the symptoms are aggravated or are not alleviated.
- Do not take the formula longer than two weeks.

Accumulation of Phlegm and Blood Type 痰瘀互結證

Main symptoms: joint deformity and pain; difficulty with and pain aggravated by movement; feeling hot or cold; fatigue; tremors of the hands or four limbs.
Tongue: purple, or with petechia; white greasy coating.
Pulse: deep and thin or choppy.
Treatment principles: tonify Blood and Qi, dissolve Phlegm, and move the Blood.

Phlegm Obstruction in the Joints

Treatment principles: moves the Blood, disperses Bruising, and opens the meridians.

Herbal Treatments
Herbal Formula
Tao Hong Yin 桃紅飲
This formula is from "Systematized Patterns with Clear-Cut 類證治裁", authored by Lin Pei Qin林佩琴 (1772–1839), in Qing dynasty 清朝 (1636–1912), and published in 1851.

Herbal Ingredients
Tao Ren桃仁9g, Hong Hua紅花 9g, Chuan Xiong川芎9g, Dang Gui Wei當歸尾 9g, Wei Ling Xian威靈仙9g

Ingredient Explanations
- Tao Ren桃仁 moves the Blood, disperses Bruising, and stops pain due to Blood stasis.
- Hong Hua 紅花 moves the Blood, releases Bruising, opens the meridians, and stops pain.

- Chuan Xiong川芎 moves the Blood, disperses Bruising, and stops pain due to Blood stasis.
- Dang Gui Wei當歸尾 moves the Blood, disperses Bruising, unblocks stagnation, opens the meridians, and stops pain.
- Wei Ling Xian威靈仙 expels Wind Damp and stops pain.

Modifications
- For low back pain, add Sang Ji Sheng桑寄生 and Gou Ji狗脊.
- For Qi deficiency, add Huang Qi黃芪 and Dang Shen党參.
- For severe Cold, add Fu Zi附子.
- For stiffness in joints, add Shen Jin Cao伸筋草 and Hai Feng Teng海風藤.

Formula Indications
- Blood stasis with a minor Invasion of Wind-Cold-Dampness.
- Stubborn Bi.

Contraindications
- Use caution if the condition of Blood Heat is present.
- It is contraindicated for pregnancy.

Blood Stagnation in Joints
***Treatment principles*:** warms Yang, resolves retained fluid, and promotes diuresis.

Herbal Treatments

Herbal Formula **Huo Luo Xiao Ling Dan**活絡效靈丹
This formula is from "Records of Chinese Medicine with Reference to Western Medicine醫學衷中參西錄", authored by Zhang Xichun張錫純" (1860~1933), and published in 2009.

Herbal Ingredients
Dang Gui當歸15g, Dan Shen丹參15g, Ru Xiang乳香15g, Mo Yao沒藥15g

Ingredient Explanations
- Dang Gui當歸 moves the Blood, disperses Bruising, and stops pain due to Blood stasis.
- Dan Shen丹參 moves the Blood, nourishes the Blood, and opens microcirculation.
- Ru Xiang乳香 moves the Blood, disperses Bruising, and stops pain due to Blood stasis.
- Mo Yao沒藥 moves Qi and the Blood and stops pain due to Qi and Blood stagnation.

Modifications
- For leg pain, add Niu Xi牛膝.

- For arm pain, add Lian Qiao連 翹.
- For Blood stasis in women, add Tao Ren桃 仁 and Wu Ling Zhi五 靈 脂.

Formula Indications
- Pain due to Qi and Blood Stagnation.

Contraindications
- Those without Blood Stasis condition should not use.
- It is contraindicated for pregnancy.

Acupuncture Treatments

Local acupuncture points:

- Shoulder – Ashi acupoints 阿 是 穴, LI15肩 髃Jianyu, SJ14肩 髎Jianliao, SI10臑 俞Naoshu.
- Elbow – Ashi acupoints 阿 是 穴, LI11曲 池Quchi, LU5尺 澤Chize, SJ10天 井Tianjing, SI8小 海Xiaohai, HT3少 海Shaohai.
- Wrist – Ashi acupoints 阿 是 穴, SJ4陽 池Yangchi, LI5陽 溪Yangxi, SJ5外 關Waiguan, SI4腕 骨Wangu.
- Spine – Ashi acupoints 阿 是 穴, DU14大 椎DaiZhui, DU12身 柱Shenzhu, DU3腰 陽 關Yaoyangguan, Jiaji夾 脊.
- Hip – Ashi acupoints 阿 是 穴, GB30環 跳Huantiao, GB29居 髎Juliao, UB54秩 邊Zhibian.
- Thigh – Ashi acupoints 阿 是 穴, ST32伏 兔Futu, UB36承 扶Chengfu, UB37殷 門Yinmen, GB34陽 陵 泉Yanglingquan, GB31風 市Fengshi.
- Knee – Ashi acupoints 阿 是 穴, Xiyan膝 眼, ST34梁 丘Liangqiu, GB34陽 陵 泉Yanglingquan, GB33膝 陽 關Xiyangguan.
- Ankle – Ashi acupoints 阿 是 穴, UB62申 脈Shenmai, UB60崑 崙Kunlun, KD6照 海Zhaohai, GB40丘 墟Qiuxu.

Modifications

- For Wind Bi Syndrome, add GB20風 池Fengchi, LI4合 谷Hegu, and LIV3太 沖Taichong.
- For Cold Bi Syndrome add moxibustion Therapy on UB23腎 俞Shenshu, DU20百 會Baihui, and DU4命 門Mingmen.
- For Damp Bi Syndrome add SP6三 陰 交Sanyinjiao, SP9陰 陵 泉Yinlingquan, ST36足 三 里Zusanli, ST40豐 隆Fenglong, UB13肺 俞Feishu, UB20脾 俞Pishu, and UB23腎 俞Shenshu.
- For Heat Bi Syndrome add DU14大 椎DaiZhui, LI4合 谷Hegu, LI11曲 池Quchi, SJ5外 關Waiguan, and ST44內 庭Neiting.
- For Kidney Yin deficiency, add KD1湧 泉Yongquan, KD3太 溪Taixi, KD6照 海Zhaohai, SP6三 陰 交Sanyinjiao, and UB23腎 俞Shenshu.
- For Kidney Yang deficiency, add moxibustion Therapy on GB34陽 陵 泉Yanglingquan, UB23腎 俞Shenshu, DU20百 會Baihui, DU4命 門Mingmen, and GB39懸 鐘Xuanzhong.

- For Blood stasis, add UB17膈俞Geshu, UB39委陽Weiyans, and SP10血海Xuehai. Add cupping Therapy after Bleeding Therapy.

PATIENT ADVISORY

1. Bone TB and bone cancer should be carefully differentiated from Bi Syndrome.
2. Patients should keep warm and avoid the pathogenic factors.
3. Limit consumption of the foods that may create Dampness, such as excess meat or protein, coffee, dairy products, nuts, oil-rich seeds, refined sugar, sweets, and salt; limit the use of alcohol, tobacco, and marijuana.
4. Limit consumption of the foods that may create Heat, such as plum, chard, beet greens, and rhubarb.
5. Limit consumption of the foods that may create Damp-Heat, such as bell peppers, eggplants, tomatoes, and potatoes.
6. Increase consumption of the foods that may drain Damp, such as celery, tea, napa cabbage, scallion, and soybean sprouts.
7. Increase consumption of the foods that may warm and tonify, such as chives and cherries.

THORACIC OUTLET SYNDROME 胸廓出口症候群

INTRODUCTION

The thoracic outlet is the space between the collarbone (or clavicle) and the first rib. It is a narrow space that encompasses the tissues including blood vessels, nerves, and muscles. When there is a compression on the tissues it can cause tingling, numbness, and coldness in the shoulders, arms, and fingers, especially when raising the arms.

Common causes include poor sleep postures, emotional stress, use of crutches; repetitive injuries from work or sports; trauma to the neck, shoulders, chest, and upper back, such as whiplash or a broken collarbone; adhesions and scarring of the soft tissues near the thoracic outlet; occupational stresses, such as dental hygienists whose work involves carrying heavy shoulder loads; and anatomical defects, such as having an extra rib in the neck.

ANATOMY

The thoracic outlet, also known as thoracic inlet and thoracic aperture, anatomically has two apertures: the superior thoracic aperture and the inferior thoracic aperture. Clinically, in the case of thoracic outlet syndrome, it refers to the superior thoracic aperture. The inferior thoracic aperture is located near the twelfth thoracic vertebra, which is not related to the symptoms of thoracic outlet syndrome.

The superior thoracic aperture lies posterosuperiorly anteroinferiorly in an oblique transverse plane to the lower portion of the neck and above the thorax. The superior portion of the outlet begins just above and behind the clavicle, and the inferior portion of it is at the upper part of the arm.

The outlet is marked by the posterior edge of the anterior scalene muscle, the anterior edge of the middle scalene muscle, and the superior aspect of the first rib; the space between these two muscles and the first rib is called the interscalene triangle.

- **The anterior scalene muscle** originates from the anterior tubercles of the transverse processes of C3-C6 and attaches onto the scalene tubercle on the inner border of the first rib. Anterior rami of C5-C7 innervates the muscle. The functions of the muscle are ipsilateral lateral flexion of the neck by its ipsilateral contraction; anterior flexion of the neck by its bilateral contraction; elevation of the first and second ribs. It is also a breathing muscle; it lifts the first rib to expand inhalation.
- **The middle scalene muscle** is the longest and biggest among all three scalene muscles, originating from the posterior tubercles of the transverse processes of C2-C7 and attaching to the scalene tubercle of the first rib. Anterior rami of C2-C7 innervates the muscle.

  The functions of the muscle are ipsilateral lateral flexion of the neck by its ipsilateral contraction; anterior flexion of the neck by its bilateral contraction; expansion of inhalation by elevation of the first rib.

  Many anatomical structures are near or pass through the interscalene triangle:
- The superior, middle, and inferior trunks of the brachial plexus are at the base of the triangle; these nerves are from the interior rami of C5 and C6 (superior trunk), the interior rami of the seventh cervical vertebra (C7) (middle trunk); the interior rami of C8 and the interior rami of T1 (interior trunk). These three trunks are involved with the motor and sensory function of the arm from the shoulder to the fingertips.
- The anterior rami of the third, fourth, and fifth cervical spinal nerves are at the upper part of the triangle.
- The subclavian artery passes directly through the triangle.
- The anterior rami of the C7 cervical spinal nerve from the long thoracic nerve (arising from the anterior rami of the C5-C7) is within the triangle. It is a motor nerve that innervates the Serratus anterior muscle, which is responsible for shoulder protraction and upward rotation of the scapular during lifting.
- The subclavian vein is at the base of the interscalene triangle. Although it is not within the triangle, it is relatively close to the triangle and is easily prone to compromise.
- The phrenic nerve is located anteriorly to the subclavian artery and oblique to the anterior scalene. Although it is not within the triangle, it is relatively close to the triangle and is also easily prone to compromise.
- The subclavius posticus muscle originates from the medial aspect of the first rib and inserts on the superior border of the scapular.

- The subclavius muscle originates from the medial aspect of the first rib and inserts on the undersurface of the middle of the clavicle. C5 and C6 innervate the muscle. The functions of the subclavius muscle are to anchor the clavicle at the sternoclavicular joint, depress the clavicle, and elevate the first rib by its contraction. The subclavius muscle is located at the anterior to the costoclavicular space. The brachial plexus and the subclavian artery and vein are also in the costoclavicular space before entering the subcoracoïd space.

WESTERN MEDICINE ETIOLOGY AND PATHOLOGY

1. Based on the pathophysiology of the symptoms, the etiology is classified into three types: neurogenic, arterial, and venous types.
   - The brachial plexus is a group of nerves from C5-T1 that innervate the muscles of the upper limbs and are involved with motor and sensory functions. Neurologic symptoms of Thoracic Outlet syndrome occur after trauma to the neck that causes edema at the interscalene triangle where the brachial plexus passes.
     - Compression can occur within the interscalene triangle space but can also occur in the subarachnoid space. Compression on the C5-T1 brachial plexus nerve roots are most often seen in young active individuals who participate in athletic activities or jobs that require repetitive above-the-shoulder upper extremity motion and heavy lifting.
     - Tumors in the upper chest and under the arm, such as Pancoast tumors, (also known as superior pulmonary sulcus tumors), can compress the brachial plexus.
   - Subclavian artery passes through the interscalene triangle: it is in the costoclavicular space and also in the subcoracoid space.[1] A compression to the artery leads to progressive stenosis and occlusion, resulting in aneurysm formation.
     - The cervical rib can compress the subclavian artery, causing stenosis or aneurysm.
     - An elongated or an enlarged transverse process of the C7 can distort the path of the brachial plexus and the subclavian artery.
   - The subclavian vein is not within the interscalene triangle, as it passes through the costoclavicular space. Upper limb movements, such as sagging the shoulders posteriorly, cervical extension, upper limb abduction, and turning the face to the opposite side with deep inhalation, decrease the costoclavicular space, resulting in brachial plexus and subclavian vein compression.
2. Common causes of the thoracic outlet syndrome by the neurovascular compression related to anomalous bones are:
   Congenital causes
     - The cervical rib

In humans, the cervical rib is a congenital, abnormal rib that arises from the C7. Its compression of the lower trunk of the brachial plexus or subclavian artery leads to weakness of the muscles around the hand near the base of the thumb and hand pain with cold and pale fingers.
- An elongated or an enlarged transverse process of C7.
- An anomalous first rib.

Traumatic causes
- Whiplash injuries, often caused by seat belts, can strain or tear the shoulder muscles. Scar tissue on the muscles can build up afterward and put pressure on the nerves and blood vessels at the thoracic outlet.[2]
- Falls cause hemorrhage, hematoma, or displaced fractures that can directly compress the nerves or vasculature. Fibrosis in the chronic condition is another consideration in the development of compression that produces symptoms.
- Fracture of the clavicle or the first rib due to trauma.
- Dislocation of the humeral head.
- Repetitive activity associated with sports or work can contribute muscle hypertrophy, swelling, and small hemorrhages, which could develop fibrosis in the chronic condition.
- Forward head posture: when the head is bent downward for long periods of time, typically observed in students, laptop computer workers, and cellphone users.
- Poor shoulder posture: people who often droop the shoulders or who have very large breasts may experience poor alignment of the torso. Such slouching posture can compress the nerves and blood vessels near the thoracic inlet.
- Heavily loaded shoulders: people who carry heavy weight on the shoulders (such as students with backpacks) or a heavy bag on one shoulder for a long period of time. Obesity also puts an undue amount of stress and pressure on the joints (CCOHS 2017).
- Pregnancy may lead to loosening of the joints (CCOHS 2017).
3. The other common causes to the compression of the subclavian blood vessels and brachial plexus are scalene muscle spasm, fibrosis, sagging shoulder girdle and excessive abduction of the upper limbs.

CHINESE MEDICINE ETIOLOGY AND PATHOLOGY

Acute State

Acute traumatic injury from many activities – such as a car accident, fracture of the clavicle, fracture of the first rib, and dislocation of the humeral head – can entrap the affected nerves and blood vessels, leading to Blood stasis, swelling, and blockage in the meridians, often resulting in pain.

Chronic State

Yellow Emperor's Inner Classic《金匮要略》 – *Sù Wèn* 素問 (*Plain Questions*), Volume I Chapter 43, on the topic of "Discussion of Blockages" 《痹論篇第四十三》:

> Yellow Emperor: "Good! Some blockages have pain, some don't have pain; some patients experience numbness; some have the condition of Cold, some have Hot; some patients experience Dryness; some experience Dampness. What are the reasons?"
> 帝曰：善。痹或痛，或不痛，或不仁，或寒，或熱，或燥，或濕，其故何也？
> Qi Bo: In the case of pain, excessive Cold, evil Qi, is present. Because there is Cold, hence, there is pain. In case there is no pain but there is numbness, the disease has persisted for a long time and it has penetrated the body deeply. The circulation of the Blood and Qi is rough; the meridians and the collaterals are deficient; hence, the patient does not feel the pain. The skin is not provided with the nutrients. Hence, it causes numbness.
> 岐伯曰：痛者，寒氣多也，有寒故痛也。其不痛不仁者，病久入深，榮衛之行澀，經絡時疏，故不通，皮膚不營，故為不仁。

Injury to the nerves and blood vessels is called "Jin Shang筋 傷" or "traumatic injury on Tendons" (Apichai 2020). In traditional Chinese medicine (TCM) theory, the Liver governs the Tendons. "Traumatic injury on Tendons" consumes Liver energy, in which mostly Yin and Blood are consumed. As the injury becomes chronic, the Liver Yin and Blood become deficient. The Kidney then follows because the element is "Wood" and the Kidney's element is "Water" – Wood and Water always affect each other. The body lacks circulation of Qi and Blood to the affected region, resulting in deficiency; once there is an invasion of Wind, Dampness, and Coldness, or there is a Blood stasis from a traumatic injury, it is prone to forming blockages.

MANIFESTATIONS

In Western Medicine

Arm pain particularly is in the inner aspect of the arm. The pain may radiate to the forearm and the last two digits; at night, the pain may turn into numbness. The arm is swollen, cold, and weak, and it is difficult to raise above the shoulder. Some patients may experience headache and neck pain; the pain may radiate to the chest causing chest distention, shortness of breath, and some patients describe the chest pain as angina pectoris. In severe cases, muscle atrophy may occur.

In Chinese Medicine

Qi and Blood Stagnation 氣滯血瘀

Main symptoms: cyanosis of the arm, finger stiffness, swelling, and cold and weak limbs.
Tongue: purple with ecchymosis.
Pulse: tight, choppy.

Liver Blood Deficiency 肝血虧虛
Main symptoms: weak, numb, and painful limbs and exhaustion after exertion. Pale face, spasm, and atrophic muscles.
Tongue: pale.
Pulse: thin, wiry.

Wind Invading Tendons 風邪侵筋

Main symptoms: painful upper limbs, it is wandering, aversion to wind, and fever. Soft or atrophic muscle. Tingling sensation in the neck and arm.
Tongue: thin white coating.
Pulse: wiry and tight.

PHYSICAL EXAMINATION

- Begin with obtaining medical history
  - Cervical rib is a predisposition to developing thoracic outlet syndrome following neck trauma, most often a whiplash injury. Thus, a history of neck injury before the thoracic outlet syndrome occurs may suggest the etiology of Neurogenic Thoracic Outlet syndrome. This, in turn, may cause Arterial Thoracic Outlet syndrome, in some cases, when the cervical ribs press against the subclavian artery.
  - Inquiring about occupation and physical activities may suggest causative factors, for instance, with dental hygienists, whose work involves carrying heavy shoulder loads.
  - Inquiring about the location of the pain may suggest the location of the associated plexus. When the upper plexus (C5,6,7) is involved, there is a pain in the side of the neck, and this pain may radiate to the ear and face. When the lower plexus (C8, T1) is involved, typically there is a pain in the anterior and posterior shoulder region, which radiates down the ulnar side of the forearm, into the hand and the ring and small fingers.
- Inspection: swelling and cyanosis of the arm often occur.
- Palpation: palpation of the shoulder is required to look for a depression, swelling, abnormal pulse, or a bony abnormality above the clavicle.
- Range of motion: generally, the pain of thoracic outlet syndrome does not decrease the range of motion of the neck and shoulder. Testing the range of motion may reproduce the symptoms; this helps to understand which positions and movements trigger the symptoms.
- Special test: ultrasound is used to detect vascular thoracic outlet syndrome. X-rays can reveal a cervical rib. Computerized tomography (CT) scan may identify the location and cause of blood vessel compression. Magnetic Resonance Imaging (MRI) can also identify the location and cause of blood vessel compression; it also can reveal congenital anomalies, such as a cervical rib.

DIFFERENTIAL DIAGNOSIS

Carpal Tunnel Syndrome

Median nerve compression produces sensation from the wrist to the hand. They include tingling, numbness or an electric shock in the thumb, index, middle fingers, and the radial half of the ring finger. The fifth digit is not affected. Shaking the hand and fingers may temporarily relieve the symptoms.

Lateral Epicondylitis

With lateral epicondylitis, the location of the pain is at the anterior and distal from the lateral epicondyle of the elbow. There is usually no specific injury history associated with the start of symptoms.

Medial Epicondylitis

In medial epicondylitis, the location of the pain is at the medial aspect of the elbow and volar forearm. Pain can be felt in the flexor pronator tendons. It is worse when the wrist is flexed or pronated against resistance.

Raynaud's Disease

With Raynaud's disease, besides cold fingers, there is also color change in the skin, usually first turning white, then blue.

TREATMENTS

Qi and Blood Stagnation 氣滯血瘀

Treatment Principle

Moves Blood, disperses Bruise, nourishes Qi, opens obstruction

Herbal Formula

Tao Hong Yin 桃紅飲

This formula is from "Systematized Patterns with Clear-Cut 類證治裁", authored by Lin Peiqin 林佩琴 (1772–1839), in Qing dynasty 清朝 (1636–1912), and published in 1851.

Formula Ingredients

Tao Ren桃仁9g, Hong Hua紅花9g, Chuan Xiong川芎9g, Dang Gui Wei當歸尾9g, Wei Ling Xian威靈仙9g

Ingredient Explanations

- Tao Ren桃仁 moves the Blood, disperses Bruising, and stops pain due to Blood stasis.
- Hong Hua紅花 moves the Blood, disperses Bruising, and stops pain due to Blood stasis.
- Chuan Xiong川芎 moves the Blood, disperses Bruising, and stops pain due to Blood stasis.

- Dang Gui Wei當 歸 尾 moves the Blood, disperses Bruising, unblocks stagnation, opens the meridians, and stops pain.
- Wei Ling Xian威 靈 仙 expels Wind-Damp and stops pain.

Modifications
- For severe Blood stasis, remove Wei Ling Xian威 靈 仙 and add Cao Wu草 鳥, Di Long地 龍, and San Qi三 七.

Formula Indications
- Numbness and tingling in four limbs.
- Blood stasis with early-stage Wind-Cold-Damp Bi Syndrome.

Contraindications
- It is contraindicated for pregnancy.
- It is contraindicated for Yin deficiency.
- It is contraindicated for Blood Heat condition.

Liver Blood Deficiency
Herbal Treatments

Treatment principles: tonifies the Liver and Kidney, opens the meridians, and soothes the Tendons.

Herbal Formula

Bu Shen Zhuang Jin Tang補 腎 壯 筋 湯

This formula is from "Supplement to Traumatology 傷 科 補 要", authored by Qian Xiuchang 錢 秀 昌 (unknown date), in Qing dynasty 清 朝 (1636–1912), and published in 1808.

Formula Ingredients

Sheng Di Huang 生 地 黃15g, Shan Zhu Yu 山 茱 萸15g, Qing Pi 青 皮6g, Bai Shao白 芍10g, Xu Duan 續 斷10g, Du Zhong杜 仲10g, Dang Gui當 歸10g, Fu Ling茯 苓10g, Wu Jia Pi 五 加 皮10g, Niu Xi牛 膝6g

Ingredient Explanations
- Sheng Di Huang生 地 cools the Blood and tonifies the Kidney Yin.
- Shan Zhu Yu茱 萸 tonifies the Liver and Kidney Yin and strengthens Kidney Yang.
- Qing Pi青 皮 soothes Liver Qi and breaks up stagnant Qi, dries Dampness, and transforms Phlegm.
- Bai Shao白 芍 nourishes the Blood, preserves Yin, calms Liver Yang, softens the Tendons, and alleviates spasm and pain.
- Xu Duan續 斷 nourishes the Liver and Kidney, strengthens the Bones and Tendons, and alleviates pain
- Du Zhong杜 仲 tonifies the Kidney and Liver and strengthens the Bones and Tendons.
- Dang Gui當 歸 nourishes the Blood, reduces edema, and stops pain.
- Fu Ling茯 苓 drains Damp and tonifies the Spleen.

- Wu Jia Pi五 加 皮 expels Wind Damp, nourishes and warms the Liver and Kidney, and strengthens the Bones and Tendons.
- Niu Xi牛 膝 expels Wind Dampness, tonifies the Liver and Kidney, and directs herbs downward.

Modifications

For strengthening the Tendons and Bones, add Gui Ban龜 膠 and Gou Qi Zi枸 杞.
For tonifying Qi, add Dang Shen黨 參, Huang Qi黃 芪, and Bai Zhu白 朮.

Formula Indications

- Late-stage trauma.
- Liver and Kidney deficiency.

Wind Invading Tendons

Herbal Treatments

Treatment principles: tonifies the Liver and Kidney, opens the meridians, and soothes the Tendons.

Herbal Formula

Fang Feng Tang 防 風 湯

This formula is from "Elucidated Prescriptions and Expositions" of Huangdi's *Plain Questions* 黃 帝 素 問 宣 明 論 方, authored by Liu Wansu 劉 完 素 (1100–1180), in Jin dynasty 金 朝 (1115–1234), and published in 1172.

Herbal Ingredients

Fang Feng防 風30g, Gan Cao甘 草30g, Dang Gui當 歸30g, Chi Fu Ling赤 茯 苓30g, Xing Ren杏 仁30g, Gui Zhi桂 枝30g, Huang Qin黃 芩9g, Qin Jiao秦 艽9g, Ge Gen葛 根9g, Ma Huang麻 黃15g

Ingredient Explanations

- Fang Feng防 風 expels Wind and drains Damp.
- Gan Cao甘 草 harmonizes the herbs, tonifies the Middle Jiao Qi, clears Heat, expectorates Phlegm, and stops cough.
- Dang Gui當 歸 moves the Blood, disperses Bruising, and stops pain due to Blood stasis.
- Chi Fu Ling赤 茯 苓 drains Damp and clears Damp-Heat.
- Xing Ren杏 仁 stops cough, calms wheezing, and moistens the Intestines.
- Gui Zhi桂 枝 warms the meridians and relieves pain.
- Huang Qin黃 芩 clears Heat, dries Damp, cools the Blood, stops bleeding, and calms Liver Yang.
- Qin Jiao秦 艽 expels Wind-Cold-Dampness, unblocks painful obstruction, alleviates pain, and benefits the joints.
- Ge Gen葛 根 relieves Heat and releases muscles, especially of the neck and upper back.
- Ma Huang麻 黃 induces sweating, releases the exterior, and warms and disperses Cold pathogens.

Modifications

For severe Wind invasion, add Sheng Jiang生 薑, Shen Jin Cao伸 筋 草, and Ren Dong Teng忍 冬 藤.

For Damp condition, remove Huang Qin黃 芩 and Ge Gen葛 根 and add Yi Yi Ren薏 苡 仁 and Wei Ling Xian威 靈 仙.

Formula Indications

• Wind-Damp Bi Syndrome, as Wind is predominant.

1. Acupuncture Treatments

Acupuncture Point Formula

Basic Points

LI17天 鼎Tianding,　ST12缺 盆Quepen,　LU1中 府Zhongfu,　UB41附 分Fufen, UB42魄 戶Pohu, UB43膏 肓Gaohuang, UB44神 堂Shentang, HT1極 泉Jiquan.

Modifications

For pain in hand, add SI5陽 谷Yanggu, LI4合 谷He Gu, and SI3後 溪Houxi.

2. Tui Na Treatments

Tui Na Techniques

Pressing, kneading, rotating, stretching, and plugging.

Tui Na Procedures

The patient is seated, the right side is an example. The therapist performs the following:

1) Presses LI17天 鼎Tianding, ST12缺 盆Quepen, LU1中 府Zhongfu for 30 seconds on each point.
2) Stands behind the patient's right side, places their left forearm under the patient's right armpit, then lifts up the affected shoulder for 1 minute.
3) Presses HT1極 泉Jiquan for 30 seconds.
4) Places their left hand on the right shoulder, holds the patient's right wrist by the right hand, and stretches the right shoulder at 45° abduction for one minute.
5) Hyperextends the right shoulder to the greatest range of motion while stretching.
6) Continuing the last step, the therapist rotates the right shoulder in flexion while stretching for one minute.
7) Rotates the right shoulder in a clockwise circle for several rounds.
8) Rotates the right shoulder in a counterclockwise circle for several rounds.
9) Repeats step 4 for one minute.
10) Presses and plugs the tissues on the right supraspinatus muscle for a minute.

11) Places their left hand on the patient's head and their right hand on the right acromion; stretches the neck by pushing the head to the left side while pressing the right acromion with the right hand for one minute. Figure 1.2 from shutterstock.com #1667170588

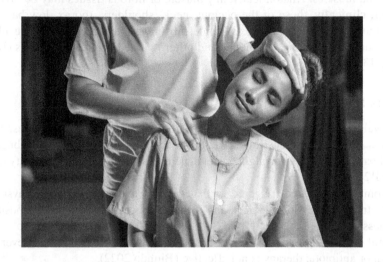

12) Kneads the posterior aspect of the neck, particularly the right side, for one minute.

Patient Advisory

- Keep the keyboard and the mouse at a level that does not cause the wrist to flex.
- Take a break from typing for 1–2 minutes every half-hour.

FIBROMYALGIA SYNDROME 纖維肌痛綜合征

INTRODUCTION

Fibromyalgia syndrome presents a widespread achy painful condition. The pain occurs at multiple locations in the muscles, tendons, and joints and generally involves both sides of the body, affecting the neck, shoulders, upper back, chest, and buttocks. Fibromyalgia typically affects young or middle-aged females (Jahan et al. 2012), and the majority of fibromyalgia patients report chronic fatigue, restless sleep, and disturbances in bowel functions; some patients also experience mood disorders, such as depression and anxiety.

Because the disease was first believed to be inflammation of the fibrous connective tissues, it was termed fibrositis by Sir William Gowers in 1904. As there was no evidence of inflammation, in 1976, the term fibromyalgia, which means pain in the fibrous connective tissues and muscles, replaced fibrositis.[3]

ANATOMY

The term fibromyalgia comes from three Latin words, "fibra" (fibrous tissue), "myo" (muscle), and "algia" (pain). Therefore, the disease refers to pain in fibrous tissues and muscles. Anatomically, any muscle or fibrous tissues may be involved; however, clinically, those of the occiput, neck, shoulders, thorax, low back, and thighs more commonly manifest. Hence, fibromyalgia indicates pain in fibrous tissues, muscles, tendons, ligaments, and other "white" connective tissues (Bondy 1992, p.1369).

WESTERN MEDICINE ETIOLOGY AND PATHOLOGY

Fibromyalgia can occur in men, women, children, and adolescents. The cause is still unknown. The painful tissues involved are not accompanied by histologic abnormality, and patients with fibromyalgia do not develop tissue damage or deformity (Jahan et al. 2012).

Fibromyalgia may be triggered or exacerbated and intensified by physical or mental stress, depression, anxiety and striving, poor sleep, trauma, exposure to Dampness or Cold,

a viral or other systemic infection (e.g., Lyme disease), or a traumatic event; but antiviral or antibiotic therapy is not effective (Biundo 2022).

Some believe it is a malfunction of the central nervous system.

CHINESE MEDICINE ETIOLOGY AND PATHOLOGY

Liver Qi Stagnation 肝氣鬱結

- In TCM, emotional upset is the main causative factor affecting the free flow of Qi in the Liver.
- The functions of the Liver include regulation of emotion, maintenance of the circulation of Qi, Blood, and body fluids, as well as the promotion of digestion and absorption.
- Liver Qi stagnation causes anger, and extreme or long-term anger impairs the Liver, which leads to more Liver Qi stagnation.
- According to TCM theory, Blood goes back to the Liver when in a lying position while in deep-stage sleep during Gallbladder and Liver functioning periods, which are between 11 p.m. and 3 a.m.

MANIFESTATIONS

In Western Medicine

Pain is the main character of fibromyalgia. It begins gradually, but becomes widespread and diffuse such that the sensation is not easily describable; patients may say, for example, that it is a stiff pain or an achy pain.

Fibromyalgia is typically a somatic symptom disorder, and pain is one of the chief complaints. Patients also experience fatigue, irritable bowel syndrome, interstitial cystitis, migraine, tension headache, paresthesia, and cognitive disturbances, such as difficulty concentrating and a general feeling of mental cloudiness. Patients tend to be stressed, tense, anxious, fatigued, ambitious, and sometimes depressed.

Symptoms can be exacerbated by environmental or emotional stress, poor sleep, trauma, exposure to Dampness or Cold, or by a physician, family member, or friend who implies that the disorder is "all in their head".

In Chinese Medicine

1. **Liver Qi stagnation symptoms** 肝氣鬱結 – anxiety, emotional upset, headaches, migraine headache, short temper, muscle stiffness in neck and shoulders, insomnia, waking frequently/difficulty falling back to sleep, dysmenorrhea, premenstrual syndrome, irritable bowel syndrome, anxiety, and depression. All symptoms may be triggered by emotional stress.
   Tongue: slightly dusky or slightly puffy.
   Pulse: wiry.

2. **Qi and Blood deficiency symptoms** 氣血兩虛 – chronic fatigue, exhaustion, dull headache, poor appetite, muscle weakness and numbness, insomnia, dream-disturbed sleep and waking up tired, palpitations, restless leg syndrome, and amenorrhea.
   Tongue: pale, thin white coating.
   Pulse: thready, weak, and deep.

3. **Kidney deficiency symptoms** 腎氣虧虛 – infertility issues for both males and females, sore lower back, irritable bladder, dysmenorrhea, amenorrhea, hot flashes and night sweats, forgetfulness, and dull headache.
   Tongue: pale or dry, with cracks.
   Pulse: thin, weak, and deep.

4. **Bi Syndrome symptoms** 痺證 – aversion to cold and rainy weather, patients present with numbness of the limbs and often complain of swollen fingers/toes.
   Tongue: big tongue body and a white, greasy coating on the tongue.
   Pulse: tight and slippery.

PHYSICAL EXAMINATION

- Begin with obtaining a medical history. Since fibromyalgia diagnosis relies on symptoms, a detailed medical history is crucial.
  - General health condition in which the patient may feel abnormally cold body temperatures, such as in the skin, legs, knees, feet, hands, or lumbar; or abnormally hot with fever, hot hands, and wet palms.
  - General symptoms include widespread pain; pain on tender points usually affects the neck, buttocks, shoulders, arms, upper back, and the chest.
  - Other symptoms include stiffness and tenderness of the muscles, tendons, and joints, involving both sides of the body. It is also characterized by restless sleep, tiredness, fatigue, anxiety, depression, and disturbances in bowel functions.

- Female patients may suffer from dysmenorrhea, female urethral syndrome, vulvar vestibulitis, and premenstrual syndrome.
- Injury history, such as falling, spraining, or direct hitting to areas of tender points. The painful tissues involved are not accompanied by tissue inflammation, tissue damage, nor deformity. Therefore, fibromyalgia may be triggered by physical injuries, even when there is no specific histologic change to the painful regions.
- The diagnostic criteria that the American College of Rheumatology (ACR) issued in 2010 excludes tender points from the process. It is replaced by two evaluative tools:
  a) The widespread pain index (WPI): pain in any one of 19 parts of the body in the past week receives 1 point for a possible maximum of 19 points.
  b) The symptom severity scale (SS): symptoms receive a scale of 0 to 3 for a possible maximum of 12 points.

A fibromyalgia diagnosis should meet all of the following criteria:
  a) Either a WPI of 7 or more, with an SS of 5 or more; or a WPI of 3–6, with an SS of 9 or more.
  b) Persistent symptoms at a similar level for at least three months.
  c) No other explanation for the symptoms.

- Careers that may be conducive to fibromyalgia, such as working in damp and/or cold environments.
- Inspection: in observing the painful sites, the tender areas are not swollen nor red
- Palpation
  - The tender points are not warm to the touch. Patients experience pain over the tender points only at the area where the examiner presses, and the pressure is only about 4 kg – or enough to cause the examiner's nail bed to blanch.
  - Tender points are defined by the ACR as 18 bilateral points on the body, nine points on each side. These 18 tender points are located on occiput (2), low cervical (2), trapezius (2), supraspinatus (2), second rib (2), lateral epicondyle (2), gluteal (2), greater trochanter (2), and knee (2). Tender points are not trigger points, rather they are where pain can be felt immediately beneath the skin when pressed. Trigger points feel like knots underneath the skin; the knots can be painful, but some people feel no pain or discomfort.
- Range of motion: the joint mobility is not affected by the pain because there is no histological change

Differential Diagnosis

Rheumatoid Arthritis

Patients with fibromyalgia syndrome report morning stiffness similar to rheumatoid arthritis, although it does not show joint swelling or deformity as in rheumatoid arthritis. Also, rheumatoid factor is present in rheumatoid arthritis, while fibromyalgia syndrome shows normal lab tests.

Myofascial Pain

Tender points occur with both fibromyalgia syndrome and myofascial pain. Myofascial pain presents localized stiffness, and patients may present fatigue, but it is not chronic fatigue.

Heart Disease

Patients with fibromyalgia syndrome present with chest pain, although it does not radiate, and the length of pain is constant.

TREATMENTS

Herbal Treatments

1. Liver Qi Stagnation 肝氣鬱結

Treatment Principle

Moves Liver Qi and stops pain.

Herbal Formula

Xiao Yao San逍 遙 散

This formula is from "Prescriptions of the Bureau of Taiping People's Welfare Pharmacy 太 平 惠 民 和 劑 局", authored by Imperial Medical Bureau 太 醫 局, in southern Song dynasty南 宋 (1127–1239), and published in 1134.

Formula Ingredients

Chai Hu柴 胡10g, Dang Gui當 歸10g, Bai Shao白 芍10g, Bai Zhu白 朮10g, Fu Ling茯 苓10g, Zhi Gan Cao甘 草5g, Sheng Jiang生 薑10g, Bo He 薄 荷5g

Ingredient Explanations

- Chai Hu柴 胡 relieves Liver Qi stagnation, pacifies the Liver, relieves the Shao Yang, and reduces fever.
- Dang Gui當 歸 nourishes the Blood, regulates menses, and invigorates and harmonizes the Blood.
- Bai Shao白 芍 calms Liver Yang, alleviates pain, nourishes the Blood, and regulates menstruation.
- Bai Zhu白 朮 tonifies the Spleen and tonifies Qi.
- Fu Ling茯 苓 strengthens the Spleen and harmonizes the Middle Jiao.
- Zhi Gan Cao甘 草 tonifies Spleen Qi, stops pain, and harmonizes the other herbs.
- Bo He薄 荷 relieves Liver Qi Stagnation, disperses stagnant Heat and enhances Chai Hu's ability to relieve the Liver.
- Sheng Jiang生 薑 harmonizes and prevents rebellious Qi and normalizes the flow of Qi at the center.

Modifications
- To make the formula stronger, add Xiang Fu香 附, Yu Jin鬱 金, and Chen Pi陳 皮.
- For Liver Fire, add Mu Dan Pi丹 皮 and Zhi Zi栀 子 to clear Heat and cool the Blood.

Formula Indications
- Liver Qi Stagnation.

Cooking Instructions
1. Grind the herbs into granules.
2. Mix 6 grams in one cup water.
3. Boil the herbs to 0.7 cup of water.

Dose
Drink the decoction when it is warm, twice daily.

2. Qi and Blood Deficiency

Treatment Principle
Tonify Qi and the Blood.

Herbal Formula
Ba Zhen Tang 八 珍 湯
This formula is from "Categorized Essentials of Repairing the Body正 體 類 要", authored by Xue Ji 薛 己 (1487–1559), in Ming dynasty 明 朝 (1368–1644), and published in 1529.

Formula Ingredients
Ren Shen人 參10g, Bai Zhu白 朮10g, Fu Ling茯 苓10g, Dang Gui當 歸10g, Chuan Xiong 川 芎10g, Bai Shao白 芍10g, Shu Di Huang 熟 地 黃10g, Zhi Gan Cao炙 甘 草5g, Sheng Jiang生 薑10g, Da Zao大 棗3 pieces

Ingredient Explanations
- Ren Shen人 參 tonifies Qi and Yang.
- Bai Zhu白 朮 tonifies the Spleen Qi and dries Dampness.
- Fu Ling茯 苓 drains Damp and tonifies the Spleen.
- Dang Gui當 歸 nourishes the Blood and moves the Blood.
- Chuan Xiong川 芎 moves the Blood and moves Qi.
- Bai Shao白 芍 nourishes the Blood, soothes the Liver, and stops pain.
- Shu Di Huang熟 地 黃 tonifies the Liver and Kidney Yin and benefits the Essence and the Blood.
- Zhi Gan Cao炙 甘 草 tonifies the Spleen, harmonizes the other herbs, and guides the herbs to all 12 meridians.

- Sheng Jiang生 薑 unblocks the pure Yang pathway and harmonizes rebellious Qi.
- Da Zao大 棗 tonifies the Spleen Qi, nourishes the Blood, and moderates and harmonizes the harsh properties of the fragrant herbs.

Modifications

- For poor appetite, add Shan Yao山 藥, Shan Zha山 楂, Mai Ya麥 芽, and Qian Shi芡 實.
- For headache due to Blood deficiency, add Man Jing Zi蔓 荊 子 and Gao Ben藁 本.

Formula Indications

- Qi and Blood deficiency.
- Liver and Spleen deficiency.

Contraindications

- It is contraindicated for those patients with Heat or excess conditions.

3. Kidney Deficiency

Treatment Principle

Warms and tonifies the Kidney Yang.

Herbal Treatments

Herbal Formula

Jin Gui Shen Qi Wan金 匱 腎 氣 丸

This formula is from "Shang Han Lun 傷 寒 雜 病 論", authored by Zhang Zhong-Jing 張 仲 景 (150–219), in eastern Han dynasty 東 漢 (25–220), and published in 200–210.

Formula Ingredients

Shu Di Huang熟 地 黃24g, Shan Yao山 藥12g, Shan Zhu Yu山 茱 萸12g, Ze Xie澤 瀉9g, Fu Ling茯 苓9g, Mu Dan Pi牡 丹 皮9g, Gui Zhi桂 枝3g, Zhi Fu Zi附 子 炮3g

Ingredient Explanations

- Shu Di Huang熟 地 黃 tonifies the Liver and the Kidney, strengthens the Bones, and benefits the Essence and the Blood.
- Shan Yao山 藥 tonifies the Kidneys and benefits the Yin and the Spleen Qi.
- Shan Zhu Yu山 茱 萸 tonifies the Liver and Kidney, strengthens the Bones, and benefits the Essence and the Blood.
- Ze Xie澤 瀉 drains Damp, especially Damp-Heat in Lower Jiao, and clears the Kidney and deficient Heat.
- Fu Ling茯 苓 drains Damp and tonifies the Spleen.

- Mu Dan Pi牡丹皮 clears Heat, cools the Blood, and moves the Blood.
- Gui Zhi桂枝 warms the meridians and relieves pain.
- Zhi Fu Zi附子炮 warms Ming Men Fire.

Modifications
- To increase the warming effect, add Rou Gui肉桂.
- For frequent, copious, clear urination, add Bu Gu Zhi補骨脂 and Lu Rong鹿茸.

Formula Indications
- Kidney Yang deficiency.

Contraindications
- Use with caution for patients with gastrointestinal weakness.
- It is contraindicated for Yin deficiency and body fluid deficiency.

Treatment Principles
Move the Blood and Qi and stop pain.
- Bi Syndrome

Herbal Formula
Shen Tong Zhu Yu Tang身痛逐瘀湯
   This formula is from "Correction of Errors in Medical Classics醫林改錯", authored by Wang Qingren 王清任 (1768–1831), in Qing dynasty 清朝 (1636–1912), and published in 1849.

Herbal Ingredients
Dang Gui 當歸9g, Chuan Xiong川芎9g, Tao Ren 桃仁9g, Hong Hua 紅花9g, Mo Yao 沒藥 6g, Wu Ling Zhi 五靈脂6g, Qin Jiao 秦艽9g, Qiang Huo 羌活9g, Xiang Fu 香附3g, Chuan Niu Xi川牛膝9g, Di Long 地龍6g, Gan Cao甘草6g

Ingredient Explanations
- Dang Gui當歸, Chuan Xiong川芎, Tao Ren桃仁 and Hong Hua紅花 moves the Blood, disperses Bruising, and stops pain due to Blood stasis.
- Qin Jiao秦艽and Qiang Huo羌活 expels Wind-Cold-Dampness, unblocks painful obstructions, and alleviates pain.
- Wu Ling Zhi五靈脂, Mo Yao沒藥and Xiang Fu香附 moves Qi and the Blood and stops pain due to Qi and Blood stagnation.
- Chuan Niu Xi川牛膝and Di Long地龍 unblocks and promotes movement in the channels and collaterals and stops spasms and convulsions.
- Gan Cao甘草 harmonizes the herbs.

Modifications
- For Heat symptoms, add Chai Hu柴胡 and Huang Bai黃柏.
- For Qi deficiency, add Dang Shen黨參 and Huang Qi黃芪.

- For pain in the lumbar area and leg, add Xu Duan續 斷, Du Zhong杜 仲, and Sang Ji Sheng桑 寄 生.
- For severe pain, add Quan Xie全 蠍 and Wu Gong蜈 蚣.
- For Cold symptoms, remove Qin Jiao 秦 芃 and add Zhi Chuan Wu制 川 烏.
- For Cold symptoms and severe pain, add Gui Zhi桂 枝 and Xi Sin細 辛.
- For Wind traveling pain, add Fang Feng防 風.
- For stiffness in joints, add Shen Jin Cao伸 筋 草, Quan Xie全 蠍, and Wu Gong蜈 蚣.

Formula Indications
- Painful obstruction due to Qi and Blood Stagnation.

Contraindications
- Contraindicated during pregnancy.

Acupuncture Treatments
Acupuncture Point Formulas

1. Liver Qi **Stagnation** 肝氣鬱結

Acupuncture Points
印 堂Yintang, LIV3太 沖Taichong, LI4合 谷Hegu, GB34陽 陵 泉Yanglingquan, Anmian, RN12中 脘Zhongwan, RN17膻 中Danzhong, UBI8肝 俞Ganshu, ST25天 樞Tianshu, SP6三 陰 交Sanyinjiao, UB19膽 俞Danshu, UB20脾 俞Pishu.

Auricular Therapy
Shenmen, Liver, Sympathetic

2. Qi and Blood Deficiency 氣血兩虛

Acupuncture Points
LI4合 谷Hegu, RN12中 脘Zhongwan, SP6三 陰 交Sanyinjiao, UB20脾 俞Pishu, ST36足 三 里Zusanli, SP10血 海Xuehai, PC6內 關Neiguan, HT6陰 郄Yinxi, HT7神 門Shenmen, KD3太 溪Taixi, UB17膈 俞Geshu.

Auricular Therapy
Adrenal, Endocrine, Heart, and Spleen

3. Qi Stagnation and Blood Stasis 氣滯血瘀

Acupuncture Points
LI4合 谷Hegu, ST36足 三 里Zusanli, SP10血 海Xuehai, PC6內 關Neiguan, UB17膈 俞Geshu, LIV3太 沖Taichong, LI10手 三 里Shousanli, SP9陰 陵 泉Yinlingquan, UBI8肝 俞Ganshu, DU20百 會Baihui.

Auricular Therapy

Shenmen, Adrenal, and Subcortex

4. Kidney Deficiency 腎氣虧虛

Acupuncture Points

RN3中 極Zhongji, RN4關 元Guanyuan, RN6 氣 海Qihai, LIV2 行 間Xingjian, LI11曲 池Quchi, GB39懸 鐘Xuanzhong, KD3太 溪Taixi, KD6照 海Zhaohai, KD7復 溜Fuliu, UB23腎 俞Shenshu, UB37殷 門Yinmen.

Auricular Therapy

Adrenal, Endocrine, Heart, and Kidney

5. Bi Syndrome 痺證

Acupuncture Points

SP10血 海Xuehai, SP3太 白Taibai, UB23腎 俞Shenshu, DU20百 會Baihui, ST36足 三 里Zusanli, UB17膈 俞Geshu, ST40豐 隆Fenglong.

Auricular Therapy

Spleen, Endocrine, and Heart
In addition to the points above, the following points are used for local tender spots according to the ACR 18.

Neck pain: SI3後 溪Houxi, UB10天 柱Tianzhu, DU16風 府Fengfu, Bailao百 勞.
Shoulder pain: LI5陽 溪Yangxi, Jianqian肩 前, Jianhou肩 後.
Arm pain: LI11曲 池Quchi, SJ5外 關Waiguan, LI4合 谷Hegu.
Upper back pain: UB11大 杼Dazhu, UB17膈 俞Geshu, SI2前 谷Qiangu.
Lower back pain: UB23腎 俞Shenshu, UB25大 腸 俞Dachangshu, UB40委 中WeiZhong.
Thigh pain: SP10血 海Xuehai, GB31風 市Fengshi, LIV8曲 泉Ququan.
Leg pain: ST36足 三 里Zusanli, GB39懸 鐘Xuanzhong, SP6三 陰 交Sanyinjiao.

Tui Na Treatments

Tui Na Techniques

Pressing, kneading, rubbing, grasping, striking, and pushing.

Tui Na Procedures

The patient is lying supine on an exam table. The pressure on muscles should be light while providing Tui Na. The therapist performs the following:

1. Applies kneading techniques on the facial muscles on the forehead and around the eyes and the temporal regions for five minutes.
2. Applies palm pressing techniques on both temporal regions for 1–2 minutes. Figure 1.3.1

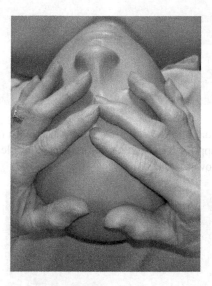

3. Applies thumb pushing techniques from the front hairline along the DU meridian to DU20 百 會 for 3–5 times.
4. Applies kneading techniques on the muscles on both arms and legs along the deltoid, biceps brachii, and triceps brachii muscles.
5. Applies rubbing techniques on the forearms from the elbows toward the wrists and ends the procedure by rubbing the fingers.
6. Applies grasping techniques on the quadriceps muscles. Figure 1.3.2 from the book "lower body pain"

The patient is lying prone on an exam table. The pressure on muscles should be light while providing Tui Na. The therapist performs the following:

1. Applies light force, striking on the Foot Yangming Stomach Meridians from ST36足 三 里Zusanli toward the ankles. Figure 1.3.3 from the book "lower body pain"; Figure 1.3.4 from the book "lower body pain"

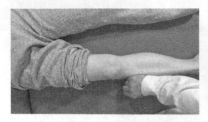 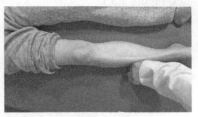

2. Places the left hand on the heel, grasps the ankle with the right hand, lifts up the leg, and applies rotating techniques on the ankle joint. Figure 1.3.5 from the book "lower body pain"

3. Ends the Tui Na procedures by rubbing the feet with both palms and pinching the toes with the fingers. Figure 1.3.6 from the book "lower body pain"; Figure 1.3.7 from the book "lower body pain"

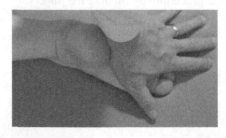

Patient Advisory

Fibromyalgia patients are advised to follow these suggestions:

1. Establish regular sleep routines. Patient bedtime is recommended not later than 11 p.m. Establish a regular time for going to bed and getting up in the morning. Maintain this schedule even on weekends and during vacations.

   Tips for good sleep habits include:
   - Avoid intense exercises approximately 6 hours before bedtime.
   - Avoid caffeine or alcohol 4–6 hours before bedtime.
   - Avoid drinking fluids right before bedtime.
   - Avoid large meals 2 hours before bedtime.
   - Avoid naps in the evening or late afternoon. Take short naps between 11 a.m. to 1 p.m., of approximately 20–30 minutes each.
   - Minimize light and maintain a comfortable, moderate temperature in the bedroom. Keep the bedroom well ventilated.
   - Use the bed only for sleep – not as an office.
   - Reduce stress. Develop a plan to avoid or limit overexertion and emotional stress.
   - Exercise regularly. At first, exercise may increase the pain. But doing it gradually and regularly often decreases symptoms. Exercise in the morning and avoid exercise in the evening. Select the type of exercise and length that the body is able to handle and avoid overdoing it.
2. Consume the correct diet. Correct diet is helpful in improving fatigue; this is particularly important for patients with Qi and Blood Deficiency. Although there is no single best food to cure fibromyalgia, patients need to know which foods are better or worse for consumption.
   - Foods that are good for nourishing Blood deficiency include eggs, yogurt, red meat, liver, beets, red dates, salmon, and molasses.
   - Foods that are good for nourishing Qi include yams, sweet potatos, rice, and oats.
   - Foods to avoid include MSG, trans fats, too much fat, fried foods, preservatives, food colors and dyes, and alcoholic beverages. Also avoid tobacco.

NOTES

1. Bhatt, D. L., *Guide to Peripheral and Cerebrovascular Intervention*. London: Remedica Medical Education and Publishing, 2004.
2. https://osteopathy.colganosteo.com/thoracic-outlet-syndrome/
3. Mandal, A., "History of fibromyalgia", News-Medical.net. February 26, 2019. https://www.news-medical.net/health/History-of-Fibromyalgia.aspx.

2 Headache and Facial Pain

HEADACHE IN CHINESE MEDICINE 頭痛（中醫）

INTRODUCTION

According to the Centers for Disease Control and Prevention's National Health Interview Survey in 2015, "approximately 20% of women and 10% of men aged ≥18 years in the United States report having a severe headache or migraine in the past 3 months".

Western medicine pain relievers are not the only solution for treating headaches and migraines; many patients find a longer period of relief from chronic headaches through traditional Chinese medicine (TCM) and without side effects, such as substance dependence.

Traditional Chinese medicine has accumulated treatment experience of headaches for thousands of years. The source of the pain could be from: an infection related to microorganisms such as bacteria, fungi, algae, or a virus; head trauma from concussion, stroke, or head surgery; internal health conditions, such as high blood pressure, anemia, dehydration, electrolyte imbalance, food allergy, and mental and/or physical stress.

Once the root of the pain is found, TCM balances it with suggestions on diet, lifestyle, exercise, as well as herbal supplements, acupuncture, and Tui Na therapeutic manipulation.

CHINESE MEDICINE ETIOLOGY AND PATHOLOGY

According to the book "The Essentials and Fundamentals of Pulse Diagnostic Chapter 17《素問脈要精微論篇第十七》", "the head is the mansion of the essential substances" "頭者精明之府". The Essence and the Marrow ascend and gather in the head. The Essence is generated by Zang Fu (organs, in Chinese medicine), and the marrow is derived from the Kidney Essence and Spleen food nutrients. The head serves as the highest center for regulating functions inside the body.

There are various causative factors, but headaches commonly occur due to exterior and/or interior causes.

1. Headache caused by the invasion of exterior factors.
2. Headache caused by the interior factors.

DOI: 10.1201/9781003203018-2

1. Exterior Factors

MECHANISM ACCORDING TO THE TYPE OF EXTERIOR FACTORS

Wind風, Heat暑, Damp濕, Dryness燥, Cold 寒, and Fire火 are six environmental elements that can influence the functions of the body. These six elements are called Six Qi六氣, when they have positive effects and are called Six Yin 六淫, if they negatively influence the body.

Six Qi 六氣

The elements exist in nature. Six Qi六氣have a positive influence, for instance, when a person dresses properly according to age, health condition, environmental seasons, and weather, such as hot or cold temperature, rain, wind, sunlight, etc.

Chinese medicine teaches that we live under the Six Qi六氣, and we need to adjust and regulate our health condition according to the changes of the Six Qi六氣.

The changes are:

- Wind diseases occur in spring.
- Summer-Heat diseases occur in summer.
- Clammy diseases occur in late summer and early autumn.
- Dry diseases occur in autumn, and Cold diseases occur in winter.

Six Yin六淫

When a person does not follow the usual pattern, or the Six Qi 六氣become extreme, these environmental conditions are able to adversely affect one's health and cause illness. In such cases, Six Qi六氣 are known as Six Yin六淫. Six Yin六淫 are the Six Qi六氣conditions that cause illness, and are also called Six Xie六邪 (Six Evil, six nefarious factors).

In among these six conditions, the most common one is Wind風. When it invades one's body, Wind travels to the head in the meridians and hinders the free flow of Qi 氣and Blood circulation.

The other five Yin 淫are able to affect one's health if they are extreme. For example, people who routinely work in a high-temperature environment are prone to be attacked by Heat暑, Dryness燥, and Fire火.

These Yin淫 conditions can work alone or in groups. Any one of the Yin 淫conditions can influence the others: for example, Wind風 can stir up Cold 寒to cause blockage in the head, resulting in a headache.

The Yin 淫conditions can also transform into another one under particular conditions: for example, Cold 寒can be transformed into Heat 暑in Yang Ming 陽明stage, the Heat goes to the forehead by the path of the Yang Ming 陽明meridian resulting in a headache in the front of the head. Heat暑 can be transformed into the Cold 寒in Jue Yin 厥陰stage as well, when the disease gets worse.

The characteristics of Wind風:

1. Wind風has a seasonal relationship with the spring months. Although Wind 風is present in all seasons, its manifestation will be stronger in the season that matches it.

2. Wind 風is a nefarious Yang 風為陽邪; its moving direction is outgoing dispersion and going upward to the head. This is stated in the text book Huangdi Neijing: "The damage inflicted by Wind affects primarily the top" 黃帝內經·素問 太陰陽明論篇第二十九.

3. Wind Evil風邪 is a corrupting influence that rarely appears alone, usually accompanied by some other external pernicious influence, such as cold and damp weather. It even helps other influences to invade the body.

4. In traditional Chinese medicine, Wind Evil 風邪is the beginning of all kinds of health conditions. Wind invasion can result in many different diseases. When a person's Yang Qi陽氣 is not sufficient, it is easy for them to catch Wind-Cold-Damp Evils; if the Yin Qi陰氣 is not sufficient, it is easy for them to catch Wind-Heat-Dryness-Fire Evils.

5. The Wind風 is characterized by constant movement 風性主動. Acute urticaria is a pattern of Wind in the skin.

6. The Wind 風is moveable and changeable 風性善行而數變; its change is rapid, and the disease processing duration is abrupt onset. Stroke is an example of a disease caused by Wind 風invasion.

7. The commonly seen diseases in spring season are typically associated with weak immune system such as allergic rhinitis and chicken pox.[1]

The characteristics of Cold 寒:

1. Cold 寒has a seasonal relationship with the winter months.
2. Cold 寒 is a Yin Evil 陰邪, it injures Yang Qi陽氣, resulting in Yang Qi 陽氣deficiency.
3. Cold 寒 contracts and tightens the soft tissues and decelerates the flow of Qi.
4. Cold 寒 is static, it likes to settle into the muscles, leading to muscle contraction.
5. The Cold Evil 寒邪 that enters the body can be transformed into the Heat Evil暑邪 when it enters Yang Ming 陽明.
6. The commonly seen diseases in winter season are common cold, flu, strep throat, norovirus.[2]

The characteristics of Heat暑:

1. Heat 暑has a seasonal relationship with the summer months.
2. It commonly affects those who undertake long trips or engage in intense physical labor outside on days of intense summer heat.
3. Heat often works together with Wind 風 to injure the exterior level of the body, and they stir upward to cause headache.
4. Heat affects the Upper Jiao上焦, and it injures the Heart Qi心氣.
5. The commonly seen diseases in the summer are sunstroke or heatstroke.

The characteristics of Damp濕:

1. Damp 濕has a seasonal relationship with the late summer months to the early autumn months.

2. The nature of the Dampness is heavy and sticky.
3. It commonly affects those who undertake outdoor activity during rainy, foggy weather. The activity can be physical work under water, sitting or lying on a moist, damp surface; sweating wets the clothes; living in a humid environment that is conducive to virus growth for a long period of time encourages the invasion of the Damp Evil 濕邪factor. The Damp Evil 濕邪penetrates the skin into the muscles and joints, meridians, and organs, such as the Lung.
4. Damp濕 can influence the others. When it influences the Wind風, it is call Wind-Damp風濕; the Wind 風carries the Damp to block the orifice, causing a headache. When the Damp 濕influences the Cold寒, it becomes Damp-Cold濕寒; when it influences the Heat熱, it becomes Damp-Heat濕熱. Damp-Cold 濕寒has a better prognosis than Damp-Heat濕熱.

The characteristics of Dryness燥:

1. Dryness 燥has a seasonal relationship with the autumn months.
2. It commonly affects the Upper Jiao 上焦, causing the common cold and headache.
3. The other common illness is seasonal flu.[3]

The characteristics of Fire火:

1. Fire is a Heat Evil 熱邪, and it is transformed from Wind風Cold 寒Heat暑, Damp濕,and Dryness燥.
2. It commonly affects all three Jiao三焦.

Infection with Six Yin 六淫may not manifest the illness right the way or in the season; it may take a considerable period of time before the symptoms appear. For example, if one suffers from Cold Evil 寒邪invasion in winter season, the Warm Disease presents in summer. If one suffers from Heat Evil熱邪 invasion in the summer, it will not appear until autumn. This is called Dormant Evil 伏邪.

Unlike the acute onset of the infection with Six Yin六淫 that presents the progression of headache from mild to severe with certain superficial symptoms under a long duration, the headache that is a symptom of the Dormant Evil 伏邪manifests as severe headache within a short duration, even the superficial symptoms did not present.

Six Yin 六淫, Six Meridians, and the Headaches

The headache in the occiput, forehead, temporal, and the vertex is called the Six Meridian Headache. It was first recorded in "Shang Han Lun 傷寒論", which was authored by Zhang Zhong-Jing 張仲景 in about 220 CE, about 1700 years ago. In Chinese medicine, the questions of locations, types, and symptoms of headache and the medical history prior to the onset of the episodes give us a sense of the pathology, etiology, and the mechanism of the headache, thereby assisting in the setup of the diagnosis and the treatment principles.

A. The occipital headache or Tai Yang headache 太陽頭痛

The Location of Tai Yang太陽

It is located on the external layer of the body, which is often the place the external pathogens first attack.

- It is commonly caused by Wind and Cold 寒 invasion. The character of Wind is Yang陽, its moving direction is ascending. The Cold and Wind reach the back side of the head by the meridian that is causing blockage, resulting in headache in the occipital region.
- Common cold or flu in western medicine belongs to this type of headache.

The Character of Tai Yang 太陽

- Tai Yang 太陽governs the exterior and is in charge of opening and dispersal. Tai Yang meridian 太陽經 is at the outermost surface, and its Qi氣 is the most prosperous among three Yang 三陽 (Tai Yang太陽, Yang Ming 陽明, and Shao Yang少陽). The Qi氣 flowing direction goes upward and outward toward the external pathogen.
- The time of day that Tai Yang disease 太陽證 tends to be cured is from 9 a.m. to 3 p.m. because this is the time period that the Yang Qi 陽氣 in the nature is strongest.

The Source of the Causes

Tai Yang disease 太陽證 is the body Yang Qi 陽氣 being injured by the Cold 寒.

The cause on its meridian: Wind and Cold 寒 invasion are the causes leading to three basic patterns of Tai Yang 太陽disease:

- Wind Strike太陽中風證.
- Cold Damage太陽傷寒證.
- Warm Disease太陽溫病.

These three patterns of disease have the same symptoms:

- Headache.
- Fever.
- Floating pulse.
    Headache. The Tai Yang meridian 太陽經is affected when the exogenous pathogenic factors attack the body surface injuring the Wei (defensive) Qi衛氣. The factors reach the back side of the head by the meridian causing the blockages, resulting in headache in the occipital region.
    Fever. The Yang Qi 陽氣is obstructed by the exogenous pathogenic factors, while Qi and Blood are struggling with the exogenous pathogenic factors at the exterior level. This results in fever.

Floating pulse. This indicates that the pathological changes are happening at the exterior level.

For treatments, please refer to **Wind-Cold Headache** 風寒頭痛.

B. The frontal headache or Yang Ming headache 陽明頭痛

The Location of Yang Ming陽明
The Yang Ming meridian is in the deep side of all three Yang三陽.

The Character of Yang Ming陽明
- Yang Ming 陽明is a Heat-producing Zang-Fu organ臟腑; it is called the "4 Greats" (great pulse, great thirst, great sweat, and great fever).
- Yang Ming 陽明governs the interior and is in charge of closing and storage of Yang Qi陽氣.
- The time of day during which Yang Ming 陽明disease tends to be cured is from 3 p.m. to 9 p.m.

The Source of Causes
Yang Ming disease 陽明證 is caused when the pathogenic Heat damages the body Yin. The exogenous pathogenic Wind and Cold 風寒 – residual from attacking Tai Yang 太陽 – progresses interiorly to attack Yang Ming 陽明, or it may be due to delayed treatment, causing body fluid consumption.

There are exterior and interior causes of this type of headache.
Exterior factor:

- Tai Yang 太陽meridian is on the surface, and it is the first place where the Xie (nefarious) Qi 邪氣Wind Cold 風寒 attacks.

When a person is exposed to an exterior Cold pathogen (called Wind Cold 風寒 in Chinese medicine), the disease stage is at the exterior body. If the exogenous pathogenic factor (called Xie (nefarious) Qi 邪氣) becomes stronger than the Wei (Defensive) Qi 衛氣, the pathogen progresses and penetrates deeper into the Yang Ming 陽明stage.

- As the Xie (nefarious) Qi 邪氣penetrates deeper into the Yang Ming 陽明 stage, the Yang Ming 陽明Heat rises because it is the ultimate stage of the struggle between the Zheng (righteous) Qi 正氣and the Xie (nefarious) Qi 邪氣. The Heat then goes to the forehead in the meridian, resulting in a headache in the front of the head.

Interior factor:

- The headache is commonly caused by the disorders of Spleen and Stomach. In western medicine, the gastrointestinal disorder is part of the Spleen and

Stomach system dysfunction. Other diseases in western medicine that are commonly seen when the system dysfunction are sinusitis, allergic rhinitis, and weak immunity.

- The Stomach is a dry Zang-Fu臟腑, and it is prone to Dryness and Heat. Poor diet or alcohol drinking results in Heat accumulation in the Stomach. The Heat then reaches the forehead by the meridian, resulting in headache in the front of the head.

C. The temporal headache or Shao Yang type of headache 少陽頭痛

The Location of Shao Yang少陽

- Shao Yang 少陽 works in between Tai Yang 太陽and Yang Ming陽明. The disease mechanism is similar to that of Jue Yin's disease 厥陰證 except Shao Yang少陽 works near exterior body while Jue Yin 厥陰works near interior body.
- Per its roles, it is located in the chest and abdomen.

The Character of Shao Yang少陽:

- It is a Fu organ腑臟. It has an exterior/interior relationship with its Zang organ Liver肝臟. Liver is an emotional Zang-Fu organ臟腑. Emotional stress, anger, anxiety, and depression easily lead to its stagnation.
- It produces Heat. Of the three Yang三陽, Tai Yang 太陽has the greatest Yang Qi陽氣, Yang Ming陽明 has the second, and Shao Yang 少陽has the weakest Yang Qi陽氣.
- Tai Yang's Yang Qi 陽明陽氣 is functioning on the exterior body, protecting us from the attack of the exogenous pathogenic factors. Yang Ming's Yang Qi 陽明陽氣 is functioning in the interior body, in charge of closing and storage of Yang Qi陽氣. Shao Yang's Yang Qi 少陽陽氣 governs the space between the exterior and interior.
- Shao Yang has a pivot dynamic; it is the driving force and catalyst for both Tai Yang 太陽and Yang Ming陽明. Since it is in between both, it enables communication between them.
- The time that Shao Yang少陽disease tends to be cured is from 3 a.m. to 9 a.m.

The Source of Causes

1. As stated in the book *Shang Han Lun*,

When the Blood is weak and the Qi is exhausted, the interstices are open, and because evil Qi enters the body and contends with the right Qi, there is binding under the ribside. The right and evil struggle by turns, so there is alternating aversion to cold and fever that stops and starts periodically, and taciturnity with no desire for food or drink. The viscera and bowels are interconnected, and so the pain is low down, hence there is retching, and Xiao Chai Hu Tang governs. 傷寒論 8-29 少陽29/宋本太陽中097前段: "血弱氣盡，腠理開， 邪氣因入， 與正氣相搏，結於脇下，正邪分爭，往來寒熱，休作有時，默默不欲 飲食，藏府 相連，其痛 必下，邪高 痛下，故使 嘔也，(一 云: 藏府相 連，其病必 下，脇膈痛 下，)小柴 胡湯主之.

When the Xie (nefarious) Qi 邪氣 is uncured in the Tai Yang 太陽exterior syndromes, it progresses into the interior, instead of transmitting to the Yang Ming 陽明interior. It stays neither on the Tai Yang 太陽exterior nor in the Yang Ming陽明 interior. This is known as Half-Exterior-Half-Interior syndrome 半表半裡證.

2. When Shao Yang syndrome 少陽證 is formed, its pathogenesis can be summarized by the three characters: stagnation, Heat, and deficiency.
   - Stagnation is derived from Qi stagnation in the meridian resulting in disharmony.
   - Heat is developed from Qi stagnation, Gallbladder Fire rises up.
   - Deficiency means the progression of Xie (nefarious) Qi 邪氣 from the exterior transmits interiorly when the Zheng (righteous) Qi 正氣 is deficient.

3. Emotional stress is also a common cause. Migraine headache in western medicine is under this type of the headache.

D. Jue Yin type of headache 厥陰頭痛or the headache on the top of the head

The Location of Jue Yin厥阴

The outcome and the mechanism of the interchanging in Jue Yin's disease 厥陰證 is similar to that of Shao Yang's disease 少陽證, except that the interchanging of Cold 寒 and Heat 熱 in Shao Yang 少陽 is more superficial, while the one on Jue Yin厥阴 is deeper.

The Character of Jue Yin厥陰

- Jue Yin厥阴 is the last meridian of the six meridians.
- The time that Jue Yin disease 厥陰證 tends to be cured is from 1 a.m. to 7 a.m.
- In Huang "Di Nei Jing Su Wen (Yellow Emperor's Internal Classic Plain Question)" Zhi Zhen Yao Da Lun, Yellow Emperor asked: "What is the meaning of Jue Yin?" Qi Bo said: "When the two Yins of Tai Yin and Shao Yin are both diminishing, it is called Jue Yin. 《素問.至真要大論》云:「帝曰:厥陰何也?岐伯曰: 兩陰交盡也。」
- "Jue厥" has three meanings:
   - 極至It is ultimate, extreme, most, virtual, absolute, and terminal. The Jue Heat 熱厥 and Jue Cold 寒厥 are two extreme characters of Jue Yin disease 厥陰證.
   - 逆It is inverse, opposite, contrary, backwards, and rebellious. The Qi, which is supposed to go downward, contrarily goes upward.
   - It also means "unconsciousness".
- Its presentation classically demonstrates as upper Heat and lower Cold. The disease condition in Jue Yin 厥陰is complex with alternating syndromes of extreme Heat and/or extreme Cold.
- The mechanism of the interchanging in Jue Yin厥陰 involves the severe deficiency and virtual exhaustion of Zheng Qi 正氣. It is the terminal stage of the illness.

- When Yin conformation goes to an extreme, it may undergo a reverse transformation into the Yang character with the proper treatment. This is the phenomenon of "things must be reversed".

The Source of the Causes

- Yin陰 and Yang陽 are disconnected when Yang Qi is floating at the superior body, causing excessive condition in the upper body and deficiency in the lower body. This is the cause of the headache on the top of the head as Huang Di Nei Jing, Su Wen Fangshengshuai growth and decline of energy 黃帝內經．素問．方盛衰論篇第八十stated: "Qi gets stuck on the superior, headache and seizure occurs on the top of the head氣上不下，頭痛巔疾".
- Other diseases in western medicine in Jue Yin disease 厥陰證 are allergic colitis, uncontrollable epilepsy, and cryptogenic tongue ulceration.

The meridians of Tai Yin太陰 and Shao Yin 少陰are not on the head, therefore, headache doesn't occur directly from these two meridians. On the other hand, Qi and Phlegm stagnation at the diaphragm blocks the pure Yang rising, leading to headache.

Interior factors

1. Chronic illnesses that injure the Zang-Fu 腑臟 leading to the disorder of Qi and Blood.
   - Long-term illness and debilitating overwork can deplete Spleen resulting Qi and Blood deficiency.
   - Traditional Chinese medicine believes that Kidney stores Essence (Jing 精), Kidney gives rise to marrow. The functions of Essence (Jing精) are controlling reproduction, growth, and development, promoting the transformation of Blood. According to Lingshu Jingmai, Yellow Emperor states: "At conception, Essence is formed. After Essence is formed, the brain and bone marrow are formed" 《黃帝內經−靈樞‧經脈第十》黃帝曰: "人始生,先成精, 精成而脑髓生".

     When one suffers from Kidney Essence deficiency, the brain and marrow lack nourishment, causing headache.
   - Chronic insomnia depletes Heart and Kidney Yin. Yin deficiency with ascendant Yang and Liver Wind causes blockage in the meridians, resulting in headaches.
   - Excessive mental work for long periods of time depletes Qi, Blood, Yin, and Yang. This is often seen in students or office workers whose work is associated with prolonged sitting and reading paper books in insufficient light and gazing at cellphones or computer monitors, especially at night.

     Chapter 23, Xuan Ming Wu Qi, in the book *Huang Di Nei Jing Su Wen* states:

Looking for a long time damages the Blood, lying down for a long time damages the Qi, sitting for a long time damages the flesh, standing for a long time damages the Bones, walking for a long time damages the Sinews.《黃帝內經·宣明五氣》：久視傷血，久臥傷氣，久坐傷肉，久立傷骨，久行傷筋，是謂五勞所傷。

- Liver Qi stagnation due to chronic stress and anger. An extended period of anger, stress, emotional frustration, or worry leads to Liver Qi stagnation (Apichai 2020); it often demonstrates the symptoms on the Foot Jue Yin Liver meridian 厥陰肝經 itself and the Fu organ of Liver, which is Foot Shao Yang Gallbladder meridian 足少陽膽經..
- Grief is the emotion of the Lung. Prolonged sadness or bereavement depletes Lung Qi. Lung's element is metal, Liver's element is wood, and metal controls wood. Weak Lung Qi loses its ability to control the Liver, leading to Liver Qi stagnation.
- Chronic Liver Qi stagnation causes chronic muscle tension in the upper back, shoulder, and neck regions, leading to headache.
- Chronic Liver Qi stagnation can overact on the Spleen, leading to Qi and Blood deficiency. Phlegm and Damp are accumulated when Spleen Qi is depleted because the functions of transformation and transportation of food, nutrition, and waste of Spleen Qi is dysfunctional.
- Worry is the emotion of the Spleen. Excessive worry will deplete the Spleen Qi. A weak Spleen is easily overacted by Liver when Liver Qi is stagnant.
- Spleen Qi also can be weakened by excessively use of certain Chinese herbs. Herbs that clear Heat or are laxatives are commonly seen in weakened Spleen Qi.
- Dysfunction of the Spleen and Stomach can also be due to long-term alcohol consumption and excessively rich diets with fatty food, sweet food, and strong-tasting food. This can result in Phlegm and Damp accumulation that blocks the Middle Jiao 中焦 and evolves over time, preventing pure Yang from rising and turbid Yin from sinking 濁陰不降. When the situation is allowed to deteriorate, turbid Yin 濁陰 mists the orifice causing headache.
- Overconsumption of some hot herbs or foods can lead to Heat in Stomach, Liver, and Gallbladder, resulting interior Heat or Fire diseases, and the Heat or Fire causes the headache.
- Headache is also a symptom of sunstroke. In Chinese medicine, there are two types of sunstroke.

    Yang Sunstroke陽暑. When one engages in intense physical labor outside on days of intense summer heat for a prolonged period of time, one's body automatically lowers the body temperature by promoting sweat and rapid breathing. If one loses too much sweat and the body temperature keeps rising, and the environment lacks fresh air; this can lead to sunstroke. This is Heat 暑invasion, and Heat暑 often works together with Wind 風to injure the exterior level of the body, and they stir upward to cause a headache.

Yin Sunstroke陰暑. This is another type of sunstroke caused by Yin 陰Cold寒 invasion.

- When one sleeps on a cold floor while dressed inappropriately, the Yin 陰 invades the body when Zheng Qi 正氣 is insufficient.
- When one rests or resides in areas of heavy moisture or dampness, such as a river, a creek or a lake, one is easily invaded by Cold 寒邪and Damp 濕邪, resulting in blockages in the meridians.
- To cool off the heat in summer, many people drink ice water and eat cold and raw food. It may not injure your health if you just eat a little occasionally. But if you eat and drink cold regularly, it easily damages the Yang Qi 陽氣of the Spleen 脾and Stomach 胃, causing Wei Qi衛氣 insufficiency against the pathogenic Cold寒邪 and Damp濕邪 invasion.
- In hot weather, many people like to be in an aired conditioned room or have an electric fan blowing constantly and directly. These conditions allow easy invasion by Cold 寒邪 and Wind風邪, resulting in blockages in the meridians.

2. Stagnation in the meridians by blood clots.

1. Chronic Liver Qi stagnation develops Blood stasis because Qi is the commander of Blood.

2. Blood stagnation headache can also be the result of a direct head trauma. A fall or a direct blow to the head is a common cause of headache resulting in Blood stasis. The headache may not present immediately following the trauma, but it can occur months or years later, under some circumstances, or it is initiated by another illness or a decline in general health.

MANIFESTATIONS

Traditional Chinese medicine makes diagnosis for headaches based on three groups of information:

1. The headache locations.
2. The types of pain.
3. The headache under certain circumstances.

A. The headache locations

The headache differentiation according to locations by Lu Yihu陸以湉, a physician in Qing dynasty (1644–1911), are explained in his book "The Medical Talks from the Deserted Cottage" 《 冷廬醫話 》published in 1858:

The Tai Yang type of headache affects the occiput and reaches the top of the head, the pain also affects the nape; the Yang Ming type of headache, the meridian connects the eyeballs, the pain is on the forehead; those who suffer from the Shao Yang type of headache, the meridian goes to the sides of the head, the pain is on the sides of the head. Tai Yang meridians are on the posterior body, Yang Ming meridians are on the anterior body, Shao Yang meridians are on the side of the body. Jue Yin meridians meet on the top of the head, therefore the pain is on the top of the head. Tai Yin meridians and Shao Yin meridians do not ascend to the head, but Phlegm and Qi are constrained in the diaphragm from descending, and can hinder the pure Yang Qi from ascending freely to the head resulting the headache.

"頭痛屬太陽者，自腦後上至巔頂，其痛連項；屬陽明者，上連目珠，痛在額前；屬少陽者，上至兩角，痛在頭角。 以太陽經行 身之後，陽 明經行身之 前，少陽經行身之側。 厥陰之脈會 於巔頂，故 頭痛在巔頂。太陰、少 陰二經雖不 上頭，然痰與氣逆壅於 膈，頭上氣 不得暢而亦 痛。

THE OCCIPITAL HEADACHE OR TAI YANG HEADACHE太陽頭痛

* Occipital headache is called Tai Yang Headache太陽頭痛in traditional Chinese medicine because the exterior pathogenic factors affect the Foot Tai Yang Bladder Meridian leading to the occipital headache; Wind 風and Cold 寒are the most commonly seen factors.
* The headache is on the back side of the head and also affects the neck, shoulders, and the upper back.

The location of the Foot Tai Yang Bladder Meridian 足太陽膀胱經 in the original text of the Huangdi Neijing (the Yellow Emperor's Internal Classic):

Meridian Diagnostics of the Huangdi Neijing 黃帝內經 (the Yellow Emperor's Internal Classic) Lingshu 靈樞 (Miraculous Pivot):

Foot Taiyang Bladder meridian足太陽膀胱經 starts from the inner canthus (Jingming睛明, UB1), ascends to the forehead, and joins the opposite meridian at the vertex (Baihui 百會, DU20).

The branch emerging from the vertex reaches to the upper area of the ear.

The straight branch of the meridian leaves from the vertex, turns back to the occipital bone, and enters the cranial cavity to connect with the brain; then it reemerges and descends to the nape (Tianzhu天柱, UB10) and joins at Dazhui 大椎 (DU14) along the medial side of the scapula and parallel to the vertebral column (1.5 Cun 寸 lateral to the dorsal midline); then it descends to the lumbar region (Shenshu 腎俞, UB23) and enters the abdominal cavity from the deep layers of muscles along the lumbar region to link with the Kidney and pertains to the Urinary Bladder.

The branch stemming from the lumbar region runs downward parallel to the spine. Passing through the buttocks, it descends along the posterior border of the lateral side of the thigh into the popliteal fossa (Weizhong 委中, UB40).

The branch emerging from the nape runs downward along the medial border of the scapula, from Fufen 附分 (UB41); then it runs alongside the region 3 Cun 寸 lateral to the spine and reaches the hip joint. From there, it goes downward along the lateral aspect of the thigh and meets the previous branch from the loin in the popliteal fossa (Weizhong 委中, UB40); then it descends through the gastrocnemius, reaching the posterior aspect of the external malleolus (Kunlun崑崙, UB60), where it runs along the lateral side of the foot dorsum to the lateral side of the tip of the little toe (Zhiyin至陰, UB67) to connect with Foot Shaoyin Kidney Meridian足少陰腎經.

《靈樞·經脈》：

膀胱足太陽之脈，起於目內眥，上額，交巔；
其支者，從巔至耳上角；
其直者，從巔入絡腦，還出別下項，循肩髆內，挾脊，抵腰中，入循膂，絡腎，
　　屬膀胱；
其支者，從腰中下挾脊，貫臀，入膕中；
其支者，從髆內左右，別下，貫胛，挾脊內，過髀樞，循髀外，從後廉，下合膕
　　中，以下貫踹內，出外踝之後，循京骨，至小趾外側。

Characteristics of Tai Yang Headache

Wind-Cold Headache 風寒頭痛

- The headache is commonly located on the back side of the head and the neck; the pain also may spread to the entire head.
- The onset is acute, and the headache generally occurs without a warning sign.
  1. The pain is intermittent.
  2. The duration of the headache is generally short.
  3. The sensation of the headache is stiff and tight. This is because Cold 寒氣 contracts the soft tissues on the neck. The external Wind 風 brings the Cold 寒邪 to settle at the upper back, neck, and occipital regions.
  4. The headache usually is accompanied by whole body ache and aversion to cold and wind; hence, it is aggravated by wind exposure. Other possible symptoms include fever, sneezing, cough with white sputum, a runny nose with a white nasal discharge or a blocked nose, absence of thirst with no desire to drink or maybe with a desire for a warm drink; there is no sweat.
  5. Tongue: thin white coating or no coating.
  6. Pulse: floating and tight.

TREATMENTS

Treatment Principle

Expel Wind and dissipate Cold; regulate Blood and stop pain.

Herbal Treatments

Herbal Formula

Chuan Xiong Cha Tiao San 川芎茶調散

This formula is from "Tai Ping Hui Min He Ji Ju Fang太平惠民和劑局方", authored by Imperial Medical Bureau 太醫局in 1134AD, in southern Song dynasty 南宋 1127–1239AD.

Herbal Ingredients

Bo He薄荷12g, Fang Feng防風4.5g, Xi Xin細辛3g, Qiang Huo羌活6g, Bai Zhi白芷6g, Gan Cao甘草炙6g, Chuan Xiong川芎12g, Jing Jie荊芥12g

Ingredient Explanations

1. Bo He薄荷12g expels Wind Heat, relieves Liver Qi stagnation, disperses stagnant Heat, and enhances Chai Hu's ability to relieve the Liver.
2. Fang Feng防風4.5g expels Wind and drains Damp.
3. Xi Xin細辛3g enters Shao Yin meridian to expel Wind, Damp, and Cold.
4. Qiang Huo羌活6g expels Wind-Cold-Dampness, unblocks painful obstruction, and alleviates pain.
5. Bai Zhi白芷6g expels Wind, drains Dampness, dispels Cold, and stops pain.
6. Gan Cao甘草炙6g harmonizes the herbs.
7. Chuan Xiong川芎12g moves the Blood, disperses Bruising, and stops pain due to Blood stasis.
8. Jing Jie荊芥12g stops bleeding, dispels Wind, and relieves muscle spasm.

Modifications

- To make the formula stronger, add Sheng Jiang 生薑, Zi Su Ye紫蘇葉.
- For dizziness, add Tian Ma 天麻 and Gao Ben藁本.

Formula Actions

- Expels Wind and stops pain.

Formula Indication

- Headache due to Wind invasion, vertigo, stuffy nose, fever, and aversion to cold.

Contraindications

- Caution for patients who suffer from headache due to Liver Wind.

Cooking Instructions

1) Place the herbs into an herb pot. Traditionally, a clay cooker is used. Gently rinse the herbs once with a strainer.
2) Add water to the pot until it covers the herbs, then add one more cup of water.
3) Presoak the herbs for 15 minutes.
4) Turn on the stove to high heat and bring the herbs to a boil. Reduce the heat to medium-low to simmer and slightly bubbling. Turn off the heat when the decoction is reduced to one cup.
5) Drain the decoction into a container.
6) Repeat steps two to five one more time and drain each decoction to the same container.
7) Divide these two cup decoctions into two separate cups.

Dose

- Drink one cup of the decoction after meals twice daily.

THE FRONTAL HEADACHE OR YANG MING HEADACHE陽明頭痛

- Frontal headache is called Yang Ming Headache陽明頭痛in traditional Chinese medicine because the residual factors from attacking Tai Yang 太陽 progress interiorly to attack Yang Ming 陽明 resulting in headache in the Foot Tai Yang Bladder Meridian 足太陽膀胱經 and leading to frontal headache.
- The headache is on the front of the head, the forehead, and the eyebrows.

The location of the Foot Yang Ming Stomach Meridian in the original text:

Foot Yangming Stomach meridian足陽明胃經 starts from the side of the alae nasi (Yingxiang迎香, LI 20) and ascends alongside the nose to meet the opposite branch at the bridge of the nose. Then it runs into the inner canthus where it meets Foot Tai Yang Bladder meridian (Jingming睛明, UB1). Turning downwards along the lateral side of the nose (Chengqi 承泣, ST1 and Sibai 四白, ST2), it enters the upper gum. After reemergence, it curves around the lips and its opposite branch descends to meet with the conception vessel at the mentolabial groove (Chengjiang 承漿), where it runs posterolaterally across the lower portion of Daying大迎 (ST5). Winding along the angle of the mandible, it ascends along the jaw, passing through the front of the ear and Shangguan上關 (GB3), it runs along the anterior hair line, finally reaching the fore head.

The branch emerging front of Daying 大迎 (ST5) runs downward to Renying人迎 (ST9) where it goes along the throat and enters the supraclavicular fossa (Quepen缺盆, ST12). Then it descends through the diaphragm, entering the Stomach to which it pertains and connecting with the Spleen.

The straight branch of the meridian, starts from the supraclavicular fossa and runs downwards through the nipple. Descending along the line 2 Cun 寸 lateral to the umbilicus, it enters Qichong 氣衝 (ST30) in the groin.

The branch starting from the lower orifice of the Stomach, descends inside the abdomen and meets with the straight portion at Qichong氣衝 (ST 30). Then it runs downward along the anterior border of the lateral aspect of the tibia and reaches the knee. From there, it continuously runs downwards along the anterior border of the lateral aspect of the tibia, passes through the dorsum of the foot, reaching the lateral side of the tip of the second toe (Lidui厲兌, ST45).

The branch separating from the region 3 Cun 寸 below the knee (Zusanli 足三里, ST36) enters the lateral side of the middle toe.

The branch emerging from the dorsum of the foot (Chongyang衝陽, ST42), runs forwards into the medial side of the great toe (Yinbai隱白, SP1), where it connects with Taiyin Spleen meridian of foot足太陰脾經.

《靈樞·經脈》：

胃足陽明之 脈，起於鼻 之交頞中， 旁納太陽之 脈，下循鼻 外，入上齒 中，還
　　出挾 口環唇，下 交承漿，卻 循頤後下廉 ，出大迎， 循頰車，上 耳前，過
　　客 主人，循髮 際，至額顱 ；
其支者，從大迎前下人迎，循喉嚨，入缺盆，下膈，屬胃，絡脾；
其直者，從缺盆下乳內廉，下挾臍，入氣沖中；

其支者, 起 於胃口, 下 循腹裏, 下 至氣沖中而 合, 以下髀 關, 抵伏兔 , 下膝臏
　　中 , 下循脛外 廉, 下足跗 , 入中指內 間;
其支者, 下廉(膝#3)三寸而別下入中趾外間;
其支者, 別跗上, 入大趾間出其端。

Characteristics of Yang Ming Headache

7. The headache is on the front of the head accompanied by spontaneous sweating and fever without aversion to cold.
8. The character of the headache is dull if it is due to Spleen and Stomach deficiency.
9. The character of the headache is sharp, splitting, searing or piercing if it is due to Stomach Heat. Usually, the headache is located behind one eye. Red eyes, swollen eyes, and excessive tearing accompany the headache. Patients may also suffer from abdominal pain that dislikes pressure, dry stool. Tongue coating is yellow and dry. Pulse is big and forceful.
10. The character of the headache is heavy if it is due to Damp or Phlegm accumulation, and patients have difficulty concentrating. Pulse: floating, abnormal moderate (it is slack, loose, viscous, sticky; the rate is normal at about 60 BPM but it feels slow to the fingers at the end of the beat even though it is normal as measured by a clock).

TREATMENTS

Treatment Principle
Expel Wind and dissipate Cold, regulate Blood, and stop pain.

Herbal Treatments
Herbal Formula

Sheng Ma Ge Gen Tang升麻葛根湯
This formula is from "Tai Ping Hui Min He Ji Ju Fang太平惠民和劑局方", authored by Imperial Medical Bureau 太醫局in 1134AD, in southern Song dynasty南宋 1127–1239AD.

Herbal Ingredients
Sheng Ma升麻5g, Ge Gen葛根5g, Bai Shao白芍5g, Gan Cao甘草5g

Ingredient Explanations
1. Sheng Ma升麻 releases the exterior, clears Heat, raises pure Yang, and lifts sunken Qi.
2. Ge Gen葛根 relieves Heat, releases muscles especially of the neck and upper back.
3. Bai Shao白芍 nourishes the Blood, soothes the Liver, and stops pain.
4. Gan Cao甘草 harmonizes the herbs.

Modifications
- To make the formula stronger, add Bai Zhi白芷 and Chuan Xiong川芎.
- For persistent fever, add Chai Hu 柴胡, Huang Qin黄芩, and Fang Feng 防風.

Formula Actions
- Induces pure Yang, releases muscle, releases exterior, and stops pain 升陽 解肌，透表解毒。

Formula Indications
- Headache due to Yang Ming Wind invasion, body ache, fever, sneezing, cough, thirst, and aversion to wind.

Contraindications
- Caution for patients who suffer from headache and measles, this formula works well for the early stage; it is contraindicated for those in whom the toxic heat from measles has migrated internally.

Cooking Instructions
1) Place the herbs into an herb pot. Traditionally, a clay cooker is used. Gently rinse the herbs once with a strainer.
2) Add water into the pot until it just covers the herbs, then add one more cup of water.
3) Presoak the herbs for 15 minutes.
4) Turn on the stove to high heat and bring the herbs to a boil. Reduce the heat to medium-low to simmer but with slight bubbling. Turn off the heat when the decoction is reduced to one cup.
5) Drain the decoction into a container.
6) Repeat steps two to five one more time and drain each decoction into the same container.
7) Divide these two cup decoctions into two separate cups.

Dose
- Drink one cup of the decoction after meals twice daily.

THE TEMPORAL HEADACHE OR SHAO YANG HEADACHE少陽頭痛

- Temporal headache is called Shao Yang Headache少陽頭痛 in traditional Chinese medicine because the residual factors from attacking Tai Yang progress interiorly to attack Shao Yang, resulting headaches in Foot Shao Yang Gallbladder meridian and leading to the temporal headache.

The location of the Foot Shao Yang Gallbladder Meridian in the original text:

Shaoyang Gallbladder meridian of foot 足少陽膽經originates from the outer canthus (Tongziliao 瞳子髎, GB1) and ascends to the corner of the forehead (Hanyan 頷厭, GB4) and runs downward to the retroauricular region (Wangu 完骨穴, GB12) where it turns upwards along the forehead to the area above the eyebrow (Yangbai陽白, GB14). Curving back to Fengchi 风池 (GB20), it runs along the neck to the shoulder and meets the opposite branch at Dazhui 大椎 (GV14) where it runs forwards into the supraclavicular fossa (Quepen缺盆, ST12).

The branch originating from the retroauricular region enters the ear, emerges in front of the ear and then reaches the posterior part of the outer canthus.

Another branch from the outer canthus descends to Daying 大迎 (ST5) where it meets the branch of hand shaoyang running at the cheek. Reaching the infraorbital region, it descends through Jiache 頰車 (ST6) in the lower jaw to the neck. After meeting with the previous branch at the supraclavicular fossa (Quepen缺盆, ST12), it enters the chest and passes through the diaphragm to connect with the Liver and pertain to Gallbladder. Emerging from Qichong 氣沖 (ST30), it runs along the margin of the pubic hair and goes transversely to the hip region (Huantiao 环跳, GB30).

The straight branch descends from the supraclavicular fossa (Quepen缺盆, ST12) to the axilla. Along the Lateral side of the chest and through the hypochondrium, it continuously descends to meet with the previous branch at Huantiao 环跳 (GB30) where it descends along the lateral side of the thigh and knee. Descending along the lateral aspect of the fibula to its lower end, it reaches the anterior aspect of the external malleolus. Then it runs along the dorsum of the foot and reaches the lateral side of the tip of the fourth toe (Zuqiaoyin 足窍阴, GB44).

Another branch separating from the dorsum of the foot (Zulinqi 足临泣, GB41) runs forwards and emerges from the lateral side of the great toe. Turning back, it passes through the nail and ends at its hairy region to connect with Jueyin Liver meridian of foot足厥陰肝經.

《靈樞·經脈》：

起於目銳眥，上抵頭角下耳後，循頸行手少陽之前，至肩上卻交出手少陽之後，入缺盆；
其支者，從耳後入耳中，出走耳前，至目銳眥後；
其支者，別銳眥，下大迎，合於手少陽，抵於頔下，加頰車，下頸，合缺盆，以下胸中，貫膈，絡肝，屬膽，循脅裏，出氣沖，繞毛際，橫入髀厭中;
其直者，從缺盆下腋，循胸，過季脅下合髀厭中，以下循髀陽，出膝外廉，下外輔骨之前，直下抵絕骨之端，下出外踝之前，循足跗上，入小趾次趾之間；
其支者，別跗上，入大指之間，循大指歧骨內，出其端，還貫爪甲，出三毛。

Characteristics of Shao Yang Headache

- The headache is on the side(s) of the head and around the ears.
- The character of the headache is sharp or throbbing.
- Sometimes, the symptoms are accompanied by a bitter taste, dry throat, and blurred vision in the morning.

TREATMENTS

Treatment Principle

Harmonize Shao Yang stage exterior and interior and harmonize the Liver and Spleen.

Herbal Treatments

Herbal Formula

Chuan Xiong Cha Tiao San 小柴胡湯

This formula is from "Shang Han Lun傷寒雜病論", authored by Zhang Zhong-Jing張仲景in 200–210AD, in eastern Han dynasty東漢 25–220AD.

Herbal Ingredients

Chai Hu 柴胡24g, Huang Qin黃芩9g, Ren Shen人參9g, Ban Xia半夏洗9g, Gan Cao 甘草6 g, Sheng Jiang生薑切9 g, Da Zao大棗擘4 pieces

Ingredient Explanations

1. Chai Hu 柴胡 moves Liver Qi, moves evil outward, and releases stagnation.
2. Huang Qin黃芩 clears Heat, dries Damp, cools the Blood, nourishes Yin, and calms Liver Yang.
3. Ren Shen人參 tonifies Qi and Yang and prevents evil Qi entering interiorly.
4. Ban Xia半夏 transforms phlegm and prevents evil Qi entering interiorly.
5. Gan Cao甘草 harmonizes the herbs.
6. Sheng Jiang生薑 unblocks the pure Yang pathway and harmonizes rebellious Qi.
7. Da Zao大棗 tonifies the Spleen Qi, nourishes the Blood, and moderates and harmonizes the harsh properties of the fragrant herbs.

Modifications

- For pronounced thirst, withdraw Ban Xia半夏 and Ren Shen人參; add Tian Ma天麻 and Gao Ben藁本.
- For Yin deficiency, add Bie Jia鱉甲 and Qing Hao青蒿.

Formula Actions

- Harmonizes Shao Yang stage exterior and interior and harmonizes Liver and Spleen.

Formula Indications

- Headache due to Liver Qi stagnation and Heat in the Liver, Gallbladder, and Stomach.

Contraindications

- Caution for patients who suffer from headache due to hypertension.
- Contraindicated for those with excess above and deficiency below.

Cooking Instructions

1) Place the herbs into an herb pot. Traditionally, a clay cooker is used. Gently rinse the herbs once with a strainer.
2) Add 2.4 liters of water to the pot.
3) Turn on the stove to high heat and bring the herbs to a boil. Turn off the heat when the decoction is reduced to 1.2 liters.
4) Drain the decoction into a container, get rid of the herbs, and place the decoction back into the pot.
5) Repeat step three, turning off the heat when the decoction is reduced to 600 ml.
6) Divide the decoction into three separate cups.

Dose

• Drink one cup of the decoction after meals three times daily.

THE VERTEX HEADACHE OR JUE YIN HEADACHE厥陰頭痛

• Vertex headache is called Jueyin Headache厥陰頭痛in traditional Chinese medicine because Yang Qi is floating at the superior body.
• The headache is on the top of the head and the eyes. Foot Jueyin Liver meridian "connects with the eye system and emerges from the forehead where it goes up to the vertex (Baihui百會, GV20) to connect with the governor vessel".

The location of the Foot Shao Yang Gallbladder Meridian in the original text:

Jueyin Liver meridian of foot 足厥陰肝經starts from the dorsal hairy region of the great toe where it goes upwards along the dorsum to Zhongfeng 中封 (Liv4), 1 Cun 寸 in front of the medial malleolus. From there, it continuously ascends along the anterior border of the medial aspect of the tibia to an area 8 Cun 寸 above the medial malleolus, where it runs across and behind Taiyin Spleen meridian of foot. Passing through the medial side of the knee, it runs along the medial aspect of the thigh into the pubic hair region where it curves around the external genitalia and goes up to the lower abdomen. Ascending through Zhangmen 章門 (Liv13) and Qimen 期門 (Liv14), it enters the abdomen and runs alongside the Stomach to pertain to the Liver and link with the Gallbladder. Then it ascends through the diaphragm and distributes in the costal and the hypochondriac region. Ascending along the posterior aspect of the throat to the nasopharynx, it connects with the eye system and emerges from the forehead where it goes up to the vertex (Baihui百會, GV20) to connect with the governor vessel.

One branch from the eye system descends inside the cheek and curves around the inner surface of the lips.

Another branch starting from the Liver ascends through the diaphragm and enters into the Lung, where it connects with Taiyin Lung meridian of hand手太陰肺經.

《靈樞·經脈》：
肝足厥陰之脈，起於大趾叢毛之際，上循足跗上廉，去內踝一寸，上踝八寸，交出太陰之後，上膕內廉，循股陰，入毛中，過陰器，抵小腹，挾胃，屬肝，絡膽，上貫膈，布脅肋，循喉嚨之後，上入頏顙，連目系，上出額，與督脈會於巔；

- 其支者,從目系下頰裏,環唇內;

  其支者,復從肝,別貫膈,上注肺。

Characteristics of Jue Yin Headache
- The headache is on the top of the head.
- The headache is dull if it is due to Liver Blood deficiency, accompanied by vertigo, dizziness, pale face, and palpitation. Tongue is pale, and pulse is thin and weak.
- The headache is sharp, hot sensation on the head if the headache is due to Liver Yang rising, accompanied by irritability, red face, insomnia, forgetfulness. Tongue is dusky with yellow coating. Pulse is wiry.
- Accompanied by both cold hands and feet, retching, vomiting with saliva, poor appetite, and dizziness.
- Tongue is pale body with white wet coating.
- Pulse is weak, thin, and slow or wiry and thin but not rapid.

TREATMENTS

Treatment Principle
Warm Middle Jiao, tonify the deficiency, descend rebellious Qi, and stop vomiting.

Herbal Treatments
Herbal Formula

Wu Zhu Yu Tang吳茱萸湯

This formula is from "Shang Han Lun傷寒雜病論", authored by Zhang Zhong-Jing 張仲景in 200–210AD, in eastern Han dynasty東漢 25–220AD.

Herbal Ingredients
Wu Zhu Yu吳茱萸9g, Ren Shen人蔘9g, Sheng Jiang生薑18g, Da Zao大棗4g

Ingredient Explanations
1. Wu Zhu Yu吳茱萸 warms Middle Jiao, disperses Cold, expels Damp Cold, and alleviates pain.
2. Ren Shen人蔘 tonifies Qi and Yang.
3. Sheng Jiang生薑 unblocks the pure Yang pathway and harmonizes rebellious Qi.
4. Da Zao大棗 tonifies the Spleen Qi, nourishes the Blood, and moderates and harmonizes the harsh properties of the fragrant herbs.

Modifications
- For severe headache, add Dang Gui當歸, Chuan Xiong川芎, and Man Jing Zi蔓荊子.

- For persistent fever, add Chai Hu 柴胡, Huang Qin黃芩, and Fang Feng防風.
- For extreme Cold, add Fu Zi附子, Gan Jiang乾薑, and Xi Xin細辛.

Formula Actions

1. Warms and tonifies Middle Jiao and Lower Jiao, descends rebellious Qi, and stops vomiting.

Formula Indication

2. Jue Yin headache, cold hands and feet, Cold attacking Middle Jiao, and Jue Yin Cold in Liver.

Contraindications

1. It is contraindicated for patients who suffer from vomiting or acid reflux due to Heat in Stomach or Yin deficiency.
2. It is contraindicated for patients who suffer from headache due to Liver Yang rising.

Cooking Instructions

1) Place the herbs into an herb pot. Traditionally, a clay cooker is used. Gently rinse the herbs once with a strainer
2) Add 1 liter of water to the pot
3) Turn on the stove to high heat and bring the herbs to a boil. Turn off the heat when the decoction is reduced to 300 ml.
4) Drain the decoction into a container, get rid of the herbs, and place the decoction back to the pot.
5) Divide the decoction into three separate cups.

Dose

- Drink one cup of the decoction after meals three times daily.

THE HEADACHE IS BEHIND THE EYES眼後頭痛

- Headache behind the eyes is a very commonly seen symptom in the migraine headache.
- When headache behind the eyes is not part of the Jue Yin Headache 厥陰 頭痛 in the six meridian pattern headaches, it is due to other medical conditions of Liver because Liver externally opens into the eyes. This type of headache is often associated with a migraine; it usually occurs on one eye and one temporal only, seldom occurring on both sides.

A: The headache is dull if it is due to Liver Blood deficiency. It is typically caused by the following two causes.

1. Traditional Chinese medicine states "Liver opens into the eyes肝開竅於 目" and "extended belaboring of the vision exhausts Blood 久視傷血".

Prolonged staring at electronic devices, such as cellphones, computer monitors, and televisions drain the Liver Blood.

2. "Liver stores Blood肝藏血" at the Liver time, which is 1 a.m. to 3 a.m. everyday while supine sleeping. Western medicine has found that sleep deprivation alters liver metabolism. Chinese medicine states that lack of sleep and going to bed late are causes of Liver Blood deficiency.

3. Traditional Chinese medicine generates Blood through Spleen, Heart, and Lung. Chinese medicine holds Kidney is the origin of congenial constitution, while Spleen is the source of acquired constitution. Transformation and transportation are two of the Spleen roles. Spleen is responsible for the intake, processing of food, and distribution of nutrients. Spleen transports the nutrients upward to the Upper Jiao; Heart and Lung generate Qi, convert the nutrients to Blood, then send the nutrients to the rest of the body by the circulatory system. The impairment of Spleen, Heart, and Lung decrease the production of Blood and give rise to Liver Blood deficiency.

4. Accompanying symptoms include pale face, pale lips, dizziness, blurred vision, floaters in the eyes, pale nails, and dry and cracked nails. Some patients may also suffer from muscle weakness, cramps and spasm, and/ or numbness and tingling in the limbs. Some also suffer from gray and dry hair and scanty menstruation.
   1. Tongue: pale body with white coating.
   2. Pulse: wiry, thin.

TREATMENTS
Treatment Principle
Tonify the Blood and regulate menstruation.

Herbal Treatments
Herbal Formula

Si Wu Tang四物湯
This formula is from "Tai Ping Hui Min He Ji Ju Fang太平惠民和劑局方", authored by Imperial Medical Bureau 太醫局in 1134AD, in southern Song dynasty南宋 1127–1239AD.

Herbal Ingredients
Dang Gui當歸10g, Chuan Xiong川芎10g, Shu Di Huang熟地黃10g, Bai Shao白芍 10g

Ingredient Explanations
1. Dang Gui當歸 nourishes the Blood, benefits Liver, and regulates menstruation.
2. Chuan Xiong川芎 moves the Blood and moves Qi.

3. Shu Di Huang熟地黃 tonifies Liver and Kidney Yin and benefits Essence and Blood.
4. Bai Shao白芍 nourishes the Blood, soothes the Liver, and stops pain.

Modifications

- To make the formula stronger, add Ju Hua菊花, Gao Ben藁本, and Gou Teng鉤藤.
- For Qi deficiency, add Ren Shen人参 and Huang Qi黃耆.
- For vertigo and dizziness due to Wind in the head, add Qin Jiu秦艽、and Qiang Huo羌活.

Formula Actions

- Nourishes the Blood, benefits Liver, and regulates menstruation.

Formula Indication

- Headache due to Liver Blood deficiency and Wind and amenorrhea.

Contraindications

- Caution for patients who are pregnant.
- Caution for patients with constitutional Spleen Yang deficiency.

Cooking Instructions

1) Place the herbs in an herb pot. Traditionally, a clay cooker is used. Gently rinse the herbs once with a strainer.
2) Add water into the pot until it just covers the herbs, then add one more cup of water.
3) Presoak the herbs for 15 minutes.
4) Turn on the stove to high heat and bring the herbs to a boil. Then reduce the heat to medium-low to simmer with slight bubbling. Turn off the heat when the decoction is reduced to one cup.
5) Drain the decoction into a container.
6) Repeat steps two to five one more time, and drain each decoction to the same container.
7) Divide the decoction into two separate cups.

Dose

- Drink one cup of the decoction after meals twice daily.

B: The headache is distending with heaviness if it is due to Liver Qi stagnation. It is typically caused by emotional stress, anger, and anxiety. The headache usually occurs on one side and can travel from one side to the other. Liver Qi stagnation often affects the Spleen system; the location of the headache can also manifest on the forehead as well.

1. Accompanying symptoms include anger, hypochondriac pain, bitter taste in the mouth, belching and sighing, and insomnia.

Tongue: pink, dusky, thin coating.

Pulse: wiry, choppy.

From a Chinese medicine perspective, this pattern of pathology can lead to Liver Yang rising and Liver Fire.

TREATMENTS

Treatment Principle

Disperse Liver, regulate Qi, harmonize the Blood, and alleviate pain.

Herbal Treatments

Herbal Formula

Chai Hu Huang Qin Tang柴胡疏肝散

This formula is from "The Complete Compendium of Jingyue景岳全書", authored by Zhang Jiebin張介賓(1111–1117AD), in Ming dynasty明代 (1624 AD).

Herbal Ingredients

Chai Hu柴胡6g, Xiang Fu香附5g, Bai Shao白芍5g, Chuan Xiong川芎5g, Zhi Ke枳殼5g, Chen Pi陳皮6g, Zhi Gan Cao甘草2g

Ingredient Explanations

1. Chai Hu 柴胡 moves Liver Qi and releases stagnation.
2. Xiang Fu香附 moves Liver Qi and Blood and stops pain due to Qi and Blood stagnation.
3. Bai Shao白芍 nourishes the Blood, soothes the Liver, and stops pain.
4. Chuan Xiong川芎 moves the Blood and moves Qi.
5. Zhi Ke枳殼 regulates Qi and removes blockage.
6. Chen Pi陳皮 regulates Qi and removes blockage.
7. Zhi Gan Cao甘草 tonifies the Spleen, harmonizes the other herbs, and guides the herbs to all 12 meridians.

Modifications

- For Blood deficiency, add Dang Gui當歸, Sheng Di Huang生地, and Shu Di Huang熟地.
- For Blood stasis, add Dan Shen丹參.

Formula Actions

- Moves Liver Qi, harmonizes the Blood, and alleviates pain due to Liver Qi stagnation.

Formula Indication

- Headache due to Liver Qi stagnation, Blood stagnation, acid regurgitation, and breast distention.

Contraindications
- Caution for patients who suffer from headache due to the Yin deficient Heat.

Cooking Instructions
1) Place the herbs into an herb pot. Traditionally, a clay cooker is used. Gently rinse the herbs once with a strainer.
2) Add water into the pot until it covers the herbs, then add one more cup of water.
3) Presoak the herbs for 15 minutes.
4) Turn on the stove to high heat and bring the herbs to a boil. Then reduce the heat to medium-low to simmer with slight bubbling. Turn off the heat when the decoction is reduced to one cup.
5) Drain the decoction into a container.
6) Repeat steps two to five one more time and drain each decoction to the same container.
7) Divide the decoction into two separate cups.

Dose
- Drink one cup of the decoction after meals twice daily.

C: The headache is severe sharp and throbbing if it is due to Liver Yang rising. It is typically caused by enduring emotional stress and anger leading to Qi stagnation and causing Heat; Heat injures Liver and Kidney Yin resulting in Liver Yang rising.

1. The headache occurs on one or both sides behind the eyes, eyebrows, and temporal. It can affect the visions, as patients often describe seeing flashing light and auras.
2. Accompanying symptoms include dizziness, tinnitus, weak and sore low back and knees, irritability, elevated blood pressure, short temper, insomnia with dreams, nausea, and vomiting.
3. Tongue: red body and red edges possibly peeled with thin yellow dry coating.
4. Pulse: wiry or wiry, thin, and rapid.

TREATMENTS

Treatment Principle
Soothe Liver, suppress Yang, nourish Yin, and clear Heat. 平肝潜阳，滋阴清火

Herbal Treatments
Herbal Formula
Tian Ma Gou Teng Yin 天麻鈎藤飲
This formula is in the book "Newly Explained Diagnosis and Treatment of TCM Internal Medicine and Miscellaneous Diseases中醫內科雜病證治新義", authored by Hu Guangci 胡光慈 in 1958.

Herbal Ingredients

Tian Ma天麻6g, Gou Teng鈎藤9g, Chuan Niu Xi牛膝12g, Shi Jue Ming石決明15g, Du Zhong杜仲12g, Huang Qin黃芩6g, Zhi Zi栀子6g, Yi Mu Cao益母草9g, Sang Ji Sheng桑寄生9g, Ye Jiao Teng夜交藤9g, Fu Shen茯神9g

Ingredient Explanations

1. Tian Ma天麻 extinguishes Wind, calms the Liver and subdues rising Liver Yang, and alleviates pain.
2. Gou Teng鈎藤 expels Wind and releases spasm.
3. Chuan Niu Xi川牛膝 unblocks and promotes Blood circulation.
4. Yi Mu Cao益母草 moves the Blood, promotes urination, reduces swelling, and clears Heat.
5. Shi Jue Ming石決明 drains Liver Fire and expels Wind.
6. Du Zhong杜仲 tonifies Liver and Kidney, strengthens Bone, and benefits sinews and Blood.
7. Huang Qin黃芩 clears Heat, dries Damp, cools the Blood, stops bleeding, and calms Liver Yang.
8. Zhi Zi栀子 clears Heat, cools the Blood, resolves Damp, reduces swelling, and opens meridians.
9. Sang Ji Sheng桑寄生 tonifies Liver and Kidney, strengthens Bone, benefits Essence and Blood; expels Wind and drains Damp.
10. Ye Jiao Teng夜交藤 nourishes Heart Yin and Blood, calms Shen, and expels Wind and Damp.
11. Fu Shen茯神 calms Shen.

Modifications

- To make the formula stronger, add Ju Hua菊花, Xia Ku Cao夏枯草, Ku Ding Cha苦丁茶, and Zhen Zhu Mu珍珠母.
- For severe Liver Heat, add Xia Ku Cao夏枯草 and Long Dan Cao龍膽草.
- For Blood stasis, add Dan Shen丹參 and Chi Shao赤芍.
- For insomnia, add Long Gu龍骨, Mu Li牡蠣, Ye Jiao Teng夜交藤, and Yuan Zhi遠志.

Formula Actions

- Soothes Liver Qi, expels Wind, moves the Blood, clears Heat, tonifies Liver and Kidney.

Formula Indication

- Headache due to Liver Yang rising, vertigo, insomnia, anger, bitter taste in the mouth.

Contraindications

- Caution for patients who suffer from Blood, Yin, and body Fluid deficiency.
- Stop taking this formula if dizziness, chest distention, shortness of breath, nausea, vomiting, or itchy skin are present.
- Do not take Tian Ma overdose.

Cooking Instructions

1) Place the herbs into an herb pot. Traditionally, a clay cooker is used. Gently rinse the herbs once with a strainer.
2) Add water into the pot until it is just enough to cover the herbs, then add one more cup of water.
3) Presoak the herbs for 15 minutes.
4) Turn on the stove to high heat and bring the herbs to a boil. Reduce the heat to medium-low to simmer with slight bubbling. Turn off the heat when the decoction is reduced to one cup.
5) Drain the decoction into a container.
6) Repeat steps two to five one more time and drain each decoction to the same container.
7) Divide the decoction into two separate cups.

Dose

• Drink one cup of the decoction after meals twice daily.

D: The headache is burning and throbbing if it is due to Liver Fire, and the pain is higher in intensity than Liver Qi stagnation and Liver Yang rising. The disease process and the mechanism are due to External Fire and Internal Fire.

1. External Fire: fatty, oily, and a pungent, hot, spicy diet and alcohol consumption are attributable as the causes of the external Fire.
2. Internal Fire: chronic Liver Qi stagnation and difficulty in managing stress are attributable as the internal cause of Liver Fire.

1. The accompanying symptoms include irritability, outbursts of anger, dizziness, tinnitus, dream-disturbed sleep that commonly occurs from 1 a.m. to 3 a.m., epistaxis， thirst, bitter taste in the mouth, scanty and dark urine, dry stool or constipation, red face, and red eyes.

TREATMENTS

Treatment Principle

Drain excess Fire from the Liver and Gallbladder, clear and drain Damp-Heat from the Lower Jiao.

Herbal Treatments

Herbal Formula

Long Dan Xie Gan Tang龍膽瀉肝湯

This formula is from the book "Analytic Collection of Medicinal Formulas" 醫方集解, authored by Wang Ang汪昂 (1615–1694AD), and published in Qing dynasty清朝 (1682AD).

Herbal Ingredients

Long Dan Cao龍膽草12g, Zhi Zi栀子9g, Huang Qin黃芩9g, Chai Hu柴胡6g, Sheng Di Huang生地黃12g, Ze Xie澤瀉9g, Dang Gui當歸5g, Che Qian Zi車前子10g, Mu Tong川木通9g, Gan Cao甘草5g

Ingredient Explanations

1. Long Dan Cao龍膽草 drains Damp-Heat from the Liver and Gallbladder channels.
2. Zhi Zi栀子 clears Heat, cools the Blood, resolves Damp, reduces swelling, and opens meridians.
3. Huang Qin黃芩 clears Heat, dries Damp, cools the Blood, stops bleeding, and calms Liver Yang.
4. Chai Hu 柴胡 moves Liver Qi and releases stagnation.
5. Sheng Di Huang生地黃 nourishes Yin and clears Heat.
6. Ze Xie澤瀉 drains Damp, especially Damp Heat in Lower Jiao, clears Kidney deficient Heat.
7. Dang Gui當歸 nourishes the Blood, benefits Liver, and regulates menstruation.
8. Che Qian Zi車前子 clears Damp Heat and promotes urination.
9. Mu Tong川木通 drains Heat and promotes urination.
10. Gan Cao甘草 harmonizes the herbs, clears Heat, and expectorates Phlegm.

Modifications

- To make the formula stronger, add Ju Hua菊花 and Tian Ma天麻.
- For redness, swelling, and pain of the eyes, add Chuan Xiong川芎 and Ju Hua菊花.
- For acute glaucoma, add Xuan Shen玄參 and Qiang Huo羌活.

Formula Actions

- Drains excess Fire from the Liver and Gallbladder and drains Damp-Heat from Lower Jiao.

Formula Indication

- Headache due to Liver Fire and Damp Heat in Lower Jiao, such as migraine headache, urethritis, acute glaucoma, and acute cholecystitis.

Contraindications

- Caution for patients who suffer from Cold and Yang deficiency in Spleen and Stomach due to the Cold herbs in the formula.

Cooking Instructions

1) Place the herbs into an herb pot. Traditionally, a clay cooker is used. Gently rinse the herbs once with a strainer.
2) Add water into the pot until it covers the herbs, then add one more cup of water.

3) Presoak the herbs for 15 minutes.
4) Turn on the stove to high heat and bring the herbs to a boil. Reduce the heat to medium-low to simmer with slight bubbling. Turn off the heat when the decoction is reduced to one cup.
5) Drain the decoction into a container.
6) Repeat steps two to five one more time, and drain each decoction to the same container.
7) Divide the decoction into two separate cups.

Dose

• Drink one cup of the decoction after meals twice daily.

The Headache Is in the Whole Head 全頭痛

• Headache in the whole head is a type that is commonly seen before having a stroke.
• It is not part of the six meridian pattern headaches; it is due to the complex medical conditions of Liver and Kidney. In Lingshu Jingmai, Yellow Emperor states that: "At conception, Essence is formed. After Essence is formed, the brain and bone marrow are formed" 《黃帝內經·靈樞·經脈第十》黃帝曰: "人始生, 先成精, 精成而腦髓生".

If the headache is chronic and the duration is long, it is typically caused by insufficient Kidney Essence (Kidney Deficiency Headache 腎虛頭痛). Kidney stores Essence, produces marrow; the brain is the Sea of Marrow. When Kidney deficiency leads to depletion of Essence and Blood, the brain is then undernourished, the Sea of Marrow is deficient, resulting in headaches, dizziness, tinnitus, and fatigue.

1. The headache sensation is "empty and dull" typically on the entire head, patients usually have difficulty pinpointing the exact spot, and the vertex is more noticeable. It is aggravated by both physical and mental exertion, and it is alleviated after sleeping.
2. The accompanying symptoms are dizziness, vertigo, blurred vision, tinnitus, fatigue, sore lower back and knees, seminal emission and vaginal discharge, insomnia, and forgetfulness.
3. Tongue: red body with thin or peeled coating.
4. Pulse: thready and weak, soft.

Treatments

Treatment Principle

Strengthen Kidney Essence and tonify Qi and Blood.

Herbal Treatments

Herbal Formula

Da Bu Yuan Jian 大補元煎

This formula is from "The Complete Compendium of Jingyue景岳全書", authored by Zhang Jiebin張介賓(1111–1117AD), in Ming dynasty 明代 (1624 AD).

Herbal Ingredients

Ren Shen人參6g, Shan Yao山藥6g, Shu Di Huang熟地6g, Du Zhong杜仲6g, Dang Gui當歸6g, Shan Zhu Yu山茱萸3g, Gou Qi Zi枸杞子6g, Zhi Gan Cao甘草炙3g

Ingredient Explanations

1. Ren Shen人參 tonifies Qi and Yang.
2. Shan Yao山藥炒 tonifies the Kidney and benefits the Yin and the Spleen Qi.
3. Shu Di Huang熟地黃 tonifies Liver and Kidney Yin and benefits Essence and Blood.
4. Du Zhong杜仲 tonifies Liver and Kidney Yang, strengthens Bone, and benefits Essence and Blood.
5. Dang Gui當歸 moves the Blood, disperses Bruising, and stops pain due to Blood stasis.
6. Shan Zhu Yu山茱萸 tonifies Liver and Kidney, strengthens Bone, and benefits Essence and Blood.
7. Gou Qi Zi枸杞子 tonifies Liver and Kidney, strengthens Bone, and benefits Essence and Blood.
8. Zhi Gan Cao甘草炙 reduces swelling and stops pain.

Modifications

- For Cold symptoms due to Yang deficiency, add Fu Zi附子, Rou Gui肉桂, and Pao Jiang炮薑.
- For Blood stasis headache, remove Shan Zhu Yu山茱萸 and add Chuan Xiong川芎.

Formula Actions

- Severe depletion of Qi and Blood, Kidney Essence deficiency.

Formula Indication

- Headache due to Kidney Essence deficiency, severe loss of Yuan Qi and Blood.

Contraindications

- Caution for patients who suffer from deficient Fire Yin deficiency, Heat in Blood stage, Stomach Fire, Lung Phlegm Heat, Wind invasion, common cold, or Wind Heat.
- While taking this formula, avoid eating spicy and hot food, cold and raw food, and fatty food.

Cooking Instructions

1) Place the herbs into an herb pot. Traditionally, a clay cooker is used. Gently rinse the herbs once with a strainer.
2) Add 400 ml of water to the pot.
3) Turn on the stove to high heat and bring the herbs to a boil. Turn off the heat when the decoction is reduced to 280 ml.
4) Drain the decoction into a container.

Dose

- Drink one cup of the decoction before meals twice daily.
- Stop taking the formula if the symptoms are aggravated or are not alleviated.
- Do not take the formula longer than 2 weeks.

If the headache is acute, there is an acute onset under a chronic condition, the location of the headache is on the whole head, or the location changes to different parts of the head at different times, it could be due to two medical conditions:

- Invasion of external Wind 風 邪 and Cold 寒邪. Please refer to **Wind-Cold Headache 風寒頭痛**.
- Liver Wind, which is an acute onset under a chronic condition. It is typically caused by:
  1. Over strain and stress. This is due to long-term overwork and aging that lead to depletion and deficiency of Liver and Kidney Yin. This causes Liver Yang and Fire to rise.
  2. Liver Yang excess and Liver and Kidney Yin deficiency. These are due to long-term emotional disorder, leading to Liver Heat and resulting in Liver Yang rising, which breeds Inner Wind.
  3. Phlegm and Dampness accumulation. This is due to long-term irregular diet such as excessive consumption of alcohol, greasy and fatty foods, etc. and can lead to high blood pressure, high cholesterol, high blood sugar and Phlegm, which is then blown upward by inner Wind and blocks the orifice.
  4. Patients may experience 1–2 days of headache; it usually starts with a whole head headache, and patients may have such a headache for quite some time before the episode. The sensation of the headache this time is different: patients usually describe it as a distending, pulsating, throbbing, bursting pain, and it usually occurs suddenly with great intensity. The location of the headache changes from the whole head to a fixed area; the duration changes from an intermittent headache to a constant one.
  5. The accompanying symptoms for Liver Yang excess and Liver and Kidney Yin deficiency include nausea, vomiting, tinnitus, low back pain, Five-Center-Heat, low-grade fever, constipation, dysphagia, sudden elevated blood pressure, dizziness, semiconsciousness and then loss

of consciousness, twitching of muscles and skin, red face, trembling or stiff tongue, hemiplegia, flaccidity, weakness of lower extremities, and aphasia.

Tongue: scarlet, small and thin; peeling with cracks.
  Pulse: deep, wiry, thin, and rapid.

TREATMENTS

Treatment Principle
Nourish Liver and Kidney Yin, expel inner Wind and prevent Liver Yang rising, and tonify the Blood.

Herbal Treatments
Herbal Formula
Qi Ju Di Huang Wan 杞菊地黃丸
This formula is from the book "Analytic Collection of Medicinal Formulas 醫方集解", authored by Wang Ang汪昂 (1615–1694AD), in Qing dynasty 清朝, and published in 1682.

Herbal Ingredients
Shu Di Huang熟地黃24g, Shan Zhu Yu山茱萸12g, Shan Yao山藥12g, Ze Xie澤瀉 9g, Mu Dan Pi牡丹皮9g, Fu Ling茯苓9g, Gou Qi Zi枸杞子9g, Ju Hua菊花9g

Ingredient Explanations
  1. Shu Di Huang熟地黃 tonifies Liver and Kidney Yin and benefits Essence and Blood.
  2. Shan Zhu Yu山茱萸 tonifies Liver and Kidney, strengthens Bone, and benefits Essence and Blood.
  3. Shan Yao山藥炒 tonifies Kidney and benefits the Yin and the Spleen Qi.
  4. Ze Xie澤瀉 drains Damp, especially Damp Heat in Lower Jiao, and clears Kidney deficient Heat.
  5. Mu Dan Pi牡丹皮 clears Heat, cools the Blood, and moves the Blood.
  6. Fu Ling茯苓 drains Damp and tonifies Spleen.
  7. Gou Qi Zi枸杞子 tonifies Liver and Kidney, strengthens Bone, and benefits Essence and Blood.
  8. Ju Hua菊花 clears Heat, expels Wind, and calms Liver Yang and Wind.

Modifications
  • For hypertension with severe dizziness, add He Shou Wu何首烏, Tian Ma 天麻, Gou Teng鉤藤, and Shi Jue Ming石決明.
  • For diminished and dimmed vision, add Tu Si Zi菟絲子, Bai Shao白芍, Dang Gui當歸, and Huang Qi黃耆.

Formula Actions

1. Nourishes Liver and Kidney Yin, tonifies the Blood, and brightens the eyes.

Formula Indication

- Headache due to Liver and Kidney Yin deficiency, Liver Yang rising, and Liver Yin and Blood deficiency.

Contraindications

- Caution for patients who suffer from deficient Yang.
- During taking this formula, avoid eating sour-tasting food or cold and raw food.

Cooking Instructions

1) Place the herbs into an herb pot. Traditionally, a clay cooker is used. Gently rinse the herbs once with a strainer.
2) Add water into the pot until it just covers the herbs, then add one more cup of water.
3) Presoak the herbs for 15 minutes.
4) Turn on the stove to high heat and bring the herbs to a boil. Reduce the heat to medium-low to simmer but with slight bubbling. Turn off the heat when the decoction is reduced to one cup.
5) Drain the decoction into a container.
6) Repeat steps two to five one more time, and drain each decoction to the same container.
7) Divide the decoction into two separate cups.

Dose

- Drink one cup of the decoction after meals twice daily.

The accompanying symptoms for Phlegm and Dampness accumulation include: semiconsciousness and heavy sensation of the body; pale face, chest distension, nausea, vomiting with mucous, poor appetite, snoring with rattling sound of phlegm in throat, and loose stool.

Tongue: pale, flabby with teeth marks; thick and greasy coating.

Pulse: slippery, weak, and deep on the right side; if Damp Phlegm is more excessive, the left side will be more wiry and slippery.

TREATMENTS

Treatment Principle

Drain Damp, transform Phlegm, tonify the Spleen, soothe the Liver, and Extinguish Wind.

Herbal Treatments

Herbal Formula

Ban Xia Bai Zhu Tian Ma Tang半夏白朮天麻湯

This formula is from "Medical Insights醫學心悟", authored by Cheng Guopeng 程國彭 (1662–1735) in Qing dynasty清朝 1636–1912, and published in 1732.

Herbal Ingredients

Zhi Ban Xia半夏4.5g, Tian Ma天麻3 g, Bai Zhu白朮3 g, Ju Hong橘紅3g, Fu Ling 茯苓3g, Gan Cao甘草1.5g, Sheng Jiang生薑2 pieces, Da Zao大棗3 pieces, Man Jing Zi蔓荊子3g
水煎服。

Ingredient Explanations

1. Zhi Ban Xia半夏 transforms Phlegm.
2. Tian Ma天麻 extinguishes Wind, calms the Liver, subdues rising Liver Yang, and alleviates pain.
3. Bai Zhu白朮 tonifies the Spleen Qi and dries Dampness.
4. Ju Hong橘紅 regulates Qi and transforms Phlegm.
5. Fu Ling茯苓 drains Damp and tonifies the Spleen.
6. Gan Cao甘草 harmonizes the herbs, tonifies Middle Jiao Qi, clears Heat, expectorates Phlegm, and stops cough.
7. Sheng Jiang生薑 unblocks the pure Yang pathway and harmonizes rebellious Qi.
8. Da Zao大棗 tonifies the Spleen Qi, nourishes the Blood, and moderates and harmonizes the harsh properties of the fragrant herbs.
9. Man Jing Zi蔓荊子 expels Wind and stops pain.

Modifications

- To make the formula stronger, add Bai Ji Li白蒺藜.
- For vertigo, add Jiang Can僵蚕 and Dan Nan Xing胆南星.

Formula Actions

- Tonifies the Spleen, drains Damp Phlegm, soothes the Liver, and extinguishes Wind.

Formula Indication

- Headache due to Phlegm blocking orifice that stirs up by Wind, dizziness, sticky white coating on the tongue, wiry slippery pulse.

Contraindications

- It is contraindicated for patients who suffer from headache not caused by Wind-Phlegm 風痰.

Cooking Instructions

1) Place the herbs into an herb pot. Traditionally, a clay cooker is used. Gently rinse the herbs once with a strainer.
2) Add water into the pot until it is just enough to cover the herbs, then add one more cup of water.
3) Presoak the herbs for 15 minutes.

4) Turn on the stove to high heat and bring the herbs to a boil. Then reduce the heat to medium-low to simmer but with slight bubbling. Turn off the heat when the decoction is reduced to one cup.
5) Drain the decoction into a container.
6) Repeat steps two to five one more time, and drain each decoction into the same container.
7) Divide the decoction into two separate cups.

Dose

- Drink one cup of the decoction after meals twice daily.

HEADACHE DUE TO WIND HEAT風熱頭痛

1. The onset is acute.
2. The duration is short.
3. The sensation is distended and hot periodically with high severity that causes the head to feel as if it is being "cracked". The headache is aggravated by heat in the summer and alleviated by cold wind.
4. The Wind Heat headache is usually accompanied by aversion to wind, fever, slight thirst and desire to drink, runny nose with a yellow sticky nasal discharge, dry mouth, sore throat, possibly swollen tonsils, cough with yellow sticky sputum, sweating, red face and eyes, slightly dark urine, and dry stool.
5. Tongue: red tongue body, slightly red edges, red tip; thin, yellow, and dry coating.
6. Pulse: floating, rapid pulse.

TREATMENTS

Treatment Principle
Expel Wind and clear Heat.

Herbal Treatments
Herbal Formula
Chai Hu Huang Qin Tang芎芷石膏湯
This formula is from "Imperially Commissioned Golden Mirror of the Orthodox Lineage of Medicine醫宗金鑑", authored by the Qianlong emperor 乾隆 (1711–1799), in Qing dynasty清朝 1636–1912, and published in 1742.

Herbal Ingredients
Chuan Xiong川芎10g, Bai Zhi白芷10g, Shi Gao石膏10g, Ju Hua菊花10g, Qiang Huo羌活10g, Gao Ben藁本10g

Ingredient Explanations
1. Chuan Xiong川芎 moves the Blood and moves Qi.
2. Bai Zhi白芷 expels Wind, drains Dampness, dispels Cold, and stops pain.

3. Shi Gao石膏 clears Heat in the Qi stage and stops thirst.
4. Ju Hua菊花 clears Heat, expels Wind, and calms Liver Yang and Wind.
5. Qiang Huo羌活 expels Wind-Cold-Dampness, unblocks painful obstruction, and alleviates pain.
6. Gao Ben藁本 expels Wind and drains Damp and Cold in Tai Yang meridians.

Modifications

- To make the formula stronger in clearing Heat, add Huang Qin黄芩 and Zhi Zi栀子.
- For excessive Heat that hurts Yin, add Zhi Mu 知母, Shi Hu石斛, and Tian Hua Fen天花粉.

Formula Actions

1. Extinguishes Wind, clears Heat.

Formula Indication

- Headache due to Wind-Heat, Liver Wind stirring internally, Liver Fire generates Wind.

Cooking Instructions

1) Place the herbs into an herb pot. Traditionally, a clay cooker is used. Gently rinse the herbs once with a strainer.
2) Add water into the pot until it just covers the herbs, then add one more cup of water.
3) Presoak the herbs for 15 minutes.
4) Turn on the stove to high heat and bring the herbs to a boil. Then reduce the heat to medium-low to simmer with slight bubbling. Turn off the heat when the decoction is reduced to one cup.
5) Drain the decoction into a container.
6) Repeat steps two to five one more time, and drain each decoction to the same container.
7) Divide the decoction into two separate cups.

Dose

- Drink one cup of the decoction after meals twice daily.

Headache Due to Wind Damp风湿头痛

- The onset is acute.
- The exact location is difficult to pinpoint, patients often describe the sensation as "the head is wrapped in a damp moist cloth".
- The sensation of the pain is heavy and fuzzy so that the patients always want to shake the head to wave something away; the brain is groggy and the thinking is not clear.

- It is aggravated by damp weather, often occurring during a heavy rain prior to a sunny time.
- It is often accompanied by aversion to cold, possibly a fever, the patient feels a fever even though the thermometer shows a normal body temperature; there is a sensation of oppression in the chest and epigastrium, a feeling of heaviness of the whole body, a stuffy and runny nose with a white nasal discharge, nausea, decreased taste, poor appetite, heavy body sensation, difficult urination, and loose stool.
- Tongue: pale red tongue body; thick, white, greasy tongue coating.
- Pulse: floating, no root, and soggy; soft pulse.

TREATMENTS

Treatment Principle
Expel Wind and dissipate Cold, regulate the Blood, and stop pain.

Herbal Treatments
Herbal Formula
Qiang Huo Sheng Shi Tang羌活勝溼湯
This formula is from "Clarifying Doubts about Damage from Internal and External Causes 內外傷辨惑論", authored by Li Dongyuan李東垣 (1180–1251), in Jin dynasty 金朝 (1115–1234), and published in 1232.

Herbal Ingredients
Qiang Huo羌活3g, Du Huo獨活3g, Gao Ben藁本1.5g, Fang Feng防風1.5 g, Chuan Xiong川芎1.5g, Man Jing Zi蔓荊子0.9g, Zhi Gan Cao甘草炙1.5g

Ingredient Explanations
1. Qiang Huo羌活 expels Wind-Cold-Dampness, unblocks painful obstruction, and alleviates pain.
2. Du Huo獨活 expels Wind, Damp, and Cold in Lower Jiao and joints and removes chronic Bi stagnation in lower limbs.
3. Gao Ben藁本 expels Wind and drains Damp in Tai Yang meridians.
4. Fang Feng防風 expels Wind and drains Damp.
5. Chuan Xiong川芎 moves the Blood and moves Qi.
6. Man Jing Zi蔓荊子 expels Wind and stops pain.
7. Zhi Gan Cao甘草炙 tonifies the Spleen, harmonizes the other herbs, and guides the herbs to all 12 meridians.

Modifications
- To make the formula stronger, add Fang Ji 防己, Qin Jiao秦艽, Fu Zi附子, and Wu Tou烏頭.
- When the headache location is mostly at the sides, add Chai Hu 柴胡, Huang Qin黃芩, and Huang Qin黃芩.

Formula Actions

- Expels Wind, drains Damp.

Formula Indication

- Headache due to Damp Phlegm and Wind, heavy sensation of the head.

Contraindications

- Caution for patients who suffer from Yin deficiency and Heat in any condition, such as Wind Heat.

Cooking Instructions

1) Place the herbs into an herb pot. Traditionally, a clay cooker is used. Gently rinse the herbs once with a strainer.
2) Add water 300 ml into the pot.
3) Turn on the stove to high heat and bring the herbs to a boil. Turn off the heat when the decoction is reduced to 150 ml.
4) Drain the decoction into a container.

Dose

- Drink one cup of the decoction when it is still warm before a meal once daily.

HEADACHE DUE TO LIVER YANG RISING肝陽頭痛

- The onset is acute.
- The location is the side of the head.
- The sensation of the pain is a pulling, distending.
- The pain is aggravated by anger or emotional stress.
- The accompanying symptoms are irritability, dizziness, flushed face, red eyes, a bitter taste in mouth, throat dryness, and tinnitus, hypochondriacal pain, and insomnia.
- Tongue: red tongue with thin yellow coating.
- Pulse: wiry and forceful.

TREATMENTS

Treatment Principle

Soothe the Liver, suppress Yang, nourish Yin, and clear Heat.

Herbal Treatments

Herbal Formula

Tian Ma Gou Teng Yin 天麻鉤藤飲

This formula is in the book "Newly Explained Diagnosis and Treatment of TCM Internal Medicine and Miscellaneous Diseases中醫內科雜病證治新義", authored by Hu Guangci 胡光慈 in 1958.

Herbal Ingredients

Tian Ma天麻6g, Gou Teng鉤藤9g, Niu Xi牛膝12g, Shi Jue Ming石決明15g, Du Zhong杜仲12g, Huang Qin黃芩6g, Zhi Zi梔子6g, Yi Mu Cao益母草9g, Sang Ji Sheng桑寄生9g, Ye Jiao Teng夜交藤9g, Fu Shen茯神9g

Ingredient Explanations

1. Tian Ma 天麻 extinguishes Wind, calms the Liver, subdues rising Liver Yang, and alleviates pain.
2. Gou Teng鉤藤 expels Wind, calms the Liver, and releases spasm.
3. Niu Xi牛膝 moves the Blood downward, tonifies Liver and Kidney; opens meridians to benefit joints and Tendons.
4. Shi Jue Ming石決明 drains Liver Fire, expels Wind; assists Tian Ma 天麻 and Gou Teng鉤藤 to enhance extinguishing Wind, and calms the Liver.
5. Du Zhong杜仲 tonifies Liver and Kidney, strengthens Bone, and benefits Essence and Blood.
6. Huang Qin黃芩 clears Heat, dries Damp, cools the Blood, stops bleeding, and calms Liver Yang.
7. Zhi Zi梔子 clears Heat cools the Blood, calms Liver Yang, resolves Damp, reduces swelling, and opens the meridians.
8. Yi Mu Cao益母草 moves the Blood, promotes urination, reduces swelling, and clears Heat.
9. Sang Ji Sheng桑寄生 tonifies Liver and Kidney, strengthens Bone, benefits Essence and Blood; expels Wind and drains Damp.
10. Ye Jiao Teng夜交藤 nourishes Heart Yin and Blood, calms Shen神, and expels Wind and Damp.
11. Fu Shen茯神 calms Shen神.

Modifications

- For excessive Liver Fire, add Xia Ku Cao夏枯草 and Long Dan Cao龍膽草.
- For insomnia, add Long Gu龍骨 and Mu Li牡蠣.
- For Blood stasis, add Dan Shen丹參and Chi Shao赤芍.
- For Liver and Kidney Yin deficiency, add Liu Wei Di Huang Wan 六味地黃丸.
- For red face and constipation, add Huang Lian黃連 and Da Huang大黃.

HEADACHE DUE TO BLOOD STASIS瘀血頭痛

- The onset can be acute followed by a known physical trauma, such as a fall or blow to the head; the headache may not necessarily appear immediately following the trauma, but can occur several hours to months or years later after the trauma, initiated by another illness or a decline in general health and circulation.
- The location is usually in the same place.
- The sensation of the pain is a sharp and stabbing pain, the pain usually lasts longer and may be more severe in the evening or at night.

- The duration is long.
- The aggravation is in the cloudy day and rain or another head injury.
- The accompanying symptoms are dark complexion, irritability, insomnia, painful menstruation, and palpitations.
- Tongue: purple and bruised.
- Pulse: choppy and wiry.
- For this type of headache, the following diseases may also occur: high blood pressure, diabetes, thyroid dysfunction, irritable bowel syndrome, or depression.

TREATMENTS

Treatment Principle
Move the Blood, disperse Bruising, and open the meridians.

Herbal Treatments
Herbal Formula
Chai Hu Huang Qin Tang血府逐瘀湯
This formula is from "Correction of Errors in Medical Classics醫林改錯", authored by Wang Qingren 王清任 (1768–1831), in Qing dynasty清朝 1636–1912, and published in 1849.

Herbal Ingredients
Dang Gui當歸9g, Sheng Di Huang生地黃9g, Tao Ren 桃仁12g, Hong Hua紅花6g, Zhi Ke枳殼6g, Chi Shao赤芍6g, Chai Hu柴胡3g, Gan Cao 甘草3g, Jie Geng 桔梗 5g, Chuan Xiong川芎5g, Chuan Niu Xi牛膝9g

Ingredient Explanations
1. Dang Gui當歸 moves the Blood, disperses Bruising, and stops pain due to Blood stasis.
2. Sheng Di Huang生地黃 nourishes Yin and clears Heat.
3. Tao Ren 桃仁 moves the Blood, disperses Bruising, and stops pain due to Blood stasis.
4. Hong Hua紅花 moves the Blood, disperses Bruising, and stops pain due to Blood stasis.
5. Zhi Ke枳殼 directs the actions of the formula to the lower body.
6. Chi Shao赤芍 moves the Blood, disperses Bruising, and stops pain.
7. Chai Hu柴胡 moves Liver Qi and releases stagnation.
8. Gan Cao 甘草 harmonizes the herbs, tonifies Middle Jiao Qi, clears Heat, expectorates Phlegm, and stops cough.
9. Jie Geng 桔梗 directs the actions of the formula to the upper body.
10. Chuan Xiong川芎 moves the Blood, disperses Bruising, and stops pain due to Blood stasis.
11. Chuan Niu Xi牛膝 unblocks and promotes movement in the channels and collaterals and stops spasms and convulsions.

Modifications

- For palpitation and insomnia, add Fu Shen茯神, Wu Wei Zi五味子, and Suan Zao Ren酸棗仁.
- For persistent fever, add Jin Yin Hua金銀花 and Lian Qiao連翹.

Formula Actions

- Moves the Blood, opens the meridians, and stops pain.

Formula Indication

- Headache due to Blood stasis.

Contraindications

- It is contraindicated for patients who suffer from headache that is not caused by Blood stasis.
- It is contraindicated during pregnancy.

Cooking Instructions

1) Place the herbs into an herb pot. Traditionally, a clay cooker is used. Gently rinse the herbs once with a strainer.
2) Add water until it just covers the herbs, then add one more cup of water.
3) Presoak the herbs for 15 minutes.
4) Turn on the stove to high heat and bring the herbs to a boil. Reduce the heat to medium-low to simmer with slight bubbling. Turn off the heat when the decoction is reduced to one cup.
5) Drain the decoction into a container.
6) Repeat steps two to five one more time and drain each decoction to the same container.
7) Divide the decoctions into two separate cups.

Dose

- Drink one cup of the decoction after meals twice daily.

HEADACHE DUE TO PHLEGM-DAMPNESS OBSTRUCTION痰濁阻絡頭痛

- The onset is slow.
- The location is vague, it can be on the forehead but the patients cannot be certain to pinpoint the exact location; patients often describe it as the whole head headache.
- The sensation of the pain is numbness, heavy head, foggy, and dizzy.
- The headache is aggravated by rainy, cloudy, and dark weather.
- The accompanying symptoms are fullness in the chest, gastric region upset, nausea, excessive phlegm or saliva, no appetite, heavy limbs, and fatigue.
- Tongue: large tongue body with teeth marks around the edge and a white greasy coating.
- Pulse: slippery.

TREATMENTS

Treatment Principle
Dispels Dampness, transforms Phlegm, strengthens the Spleen, soothes the Liver, and extinguishes Wind.

Herbal Treatments
Herbal Formula
Ban Xia Bai Zhu Tian Ma Tang半夏白朮天麻湯

This formula is from "Medical Insights醫學心悟", authored by Cheng Guopeng 程國彭 (1662–1735), in Qing dynasty清朝 1636–1912, and published in 1732.

Herbal Ingredients
Zhi Ban Xia半夏4.5g, Tian Ma天麻3 g, Bai Zhu白朮3 g, Ju Hong橘紅3g, Fu Ling 茯苓3g, Gan Cao甘草1.5g, Sheng Jiang生薑2 pieces, Da Zao大棗3 pieces, Man Jing Zi蔓荊子3g

Ingredient Explanations
1. Zhi Ban Xia半夏 transforms Phlegm.
2. Tian Ma天麻 extinguishes Wind, calms the Liver and subdues rising Liver Yang, and alleviates pain.
3. Bai Zhu白朮 tonifies the Spleen Qi and dries Dampness.
4. Ju Hong橘紅 regulates Qi and transforms Phlegm.
5. Fu Ling茯苓 drains Damp and tonifies the Spleen.
6. Gan Cao甘草 harmonizes the herbs, tonifies the Middle Jiao Qi, clears Heat, expectorates Phlegm, and stops cough.
7. Sheng Jiang生薑 unblocks the pure Yang pathway and harmonizes rebellious Qi.
8. Da Zao大棗 tonifies the Spleen Qi, nourishes the Blood, and moderates and harmonizes the harsh properties of the fragrant herbs.
9. Man Jing Zi蔓荊子 expels Wind and stops pain.

Modifications
* To make the formula stronger, add Bai Ji Li白蒺藜.
* For vertigo, add Jiang Can僵蚕 and Dan Nan Xing胆南星.

Formula Actions
Tonifies the Spleen, drains Damp Phlegm, soothes the Liver, and extinguishes Wind.

Formula Indication
* Headache due to Phlegm blocking the orifice that is stirred up by Wind, dizziness, sticky white coating on the tongue, and wiry slippery pulse.

Contraindications
* It is contraindicated for patients who suffer from headache not caused by Wind-Phlegm 風痰.

Cooking Instructions

1) Place the herbs into an herb pot. Traditionally, a clay cooker is used. Gently rinse the herbs once with a strainer.
2) Add water into the pot until it just covers the herbs, then add one more cup of water.
3) Presoak the herbs for 15 minutes.
4) Turn on the stove to high heat and bring the herbs to a boil. Reduce the heat to medium-low to simmer with slight bubbling. Turn off the heat when the decoction is reduced to one cup.
5) Drain the decoction into a container.
6) Repeat steps two to five one more time, and drain each decoction to the same container.
7) Divide the decoction into two separate cups.

Dose

- Drink one cup of the decoction after meals twice daily.

THE HEADACHE DURING A MENSTRUAL CYCLE月經期頭痛

- The onset is during the menstrual cycle. The duration is not certain, it can be one day or longer. Onset can be slow.
- The locations can be on the forehead, occiput, vertex, and temporals.
- The sensation of the pain is classified into excessive and deficient conditions.

A. Excessive condition: The headache comes before menstruation. The sensations are distending, pulling, stabbing pain.

1. **Liver Fire肝火**

   Etiology and Pathology: this is commonly seen in the patients with underlying Liver Yang rising, and the Blood is unable to reach the Chong and Ren meridians during the menstrual cycle. Because Foot Jueyin Liver and Du meridians are on the vertex, Liver Fire easily reaches the top of the head, resulting in headache.

   Onset: the headache usually occurs before menstruation. The pain can be severe. The location is usually on the vertex.

   Accompanying symptoms: dizziness, vertigo, irritability, bitter taste in the mouth, and dry throat.

   Tongue: red tongue body with thin yellow coating.

   Pulse: thin, wiry, and rapid.

   Treatment Principle: nourish Yin, soothe the Liver, and expel Wind.

Herbal Formula

Tian Ma Gou Teng Yin 天麻鉤藤飲

This formula is in the book "Newly Explained Diagnosis and Treatment of TCM Internal Medicine and Miscellaneous Diseases中醫內科雜病證治新義", authored by Hu Guangci 胡光慈 in 1958.

Herbal Ingredients

Tian Ma天麻6g, Gou Teng鉤藤9g, Niu Xi牛膝12g, Shi Jue Ming石決明15g, Du Zhong杜仲12g, Huang Qin黃芩6g, Zhi Zi梔子6g, Yi Mu Cao益母草9g, Sang Ji Sheng桑寄生9g, Ye Jiao Teng夜交藤9g, Fu Shen茯神9g

Ingredient Explanations

1. Tian Ma 天麻 extinguishes Wind, calms the Liver and subdues rising Liver Yang, and alleviates pain.
2. Gou Teng鉤藤 expels Wind, calms the Liver, and releases spasm.
3. Niu Xi牛膝 moves the Blood downward, tonifies Liver and Kidney, and opens the meridians to benefit joints and Tendons.
4. Shi Jue Ming石決明 drains Liver fire, expels Wind, assists Tian Ma 天麻 and Gou Teng鉤藤 to enhance extinguishing Wind, and calms the Liver.
5. Du Zhong杜仲 tonifies Liver and Kidney, strengthens Bones and benefits Essence and the Blood.
6. Huang Qin黃芩 clears Heat, dries Damp, cools the Blood, stops bleeding, and calms Liver Yang.
7. Zhi Zi梔子 clears Heat, cools the Blood, calms Liver Yang, resolves Damp, reduces swelling, and opens meridians.
8. Yi Mu Cao益母草 moves the Blood, promotes urination, reduces swelling, and clears Heat.
9. Sang Ji Sheng桑寄生 tonifies Liver and Kidney, strengthens Bones, benefits Essence and the Blood; expels Wind and drains Damp.
10. Ye Jiao Teng夜交藤 nourishes Heart Yin and Blood, calms Shen神, expels Wind and Damp.
11. Fu Shen茯神 calms Shen神.

Modifications

- For excessive Liver Fire, add Xia Ku Cao夏枯草 and Long Dan Cao龍膽草.
- For insomnia add Long Gu龍骨 and Mu Li牡蠣.
- For Blood stasis, add Dan Shen丹參and Chi Shao赤芍.
- For Liver and Kidney Yin deficiency, add Liu Wei Di Huang Wan 六味地黃丸.
- For red face and constipation, add Huang Lian黃連 and Da Huang大黃.

2. Blood Stasis血瘀

Etiology and pathology: There should be no pain while menstruating if the flow of Qi and Blood is smooth without obstruction. Headache occurs during the menstrual cycles often seen on the patients underlying Blood stasis. During the menstrual cycle the blood clots move with the flows of the Blood and block the meridians and the collaterals, hence, the pure Yang is unable to rise to the orifice.

Onset: the headache is abrupt, the pain can be persistent stabbing, pain commonly starts 1 to 3 days before the period or starts on the first day of the period. It usually occurs before or during menstruation. When the pain is severe, there is a throbbing sensation on the temporal acupuncture point Taiyang太陽 region. The location of the headache is fixed on one area; it also expands to the whole side when it is severe.

Accompanying symptoms: Throbbing, cramping pain in the lower abdomen that can be intense; usually the pain starts 1–3 days before the period and subsides in 2–3 days. The pain may also radiate to the lower back and thighs.

Tongue: dusky with ecchymoses on the edges.

Pulse: thin, choppy or wiry, choppy.

Treatment Principle

Regulate Qi, move Blood, disperse Bruising, and open the channel.

Herbal Treatments

Herbal Formula

Tong Qiao Huo Xue Tang通竅活血湯

This formula is from "Correction of Errors in Medical Classics醫林改錯", authored by Wang Qingren 王清任 (1768–1831) in Qing dynasty清朝 1636–1912, and published in 1849.

Herbal Ingredients

Chuan Xiong川芎3g, Chi Shao赤芍3g, Tao Ren 桃仁9g, Hong Hua 紅花9g, She Xiang 麝香0.15g, Cong Bai蔥白切碎15g, Sheng Jiang生薑切碎9g, Da Zao (without seed) 紅棗去核7 pieces or 300g, rice wine 250ml

Ingredient Explanations

1. Chuan Xiong 川芎 extinguishes Wind, calms the Liver and subdues rising Liver Yang, and alleviates pain.
2. Chi Shao 赤芍 moves the Blood, disperses Bruising, and stops pain.
3. Tao Ren 桃仁 moves the Blood, disperses Bruising, and stops pain due to Blood stasis.
4. Hong Hua 紅花 moves the Blood, disperses Bruising, and stops pain due to Blood stasis.
5. She Xiang 麝香 opens orifices, revives Shen, moves the Blood, opens the 12 meridians, and stops pain.
6. Cong Bai 蔥白 unblocks the pure Yang pathway, disperses Cold, induces sweat, releases the exterior, and relieves toxicity.
7. Sheng Jiang 生薑 unblocks the pure Yang pathway and harmonizes rebellious Qi.
8. Da Zao 紅棗 tonifies the Spleen Qi, nourishes the Blood, and moderates and harmonizes the harsh properties of the fragrant herbs.
9. Rice wine黃酒 warms meridians, disperses Cold, moves the Blood, tonifies the Spleen, expels Wind, and induces other herbs into their meridians.

Modifications

- For sharp headache, add Shi Chang Pu.
- For post-concussion syndrome, add Wu Ling Zhi五靈脂 and Man Jing Zi 蔓荊子.

Cooking Instructions
- Cook the herbs except She Xiang 麝香 with the rice wine to 150ml, discard the herbs, keep the wine.
- Cook the wine with She Xiang 麝香, boiling it two times.

Dose
- For adults, drink the wine before bed once a day for three days, take one day off, then continue drinking for another three days.

Formula action
- Moves the Blood, disperses Bruising, and opens orifices and meridians.

Formula Indication
- Recalcitrant headache due to Blood stagnation, migraine, post-concussion syndrome, and hair loss.

Contraindications
- Do not use while pregnant or nursing.
- Do not use with heavy menstrual bleeding.
- Consult a physician prior to use if you are taking blood thinners.
- Not for long-term use.

B. Deficient condition: headache presents at the end or after menstruation. The sensations are empty, dull headache.
  1. **Blood Deficiency血虚**

Etiology and Pathology: This is commonly seen in patients with underlying Blood deficiency; the body is insufficient in Blood supply. The Blood becomes more deficient when menstruation occurs.

Onset: This type of headache can occur both before and after menstruation.

Accompanying symptoms: dizziness, palpitation, insomnia, fatigue, and lassitude.

Tongue: pale, thin coating.

Pulse: thin, weak.

Treatment Principle
Nourish the Blood and tonify Qi.

Herbal Formula
Si Wu Tang 四物湯
This formula is from "Tai Ping Hui Min He Ji Ju Fang太平惠民和劑局方", authored by Imperial Medical Bureau 太醫局 in southern Song dynasty南宋 (1127–1239), and published in 1134.

Herbal Ingredients
Dang Gui 當歸10 g, Sheng Di Huang 生地黃12 g, Chuan Xiong 川芎8 g, Bai Shao 白芍12 g

Ingredient Explanations

- Dang Gui 當歸 moves the Blood, disperses Bruising, and stops pain due to Blood stasis.
- Sheng Di Huang生地黄 nourishes the Blood, nourishes Liver and Kidney Yin, clears Heat, and cools the Blood.
- Chuan Xiong 川芎 moves the Blood, disperses Bruising, and stops pain due to Blood stasis.
- Bai Shao白芍 nourishes the Blood, soothes the Liver, and stops pain.

Acupuncture Treatments

Acupuncture point formula

- Patient position: The patient is positioned according to the acupuncture point selection.

Acupuncture Points

1. **Tai Yang Headache 太陽頭痛:** UB10天柱 Tianzhu is the main acupuncture point; it is assisted by DU14大椎 Dazhui, DU19 後頂 Houding, DU20 百會 Baihui, UB62 申脉 Shenmai, SI3 後溪 Houxi, UB60 昆侖Kunlun, and Ashi acupoints 阿是穴.
2. **Yang Ming Headache陽明頭痛:** ST8 頭維 Touwei is the main acupuncture point; it is assisted by Yintang 印堂, DU23上星Shangxing, 陽白 Yangbai、UB2 to Yuyao 攢竹透魚腰Cuanzhu through Yuyao, SJ23 絲竹空 Sizhukong、LI4 合谷 Hegu, and ST44 內庭Neiting.
3. **Shao Yang Headache少陽頭痛:** GB8 率谷Shuaigu is the main acupuncture point; it is assisted by Taiyang, SJ23 絲竹空 Sizhukong, SJ20 角孫 Jiaosun，GB20 風池 Fengchi, SJ5外關Waiguan, and GB41足臨泣zulinqi.
4. **Jue Yin Headache厥陰頭痛:** DU20 百會 Baihui is the main acupuncture point; it is assisted by LIV5蠡溝Ligou, UB7通天Tongtian, LIV3太衝 Taichong, LIV2行間Xingjian, KD3太溪Taixi, and KD1湧泉Yongquan.

Modifications

1. Add GB20 風池 Fengchi, UB12 風門 Fengmen, and DU16 風府 Fengfu if the headache is caused by Wind 風 invasion.
2. Add UB9玉枕 Yuzhen、DU22囟會Xinhui if the headache is caused by Cold 寒 invasion.
3. Add moxibustion Therapy on DU14大椎 Dazhui if the headache is caused by Wind Cold 風寒.
4. Add Taiyang 太陽 and DU23上星Shangxing if the headache is caused by Heat 熱 invasion.
5. Apply reducing techniques on LI10 曲池 Quchi if the headache is caused by Wind Heat 風熱 .
6. Add SJ17 翳風Yifeng and KD4大鐘 Dazhong if the headache is caused by Damp 濕 invasion.
7. Apply reducing techniques on SP6 三陰交 Sanyinjiao if the headache is caused by Wind Damp 風濕.

8. Add KD3太溪 Taixi and UB23 腎俞 Shenshu if the headache is caused by Yin 陰 deficiency.

9. Add DU20百會Baihui and Ren17膻中Danzhong if the headache is caused by Qi 氣 deficiency.

10. Add ST8 頭維Touwei and GB13本神 Benshen if the headache is caused by Qi 氣 stagnation.

11. Add UB60崑崙Kunlun and REN6 氣海 Qihai. If the headache is caused by Yang 陽 deficiency

12. Add UB17膈俞Geshu and UB20 脾俞 Pishu if the headache is caused by Blood deficiency.

13. Add REN6 氣海Qihai, SP10血海 Xuehai, and ST36足三里Zusanli if the headache is caused by Qi 氣and Blood deficiency.

14. Add LIV6 中都 Zhongdu, SP10血海 Xuehai, LIV3 太衝Taichong, DU16 風府 Fengfu, and LIV1 大敦 Dadun if the headache is caused by Blood stasis.

15. Add LI4 合谷Hegu, LIV3 太衝Taichong, and UB17 膈俞Geshu if the headache is caused by Qi 氣and Blood stagnation.

16. Add ST40豐隆 Fenglong and ST36足三里Zusanli if the headache is caused by Phlegm Dampness obstruction.

17. Add KD3太溪 Taixi, SP6 三陰交 Sanyinjiao, UB10天柱 Tianzhu, and UB23 腎俞 Shenshu if the headache is caused by Kidney deficiency.

18. Add GB5 懸顱 Xuanlu, DU20百會Baihui, LIV2 行間 Xingjian, and GB43 俠溪 Jiaxi if the headache is caused by Liver Yang rising.

19. All types of headaches should consider bleeding the Ashi acupoints 阿是穴 for better results.

Bleeding Therapy

1. Point selected is Ashi acupoints 阿是穴.
2. Treatment frequency is once a day.

Procedures

- Prior to the procedure, the tools should be prepared and ready to be used in the working area.
- Set up a clean field with paper towels.
- Place the tools in the clean field, including dry sterile cotton balls, medical alcohol wipes or cotton swabs soaked in alcohol, sterile gloves, a biohazard sharps container, a biohazard trash container, goggles and face mask, and paper towels. Other required tools include a medical lancet, or a traditional three-edge needle, or the seven-star needle (a bleeding tool). In this case, a medical lancet or a traditional three-edge needle is selected for bleeding Ashi acupoints 阿是穴.
- The patient lies prone with the legs extended so that Ashi acupoints 阿是穴 are facing upward.
- The room should be well lit.
- The therapist washes their hands thoroughly, and wipes their hands with a single-use napkin, and puts on sterile gloves.

- Wipe the acupuncture point Ashi acupoints 阿是穴 with a medical alcohol wipe or a cotton swab soaked in alcohol, moving from the center of the acupuncture point to the periphery.
    1. Place the medical lancet or the traditional three-edge needle at an angle of 90 degrees to the skin.
- In one quick move, prick the needle into the acupuncture point about 0.1 Cun 寸 deep, eliciting a few drops of blood.
- If more drops of blood are needed, absorb the blood with a sterile cotton ball.
- Lastly, the point is pressed with a sterile cotton ball until the bleeding ceases.
- Dispose of the tools in the proper containers.
- Wash hands thoroughly.

Electro-Acupuncture and Moxibustion Therapy

Warming needling techniques on GB20 風池 Fengchi, UB12 風門 Fengmen, and DU16 風府 Fengfu.

Treatment Frequency

The acupuncture needling treatment is once a day, needle retention time is 30 minutes.

Moxibustion Therapy is once every other day.

Tui Na Treatments

Tui Na Techniques

• Pressing, kneading, chafing, grasping, and one finger meditation.

Tui Na Procedures

The patient is in a sitting position; the therapist stands behind the patient.

1. Apply pressing technique with the thumb tips on UB10天柱 Tianzhu, GB20 風池 Fengchi, and UB12 風門 Fengmen for 3–5 minutes on each acupuncture point.
2. Apply "One Finger Meditation Technique" with the thumb tips along both sides of the neck where Tai Yang meridians are located 3–5 times.
3. Apply grasping technique with the first 3 fingers (one hand) on both GB20 風池 Fengchi 10–20 times. Figure 2.1.1; Figure 2.1.2

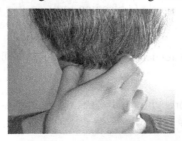

4. Apply grasping technique with all fingers (both hands) along the sides of the neck for 10–20 times. Figure 2.1.3; Figure 2.1.4

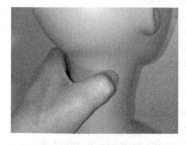 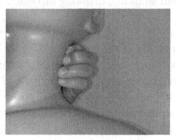

5. Apply clenching technique on GB20 肩井 Jianjing. Figure 2.1.5; Figure 2.1.6

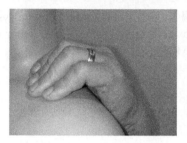 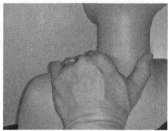

6. Applies chafing technique on the spine from DU14 大椎 Dazhui to DU4 命門 Mingmen for 10–20 times. Figure 2.1.7

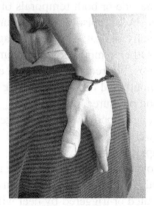

Patient Advisory

Dietary Recommendations

1. More liquid food such as congee.
2. The temperature of food should not be cold; it needs to be warm.
3. Eat more easily digested food.
4. Eat more foods that are high vitamin C.

5. Avoid fatty, sweet, spicy, curry, and raw food.
6. Avoid deep-fried food and barbecued food.
7. Avoid smoking, drinking alcohol, and coffee.

Lifestyle
1. Get plenty of rest.
2. Have more physical and mental relaxation.
3. Avoid staying up late, be on the bed, and get ready to sleep at around 10 or 10:30 p.m.
4. Routine morning exercise is recommended. Morning before 9 o'clock is the best time for physical exercise. Select exercises that promote sweat, with light sweating on the whole body. Cardio exercise is good for migraine patients; 3 times a week for 40 minutes will prevent headache.

Morning exercise is to tonify Yang Qi 陽氣, it is to boost the immune system to prevent Wind Cold/Heat invasion, it is to refresh the brain to have better energy, memory, and clearing the brain fog; it also helps to manage stress.

5. Manage mental stress.

THE TYPES OF PAIN

Throbbing Headache跳痛性頭痛
- Patients may describe the sensation as pulsating, bursting, and distending.
- It indicates an excess condition of headaches.
  1. If it affects only one side or both temporals of the head, the headache usually is preceded by an episode of emotional upset such as anger or argument. It usually indicates Liver Qi stagnation, Liver Yang rising, or Liver Fire. Other symptoms include emotional stress, irritability, dizziness, tinnitus, blurred vision, nausea and vomiting, insomnia, flushed face, dry mouth and thirst, bitter taste in the mouth, constipation, increased blood pressure, and epistaxis.
     - Tongue: red tongue body with red edges and possibly peeled. Yellow and dry coating.
     - Pulse: wiry, rapid, and big.
  2. If it affects the whole head, particularly in the forehead, it usually indicates exterior Wind Heat invasion, and the onset is usually acute. The headache is aggravated or triggered by heat wind blowing or high temperature. Other symptoms include nasal congestion, sore throat with thirst, flushed face, constipation, and yellow urine.
     1. Tongue: red tongue body with yellow coating.
     2. Pulse: rapid, floating.
  3. If it affects only a fixed area of the head, it is possibly a Blood stasis headache. The sensation of the pain is persistent and steady headache. It is accompanied by forgetfulness and palpitations.

Heaviness Headache 重痛性頭痛

1. Patients may describe the sensation as a heavy feeling in the head, a tight band around the head, or the head lifts a heavy object.
2. It commonly indicates the head is obstructed by Dampness. Heaviness headache can also indicate Liver Qi stagnation, Spleen Qi dysfunction, and turbid Yin unable to descend.
3. The headache has a certain seasonal or regional nature, especially when living in humid places for a long time or when the weather is too humid.

Stabbing Headache刺痛性頭痛

1. Patients often describe the headache as very intense pain; the sensation of the headache is like a knife stabbing or cutting.
2. It has a fixed location.
3. The duration is chronic.
4. It is accompanied by forgetfulness and insomnia. The tongue body color is purple, the pulse is choppy.
5. The patient may have a head injury, such as a punch to the head from the boxing or simply just from a head concussion years ago.
6. It commonly indicates the headache is caused by Blood stasis.

Emptiness Headache虛型性頭痛

7. Patients often describe the headache as a mild dull pain. The duration is long, it is usually tolerable.
8. Patients will press the head with their hands.
9. The location is vague, it can start at the back of the head, sometimes with neck pain, though it can be felt all over the head; patients often describe it as sitting deep inside of the head.
10. It gets worse when standing, sitting, coughing, sneezing, and exertion; it gets better when lying down.
11. It is accompanied by low back pain, tinnitus, dizziness, nausea, spontaneous sweating, shortness of breath, palpitation, fatigue, and poor appetite.

Electric Shock-Like Headache電擊樣頭痛

- Patients may describe the sensation as a continuous intense pain, stabbing, aching, burning, with intermittent electric shocking or shooting pain.
- The location is usually in one side of the head.
- The pain lasts several seconds to 1–2 minutes. It is abrupt onset, there is no aura.
- The accompanying symptoms are facial muscle twitching, tearing, salivation, facial flushing, and conjunctival hyperemia.
- It is a commonly seen headache that is related to trigeminal neuralgia and occipital neuralgia. On the other hand, acupuncture needling procedures may trigger the electric shock-like headache. The common acupuncture points that may cause the headache are DU15啞門, DU16風府, Yiming翳明, GB20風池, and Anmian安眠.

- It commonly indicates the head is obstructed by the Wind. Liver Yang rising, Liver Blood deficiency, and Blood stasis contribute the "Head Wind 頭風".

Thunderclap Headache炸裂样痛頭痛

- Patients describe the sensation as an extremely painful headache and commonly say it is worst headache they have ever experienced. It usually starts suddenly like a clap of thunder, without warning, and the intensity of pain peaks within one minute and can last for at least five minutes.
- The location and type of pain are not specific.
- The accompanying symptoms are nausea, vomiting, neck stiffness, irritability, seizures, fever, limb weakness and numbness on one side of the body, slurred speech, confusion, dizziness, light sensitivity, double vision, drooping eyelid, possible loss of consciousness, and even death.
- It is a commonly seen headache that relate to subarachnoid hemorrhage. Immediate medical attention is extremely important.
- It commonly indicates the head is obstructed by Blood.

Tension Headache緊箍样頭痛

- Patients may describe the sensation as like a band squeezing around the head; it is dull, pressure-like but not a throbbing sensation.
- The location is usually all over the head; it is not commonly in one point or one side of the head. But the pain can be more in the scalp, temporals, nape, and possibly in the shoulders as well.
- The frequency and duration varies; it may occur daily or only once, and the headache may last for 30 minutes to a week.
- The headache commonly occurs after holding the head in one position for a long time without moving such as when typing, using a microscope, or sleeping in a cold room with the neck in an abnormal position. This type of headache is often due to workday stresses and may be worse in the late afternoon.
- The accompanying symptoms are insomnia, nausea, and vomiting.
- It is a commonly seen headache that can be triggered by and occurs upon mental and physical fatigue or overexertion, emotional and physical stress, common cold, teeth grinding, jaw clenching, eye strain, excessive alcohol, or smoking use. The headache may be relieved by rest, hot pads, and massaging the painful sites.
- It commonly indicates the headache is caused by Liver Qi stagnation.

THE HEADACHE UNDER CERTAIN CIRCUMSTANCES

Morning Headache晨起頭痛

1. Patients often describe the time of the headache as worse in the morning.
2. If the location of the headache is in the occipital region, hypertension is considered.

3. If the location is in the middle of the forehead, and the sensation is dull and deep in the head, intracranial space-occupying lesions, such as tumors or abscesses, present within the cranium or skull are considered.
4. Sinusitis is commonly seen when the location is in the forehead.
5. It commonly indicates the headache is caused by the obstruction of Phlegm-Dampness.

Nighttime Headache 入夜頭痛

1. Patients report that this type of headache typically only occurs in the middle of the night, usually one to two hours after going to bed.
2. The location is usually in or around one eye on one side of the head. It may radiate to the face, neck, and other parts of the head on the affected side.
3. The intensity of the pain is extremely high that can wake up the patient.
4. Onset is abrupt, usually without warning.
5. The frequency of the headache occurs in bouts: it can be 1–2 times nightly, and it often occurs at the same time, like an alarm clock. Most patients have no headache in the daytime, and it is commonly seen more in the spring and autumn seasons.
6. The duration of the headache can last from 15–180 minutes in each episode, which usually ends suddenly. Patients usually can return to their normal life but may be exhausted. The periods can last from weeks to months, then followed by remission periods with no headache for months or sometimes even years.
7. The accompanying symptoms are red eyes on the affected side, excessive tearing, stuffed or runny nose on the affected side, forehead and facial sweating on the affected side, swelling around the eye on the affected side, drooping eyelid on the affected side and some patients may have nausea and vomiting.
8. When it becomes chronic, patients may suffer from emotional depression.
9. It commonly indicates the headache is caused by Blood stasis.

Fatigue Headache 勞累性頭痛

10. Patients often report that this type of headache occurs when fatigued; for instance, after shopping in the morning, the headache occurs in the afternoon. Rest helps. The pain is aggravated if the patient doesn't rest.
11. The onset is slow, gradual.
12. The duration may last for hours.
13. It commonly indicates the headache is caused by the deficiency of Qi, Blood, Yin, and Yang.
14. Qi deficiency headache: The sensation usually starts with a mild dull headache in bouts, the duration and pain intensity increase gradually. It is accompanied by lassitude, shortness of breath, spontaneous sweating, and poor appetite.
15. Blood deficiency headache: the headache is intermittent, accompanied with dizziness, palpitations, and pale complexion.

16. Qi deficiency headache: The sensation usually starts with a mild dull headache in bouts, the duration and pain intensity increase gradually.
17. Complexion.
18. Yin deficiency headache: empty sensation, accompanied by vertigo, dizziness, tinnitus, and sore lower back and knees. The intensity increases when the activities persist.
19. Yang deficiency headache: the intensity is mild, but the pain may increase when the patient is more fatigued. It is accompanied by cold limbs.

Hangover Headache 酒精性頭痛

1. Patients often report this type of headache after consuming alcohol.
2. It occurs within 3 hours, and it can last up to 72 hours.
3. The location of the headache is on both sides of the head.
4. The sensation is throbbing, pulsating.
5. The headache is aggravated by physical activity.
6. It commonly indicates the headache is caused by Phlegm-Dampness obstruction.

Caffeine Headache 咖啡依賴性頭痛

1. Patients often report this type of headache after drinking coffee, tea, colas, or after forgetting to drink coffee. This kind of headache is common in heavy coffee drinkers.
2. It usually occurs within hours.
3. The sensation of the headache is heavy.
4. It is accompanied by restlessness, tachycardia, arrhythmia, insomnia, tinnitus, blurred vision, increased diuresis, and muscle twitching.
5. It commonly indicates the headache is caused by Damp-Heat obstruction.

Orgasm Headache 房事頭痛

6. Patients often report that this type of headache occurs before or at the moment of sexual activity.
7. The sensation may be a dull pain in the head and neck that builds up and gets worse as sexual arousal increases. It may also be a sudden severe throbbing pain; usually these two types happen at once.
8. The duration can last several minutes to hours or even days.
9. It is accompanied by chest distention, nausea, vomiting, and lower abdominal distension.
10. It commonly indicates the headache is caused by Liver Fire flaring, Blood Heat blocking meridians, and Kidney deficiency.
11. If the headache is ameliorated by the orgasm, it also indicates Liver Fire flaring.

Weekend Headache 週末頭痛

1. Patients often report that the headache tends to occur during the weekend especially on Saturday morning, even if there was a good sleep the prior

night – there is no appointment or work today, and it is supposed to be a relaxing day.

2. It is accompanied by fatigue and yawning, moody, tight scalp and neck, even face.

3. It commonly indicates the headache is caused by Liver Yang rising.

Food Triggered Headaches 食物引起疼痛

1. Patients often report that this type of headache tends to occur after consuming certain foods or drinks.

2. The following foods are commonly reported to trigger headaches: aged cheese, alcohol, pickled foods, onions, garlic, dried fruits such as figs, raisins and dates; chocolate, chicken liver, tomato, milk; certain fresh fruits, such as orange, tangerine, ripe banana, papaya, raspberries, kiwi, red plum, pineapple; and monosodium glutamate (MSG).

3. It commonly indicates the headache is caused by Liver Yang rising.

Weather Triggered Headaches 天氣變化頭痛

1. Patients often report that this type of headache occurs related to weather changes. Different people may be sensitive to certain weather factors. For example, the amount of sunshine may be a factor for patient A, leading to headache; while for patient B, the sunshine didn't cause any headache, even they were together at the same time, same place, and same length of time. Instead of the factor of sunshine, a typhoon may be a factor causing patient B's headache, but patient A is not affected. It is also possible that the headache sufferer is predisposed to such triggers due to other factors, like fatigue, stress, wrong diet, and lack of sleep.

2. The reported weather factors that cause headaches include the following:
   1. A strong wind such as typhoon.
   2. Temperature and humidity changes, such as an abnormal heat wave or an extremely cold snow day.
   3. Strong sunshine.
   4. Dry air.
   5. Smoke from mountain fire.

3. It commonly indicates the headache is caused by Liver Yang rising 肝陽上亢.

MIGRAINE HEADACHE偏頭痛

INTRODUCTION

Migraine headache is a neurological condition that can cause multiple symptoms. Migraines can occur at any age from childhood to elderly. Women are more likely than men to experience migraines. A family history of migraines is the most potent and consistent risk factor for migraine; a positive family history tends to manifest earlier in life and with greater severity than those without a family history.

Approximately in 14% of the global population experiences migraine headaches. World Health Organization (WHO) ranks migraine as the most prevalent, disabling, long-term neurological condition.[4]

ANATOMY

The migraine anatomy and the locations of the trigeminal nerve need to be introduced.

The trigeminal nerve is the fifth cranial nerve, and it is the biggest nerve among the 12 cranial nerves. The trigeminal nerve originates from four nuclei (three sensory nuclei and one motor nucleus) in the brain stem. The sensory nuclei form a sensory root at the level of the pons, the motor nucleus continues to form a motor root.

The sensory root expands into the trigeminal ganglion at the middle cranial fossa outside the central nervous system, and the ganglion is located superiorly to the motor root at the trigeminal cave where the motor root lays on the floor.

The sensory ganglion generates three divisions: ophthalmic (V1), maxillary (V2) and mandibular (V3).

The ophthalmic branch(V1) is the smallest branch among the three terminal branches of the trigeminal nerve. It provides sensory innervation to the face and scalp above the orbits. It contains sympathetic nerve fibers responsible for pupil dilation and supplies the ciliary body, iris, lacrimal gland, conjunctiva, and cornea. It supplies the superior portion of the nasal cavity, the frontal sinus, the dura mater and portions of the anterior cranial fossa.

The maxillary branch (V2) is also a sensory branch, it covers the area below the orbits and above the mouth, including the inferior portion of the nasal cavity, the maxillary teeth, and maxillary sinus.

The mandibular branch (V3) reaches the lower jaw area and from the chin up to the area in front of the ear. It is the biggest branch (n. ophthalmicus) among the three terminal branches of the trigeminal nerve. It is also the only division of the trigeminal nerve that has both sensory and motor components. The sensory portion is responsible for pain and temperature information from the mandibular teeth, buccal mucosa, temporomandibular joint, the anterior two-thirds of the tongue, and the face below the territory of the maxillary nerve. The motor component innervates the muscles listed below.

The mandibular branch (V3) of the trigeminal nerve controls the following eight muscles of mastication:

1. **Masseter muscle** is supplied by the anterior division of the mandibular division (V3).
2. **Temporal muscle** is supplied by the anterior division of the mandibular division (V3).
3. **Medial pterygoid muscle** is supplied by the anterior division of the mandibular division (V3).
4. **Lateral pterygoid muscle** is supplied by the anterior division of the mandibular division (V3).
5. **Tensor veli palatini muscle** is supplied by the medial pterygoid nerve, a branch of the mandibular nerve (V3).
6. **Mylohyoid muscle** the mylohyoid nerve is a branch of the inferior alveolar nerve, a branch of the mandibular nerve (V3).
7. **Digastric muscles** consist of two muscular bodies which are supplied by different cranial nerves.

8. The anterior body is supplied by the trigeminal via the mylohyoid nerve, a branch of the inferior alveolar nerve, itself a branch of the mandibular division of the trigeminal nerve.[5]
9. The posterior belly is supplied by the digastric branch of the facial nerve.
10. **Tensor tympani muscle** is supplied by the tensor tympani nerve, a branch of the mandibular branch of the trigeminal nerve. As the tensor tympani is supplied by motor fibers of the trigeminal nerve, it does not receive fibers from the trigeminal ganglion, which has sensory fibers only.

WESTERN MEDICINE ETIOLOGY AND PATHOLOGY

1. Migraine is considered a neurovascular pain syndrome; it has been seen in people whose nervous system is more sensitive than others. These people's nerve cells in the brain are easily stimulated then producing electrical activity. Neuroanatomical and neurochemical studies revealed that most sensory fibers from the intracranial and the extracranial tissues originate in the fifth cranial nerve ganglion, also called the trigeminal ganglion[6]. The fifth cranial nerve, the trigeminal nerve, sends impulses from the eyes, upper eyelids, mouth, jaw, scalp, and forehead to the brain. When the nerve is stimulated, it releases substances that cause painful inflammation in the blood vessels of the brain and the layers of tissues that cover the brain.
2. Migraines can be triggered when estrogen levels increase or fluctuate.

Migraines become common during puberty when estrogen levels increase; some female patients have migraines before, during, or after menstrual periods.

Migraines often occur less in the last trimester of pregnancy when estrogen levels are relatively stable.

Migraines become particularly difficult to control during menopause when estrogen levels are fluctuating.

CHINESE MEDICINE ETIOLOGY AND PATHOLOGY

Liver Yang Rising 肝陽上亢

1. **Indulgence**

Long-term over-physical and -mental work lead to depletion and deficiency of Kidney Yin, and Kidney deficiency fails to nourish the Liver Blood. One of the functions of Liver Blood is to manage and govern Liver Qi. When Liver Blood is insufficient, Liver Qi is excessive; Qi is Yang energy, the moving direction of it is going upward, this results in the rising of Liver Yang.

2. **Emotional Disorder**

Emotional disorder such as depression, anxiety, short temper, and fear lead to Liver Qi stagnation. A famous physician Zhu Dan Xi in Yuan Dynasty said "excessive Qi causes Fire" 元代医学家朱丹溪说 "气有余便是火". Fire drains Liver Blood and Yin, Liver Blood and Yin deficiency cause Uprising of Liver Yang due to Yin failing to control Yang.

3. **Old Age**
    Older people are more seen for Kidney and Liver insufficiency. Kidney Yin fails to nourish Liver Yin, then causes Liver Yang rising.
4. Foot Yue Yin Liver meridian 足厥陰肝經 and Foot Shao Yang Gallbladder meridian 足 少陽膽經 are in the interior–exterior relationship 表里關係. Therefore, Liver Yang rising headache is at the temporal region where Gallbladder meridians are located.

Blood Stasis 瘀阻腦絡

1. Chinese medicine holds that "Prolonged illness is more Blood stasis久病多瘀".

"The first illness is in the meridians, the long-term illness is in the collaterals 初病在經，久病在絡". "The meridian governs the Qi, and the collateral governs the Blood 經主氣，絡主血". "Qi is stagnant in a meridian at first, then the illness enters the collaterals when Blood is injury in chronic stage 初為氣結在經，久則血傷入絡". These statements are from the "Clinical Guide to Medical Records" by Ye Tianshi葉天士, a famous doctor in the Qing Dynasty清朝.

If a headache lasts a long time, whether it is the cause of lack of righteous Qi or the entrenchment of evil Qi, it will eventually lead to a result; that is, the movement of Qi and Blood will slow down in Shao Yang Meridians少陽經. Because Qi is a commander of Blood, due to insufficient Qi and weak pushing force, speed will slow down; insufficient Blood, unable to flow, speed will also slow down, leading to Blood clotting in the meridians.

2. Heat evil can refine the body fluid into a Phlegm, Blood is a kind of body fluid, that when it is hot, it becomes thick and causes blockage.
3. If Cold evil 風 寒邪 injures Yang energy 寒邪傷陽 and enters the collaterals, the cold temperature cools down and thickens the Blood to a Phlegm to cause a stagnation.
4. Emotional disorders such as depression, anxiety, short temper, and fear lead to Liver Qi stagnation. Chronic Liver Qi stagnation develops Blood stasis because Qi is the commander of Blood.
5. Blood stasis migraine can also develop from a direct head trauma. A fall or a direct blow to the head is a common cause of headache resulting in Blood stasis. The headache may not present immediately following the trauma, but it can occur months or years later under some circumstances or it can be triggered by another illness such as stress, anger, or a decline in general health.

Wind Cold Invasion 風寒外襲

1. Temporal headache is called Shao Yang Headache少陽頭痛 in traditional Chinese medicine because the residual factors from attacking Tai Yang progressing interiorly to attack Shao Yang少陽 resulting in headaches in Foot Shao Yang Gallbladder meridian 足 少陽膽經 leading to the temporal headache.
2. Patients have a history of exposing exterior Wind Cold pathogen.

Liver Qi Stagnation 肝氣鬱結

- Liver Qi stagnation is due to chronic stress and anger. An extended period of anger, stress, emotional frustration, and worry lead to Liver Qi stagnation (Apichai 2020); it often demonstrates the symptoms on the Foot Jue Yin足厥陰 Liver meridian 足厥陰肝經 itself and the Fu 腑 organ of Liver, which is Foot Shao Yang 足少陽Gallbladder meridian 足少陽膽經.
- Grief is the emotion of the Lung. Prolonged sadness or bereavement depletes Lung Qi. Lung's element is metal, Liver's element is wood, metal controls wood. Weak Lung Qi loses its ability to control the Liver, leading to Liver Qi stagnation.
- Chronic Liver Qi stagnation causes chronic muscle tension in the upper back, shoulder, and neck regions, leading to the headache, and is the reason that chronic Liver Qi stagnation develops Blood stasis – because Qi is the commander of Blood.

Liver Blood Deficiency 肝血虧虛

"Liver is the root on women". This sentence is from the "Clinical Guide to Medical Records" by Ye Tianshi葉天士, a famous doctor in the Qing Dynasty清朝.

"Liver stores Blood". The body requires Blood for physical activity. During menstruation Blood is lost, which may cause Liver Blood deficiency. Wind stirs upward and causes blockage in the Gallbladder meridians, causing migraine.

Phlegm Blocking Orifice 痰濁蒙竅

Phlegm Blocking Orifice is due to long-term irregular diet, such as excessive consumption of alcohol, rich food, greasy and fatty foods, sweet food, strong-tasting food, etc., causing dysfunction of the Spleen and Stomach, which leads to Phlegm Damp Heat accumulation. Heat injures Yin, Yin fails to control Yang, causing high blood pressure, high cholesterol, and high blood sugar. Damp濕 can influence inner Wind風 – it is called Wind-Damp風濕. Wind is Yang, the direction of Yang is going upward; Phlegm is then blown upward by inner Wind and blocks the orifice. Liver overreacts on Spleen when it is deficient, and Wind from Liver blows Phlegm upward to block the orifice.

Liver and Kidney Yin Deficiency 肝腎陰虛

More often, older people present with some Kidney and Liver insufficiency. Kidney Yin fails to nourish Liver Yin, then both Kidney and Liver Yin are deficient. Dried wood is due to lack of water moistening, causing Heat in Liver. Heat generates Wind, and Wind blocks meridians in the head, resulting in migraine.

MANIFESTATIONS

In Western Medicine

There are two major types of migraine.

a. Migraine without aura, formerly known as "common migraine", is a clinical syndrome characterized by headache with specific features and associated symptoms.

b. Migraine with aura usually includes visual symptoms such as lines, flashing lights, and geometric patterns. These symptoms usually precede or sometimes accompany the headache. Approximately only 1 in 4 migraine patients will experience auras.

Some patients who have frequent attacks with aura may also experience some attacks without aura.

Some patients also experience a prodrome phase that occurs hours or days before the headache, and/or a post-drome phase following headache resolution.

A migraine attack usually occurs in four phases:

1. **Prodrome Phase:** patients experience irritability, depression, stiff neck, thirst, lassitude, yearning, more urination, and constipation. Usually, these symptoms occur several hours to 1–2 days before a migraine attack. Not all migraine patients experiencing this stage.
2. **Aura Phase:** patients experience mainly visual disturbances, such as seeing bright sparks, floaters, or colored lines. Some patients also suffer from hallucination, anxiety, and agitation. These symptoms occur 10–20 minutes before a migraine attack. Not all migraine patients experience this stage.
3. **Attack Phase:** patients experience a high intensity throbbing headache with accompanying symptoms such as nausea, vomiting, dizziness, blurred vision, confusion, and hypersensitivity to light (photophobia) and sound (phonophobia). An attack can last 4–72 hours. The headache is commonly about 60% unilateral location and 40% bilateral location. Bilateral location is commonly seen in children and adolescents (aged under 18 years). The attack can occur from once several days to 1–2 times a year.
4. **Post-drome Phase:** After the above symptoms subside, migraine patients may suffer from confusion, moodiness, elation, and fatigue. These symptoms may last up to 24 hours.

In Chinese Medicine

Liver Yang Rising 肝陽上亢

The migraine is pulling distending, and it is aggravated by anger or emotional stress. The location can be on one or both sides. The accompanying symptoms are irritability, short temper, insomnia, hypochondriac pain, dry mouth, and bitter taste in mouth.

Tongue: red tongue with thin yellow coating.

Pulse: wiry and rapid.

Blood Stasis 瘀阻腦絡

The migraine is stabbing; the location can be one or both sides, and it is fixed without traveling. The duration is long.

Tongue: dusky with ecchymoses, thin white coating.

Pulse: thin and choppy.

Wind Cold Invasion 風寒外襲

The migraine is pulling, heavy, sharp or throbbing pain. The patient is aversion to wind because wind can aggravate the migraine. The episode is usually acute onset without warning. The location is on one or both temporal regions closer to the forehead. Sometimes the symptoms are accompanied by a stiff neck and the occipital regions, absence of thirst, no sweat and no desire to drink, or with a desire for a warm drink. Patients usually have a history of wind and cold air blowing to the head, neck and upper back such as when working in an office where the air conditioning vents are blowing directly at the person.

Tongue: thin white coating.

Pulse: floating, tight.

Liver Qi Stagnation 肝氣鬱結

The migraine is sharp and throbbing with tension or distending headaches. Often, these types of headaches are located at the temporal region near the ear, and often it is behind the eye. It is usually on one side only and it often alternates between sides; when it occurs on the left side, the migraine is usually due to poor blood circulation; if it is on the right side, it is usually due to Qi stagnation or deficiency. The migraine seldom occurs on both sides simultaneously. The migraine is usually aggravated by emotional disorder such as anger, stress, depression, and anxiety; meanwhile, it is improved with relaxation or sexual activity. Sometimes, the symptoms are accompanied by a bitter taste in the mouth in the morning, dried throat, and blurred vision. Patients may have pain, tightness, or muscle spasms on the neck, shoulder and upper back This is usually due to long-term stress, for instance, leaning the head and neck to one side for a long time, such as holding a telephone between the shoulder and the chin.

Tongue: thin white coating.

Pulse: wiry.

Liver Blood Deficiency 肝血虧虛

The migraine is dull, empty, achy. The pain is intermittent, and it is worsened by overexertion or when the patient is fatigued. Onset is slow. The symptoms and signs are accompanied by pale complexion, dizziness, vertigo, palpitations, insomnia, dream-disturbed sleep, spontaneous sweating, shortness of breath, and aversion to wind.

Tongue: pale body, thin white coating.

Pulse: thin, thready, weak.

Phlegm Blocking Orifice 痰濁蒙竅

The migraine is dull and heavy. The location is on one side. The symptoms are accompanied by chest and abdominal distension, poor appetite and nausea, vomiting phlegm.

Tongue: dusky, large body with teeth marks and white think greasy coating.

Pulse: wiry and deep, or wiry and slippery.

Liver and Kidney Yin Deficiency 肝腎陰虛

The migraine is dull and empty. the location is on one side. The migraine is aggravated by both physical and mental exertions and it is alleviated after sleeping. The symptoms are accompanied by tinnitus, deafness, soreness and weakness in lower back and knees, Five-Center Heat, nigh sweat, irritability, insomnia, seminal emission, dry mouth and forgetfulness, underweight.

Tongue: red, thin and small body, peeled or no coating; cracks.

Pulse: thin, rapid.

PHYSICAL EXAMINATION

1. Begin with obtaining medical history.
    1. Migraine history: what the pain feels like – is the pain pulsing, pressing, throbbing, stabbing or heavy, with or without aura; where is the pain located, is it one or both sides, where does the pain start, is it going toward the ear or behind the eye, any pain on the neck and shoulder; how severe is the pain, how long does the pain last, does the pain get worse with physical activity or get better with sleep, is it abrupt or slow onset; what other accompanying symptoms, such as nausea, vomiting, appetite changes, attitude, or behavior changes, have presented; what medications or try other treatments were used, and did they work; at what age was the first episode and under what circumstances that could be causative factors, such as stress, medications, or foods (such as MSG or coffee); how many days or hours has this attack been ongoing; is it a single type of multiple types, how often do the attacks occur. Who else in the family has migraines?
    2. Work schedule: does the patient work the night shift, more than eight hours a day, or more than 40 hours a week.
    3. Work environment: does the patient work in a cold, windy, wet, or rainy place; do they work in a high temperature environment that causes sweating.
    4. Stress: from work or life in both emotional and physical.
    5. Sport activities that often cause physical trauma or excessive sweating.
    6. Sleep schedule that doesn't match the natural clock, such as staying up too late (later than 11 p.m.).
    7. Does the patient consume foods that may damage the Spleen and Stomach, such as fatty, oily, cold (icy), raw, too-sweet foods, including alcohol consumption; are meals scheduled (breakfast, lunch, and dinner), and how long is the gap between dinner and the bedtime. Has the patient had any weight loss or gain.
    8. Illnesses: insomnia, night sweat, nocturia, depression, menstrual period, sexual activities, high blood pressure, high blood glucose, high blood cholesterol, digestive disorders, kidney failure, head injury, and other types of headaches.
        a. Inspection
    9. Appearance: alertness, coordination and gait, fatigue, painful facial expressions, sighing, moaning, asking for help.

10. Vital signs: tachycardia or bradycardia, hypertension or hypotension.
11. Tongue condition, such as red or pale, large or small size, thin or thick coating.
    a. Palpation
    1. Head and neck: is the skin temperature cold or hot; muscle tenderness on neck muscles, shoulder trapezius muscle or sternocleidomastoid muscle.
    2. Pulse types such as wiry, slippery.
    a. Range of Motion
    3. Cervical and shoulder range of motion.
    1. Special Tests

X-rays, electroencephalogram (EEG) or MRI, CT scan or eye exam, and spinal tap are not helpful in diagnosing migraine, but they might be needed to rule out other medical conditions. Sinus x-ray may be used to rule out a sinus problem; eye exam may be used to rule out glaucoma or pressure on the optic nerve; spinal tap may rule out brain or spinal cord infection; EEG may rule out seizures. And an eye exam may rule out Adie-type pupil. Laboratory tests may rule out any suspected coexistent metabolic problems.

DIFFERENTIAL DIAGNOSIS

Subarachnoid Hemorrhage
4. Same: headache, nausea, and vomiting.
5. Difference: loss of consciousness.
6. Onset is sudden, the pain is severe, the type of headache is like a thunder-clap that patients usually describe as the worst headache of their life. CT shows hyperattenuating material filling the subarachnoid space. Lumbar puncture (LP) shows elevated opening pressure, elevated red blood cell (RBC), and xanthochromia.

Meningitis
1. Same: headache, photophobia, phonophobia, nausea, and vomiting.
2. Difference: high fever and neck rigidity.
3. Onset is sudden with seizures, sleepiness, or difficulty waking, skin rash may be present, such as in meningococcal meningitis. CT scan and lumbar puncture can determine the risk of herniation. Cerebrospinal fluid level (CSF) analysis shows increase in leukocytes, increase in protein, and decrease in glucose.

Cerebral Hemorrhage
1. Same: headache and vomiting.
2. Difference: increase intracranial pressure (ICP) and loss of consciousness.
3. Onset is sudden. Problems with balance and coordination, seizure, weakness on one side of the body, and coma are symptoms. CT scan shows a hyperattenuating clot.

TREATMENTS

1. Liver Yang Rising 肝陽上亢

Treatment Principle

1. Soothe the Liver, suppress Yang, nourish Yin, and clear Heat.

Herbal Treatments

Herbal Formula

Tian Ma Gou Teng Yin 天麻鉤藤飲

This formula is in the book "Newly Explained Diagnosis and Treatment of TCM Internal Medicine and Miscellaneous Diseases中醫內科雜病證治新義", authored by Hu Guangci 胡光慈 in 1958.

Herbal Ingredients

Tian Ma天麻6g, Gou Teng鉤藤9g, Chuan Niu Xi牛膝12g, Shi Jue Ming石決明15g, Du Zhong杜仲12g, Huang Qin黃芩6g, Zhi Zi梔子6g, Yi Mu Cao益母草9g, Sang Ji Sheng桑寄生9g, Ye Jiao Teng夜交藤9g, Fu Shen茯神9g

Ingredient Explanations

1. Tian Ma天麻 extinguishes Wind, calms the Liver and subdues rising Liver Yang, and alleviates pain.
2. Gou Teng鉤藤 expels Wind and releases spasm.
3. Chuan Niu Xi川牛膝 unblocks and promotes Blood circulation.
4. Yi Mu Cao益母草 moves the Blood, promotes urination, reduces swelling, and clears Heat.
5. Shi Jue Ming石決明 drains Liver Fire and expels Wind.
6. Du Zhong杜仲 tonifies the Liver and Kidney, strengthens Bone, and benefits Essence and Blood.
7. Huang Qin黃芩 clears Heat, dries Damp, cools the Blood, stops bleeding, calms Liver Yang.
8. Zhi Zi梔子 clears Heat, cools the Blood, resolves Damp, reduces swelling, and opens the meridians.
9. Sang Ji Sheng桑寄生 tonifies the Liver and Kidney, strengthens Bone, benefits Essence and Blood; expels Wind and drains Damp.
10. Ye Jiao Teng夜交藤 nourishes Heart Yin and the Blood, calms Shen, expels Wind and Damp.
11. Fu Shen茯神 calms Shen.

Modifications

1. For severe Liver Heat from Liver Qi stagnation transforming into Fire that presents dry stool and dark yellow urination, add Da Huang大黃, Niu Bang Zi牛蒡子, and Long Dan Cao龍膽草.

Formula Actions

2. Soothes Liver Qi, expels Wind, moves the Blood, clears Heat, tonifies the Liver and Kidney.

Formula Indication

- Headache due to Liver Yang rising, vertigo, insomnia, anger, and bitter taste in the mouth.

Contraindications

- Caution for patients who suffer from Blood, Yin, and body Fluid deficiency.
- Stop taking this formula if dizziness, chest distention, shortness of breath, nausea, vomiting, or skin itch are present.
- Do not take Tian Ma overdose.

Cooking Instructions

1) Place the herbs into an herb pot. Traditionally, a clay cooker is used. Gently rinse the herbs once with a strainer.
2) Add water into the pot until it is just enough to cover the herbs, then add one more cup of water.
3) Presoak the herbs for 15 minutes.
4) Turn on the stove to high heat and bring the herbs to a boil. Reduce the heat to medium-low to simmer but with slight bubbling. Turn off the heat when the decoction is reduced to one cup.
5) Drain the decoction into a container.
6) Repeat steps two to five one more time, and drain each decoction to the same container.
7) Divide the decoctions into two separate cups.

Dose:

- Drink one cup of the decoction after meal twice daily.

Blood Stasis 瘀阻腦絡

Treatment Principle

3. Move the Blood, disperse Bruising, open the orifice, stop pain.

Herbal Treatments

Herbal Formula

Xue Yu Zhu Yu Tang血府逐瘀湯

This formula is from "Correction of Errors in Medical Classics醫林改錯", authored by Wang Qingren 王清任 (1768–1831), in Qing dynasty清朝 1636–1912, and published in 1849.

Herbal Ingredients

Dang Gui當歸9g, Sheng Di Huang生地黃9g, Tao Ren 桃仁12g, Hong Hua紅花6g, Zhi Ke枳殼6g, Chi Shao赤芍6g, Chai Hu柴胡3g, Gan Cao 甘草3g, Jie Geng 桔梗 5g, Chuan Xiong川芎5g, Chuan Niu Xi牛膝9g

Ingredient Explanations

4. Dang Gui當歸 moves the Blood, disperses Bruising, and stops pain due to Blood stasis.

5. Sheng Di Huang生地黃 nourishes Yin and clears Heat.
6. Tao Ren 桃仁 moves the Blood, disperses Bruising, and stops pain due to Blood stasis.
7. Hong Hua紅花 moves the Blood, disperses Bruising, and stops pain due to Blood stasis.
8. Zhi Ke枳殼 directs the actions of the formula to the lower body.
9. Chi Shao赤芍 moves the Blood, disperses Bruising, and stops pain.
10. Chai Hu柴胡 moves Liver Qi and releases stagnation.
11. Gan Cao 甘草 harmonizes the herbs, tonifies Middle Jiao Qi, clears Heat, expectorates Phlegm, and stops cough.
12. Jie Geng 桔梗 directs the actions of the formula to the upper body.
13. Chuan Xiong川芎 moves the Blood, disperses Bruising, and stops pain due to Blood stasis.
14. Chuan Niu Xi牛膝 unblocks and promotes movement in the channels and collaterals and stops spasms and convulsions.

Modifications
- For palpitation and insomnia, add Fu Shen茯神, Wu Wei Zi五味子, and Suan Zao Ren酸棗仁.
- For persistent fever, add Jin Yin Hua金銀花 and Lian Qiao連翹.

Formula Actions
- Moves the Blood, opens the meridians, and stops pain.

Formula Indication
- Headache due to Blood stasis.

Contraindications
- It is contraindicated for patients who suffer from headache that is not caused by Blood stasis.
- It is contraindicated during pregnancy.

Cooking Instructions
1) Place the herbs into an herb pot. Traditionally, a clay cooker is used. Gently rinse the herbs once with a strainer.
2) Add water into the pot until it is just enough to cover the herbs, then add one more cup of water.
3) Presoak the herbs for 15 minutes.
4) Turn on the stove to high heat and bring the herbs to a boil. Reduce the heat to medium-low to simmer but with slight bubbling. Turn off the heat when the decoction is reduced to one cup.
5) Drain the decoction into a container.
6) Repeat steps two to five one more time, and drain each decoction to the same container.
7) Divide the decoction into two separate cups.

Dose
- Drink one cup of the decoction after meals twice daily.

Wind Cold Invasion 風寒外襲

Treatment Principle
1. Expel Wind and dissipate Cold, release exterior, regulate the Blood, and stop pain.

Herbal Treatments

Herbal Formula

Chuan Xiong Cha Tiao San 川芎茶調散

This formula is from "Tai Ping Hui Min He Ji Ju Fang太平惠民和劑局方", authored by Imperial Medical Bureau 太醫局in 1134AD, in southern Song dynasty 南宋 1127–1239AD.

Herbal Ingredients

Bo He薄荷12g, Fang Feng防風4.5g, Xi Xin細辛3g, Qiang Huo羌活6g, Bai Zhi白芷6g, Gan Cao甘草炙6g, Chuan Xiong川芎12g, Jing Jie荊芥12g

Ingredient Explanations
2. Bo He薄荷expels Wind Heat, relieves Liver Qi stagnation, disperses stagnant Heat, and enhances Chai Hu's ability to relieve the Liver.
3. Fang Feng防風expels Wind and drains Damp.
4. Xi Xin細辛enters the Shao Yin meridian to expel Wind, Damp, Cold.
5. Qiang Huo羌活expels Wind-Cold-Dampness, unblocks painful obstruction, and alleviates pain.
6. Bai Zhi白芷expels Wind, drains Dampness, dispels Cold, and stops pain.
7. Gan Cao甘草炙harmonizes the herbs.
8. Chuan Xiong川芎moves the Blood, disperses Bruising, and stops pain due to Blood stasis.
9. Jing Jie荊芥stops bleeding, dispels Wind, and relieves muscle spasm.

Modifications
- To make the formula stronger, add Sheng Jiang生薑 and Zi Su Ye紫蘇葉.
- To warm, add Gui Zhi桂枝.
- For dizziness, add Tian Ma天麻 and Gao Ben藁本.

Formula Actions
- Expels Wind and stops pain.

Formula Indication
- Headache due to Wind invasion, vertigo, stuffy nose, fever, and aversion to cold.

Contraindications

- Caution for patients who suffer from headache due to Liver Wind.

Cooking Instructions

1) Place the herbs into an herb pot. Traditionally, a clay cooker is used. Gently rinse the herbs once with a strainer.
2) Add water into the pot until it is just enough to cover the herbs, then add one more cup of water.
3) Presoak the herbs for 15 minutes.
4) Turn on the stove to high heat and bring the herbs to a boil. Reduce the heat to medium-low to simmer but with slight bubbling. Turn off the heat when the decoction is reduced to one cup.
5) Drain the decoction into a container.
6) Repeat steps two to five one more time, and drain each decoction to the same container.
7) Divide the decoction into two separate cups.

Dose:

- Drink one cup of the decoction after meals twice daily.

Liver Qi Stagnation 肝氣鬱結

Treatment Principle

1. Tonify and move the Blood, soothe the Liver, suppress Yang, nourish Yin, and clear Heat.

Herbal Treatments

Herbal Formula

Tao Hong Si Wu Tang and Tian Ma Gou Teng Yin桃紅四物湯合天麻鉤藤飲

Herbal Ingredients

Tao Ren桃仁 10g, Hong Hua 紅花5g, Dang Gui當歸10g, Chuan Xiong川芎6g, Shu Di Huang熟地黃10g, Bai Shao白芍10g, Tian Ma天麻6g, Gou Teng鉤藤9g, Chuan Niu Xi牛膝12g, Shi Jue Ming石決明15g, Du Zhong杜仲12g, Huang Qin黃芩6g, Zhi Zi栀子6g, Yi Mu Cao益母草9g, Sang Ji Sheng桑寄生9g, Ye Jiao Teng夜交藤9g, Fu Shen茯神9g

Ingredient Explanations

1. Tao Ren 桃仁 moves the Blood, disperses Bruising, and stops pain due to Blood stasis.
2. Hong Hua紅花 moves the Blood, disperses Bruising, and stops pain due to Blood stasis.
3. Dang Gui當歸 nourishes the Blood, benefits the Liver, and regulates menstruation.
4. Chuan Xiong川芎 moves the Blood, moves Qi.

5. Shu Di Huang熟地黃 tonifies the Liver and Kidney Yin and benefits Essence and the Blood.
6. Bai Shao白芍 nourishes the Blood, soothes the Liver, and stops pain.
7. Tian Ma天麻 extinguishes Wind, calms the Liver and subdues rising Liver Yang, and alleviates pain.
8. Gou Teng鉤藤 expels Wind and releases spasm.
9. Chuan Niu Xi川牛膝 unblocks and promotes Blood circulation.
10. Yi Mu Cao益母草 moves the Blood, promotes urination, reduces swelling, clears Heat.
11. Shi Jue Ming石決明 drains Liver Fire and expels Wind.
12. Du Zhong杜仲 tonifies the Liver and Kidney, strengthens Bone, and benefits Essence and the Blood.
13. Huang Qin黃芩 clears Heat, dries Damp, cools the Blood, stops bleeding, calms Liver Yang.
14. Zhi Zi栀子 clears Heat, cools the Blood, resolves Damp, reduces swelling, and opens meridians.
15. Sang Ji Sheng桑寄生 tonifies the Liver and Kidney, strengthens Bone, benefits Essence and the Blood; expels Wind, drains Damp.
16. Ye Jiao Teng夜交藤 nourishes Heart Yin and the Blood, calms Shen, expels Wind, Damp.
17. Fu Shen茯神 calms Shen.

Modifications

- To make the formula stronger, add Ju Hua菊花, Gao Ben藁本, and Chai Hu 柴胡.
- For Qi deficiency, add Ren Shen人參 and Huang Qi黃耆.
- For vertigo and dizziness due to Wind in the head, add Qin Jiu秦艽 and Qiang Huo羌活.

Formula Actions

- Nourishes and moves the Blood, benefits the Liver, expels Wind, clears Heat, and tonifies Liver and Kidney.

Formula Indication

- Headache due to Liver Qi stagnation and Blood stasis.

Contraindications

- Caution for patients who are pregnant.
- Caution for patients with constitutional Spleen Yang deficiency.

Cooking Instructions

1) Place the herbs into an herb pot. Traditionally, a clay cooker is used. Gently rinse the herbs once with a strainer.
2) Add water into the pot until it is just enough to cover the herbs, then add one more cup of water.

3) Presoak the herbs for 15 minutes.
4) Turn on the stove to high heat and bring the herbs to a boil. Reduce the heat to medium-low to simmer but with slight bubbling. Turn off the heat when the decoction is reduced to one cup.
5) Drain the decoction into a container.
6) Repeat steps two to five one more time, and drain each decoction to the same container.
7) Divide the decoction into two separate cups.

Dose:
- Drink one cup of the decoction after meals twice daily.

Liver Blood Deficiency 肝血虧虛

Treatment Principle
Tonify the Blood.

Herbal Treatments
Herbal Formula
Si Wu Tang四物湯
This formula is from "Tai Ping Hui Min He Ji Ju Fang太平惠民和劑局方", authored by Imperial Medical Bureau 太醫局in 1134AD, in southern Song dynasty 南宋 1127–1239AD.

Herbal Ingredients
Dang Gui當歸10g, Chuan Xiong川芎10g, Shu Di Huang熟地黃10g, Bai Shao白芍 10g

Ingredient Explanations
1. Dang Gui當歸 nourishes the Blood, benefits the Liver, and regulates menstruation.
2. Chuan Xiong川芎 moves the Blood and moves Qi.
3. Shu Di Huang熟地黃 tonifies the Liver and Kidney Yin and benefits Essence and the Blood.
4. Bai Shao白芍 nourishes the Blood, soothes the Liver, and stops pain.

Modifications
- For Yang deficiency, add Shu Di Huang熟地黃, Lu Jiao Jiao鹿角膠, Shan Zhu Yu山茱萸, Zhi Fu Zi附子, and Rou Gui 肉桂.
- For Qi deficiency, add Ren Shen人參 and Huang Qi黃耆.
- For vertigo and dizziness due to Wind in the head, add Qin Jiu秦艽, Qiang Huo羌活.

Formula Actions
- Nourishes the Blood, benefits the Liver, improves Blood circulation.

Formula Indication
- Headache due to Liver Blood deficiency.

Contraindications
- Caution for patients who have weak digestion.
- Caution for patients with diarrhea.

Cooking Instructions
1) Place the herbs into an herb pot. Traditionally, a clay cooker is used. Gently rinse the herbs once with a strainer.
2) Add water into the pot until it just covers the herbs, then add one more cup of water.
3) Presoak the herbs for 15 minutes.
4) Turn on the stove to high heat and bring the herbs to a boil. Then reduce the heat to medium-low to simmer but with slight bubbling. Turn off the heat when the decoction is reduced to one cup.
5) Drain the decoction into a container.
6) Repeat steps two to five one more time, and drain each decoction to the same container.
7) Divide the decoction into two separate cups.

Dose:
- Drink one cup of the decoction after meals twice daily.

Phlegm Blocking Orifice 痰濁蒙竅

Treatment Principle
Tonify Spleen Qi, drain Damp, and extinguish Wind.

Herbal Formula
Ban Xia Bai Zhu Tian Ma Tang半夏白朮天麻湯
This formula is from "Medical Insights 醫學心悟", authored by Cheng Guopeng 程國彭 (1662–1735), in Qing dynasty 清朝 (1636–1912), and published in 1732.

Herbal Ingredients
Ban Xia半夏4.5g, Bai Zhu白朮3g, Tian Ma天麻3g, Ju Hong橘紅3g, Fu Ling茯苓3g, Gan Cao甘草1.5g, Sheng Jiang生薑2 pieces, Da Zao大棗3 pieces, Man Jing Zi蔓荊子3g

Ingredient Explanations
1. Ban Xia半夏 transforms phlegm and prevents evil Qi entering interiorly.
2. Bai Zhu白朮 tonifies the Spleen Qi, dries Dampness.
3. Tian Ma天麻 extinguishes Wind, calms the Liver and subdues rising Liver Yang, and alleviates pain.
4. Ju Hong橘紅 regulates Qi and transforms phlegm.

5. Fu Ling茯苓 drains Damp and tonifies the Spleen.
6. Gan Cao甘草 harmonizes the herbs, tonifies Middle Jiao Qi, clears Heat, expectorates phlegm, and stops cough.
7. Sheng Jiang生薑 unblocks the pure Yang pathway, and harmonizes rebellious Qi.
8. Da Zao大棗 tonifies the Spleen Qi, nourishes the Blood, and moderates and harmonizes the harsh properties of the fragrant herbs.
9. Man Jing Zi蔓荊子 expels Wind and stops pain.

Modifications

- For Qi deficiency, add Ren Shen人參 and Huang Qi黃耆.

Formula Actions

- Tonifies the Spleen, drains Damp Phlegm, soothes the Liver, and extinguishes Wind.

Formula Indication

- Migraine due to Phlegm blocking orifice that stirs up by Wind, dizziness, tongue has a sticky white coating, and pulse is wiry and slippery.

Contraindications

- It is contraindicated for patients who suffer from headache not caused by Wind-Phlegm 風痰.

Cooking Instructions

1) Place the herbs into an herb pot. Traditionally, a clay cooker is used. Gently rinse the herbs once with a strainer.
2) Add water into the pot until it is just enough to cover the herbs, then add one more cup of water.
3) Presoak the herbs for 15 minutes.
4) Turn on the stove to high heat and bring the herbs to a boil. Reduce the heat to medium-low to simmer but with slight bubbling. Turn off the heat when the decoction is reduced to one cup.
5) Drain the decoction into a container.
6) Repeat steps two to five one more time, and drain each decoction to the same container.
7) Divide the decoction into two separate cups.

Dose

- Drink one cup of the decoction after meals twice daily.

Liver and Kidney Yin Deficiency 肝腎陰虛

Treatment Principle

Tonify Liver and Kidney Yin, clear Heat, and extinguish Wind.

Herbal Formula
Liu Wei Di Huang Wan六味地黃丸

This formula is from "Key to Therapeutics of Children's Diseases 小兒药证直诀", authored by Qian Yi 钱乙 (1032–1113), a physician of the Song dynasty 北宋 (960–1127), and published in 1119–1125.

Herbal Ingredients

熟地黃24g, 山茱萸12g, 山藥12g, 澤瀉9g, Fu Ling茯苓9g, 牡丹皮9g

Ingredient Explanations

1. Shu Di Huang熟地黃 tonifies Liver and Kidney Yin, and benefits Essence and the Blood.
2. Shan Zhu Yu山茱萸 tonifies the Liver and Kidney, strengthens Bones, and benefits Essence and the Blood.
3. Shan Yao山藥 tonifies the Kidney and benefits the Yin and the Spleen Qi.
4. Ze Xie澤瀉 drains Damp, especially Damp Heat in Lower Jiao and clears Kidney deficient Heat.
5. Fu Ling茯苓 drains Damp and tonifies the Spleen.
6. Mu Dan Pi牡丹皮 clears Heat, cools the Blood, moves the Blood.

Modifications

For Liver Blood deficiency, add Dan Gui當歸 and Bai Shao白芍.
1. For frequent urination, remove Ze Xie澤瀉 and add Yi Zhi Ren益智仁 and Fu Pen Zi覆盆子.
2. For pain in low back and knee, add Niu Xi牛膝, Du Zhong杜仲, and Sang Ji Sheng桑寄生.

Formula Actions

• Nourishes Liver and Kidney Yin.

Formula Indication

• Migraine due to Liver and Kidney Yin deficiency.

Contraindications

It is contraindicated for patients who suffer from headache caused by Yang deficiency.
1. Caution for patients who suffer from diarrhea and indigestion due to Spleen deficiency.

Cooking Instructions

1) Place the herbs into an herb pot. Traditionally, a clay cooker is used. Gently rinse the herbs once with a strainer.
2) Add water into the pot until it just covers the herbs, then add one more cup of water.

3) Presoak the herbs for 15 minutes.

4) Turn on the stove to high heat and bring the herbs to a boil. Reduce the heat to medium-low to simmer but with slight bubbling. Turn off the heat when the decoction is reduced to one cup.

5) Drain the decoction into a container.

6) Repeat steps two to five one more time, and drain each decoction to the same container.

7) Divide the decoction into two separate cups.

Dose:

- Drink one cup of the decoction about 15 minutes after meals twice daily.

Acupuncture Treatments

Acupuncture Point Formula

Local Points

Tai Yang太陽, Yin Tang印堂, SJ3中渚ZhongZhu, SJ17翳風Yifeng, ST10水突Shuitu, GB20風池Fengchi, GB21肩井Jianjing, Ashi阿是穴

Distal Points

GB39懸鍾Xuanzhong, LI4合谷Hegu, LIV3太衝Taichong, UB63金門Jinmen

Modifications

1. For Liver Yang rising, add GB5 懸顱 Xuanlu, DU20百會 Baihui, Sishencong四神聰, GB4頷厭 Hanyan, GB41足臨泣Zulinqi, SJ23絲竹空 Sizhukong, LIV2 行間 Xingjian, GB34陽陵泉Yanglingquan, ST44內庭 Neiting, GB43 俠溪 Jiaxi, and GB5 through GB8懸顱透率谷.

2. For Blood stasis, add LIV6 中都 Zhongdu, SP10血海 Xuehai, LIV3 太衝 Taichong, DU16 風府 Fengfu, UB17 膈俞Geshu, LIV1 大敦 Dadun, and bleeding Tai Yang太陽.

3. For Qi 氣 stagnation, add ST8 頭維Touwei and GB13本神 Benshen.

4. For Qi and Blood stagnation, add SP10血海Qihai, SP6三陰交Sanyinjiao, LIV3太衝Taichong, and UB17膈俞Geshu.

5. For Qi and Blood stagnation, add ST36足三里Zusanli and REN6 氣海 Qihai.

6. For Cold 寒 invasion, add UB9玉枕 Yuzhen, GB20風池Fengchi, LI4合谷Hegu, LI11曲池Quchi, QuchiDU22囟會Xinhui, and Tai Yang through SJ20Jiaosun太陽透角孫.

7. For Blood deficiency, add UB17膈俞Geshu, ST36足三里Zusanli, REN6 氣海Qihai, and UB20 脾俞Pishu.

8. For Phlegm Dampness obstruction, add ST40豐隆Fenglong, SP6三陰交 Sanyinjiao, DU23上星 Shangxing, ST8頭維Touwei, SP9 陰陵泉, REN12 中脘 Zhongwan, and ST36足三里Zusanli.

9. For Yin 陰 deficiency, add KD3太溪 Taixi, SP6 三陰交 Sanyinjiao, and UB23 腎俞 Shenshu.

Bleeding Therapy

10. Point selected is Tai Yang太陽.
11. Remove 4–8 drops with a cotton wool swab.
12. Treatment frequency is once a week for six weeks.

Procedures

- Prior to the procedure, the tools should be prepared and ready to be used in the working area.
- Set up a clean field with paper towels.
- Place these tools in the clean field, including dry sterile cotton balls, medical alcohol wipes or cotton swabs soaked in alcohol, sterile gloves, a biohazard sharps container, a biohazard trash container, goggles and face mask, and paper towels. Also, a medical lancet or a traditional three-edge needle or the seven-star needle (a bleeding tool); a medical lancet or a traditional three-edge needle is selected for bleeding Tai Yang太陽 in this case.
- The room should be well lit.
- The patient lies supine.
  1. The therapist washes and dries their hands thoroughly and wipes their hands with a single-use napkin. Put on sterile gloves.
  2. The therapist sits facing the acupuncture point.
- Wipe the acupuncture point Tai Yang太陽 with a medical alcohol wipe or a cotton swab soaked in alcohol, moving from the center of the acupuncture point to the periphery.
  2. Place the medical lancet or the traditional three-edge needle at an angle of 90 degrees to the skin.
- In one quick move, prick the needle into the acupuncture point about 0.1 Cun 寸 deep, eliciting a few drops of blood.
- If more drops of blood are needed, absorb the blood with a sterile cotton ball.
- Lastly, the point is pressed with a sterile cotton ball until the bleeding ceases.
  3. Apply a bandage on the prick wound.
- Dispose of the tools in the proper containers.
- Wash the hands thoroughly.

Auricular Therapy

Forehead, Occiput, Shenmen, Neck, Heart, Liver, Ear Apex, and Helix 6.

Aquapuncture Therapy

Vitamin B12 on an Ashi point on the painful site.

Tui Na Treatments

Tui Na techniques

- Pressing, kneading, percussing, grasping, one finger meditation, rubbing and dry-cleaning-hair.

Tui Na procedures

The patient is in a sitting position; the therapist stands in front of the patient.

1. Applies "One Finger Meditation Technique" with a thumb tip on Yin Tang 印堂 for 20–30 times.
2. Applies pressing technique with a thumb tip on UB2 攢竹Zanzhu for a minute.
3. Applies rubbing technique with both thumb pads along both sides of the nasolabial groove where LI20 迎香 Yingxiang and Bitong 鼻通 are located for 1 minute.

The patient is in a sitting position; the therapist stands behind the patient.

1. Applies grasping technique with the first 3 fingers (one hand) on both GB20 風池 Fengchi for 3 minutes. Figure 2.2.1

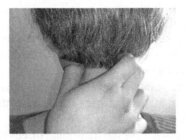

2. Applies grasping technique with all fingers (both hands) along the sides of the neck for 3 minutes. Figure 2.2.2; Figure 2.2.3

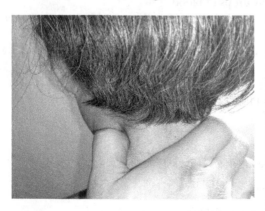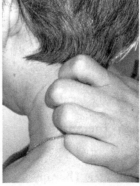

3. Applies clenching technique on GB20 肩井 Jianjing.
4. Applies percussing technique with all five fingertips on DU20 百會 Baihuifor for 10–20 times alternating with "Dry-Cleaning-Hair" technique on the scalp 5 times after every 10 times of fingertip percussing.

PATIENT ADVISORY

Dietary Recommendations

1. Watch what you eat and drink. If you get a migraine, review what food and drink you have had in the last several meals even from a day ago, write them down, and track them to find out a pattern that you repeatedly have. The drinks that may trigger migraine are alcohol, especially red wine; caffeine, including coffee, tea, chocolate, sodas, colas – do not cut back suddenly, instead try to slow down drinking caffeine. The foods that may trigger migraine are foods that contain monosodium glutamate (MSG), chicken powder (used in many restaurants and also contains MSG), artificial sweeteners, aged cheeses, pickled and fermented foods, baked foods, bread, dried fruits, potato chips, pizza, peanuts, chicken livers, foods that contain preservatives (such as nitrates and including pepperoni), hotdogs (contain nitric oxide and can cause or contribute to migraine). Eating frozen foods and drinks, like ice cream or slushies, can trigger severe, stabbing pains in the head. Foods that contain high levels of sodium can increase blood pressure, causing headaches or migraine attacks.

   1. Eat meals on time. Spleen Qi needs food and drink regularly. Migraine headache patients are highly recommended to focus on the foods that contain calcium, magnesium, fish oil, B vitamins, C vitamins, and turmeric.
   2. Foods that may release migraine as below:

   Fruits and vegetables: banana, star fruit, spinach, seaweed, kelp, cauliflower, and cabbage.

   Whole grains: brown rice, whole oats, whole buckwheat, and coarse coix seed.

   Nuts: sunflower seeds, cashews, and almonds.

   Beans: black beans, soybeans, peas, and broad beans.

   3. Drink more water. Prior to and during menstruation, drinking more water will help to reduce headache. Warm temperature water is recommended.
   4. Avoid eating or drinking before sleep. To prevent Phlegm and Damp accumulation, avoid having the stomach working hard about 2 hours before bed.

Lifestyle

5. Be careful with exercise. Appropriate exercise is needed for migraine patients, but over exercise, wrong types of exercise, exercising at the wrong time, or exercising in certain weather may trigger migraine. The morning before 9 o'clock is the best time for physical exercise. Select types of exercise that promote sweat; the sweating should be light and all over the whole body. Cardio exercise is a good one for migraine patients, 3 times a week for 40 minutes will help to prevent headache.

Morning exercise tonifies Yang Qi 陽氣, boosts the immune system to prevent Wind Cold/Heat invasion, refreshes the brain for better energy, memory and clearing brain fog; it also helps manage stress. Evening is appropriate for mental exercise. Patients who suffer from the Phlegm Blocking Orifice type of migraine should be cautioned against exercising in the rain, fog, and wind.

6. Avoid staring at television, computer monitors, cellphones, and/or tablets for extended periods.
7. Avoid going to bed later than 11 p.m. because it is considered part of the next day. Be on the bed and ready to sleep at around 10 or 10:30 p.m. Getting 7–8 hours of sleep and going to bed early will help to store energy for use on the following day.
8. Take a warm bath. During the migraine attack and before bed, taking a warm bath will relax tight muscles, help to stop headache, and improve sleep.
9. Reduce stress. Stress is a common cause of Liver Qi stagnation and Liver Yang rising. Exercise, yoga, meditation, praying, joining a party, reading funny books, and watching comedy can ease stress.
10. Heating pad. Applying a heating pad on the neck is helpful when the migraine is from Wind Cold Invasion, the type of migraine which commonly begins at the occipital region.

POST-CONCUSSION SYNDROME 脑震荡症候群

INTRODUCTION

Concussion occurs when there is a traumatic brain injury in which the brain rapidly shifts or turns. It is the most common type of brain injury, accounting for hundreds of thousands of emergency room visits each year.[7]

Many physical activities are prone to causing head injury, including falls, blows, shaking, and sports, like wresting, football, bicycling, horseback riding, hockey, rugby, soccer, and basketball.

After a concussion, the majority of people recover from the initial symptoms within 2–4 weeks; a minority of people who experience post-concussion syndrome have the symptoms of concussion longer than six weeks.

People who suffer multiple concussions develop post-concussion syndrome, which may cause lifelong brain damage.

ANATOMY

The brain is surrounded by meninges and cerebrospinal fluid, which cushion the brain. Meninges are many arteries, veins, and cranial nerves. Cerebrospinal fluid surrounds the brain and cushions it from direct contact with the skull. During an impact, the brain is forcibly rotated and battered and shifted to the inner side of the skull Unfortunately, the skull doesn't absorb all the impact of a violent force, it absorbs only part of it; different parts of the brain can move at different speeds, such movements can alter the balance of ions and chemicals in the brain and stretch or even rip nerve tissues and blood vessels, bruising the brain in the worse concussions.[8,9,10]

WESTERN MEDICINE ETIOLOGY AND PATHOLOGY

A concussion is often referred to as a traumatic brain injury (TBI) which is relatively minor, hence, sometimes it is called a mild traumatic brain injury (mTBI). A concussion is a type of closed head injury which is caused by external factors directly hitting the head, jolting the head, and shaking the brain.

It is a commonly seen head injury from certain sport activities particularly football, wresting, and boxing.

Following concussion, cerebral pathophysiology can be adversely affected for weeks, but not all people who suffer from concussion develop the symptoms. Abrupt release of neurotransmitters and unchecked ionic fluxes occur immediately after the brain injury. The binding of glutamate to the N-methyl-D-aspartate (NMDA) receptor leads to efflux of potassium and influx of calcium. The sodium–potassium (Na^1–K^1) pump works to maintain the negatively charged membrane inside the cell and restores the neuronal membrane potential.[11] The pump requires increasing amounts of adenosine triphosphate (ATP), triggering an increase in glycolysis which is to increase in glucose use. The increase in glycolysis results in an increase of lactate production, with a decrease in lactate metabolism, resulting in lactate accumulation. Elevated lactate levels result in cell swelling and a reduction in cerebral blood flow, leading to ischemia in the brain. Decreased cerebral blood flow causes reduction in glucose level, leading to reduced ATP levels, resulting in elevated intracellular calcium.

Magnesium is necessary for maintaining the cellular membrane potential and imitating protein synthesis. Magnesium levels in the brain also decrease following concussion. A low level of magnesium may lead to greater influx of calcium.

Increased intracellular calcium may lead to the production of harmful chemicals called free radicals, cytoskeletal reorganization, and activation of apoptotic genetic signals and may lead to cell death.

Concussion also impairs choline acetyltransferase activity resulting in learning and spatial memory deficits.

Concussion may subsequently lead to the development of seizures when the inhibition of hippocampal dentate granule cells is compromised because of a loss of g-aminobutyric acid-producing (GABAergic) hilar neurons.

CHINESE MEDICINE ETIOLOGY AND PATHOLOGY

Although headache is a main symptom of concussion, it is not part of the common headache. The common headaches are mostly caused by Wind, Heat, Phlegm, and deficiency of Zang-Fu (the internal organs in Chinese medicine), and the main cause is related to Zang-Fu. Concussion is the cause of head trauma that is directly related to the brain. The pathogenesis is mainly failure of Qi and Blood flow in the meridians.

Traditional Chinese medicine defines post-concussion syndrome into five types:

1. Blood Stagnation 瘀血阻滯型

The head is where the Yang energy gathers. The brain is the palace of the soul, which controls the whole body. If the head is shocked by an external force, the Qi will inevitably be damaged. The Qi and Blood will be chaotic stagnant.

1. Qi and Blood Xu 氣血虧損型

"Shen" is a mandarin pronunciation, it can be translated as "Spirit" or "Mind", and it implies the consciousness, mental functions, mental health, vitality, and presence of mind. Healthy "Shen 神" of the brain depends on the strength of the Kidney Essence Qi, and Blood.

"Shen 神" lives in the Blood vessels, and it is a part of the Heart system in traditional Chinese medicine. Heart produces Blood, Blood is the mother of Qi; it is formed from Spleen food and fluid nutrients. The Marrow is derived from the Kidney Essence Qi, and Blood; the Essence is generated by Zang-Fu 臟腑 (organs in Chinese medicine). Both Marrow and Essence ascend and gather in the head.

Disturbed "Shen 神" often is a result of long-term emotional problems or after a serious shock, especially if it occurred a long time ago. Shen disturbance depletes Qi and Blood.

1. Kidney Essence Deficiency 腎精虧損型

In Chinese medicine, the brain function is related to the Kidney system, and Kidney sends Essence to the brain to support its functions. Memory, concentration, thinking and intelligence, mental activity, physical movement and sensation, daytime energy and nighttime sleeping are functions of the brain. Any malfunction of the brain requires more Kidney Essence to repair, but over time, Kidney Essence is depleted.

1. Phlegm Damp Misting the Orifice 痰浊上蒙型

Constant worry knots Spleen Qi, and a weak spleen creates more worry. One of the roles in Spleen is "Spleen Houses Thought 脾 藏 意": Spleen is directly related to the capacity for thinking, concentration, discernment, and intention; it is responsible for analytical thinking, memory, cognition, intelligence, and ideas. Spleen is affected once the brain is injured resulting in Spleen Qi deficiency. The Spleen also "transforms and transports foods and fluids 脾主運化". Spleen deficiency leads to an accumulation of moisture within the body, the moisture is called "Phlegm" which is distinguished into "substantial" and "insubstantial". Substantial phlegm is visible such as sputum that cannot penetrate the meridian, presenting in the lungs and stomach; insubstantial phlegm is invisible that can penetrate and stay in the meridians, it combines with other pathological influences, such as Wind, Fire, and Damp, then is distributed to various parts of the body; it results in loss of consciousness or dizziness and vertigo when it goes to the Heart and brain.

1. Liver Yang Uprising 肝陽上亢型

After concussion, the persisting psychological symptoms such as personality changes, restlessness, irritability and depressive mood, and anxiety cause Liver Qi stagnation. Liver overacts on Spleen, leading to insubstantial phlegm accumulation. Kidney Yin deficiency is followed by long-term Liver Qi stagnation. Wind stirs upward together with the Yin deficient Fire and insubstantial Phlegm to the brain.

MANIFESTATIONS

Post-concussion symptoms vary depending on the person. They typically begin to manifest within days to weeks of an accident. Most symptoms are transient and typically resolve within two weeks. Persistent post-concussive syndrome occurs when the symptoms persist beyond three months (Permenter, Fernández-de Thomas, and Sherman 2022).

The common signs and symptoms of a concussion include:

1. Headache.
2. Vertigo, poor balance, and problems walking.
3. Fatigue.
4. Amnesia.
5. Fogginess.
6. Poor concentration.
7. Insomnia, trouble falling asleep.
8. Irritability.
9. Anxiety and depression.
10. Tinnitus.
11. Blurry vision.
12. Sensitivity to noise and light.

These symptoms may not present on the same day, but can show up days later after an accident:

1. Trouble learning and focusing.
2. Poor memory.
3. Confusion.

These symptoms may indicate a serious health issue. Contact the medical provider or go to the nearest emergency room or an urgent care right away:

1. Neck pain or pressure pain.
2. Double vision.

3. Weakness or stabbing, burning pain in the arms or legs.
4. Severe headache that is different than the ones before, does not seem to get better, and gets progressively worse.
5. Seizure or convulsion.
6. Increasingly restless, agitated or combative.
7. Unable to actively open eyes.
8. Repeated vomiting that does not seem to stop.
9. Paralysis or numbness of the face, arm, or leg.
10. Problems seeing in one or both eyes.
11. Loss of consciousness.
12. Increased confusion, such as difficulty recognizing well-known people or places.
13. More than one concussion.
14. Clear fluid or blood draining from the nose or ears.

PHYSICAL EXAMINATION

1. Cognitive Ability/Memory Function Check

The initial evaluation begins with a quick cognitive ability/memory function check. The initial evaluation can be performed on the scene of injury or during the emergency room visit.

Both cognitive testing and physical examination are recommended for at least the first 24–48 hours post-injury as outlined by Zurich guidelines (Permenter, Fernández-de Thomas, and Sherman 2022).

Cognitive testing evaluates the patient's ability in thinking, memorizing, and concentration.

Ask the following questions and record the responses. Failure to answer any one of these questions indicates a suspected concussion.
1. Which city or what room are we in right now?
2. What activity/sport/game are we playing now?
3. Is now before or after lunch/dinner?
4. What is the name of your coach?
5. What school do you go to?

1. Vital Signs

Autonomic dysregulation is common after a head injury. It may present as abnormal blood pressure and heart rates.

1. Orthostatic hypotension which it is defined as a 20 mm Hg or greater reduction in systolic blood pressure or a 10 mm Hg reduction in diastolic blood pressure after 3 minutes of standing from the supine position with or without symptoms (Matuszak et al. 2016; Haider et al. 2020).

Orthostatic vital sign procedures:

1. Patient lies supine for 2 minutes and documents the blood pressure and heart rate.
2. Patient stands and documents the blood pressure and heart rate after 1 minute and after 3 minutes.
    1. Postural orthostatic tachycardia syndrome is a condition that affects blood flow to the brain, resulting in syncope, lightheadedness, and rapid increase in heart rates. These symptoms occur when a head injury patient stands up from a reclining position and are relieved by sitting or lying back down.

A head-up tilt test (HUTT) is performed to determine the cause of the postural orthostatic tachycardia syndrome. The test shows how different positions affect the blood pressure, heart rhythm, and heart rate. The test begins with the patient in an upright position and rests 15 minutes on a supine table. The table is then tilted to 30 degrees for 2–3 minutes, 45 degrees for 2–3 minutes, and 70 degrees for up to 45 minutes, while the blood pressure and ECG are monitored throughout at different table angles. The table will be returned to the supine position and the patient stays on the table for 5–10 minutes. A negative HUTT is defined as slight fluctuations in systolic and diastolic blood pressure and in heart rate, without any other abnormalities. A positive HUTT result is characterized by a loss of consciousness following various hemodynamic patterns, increased heart rate 30 bpm in the first 10 minutes of tilting, heart rate 120 bpm in the first 10 minutes of tilting, and increased heart rate 30 bpm when isoprenaline is infused at a rate of 1 mg/min (Barón-Esquivias, Gonzalo, and Martínez-Rubio 2003).
    1. Neurological Examination
        2. A screening neurological examination is a series of procedures that helps the physician to confirm or rule out a neurological disorder or medical condition of mental status, mood, behavior, cranial nerves, motor system, sensory system, coordination, vision, hearing, speech, coordination, gait, and balance (NINDS 2019).

1. The common tests for upper motor neuron disorders include hyperreflexia such as brisk reflexes or clonus, finger rolling, pronator drift, Hoffman sign, and Babinski sign.

1. Head and Cervicothoracic Examination
    1. Spurling's test is the assessment of cervical nerve root entrapment.
    2. Cervical proprioception abnormalities may delay recovery and lead to persistent symptoms. To access cervical proprioception, common tests used in evaluation include cervical range of motion device, head reposition accuracy tests, position-matching tasks.
2. Balance/Coordination Examination
    3. In screening head injury patients with vestibular disorders, Tandem Gait is recommended (Cohen et al. 2019). The test is to ask the patient

to walk in a straight line heel to toe, with the arms down by the body. If the gait is unsteady, jerky, the test is positive, and it indicates a cerebellar disease, such as ataxia, dysarthria, nystagmus, and vertigo (Ataullah 2021). Usually, Romberg's sign, the finger-to-nose test, and the heel to shin test are applied with Tandem Gait; if the patient cannot complete the tests, it suggests a cerebellar disorder.

1. Vestibulo-Ocular Examination

In screening head injury patients with vestibular disorders, the head thrust test is an effective tool. With the patient in a sitting position, and the examiner sitting facing them, the patient gazes at the examiner's nose, without blinking for two minutes, and ensuring that their eyes don't stray from the examiner's. The examiner then holds the patient's head and moves it quickly and unpredictably to 10–15 degrees of the neck rotation. A positive test result is when the patient's gaze strays from the examiner's nose.[5]

1. Ocular/Ophthalmologic Examination

Vision difficulties or dysfunction followed by a head trauma may children returning to school or office workers whose work is mostly visual from returning to their jobs. The King–Devick test is an accurate and reliable method that requires less than 2 minutes to administer. It can evaluate eye movements, and it captures impairments of eye movement.

In this test, the patient reads numbers on 3 cards from left to right as quickly as possible, without making any errors. Errors corrected before going on to the next number are not counted (Galetta et al. 2011).

2. Imaging tests with MRI or CT are not always needed in the early evaluation of concussion; this is because most of the pathogenic structural defects in the brain are not seen on the imaging that often presents normally on neuroimaging (Than 2022). In an emergency department, physicians might order CT and MRI imaging tests if more serious issues are suspected such as brain swelling, bleeding in the skull, cervical spine or spinal cord injury, or if symptoms are worsening. Physicians also use CT and MRI in patients, whose symptoms last more than one month after the initial head injury, to rule out other etiologies.

DIFFERENTIAL DIAGNOSIS

Acute concussion needs to differentiate from other potentially severe injuries, including cervical spine injury, acute epidural hematoma, acute subdural hematoma, and intracranial hematoma.

Cervical Spine Injury

C1–C4 Injury

1. Potential paralysis to arms, hands, trunk, and legs.
2. Potential inability to breath independently, cough, or control bowel movements and/or bladder.
3. Impaired or reduced ability to speak.

C5 Injury

1. Potential paralysis to wrists, hands, trunk, and legs.
2. Potential weakened breathing.
3. Little or no voluntary control of bowel or bladder.
4. Able to raise arms and bend elbows.
5. Able to speak and use diaphragm.

C6 Injury

1. Potentially affecting the nerves on the wrist extension.
2. Potential paralysis to trunk, hands, and legs; a false case of carpel tunnel syndrome.
3. Numbness and tingling on arms, hands, and fingers.
4. Breathing may be taxed.
5. No voluntary control of bowel or bladder.
6. Able to speak.

C7 Injury

1. Has full neck movement.
2. May experience numbness and tingling on hands and fingers and a referred burning pain in the scapula and triceps.
3. Decreased dexterity on the hands and fingers.
4. No voluntary control of bowel or bladder.
5. Breathing may be taxed, but the patient usually doesn't need a ventilator.

C8 Injury

1. Potential paralysis to the legs, trunk, and hands.
2. Potentially affecting movements of the hands, finger flexion, and the forearm, but shoulder and arm movements are not affected.

Epidural Hematoma

An epidural hematoma (EDH) occurs when blood accumulates between the skull and the dura mater.

1. With or without a brief loss of consciousness. The typical pattern of the loss of consciousness indicating an epidural hematoma is a loss of consciousness

followed by a period of alertness, that may last several hours, then rapid deterioration, sometimes leaving the patient in a coma. But this pattern may NOT appear in all people.
2. Confusion.
3. Dizziness.
4. Drowsiness.
5. Seizures.
6. Nausea or vomiting.
7. Enlarged pupil in the ipsilateral eye.
8. Severe headache.
9. Weakness in part of the body, usually on the contralateral side.

Subdural Hematoma

1. With or without a loss of consciousness. The person may appear normal for days after a head injury but slowly become confused, then lose consciousness, or even go into a coma several days later.

Intracranial Hematoma
1. Persistent increasing headache.
2. Vomiting.
3. Drowsiness.
4. Dizziness.
5. Confusion.
6. Unequal pupil size.
7. Slurred speech.
8. Lethargy.
9. Seizures.
10. Paralysis on the contralateral side.
11. Unconsciousness.

TREATMENTS

Herbal Treatments
Blood Stagnation瘀血阻滯型 .

Treatment Principle
Move the Blood and disperse Bruising.

Herbal Formula
Tong Qiao Huo Xue Tang通竅活血湯
This formula is from "Correction of Errors in Medical Classics醫林改錯", authored by Wang Qingren 王清任 (1768–1831), in Qing dynasty 清朝 (1636–1912), and published in 1849.

Herbal Ingredients

Chi Shao赤芍3g, Chuan Xiong川芎3g, Tao Ren 桃仁9g, Hong Hua紅花9g, She Xiang麝香0.5g, Cong Bai老蔥15g, Sheng Jiang生薑9g, Da Zao紅棗7 pieces

Ingredient Explanations

1. Chi Shao赤芍 moves the Blood, disperses Bruising, and stops pain.
2. Chuan Xiong川芎 moves the Blood, disperses Bruising, and stops pain due to Blood stasis.
3. Tao Ren 桃仁 moves the Blood, disperses Bruising, and stops pain due to Blood stasis.
4. Hong Hua紅花 moves the Blood, disperses Bruising, and stops pain due to Blood stasis.
5. She Xiang麝香 opens orifices, revives Shen, moves the Blood, opens the 12 meridians, and stops pain.
6. Cong Bai老蔥 unblocks the pure Yang pathway, disperses Cold, induces sweat, releases the exterior, and relieves toxicity.
7. Sheng Jiang生薑 unblocks the pure Yang pathway and harmonizes rebellious Qi.
8. Da Zao紅棗 tonifies the Spleen Qi, nourishes the Blood, moderates and harmonizes the harsh properties of the fragrant herbs.

Formula Actions

1. Moves the Blood.
2. Opens the orifices.

Contraindications

• Contraindicated during pregnancy.

Modifications

1. To make the formula stronger, add Wu Ling Zhi 五靈脂, Man Jing Zi蔓荊子, Tian Ma天麻, Gou Teng鉤藤, and Da Huang大黃.
2. For severe headache, add Wu Ling Zhi 五靈脂, Man Jing Zi蔓荊子, Quan Xie全蠍, Wu Gong蜈蚣, and Tu Bie Chong土鱉蟲.

Cooking Instructions

1) Place the herbs into an herb pot. Traditionally, a clay cooker is used. Gently rinse the herbs once with a strainer.
2) Add 250 ml cooking wine to the pot.
3) Turn on the stove to high heat and bring the herbs to a boil. Turn off the heat when the decoction is reduced to 150 ml.
4) Drain the decoction into a container. Keep only the herbal wine.
5) Pour the herbal wine back into the herb pot and add She Xiang麝香.
6) Bring the herbal wine to the boiling point two times.
7) Drink the herbal wine before bed.

Qi and Blood Xu氣血虧損型

Treatment Principle
Tonify Qi and nourish the Blood.

Herbal Formula
Gui Pi Tang歸脾湯
This formula is from "Categorized Essentials of Repairing the Body 正體類要", authored by Xue Ji 薛己 (1487–1559), in Ming dynasty 明朝 (1368–1644), and published in 1529.

Herbal Ingredients
Ren Shen人參3g, Zhi Huang Qi 黃耆炙15g, Bai Zhu白朮9g, Fu Ling茯苓9g, Dang Gui 當歸9g, Suan Zao Ren酸棗仁炒9g, Long Yan Rou龍眼肉9g, Zhi Yuan Zhi遠志6g, Mu Xiang木香3g, Zhi Gan Cao甘草炙4.5g, Sheng Jiang生薑2 slices, Da Zao 大棗3 pieces

Ingredient Explanations
1. Ren Shen人參 tonifies Qi and Yang.
2. Zhi Huang Qi黃耆炙 tonifies Qi and strengthens the Spleen.
3. Bai Zhu白朮 tonifies the Spleen Qi and dries Dampness.
4. Fu Ling茯苓 drains Damp and tonifies the Spleen.
5. Dang Gui當歸 moves the Blood, disperses Bruising, and stops pain due to Blood stasis.
6. Suan Zao Ren酸棗仁 nourishes Heart Yin and tonifies Liver Blood.
7. Long Yan Rou龍眼肉 drains Damp-Heat from the Liver and Gallbladder channels.
8. Yuan Zhi遠志 expels Phlegm, calms Shen, and reduces abscesses.
9. Mu Xiang木香 promotes the movement of Qi and alleviates pain.
10. Zhi Gan Cao甘草炙 tonifies the Spleen, harmonizes the other herbs, and guides the herbs to all 12 meridians.
11. Sheng Jiang生薑 unblocks the pure Yang pathway, harmonizes rebellious Qi.
12. Da Zao紅棗 tonifies the Spleen Qi, nourishes the Blood, and moderates and harmonizes the harsh properties of the fragrant herbs.

Formula Actions
13. Nourishes the Blood.
14. Tonifies Qi.
15. Strengthens the Spleen.

Contraindications
- Contraindicated for those who are bleeding with interior Heat or insomnia.

Modifications
16. To make the formula stronger, add Shu Di Huang 熟地and E Jiao阿膠.

17. For severe insomnia and forgetfulness, add He Huan Pi 合歡皮, Ye Jiao Teng夜交藤, and Shi Chang Pu 石菖蒲.

Cooking Instructions
1) Place the herbs into an herb pot. Traditionally, a clay cooker is used. Gently rinse the herbs once with a strainer.
2) Add water into the pot until it just covers the herbs, then add one more cup of water.
3) Presoak the herbs for 15 minutes.
4) Turn on the stove to high heat and bring the herbs to a boil. Reduce the heat to medium-low to simmer but with slight bubbling. Turn off the heat when the decoction is reduced to one cup.
5) Drain the decoction into a container.
6) Repeat steps two to five one more time, and drain each decoction to the same container.
7) Divide the decoction into two separate cups.

Dose
• Drink one cup of the decoction after meals twice daily.

Kidney Essence Deficiency肾精虧損型

Treatment Principle
Tonify the Kidney and Lung, nourish the Yin, anchor the Yang, and drain Heat and Fire.

Herbal Formula
He Che Da Zao Wan河車大造丸

This formula is from the book "Analytic Collection of Medicinal Formulas醫方集解", authored by Wang Ang汪昂 (1615–1694AD), in Qing dynasty 清朝, and published in 1682.

Herbal Ingredients
Zi He Che紫河車1 piece, Gui Ban 龜板 60g, Huang Bai黃柏45g, Du Zhong杜仲 45g, Niu Xi牛膝36g, Sheng Di Huang生地黃75g, Sha Ren砂仁18g, Fu Ling茯苓60g, Tian Men Dong天冬36g, Mai Men dong麥冬36g, Ren Shen人參36g

Ingredient Explanations
1. Zi He Che紫河車 tonifies the Liver and Kidney and assists the Essence; tonifies the Qi and Blood.
2. Gui Ban 龜板 nourishes the Yin and holds down the Yang; strengthens the Kidney and the Bones; cools the Blood and stops uterine bleeding; nourishes the Heart 败龟版阴气最全.
3. Huang Bai黃柏 expels Damp-Heat in the Lower Jiao; clears Kidney Yin Deficient Heat and toxic Fire.

4. Du Zhong杜仲 tonifies Liver and Kidney, strengthens Bone, and benefits Essence and Blood.
5. Huai Niu Xi牛膝 tonifies the Liver and Kidney and directs herbs downward.
6. Sheng Di Huang生地黃 nourishes Yin and clears Heat.
7. Sha Ren砂仁 warms the Spleen and transforms Dampness, promotes the movement of Qi for Damp of the Stomach and Spleen; settles a restless fetus and stops morning sickness.
8. Fu Ling茯苓 promotes urination and drains Dampness; tonifies Spleen and Stomach; assists the Heart and calms Shen.
9. Tian Men Dong天冬 nourishes Yin of the Lungs and Kidney and expectorates Phlegm.
10. Mai Men dong麥冬 replenishes Yin Essence and promotes secretions; lubricates and nourishes the Stomach; soothes the Lung; nourishes the Heart.
11. Ren Shen人參 tonifies Qi and Yang.

Formula Actions

12. Nourishes Liver and Kidney.
13. Tonifies Kidney Yang.
14. Nourishes the Blood, Qi, and Essence.
15. Tonifies Yin and empties deficient Fire due to deficiency of Yin.
16. Regulates Chong and Ren meridians.

Caution and Contraindications

17. While taking the formula, patients should avoid greasy food.
18. This formula is taken before meals.
19. Patients who are vomiting or who have diarrhea, abdominal distension, and cough with phlegm should take the formula with caution.
20. Patients who suffer from common cold should not take the formula.
21. Patients who are pregnant should take the formula with caution.
22. Patients who suffer from hypertension, diabetes mellitus should take the formula under the guidance of the physician.
23. If symptoms persist after two weeks, the patient should stop taking the formula.
24. Keep all herbs out of the reach of children.

Modifications

25. When taking this formula in summer, add Wu Wei Zi 五味子.
26. Female patients should remove Gui Ban 龜板but add Dang Gui當歸.

Cooking Instructions

Making Di Huang paste地黃膏

1) Place Sha Ren砂仁and Fu Ling茯苓 in a cotton soup bag.
2) Place Sheng Di Huang生地黃 and the soup bag into an herb pot. Traditionally, a clay cooker is used.

3) Gently rinse the herbs once with a strainer.
4) Add enough cooking wine to the pot to just cover the herbs.
5) Turn on the stove to high heat and bring the herbs to a boil until dry, then add more cooking wine.
6) Repeat step 5 for seven times.
7) Remove the cotton soup bag and its herbs.
8) Pestle Di Huang地黃 to a paste.

Making the pills

1) Grind the herbs to a fine powder.
2) Pour the powder and Di Huang paste地黃膏 into a bowl and then mix them to a dough (avoid using a metal spoon).
3) When they start to take shape, add a little wine if needed, the dough should feel like clay or play-doh but harder, transfer it from the bowl to a clean board
4) Divide the dough evenly into 10 grams each and then divide it into 100 small pieces evenly.
5) Begin shaping each small piece dough with your hands to mold the dough to small balls the size of soybeans.

Drying the pills

1) Preheat a dehydrator with the thermostat set to 95°F to 115°F. In areas with higher humidity, temperatures as high as 125°F may be needed.
2) Place the herbal doughballs in a single layer on dehydrator trays, drying times may vary from 1 to 4 hours. Check the condition periodically to avoid over-drying. Check your dehydrator instruction booklet for specific details.

Storing the pills: https://food.ndtv.com/food-drinks/the-right-way-to-store-fresh-and-dried-herbs-expert-tips-1783908

1) Store the pills in an airtight black colored or dark container so that exposure to oxygen does not spoil them. Avoid using plastic and metal containers as they may leach chemicals and moisture into the pills; instead use glass bottles.
2) Keep these jars away from sunlight; instead store them in a dark and dry place such as refrigerator. Sunlight may ruin the property of the herbs.

Dose
Take 6 grams twice daily with warm wine in winter or with warm soup in the other seasons.
    Phlegm Damp Misting the Orifice痰浊上蒙型

Treatment Principle
1. Transform Phlegm.
2. Descend rebellious Qi.

Herbal Formula

Ban Xie Bai Zhu Tian Ma Tang半夏白朮天麻湯

This formula is from "Medical Insights 醫學心悟", authored by Cheng Guopeng 程國彭 (1662–1735), in Qing dynasty清朝 1636–1912, and published in 1732.

Herbal Ingredients

Zhi Ban Xia半夏4.5g, Bai Zhu白朮3g, Tian Ma天麻3g, Ju Hong橘紅3g, Fu Ling茯苓3g, Gan Cao甘草1.5g, Sheng Jiang生薑2pieces, Da Zao大棗3g, Man Jing Zi 蔓荊子3g

Ingredient Explanations

1. Zhi Ban Xia半夏 transforms Phlegm.
2. Tian Ma天麻 extinguishes Wind, calms the Liver, subdues rising Liver Yang, and alleviates pain.
3. Bai Zhu白朮 tonifies the Spleen Qi and dries Dampness.
4. Ju Hong橘紅 regulates Qi and transforms Phlegm.
5. Bai Zhu白朮 tonifies the Spleen Qi and dries Dampness.
6. Fu Ling茯苓 drains Damp and tonifies the Spleen.
7. Gan Cao甘草 harmonizes the herbs, tonifies Middle Jiao Qi, clears Heat, expectorates Phlegm, and stops cough.
8. Sheng Jiang生薑 unblocks the pure Yang pathway and harmonizes rebellious Qi.
9. Da Zao大棗 tonifies the Spleen Qi, nourishes the Blood, and moderates and harmonizes the harsh properties of the fragrant herbs.
10. Man Jing Zi 蔓荊子 expels Wind and stops pain.

Modifications

11. To make the formula stronger, add Shen Qu神曲, Mai Ya麥芽, Ze Xie澤瀉, Ren Shen人參, Huang Qi黃芪, Gan Jiang乾薑, and Huang Bai黃柏.
12. For severe headache, add Bai Ji Li白蒺藜.
13. For headache due to Blood deficiency, add Dang Gui當歸 and Chuan Xiong 川芎.
14. For severe Qi deficiency, add Ren Shen 人參, Dang Shen黨參, and Huang Qi黃芪.
15. For severe vertigo, add Jiang Can殭蠶 and Dan Nan Xing膽南星.
16. For arteriosclerosis in the brain, add Gou Teng鉤藤.
17. For Liver Yang rising, add Gou Teng鉤藤 and Dai Zhe Shi代赭石.
18. For severe dizziness due to Liver excess, add Jiang Can殭蠶, Gou Teng鉤藤, and Dan Nan Xing膽南星.

Formula Actions

1. Tonifies Spleen Qi.
2. Dissolves Dampness.
3. Transforms Phlegm.

4. Soothes the Liver.

5. Expels Wind.

Contraindications

- Contraindicated for those with interior Heat due to Kidney and Liver Yin deficiency or dizziness and vertigo due to Qi and Blood deficiency.

Cooking Instructions

1) Place the herbs into an herb pot. Traditionally, a clay cooker is used. Gently rinse the herbs once with a strainer.
2) Add water into the pot just enough to cover the herbs, then add one more cup of water.
3) Presoak the herbs for 15 minutes.
4) Turn on the stove to high heat and bring the herbs to a boil. Reduce the heat to medium-low to simmer but with slight bubbling. Turn off the heat when the decoction is reduced to one cup.
5) Drain the decoction into a container.
6) Repeat steps two to five one more time, and drain each decoction to the same container.
7) Divide the decoction into two separate cups.

Dose

- Drink one cup of the decoction after meals twice daily.

Liver Yang Uprising肝陽上亢型

Treatment Principle

Soothe the Liver, suppress Yang, nourish Yin, and clear Heat.

Herbal Treatments

Herbal Formula

Tian Ma Gou Teng Yin 天麻鉤藤飲

This formula is in the book "Newly Explained Diagnosis and Treatment of TCM Internal Medicine and Miscellaneous Diseases中醫內科雜病證治新義", authored by Hu Guangci 胡光慈 in 1958.

Herbal Ingredients

Tian Ma天麻6g, Gou Teng鉤藤9g, Chuan Niu Xi牛膝12g, Shi Jue Ming石決明15g, Du Zhong杜仲12g, Huang Qin黃芩6g, Zhi Zi栀子6g, Yi Mu Cao益母草9g, Sang Ji Sheng桑寄生9g, Ye Jiao Teng夜交藤9g, Fu Shen茯神9g

Ingredient Explanations

1. Tian Ma天麻 extinguishes Wind, calms the Liver and subdues rising Liver Yang, and alleviates pain.

2. Gou Teng鉤藤 expels Wind and releases spasm.
3. Chuan Niu Xi川牛膝 unblocks and promotes Blood circulation.
4. Yi Mu Cao益母草 moves the Blood, promotes urination, reduces swelling, and clears Heat.
5. Shi Jue Ming石決明 drains Liver fire and expels Wind.
6. Du Zhong杜仲 tonifies Liver and Kidney, strengthens Bones, and benefits Essence and Blood.
7. Huang Qin黃芩 clears Heat, dries Damp, cools the Blood, stops bleeding, and calms Liver Yang.
8. Zhi Zi栀子 clears Heat, cools the Blood, resolves Damp, reduces swelling, and opens the meridians.
9. Sang Ji Sheng桑寄生 tonifies Liver and Kidney, strengthens Bones, benefits Essence and Blood; expels Wind and drains Damp.
10. Ye Jiao Teng夜交藤 nourishes Heart Yin and Blood, calms Shen, expels Wind and Damp.
11. Fu Shen茯神 calms Shen.

Modifications

- To make the formula stronger, add Ju Hua菊花, Xia Ku Cao夏枯草, Ku Ding Cha苦丁茶, and Zhen Zhu Mu珍珠母.
- For severe Liver Heat, add Xia Ku Cao夏枯草 and Long Dan Cao龍膽草.
- For Blood stasis, add Dan Shen丹參 and Chi Shao赤芍.
- For insomnia, add Long Gu龍骨, Mu Li牡蠣, Ye Jiao Teng夜交藤, and Yuan Zhi遠志.

Formula Actions

- Soothes Liver Qi, expels Wind, moves the Blood, clears Heat, tonifies the Liver and Kidney.

Formula Indication

- Headache, vertigo, insomnia, anger, bitter taste in the mouth due to Liver Yang uprising.

Contraindications

- Caution for patients who suffer from Blood, Yin, and body Fluid deficiency.
- Stop taking this formula if there is dizziness, chest distention, shortness of breath, nausea, vomiting, or skin itch.
- Do not take Tian Ma overdose.

Cooking Instructions

1) Place the herbs into an herb pot. Traditionally, a clay cooker is used. Gently rinse the herbs once with a strainer.
2) Add water into the pot until it just covers the herbs, then add one more cup of water.
3) Presoak the herbs for 15 minutes.

4) Turn on the stove to high heat and bring the herbs to a boil. Reduce the heat to medium-low to simmer but with slight bubbling. Turn off the heat when the decoction is reduced to one cup.

5) Drain the decoction into a container.

6) Repeat steps two to five one more time and drain each decoction to the same container.

7) Divide the decoction into two separate cups.

Dose

- Drink one cup of the decoction after meals twice daily.

Acupuncture Treatments

Acupuncture Point Formula

Local Points

In the coma period, needling DU26 to promote awakening.

Acupuncture point formula

SJ5外關 Waiguan + GB20風池Fengchi, LI4合谷Hegu+ DU23上星Shangxing, DU15 啞門Yamen + ST36足三里Zusanli, SI3後溪Houxi + UB60崑崙Kunlun, SJ9四瀆Sidu, Tai Yang太陽, LU7列缺Lieque, LI11曲池Quchi, KD1涌泉Yongquan, LI4合谷Hegu

Modifications

For persistent hiccups, add REN12中脘Zhongwan.

For severe insomnia, add HT7神門Shenmen and SP6三陰交Sanyinjiao.

Needling Technique

Even

Treatment Frequency

Once daily

1. **Tui Na Treatments**

Tui Na Techniques

Rubbing, grasping, pushing

Tui Na Procedures

Step One

1. Patient is in sitting position; the therapist stands behind.

Rubbing Techniques

2. Rubbing region: the sternocleidomastoid (SCM) muscle.
3. Place the second to the fifth fingers on both hands on the SCM muscle.

4. Rub the muscles with the pads of the fingers 160 times per minutes for 7–10 times.

Grasping Techniques

1. Points: both sides GB19腦空Naokong and GB20風池Fengchi.
2. Apply one hand to grasp both points with the thumb tip and the index fingertip approximately 20 times on each point.

Grasping Techniques

3. Grasping region: occipital region.
4. Place one hand on the patient's forehead; place the other hand on the patient's occipital region.
5. Apply grasping with all fingertips on both sides of the occiput for 3–5 times.

Step Two

1. Patient is in sitting position; the therapist stands in front and faces the patient.

Pushing Techniques

2. Points: Yin Tang印堂, LI20迎香Yingxiang, SJ20角孫Jiaosun. Figure 2.3.1

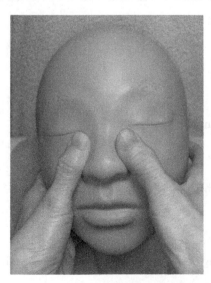

3. Pushing the points with the thumb pad. Pushing Yin Tang印堂 upward toward the hairline, pushing LI20迎香Yingxiang downward along the ala nasi, and pushing SJ20角孫Jiaosun toward the occipital region. Figure 2.3.2

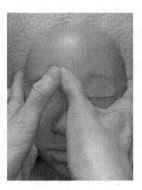

4. Repeat for 7–10 times.

PATIENT ADVISORY

1. Foods that are high in phosphorus, such as fish and eggs, are good for brain edema.
2. Foods that are good for brain injury include skim milk, soymilk, noodles, vegetable oil, cane sugar, and black fungus,
3. Post-concussion patients are recommended to avoid fatty foods, high cholesteral foods, and caffeine, such as coffee, tea, etc.
4. Avoid cold and raw foods such as cold water or drink, munkbean, winter melon, bittermelon, and water chestnuts.
5. Avoid spicy foods, such as peppers, mustard, and chives.

TEMPOROMANDIBULAR JOINT DISORDERS 顳頜關節功能紊亂癥

INTRODUCTION

Among 360 joints in the adult human body, the temporomandibular joint (TMJ) is the busiest joint, opening and closing approximately 1500–2000 times a day (Saito et at. 2009).

The temporomandibular joint is where the jawbone connects to the head bone. The joint allows rotational movements when opening and closing the mouth. The movements are constrained by various passive tension of the soft tissues such as the ligaments and muscles.

Pain that occurs at the temporomandibular joint (TMJ) can cause lifestyle limitation such as in eating, chewing, yawning, laughing and speaking, and even breathing.

ANATOMY

The temporomandibular joint is a synovial joint located anterior to the ear tragus on both sides of the face. It is one of the most complex joints in the whole body regardless of its anatomical structure or the physiological functions. And it is the only joint in the whole body that performs the movements simultaneously.

The TMJ has a rotational movement in the sagittal plane and a translation movement on its own axis. Every point of the moving object simultaneously has the same direction and velocity: when the mandible moves forward, the teeth, condyles, and rami all move in the same direction and to the same degree of direction and speed.[12]

The joint is formed by the condylar process and the glenoid fossa of the temporal bone of the zygomatic arch. There are the joint capsule, articular disc, mandibular condyles, articular surface of the temporal bone, temporomandibular ligament, stylomandibular ligament, sphenomandibular ligament, and lateral pterygoid muscle.

The TMJ is the only joint in the human body that can be dislocated without being injured, and the joint capsule does not tear during dislocation. The capsule is inferior to the mandibular fossa, posterior to the articular tubercle, and superior to the condyle of the mandible.

The TMJ is one of the few synovial joints in the human body with an articular disc. The dense fibrocartilaginous articular disc is located between the head of the mandibular condyle and the glenoid fossa of the temporal bone. The posterior portion of the disc splits in the vertical dimension; the area between the split continues posteriorly and is called "retrodiscal tissue". Unlike the disc, the retrodiscal tissue is vascular and innervated. The disc gets its nutrients from the surrounding synovial fluid because it is avascular and lacks innervation.

Ligaments that support the joint are the temporomandibular, sphenomandibular, and stylomandibular ligaments. The temporomandibular ligament is the major ligament among the three ligaments and is a thickened lateral portion of the articular capsule. The base of this fan-shaped ligament is attached to the zygomatic process of the temporal bone and the articular tubercle; its apex is fixed to the lateral side of the neck of the mandible. It is responsible for synchronizing the condyle and articular disc and prevents excessive retraction or backward movement of the mandible. The other two ligaments are involved in controlling the mandibular movements but are not directly attached to any part of the joint.

The mandibular nerve is a terminal branch of the trigeminal nerve. It is a sensory innervation. The motor innervation is to the muscles. Other nerves are the facial nerves (CN VII), C1, C2, and C3.

The joint's arterial blood supply is mainly provided by the superficial temporal branch of the external carotid artery.

The temporalis and masseter muscles are responsible for elevating the mandible; the lateral pterygoid muscle is responsible for protracting the mandible, depressing the chin, and lateral deviation of the mandible; the medial pterygoid muscle is responsible for aiding the masseter muscle to elevate mandible; the platysma and the suprahyoid muscles (digastric, stylohyoid, geniohyoid, and mylohyoid muscles) are responsible for depressing the mandible.

The TMJ is one of the few synovial joints in the human body with an articular disc, another being the sternoclavicular joint.

Translation occurs within the superior cavity of the joint between the superior surface of the articular disc and the inferior surface of the articular fossa (i.e., between the disc-condyle complex and the articular fossa).

WESTERN MEDICINE ETIOLOGY AND PATHOLOGY

Stress Factors

Physical stress, mental stress, and emotional stress can lead to the overuse of the muscles of mastication (such as clenching or grinding teeth), triggering the onset or aggravation of TMJ dysfunction.[13]

Traumatic Factors

1. Inflammatory Conditions
   1. Inflammation is often caused by trauma. The retrodiscal tissue is the common site of severe pain when it is inflamed.
2. Internal Derangement
   2. Internal derangement is caused by the anatomic structural changes within the joint. This can be caused by bruxism, excessive chewing of hard food, dragging heavy objects with the teeth beyond physiological limits, direct trauma, and long-term unilateral chewing.
   3. The articular disc can become displaced by itself; commonly it will displace anteriorly. The disc will be pushed anteriorly when opening. If the disc reduces itself, it causes an audible, palpable click; if the displaced disc cannot reduce itself and is in a locked position, it causes pain, and the jaw will not produce an audible click even patients sense a click.
3. Arthritis
   4. When a plain x-ray reveals a flattening of the condylar head, often it is degenerative arthritis; if it is not age-related degeneration, it commonly is a secondary to trauma occurring at a younger age.[14]
4. Muscle Spasm
   5. Prolonged dental procedures, in which the mouth is held open for extended periods, affect the muscles of mastication – especially the masseter, temporalis, and the pterygoid muscles – leading to muscle spasm.
   6. Sudden cold irritation, such as using an ice pack on the jaw, decreases the temperature, putting extra stress on the jaw muscles.
5. Fractures
   7. A direct blow to the jaw or a fall onto the chin can lead to a fracture on the mandibular symphysis or the condylar neck. The fracture often is combined with a dislocation of one or both condyles.

CHINESE MEDICINE ETIOLOGY AND PATHOLOGY

1. Liver Qi Stagnation 肝氣鬱結

In Chinese philosophy, Qi is a life force that brings the blood to flow to all organs and parts of the body through meridians to maintain health and balance.

Stress is the most common cause of blockages in the meridians. The flow of Liver Qi is stagnant. Liver Qi stagnation results in anger, frustration, depression, sadness, and stress. And emotional states can reversely cause Liver Qi to stagnate further and lead to other symptoms.

Bruxism and jaw clenching are symptoms of stress. Emotional stress can lead to overuse of the muscles of mastication resulting in misalignment of the jaw.

2. Local Qi and Blood Stagnation 氣滯血瘀

Patients usually have a history of trauma, infection, or inflammation.

1. Trauma: a direct blow to the jaw such as the motor vehicle accidents, a slip and fall, or sports injuries leading to a broken or dislocated jaw.
2. Infection: an infected salivary gland can lead to a temporomandibular disorder (TMD). The swelling around a wisdom tooth infection that prevents the jaw from opening can also lead to TMD.
3. Inflammation: both rheumatoid arthritis (RA) and osteoarthritis (OA) can break down the cartilage of the TMJ over time, reducing jaw movement support in the joint socket.

The adhesions and scarring that occur in trauma, strains, sprains, bruises, inflammation, and infection are Blood stagnation which can lead to a decrease in local Blood circulation.

Mental stress can lead to mastication muscle tension, which slows Blood circulation around the jaw and can result in Qi and Blood stagnation.

3. Wind-Damp-Cold Bi Syndrome 風濕寒痺

The cold temperature in the cold weather brings on the teeth chattering and shivering, the jaw becomes tighter. The seasonal stress from being in the short daytime, less sunlight, cold, and rain increase the stagnation of Liver Qi. Wind that is the elemental factor of Liver stirs up the Cold and Damp to block meridians flowing to the jaw, blockage manifests pain. Tendon (the soft tissues around the jaw) that is the tissue of Liver becomes tighter and manifests pain.

4. Wind-Damp-Heat Bi Syndrome 風濕熱痺

It is common called Heat Bi Syndrome. If a patient has excess Yang by nature or has underlying Yin deficiency that accumulates deficient Fire, then invasion by Wind-Damp-Cold pathogenic factors allows the body constitution to transfer the factors into the Heat type of Bi Syndrome. Wind Damp Heat block Qi and Blood in the jaw joint, leading to stagnation.

The stagnation is combined with three components: Wind, Damp, and Cold. Different patient's symptoms may manifest differently depending on the person's body constitution.

5. Liver and Kidney Yin Deficiency 肝腎陰虛

Yin represents the moistening and nourishing fluids of the body, which are responsible for the lubrication of joints and muscles. The major clinical manifestation of Yin deficiency is dryness in the jaw joint. Yin deficiency and the dryness can thicken the body fluid; therefore, Blood is difficult to flow to the jaw joint leading to Blood deficiency or stasis and resulting in pain.

Kidney Yin deficiency is caused by:

1. Protracted illnesses in any organ system will impair Kidney.
2. Intemperate sexuality.
3. Multiple pregnancies.
4. Loss of large quantities of blood (such as prolonged or very heavy menstrual periods).
5. Drug abuse, such as taking the herbs that are hot in nature in large quantities for a long time.

Liver Yin deficiency is caused by:

1. Excessive physical exertion.
2. Long term Liver Blood deficiency, such as caused by long hours staring at an electronic monitor or cellphone.
3. Long-term Kidney Yin deficiency.
4. Prolonged or very heavy menstrual periods.
5. High fever.
6. Emotional attack such as anger, depression, sadness, grieving, and anxiety.
7. Multiple childbirths.

Kidney and Liver in Chinese medicine have the same source, as they mutually promote each other. The element of Kidney is water and the element of Liver is wood; insufficiency of the water will fail to nourish the wood causing insufficiency of Liver Yin and leading to hyperactivity of Liver Yang; therefore, the hot in nature injures Kidney Yin.

MANIFESTATIONS

1. Liver Qi Stagnation 肝氣鬱結

Inappropriate anger, frustration, anxiety, depression, melancholy, tendency to sigh, headaches (especially temporal headaches if the Gallbladder is involved), facial muscle tension, chest distention, hypochondriac pain, earache, waking up between 1 a.m. and 3 a.m., and tinnitus. Women: irregular menses, dysmenorrhea, breast distention, and pain before and during menses. Tongue: normal body; thin, white coating; distended slim sublingual veins. Pulse: wiry.

2. Local Qi and Blood Stagnation 氣滯血瘀

Patients usually have a history of trauma, infection, and inflammation. The injury can be a direct blow to the jaw. Symptoms are bruising, swelling facial pain, ear pain, tinnitus, deviated jaw, and limitation of movement.

Tongue: purple tongue body, may have bruise marks; thin, white coating; distended ipsilateral sublingual vein like an earthworm.

Pulse: choppy or wiry.

3. Wind-Damp-Cold Bi Syndrome 風濕寒痹

Pain made worse when exposed to cold; pain is better with warmth; heavy sensation of the jaw joint, and body ache.

Tongue: normal body; white coating, thick and white coating.

Pulse: superficial and tight and/or wiry and slow

Wind symptoms行痹: acute onset, pain that changes position, the pain intensity increases when exposed to wind or cold; patients may have fever or slight chills.

Tongue: thin, white or thin greasy.

Pulse: floating, tight.

Damp symptoms着痹: jaw joint swelling, severe pain, difficulty moving the jaw, numbness on the skin, pain is worse when exposed to rain, wind, cold, and darkness.

Tongue: white, greasy coating.

Pulse: slow, deep, soft, and thin.

Cold symptoms痛痹: severe pain, pain location is more fixed but can be wandering, difficulty moving the jaw; it is worse when exposed to cold, better when exposed to warm. The skin color is not red, and it is not hot when touched.

Tongue: thin, white coating.

Pulse: wiry, tight.

4. Wind-Damp-Heat Bi Syndrome 風濕熱痹

Pain is worse when exposed to cold temperature and it is better with warmth, numbness, heaviness of the jaw joint, body aches, and difficulty moving the jaw.

The stagnation is a combination of three components: Wind, Damp, and Cold. Different patients' symptoms may manifest differently depending on the person's body constitution.

Wind symptoms行痹: acute onset, pain that changes position, the pain intensity increases when exposed to wind or cold; patients may have fever or slight chills.

Tongue: thin, white, or thin, greasy coating.

Pulse: floating, tight.

Damp symptoms着痹: jaw joint swelling, severe pain, difficulty moving the jaw, numbness on the skin, pain is worse when exposed to rain, wind, cold, and darkness.

Tongue: white, greasy coating.

Pulse: slow, deep, soft, and thin.

Heat symptoms痛痹: severe pain, pain location is more fixed but can be wandering, difficulty moving the jaw, it is worse when exposed to cold, better when exposed to warm. The skin color is not red, and it is not hot when touched.

Tongue: thin, white coating.

Pulse: wiry, tight.

5. Liver and Kidney Yin Deficiency 肝腎陰虛

In both Liver and Kidney Yin deficiency, there is pain in the jaw joint, difficulty in opening and closing the mouth, loose teeth, slightly red cheeks, dizziness, dry painful throat, Five-Center-Heat (hot palms, soles, and chest), afternoon fever, night sweats, irritability, insomnia, vivid dreams, seminal emission, irregular or scanty menses or amenorrhea, and vaginal discharge.

Liver Yin Deficiency

Dry eyes, blurred vision, floaters in the eyes; hypochondriac pain, numbness of limbs, Wei syndrome, limb weakness.

Kidney Yin Deficiency

Tinnitus, deafness, forgetfulness, sore and painful low back and knees, dry stool, and constipation.

Tongue: red or pale red body, peeled coating.

Pulse: thready and rapid.

PHYSICAL EXAMINATION

1. Begin with obtaining medical history.
   1. General health conditions such as the Shen, cheek color, and any previous fractures or trauma, dental history, bruxism, or rheumatoid arthritis.
   2. Injury history to the temporomandibular joint, such as car accident, falls, or direct blows to the joint.
   3. Related symptoms such as earache, headaches, migraines, neck pain, and dizziness.
   4. Career activities that may be prone to injury, such as football players and boxers.
   5. Sports injury that falling and directly attack the temporomandibular joint.

   a. Inspection

1. Observe the region to rule out swelling, erythema, facial deformation, facial symmetry; the mandible often deviates toward the affected side during opening when it is dislocated.
2. Ask patients to slowly open and close the mouth to observe the difficulty of the movements and listen for the audible clicking noise or joint crepitus;

also observe the timing of locking and the level of irritability to determine
the severity of symptomology. The clicking noise usually occurs at the
beginning of opening the mouth and at the end of closing the mouth, but
sometimes it occurs at the end of opening the mouth and the beginning of
closing the mouth. There is a discomfort with the clicking, the mouth opening and closing may be limited with it.

3. Measure the distance between the incisal edges of upper and lower anterior
   teeth when opened; an opening of less than 35 mm is considered abnormal
   in an adult. There is no upper limit of normal but few patients can exceed
   60 mm comfortably (Meyer 1990).
   a. Palpation

1. Palpate directly over the joint for tenderness while the patient opens and
   closes the mandible joint.
2. Skin temperature, such as warm to touch.
3. Palpable clicking when opening or closing the mouth.
1. Range of Motion
   1. Compare the motions of the joint anatomy on both left and right joint in
      active movement.

1. Special Tests
   a. Tongue Blade Test

1. The patient is seated, the therapist stands in front.
2. Place a wooden tongue depressor in the patient's mouth between the teeth
   then ask the patient to bite it tightly.
3. The therapist tries to twist the tongue blade. The test is positive if the blade
   is twisted and broken.
      X-rays or CT, MRI may be ordered to rule out mandibular fracture and
   to display the TMJ anatomic structures.

DIFFERENTIAL DIAGNOSIS

Temporomandibular Joint Disorders need to be differentiated from migraine because
they both have pain unilaterally.

1. The pain quality of migraine headache is pulsating.
2. Migraine headache may be accompanied by photophobia and phonophobia
   and may present with nausea and/or vomiting.

Temporomandibular Joint Disorders need to be differentiated from mandible dislocation because the pain can be at the TMJ, and palpation may reveal tenderness over
the location of the TMJ.

1. Mandible dislocation presents with jaw deviation in unilateral dislocation.
2. The patient's mouth will often be slightly open.
3. If the dislocation occurs as a superior dislocation, the trigeminal nerve (fifth cranial nerve), facial nerve (seventh cranial nerve), and the vestibulocochlear nerve (eighth cranial nerve) may be damaged (Hillam and Isom 2021).

TMD needs to be differentiated from mandible fracture because both medical conditions may have an injury history of a traumatic force directly to the TMJ. The pain in both conditions can be at the TMJ, both presenting with redness, swelling, hot sensation, and possibly numbness.

1. Chin deviation with mandibular opening and closing in unilateral mandibular fracture.
2. Mandible fracture presents with teeth misalignment.
3. Bleeding from the mouth.
4. Ecchymosis of the floor of the mouth is commonly seen in mandibular fractures in the acute stage (Pickrell, Serebrakian, and Maricevich 2017).

TREATMENTS

1. Liver Qi Stagnation肝氣抑鬱型

Treatment Principle
Move Liver Qi and stop pain.

Herbal Formula
Xiao Yao San逍遙散
This formula is from "Prescriptions of the Bureau of Taiping People's Welfare Pharmacy 太平惠民和劑局方", authored by Imperial Medical Bureau 太醫局, in southern Song dynasty南宋 (1127–1239), and published in 1134.

Herbal Ingredients
Chai Hu柴胡10g, Dang Gui當歸10g, Bai Shao白芍10g, Bai Zhu白朮10g, Fu Ling茯苓10g, Zhi Gan Cao甘草5g, Sheng Jiang生薑10g, Bo He 薄荷5g

Ingredient Explanations
5. Chai Hu 柴胡 relieves Liver Qi stagnation, pacifies the Liver, relieves the Shao Yang, and reduces fever.
6. Dang Gui 當歸 nourishes the Blood, regulates menses, and invigorates and harmonizes the Blood.
7. Bai Shao 白芍 calms Liver Yang, alleviates pain, nourishes the Blood, and regulates menstruation.
8. Bai Zhu 白朮 tonifies the Spleen and tonifies Qi.
9. Fu Ling 茯苓 strengthens the Spleen and harmonizes the Middle Jiao.

10. Zhi Gan Cao 甘草 tonifies Spleen Qi, stops pain, and harmonizes the other herbs.
11. Bo He 薄荷 relieves Liver Qi stagnation, disperses stagnant Heat, and enhances Chai Hu's ability to relieve the Liver.
12. Sheng Jiang 生薑 harmonizes and prevents rebellious Qi and normalizes the flow of Qi at the center.

Modifications

1. To make the formula stronger, add Xiang Fu香附, Yu Jin鬱金, and Chen Pi 陳皮.
2. For Liver Fire, Add Mu Dan Pi丹皮 and Zhi Zi栀子 to clear Heat and cool the Blood.

Formula Actions

3. Soothes the Liver.
4. Releases stagnation.
5. Strengthens Spleen
6. Harmonizes Liver and Spleen.

Cooking Instructions

1. Grind the herbs into granules.
2. Mix 6 grams in one cup water.
3. Boil the herbs with 0.7 cup water.

Dose

Drink the decoction when it is warm, twice daily.

2. Local Qi and Blood Stagnation 淤血痹阻型

Treatment Principle

Tonify and move the Blood, soothe the Liver, regulate Qi, and stop pain.

Herbal Formula

Tao Hong Si Wu Tang桃紅四物湯

This formula is from "Imperially Commissioned Golden Mirror of the Orthodox Lineage of Medicine醫宗金鑑", authored by the Qianlong emperor 乾隆 (1711–1799), in Qing dynasty 清朝 (1636–1912), and published in 1742.

Herbal Ingredients

Tao Ren桃仁 25 pieces, Hong Hua 紅花2.4g, Chuan Xiong 川芎3g, Dang Gui 當歸 3g, Bai Shao白芍3g, Shu Di Huang熟地黃2.4g

Ingredient Explanations

1. Tao Ren桃仁 breaks up stubborn Blood stasis.
2. Hong Hua紅花 moves the Blood, releases Bruising, opens the meridians, and stops pain.

3. Chuan Xiong川芎 moves the Blood and Qi, expels Wind, and stops pain.
4. Bai Shao白芍 nourishes the Blood, preserves Yin, calms Liver Yang and Liver Wind, softens Tendons, and alleviates spasm and pain.
5. Dang Gui當歸 moves and tonifies the Blood, disperses Cold, and stops pain due to Blood stasis.
6. Shu Di Huang熟地黃 nourishes the Blood and nourishes Liver and Kidney Yin.

Formula Action

1. Tonifies and moves the Blood, breaks up stubborn Blood stasis, and regulates the Blood circulation.

Modifications

- To make the formula stronger, add Wu Wei Xiao Du Yin五味消毒飲.
- For Qi deficiency, add Ren Shen人參 and Huang Qi黃耆.

Formula Indication

- Temporomandibular joint disorders due to local Qi stagnation and Blood stasis.

Contraindications

- Caution for patients who are pregnant.
- Caution for patients with constitutional Spleen Yang deficiency.

Cooking Instructions

1. Place the herbs into an herb pot. Traditionally, a clay cooker is used. Gently rinse the herbs once with a strainer. You may notice some red flowers in the strainer; place them back into the herb pot.
2. Add water into the pot until it just covers the herbs, then add one more cup of water.
3. Presoak the herbs for 15 minutes.
4. Turn on the stove to high heat and bring the herbs to a boil. Then reduce the heat to medium-low to simmer but with slight bubbling. Turn off the heat when the decoction is reduced to one cup.
5. Drain the decoction into a container.
6. Repeat steps 2–5 for two times and drain each decoction to the same container.
7. Divide the decoction into three separate cups.

Dose

Drink the decoction one cup, three times a day.

3. Wind-Damp-Cold Bi Syndrome風寒侵襲型

Treatment Principle
Expel Wind, disperse Cold, opens the meridians, and stop pain.

Herbal Treatments
Herbal Formula
Juan Bi Tang蠲痹湯
This formula is from "A Book of Formulas to Promote Well-Being 嚴氏濟生方", authored by Yan Yonghe 嚴用和 (1200~1268), in Song dynasty 南宋 (1127–1239), and published in 1253.

Herbal Ingredients
Dang Gui當歸9g, Chi Shao赤芍9g, Jiang Huang薑黃9g, Huang Qi黃耆9g, Qiang Huo羌活9g, Gan Cao甘草3g, Sheng Jiang生薑15g, Da Zao大棗3 pieces

Ingredient Explanations
8. Dang Gui當歸 nourishes the Blood, benefits the Liver, and regulates menstruation.
9. Chi Shao赤芍 moves the Blood, disperses Bruising, and stops pain.
10. Jiang Huang薑黃 moves the Blood, opens meridians, expels Wind, and reduces swelling.
11. Huang Qi黃耆 tonifies Qi, strengthens the Spleen, raises the Yang Qi of the Spleen and Stomach, tonifies Wei Qi, stabilizes the exterior, and tonifies the Blood.
12. Qiang Huo羌活 expels Wind-Cold-Dampness, unblocks painful obstruction, and alleviates pain.
13. Gan Cao甘草 harmonizes the herbs, tonifies Middle Jiao Qi, clears Heat, expectorates Phlegm, and stops cough.
14. Sheng Jiang生薑 unblocks the pure Yang pathway and harmonizes rebellious Qi.
15. Da Zao大棗 tonifies the Spleen Qi, nourishes the Blood, and moderates and harmonizes the harsh properties of the fragrant herbs.

Modifications
- For Cold symptoms, add Fu Zi附子.
- For Damp symptoms, add Cang Zhu蒼朮, Fang Ji防己, and Yi Yi Ren薏苡仁.
- For Wind symptoms, add Fang Feng防風.

Formula Actions
- Nourishes and moves the Blood, drains Damp, and expels Wind.

Formula Indication
- Temporomandibular Joint Disorders due to Wind-Damp-Cold Bi Syndrome.

Cooking Instructions

1) Place the herbs into an herb pot. Traditionally, a clay cooker is used. Gently rinse the herbs once with a strainer.
2) Add water into the pot until it just covers the herbs, then add one more cup of water.
3) Presoak the herbs for 15 minutes.
4) Turn on the stove to high heat and bring the herbs to a boil. Reduce the heat to medium-low to simmer but with slight bubbling. Turn off the heat when the decoction is reduced to one cup.
5) Drain the decoction into a container.
6) Repeat steps 2–5 one more time and drain each decoction to the same container.
7) Divide the decoction into two separate cups.

Dose

• Drink one cup of the decoction after meals twice daily.

4. Wind-Damp-Heat Bi Syndrome風熱侵襲型

Treatment Principle

Expel Wind, clear Heat, open the meridians, and stop pain.

Herbal Treatments

Herbal Formula

Xuan Bi Tang宣痹湯

This formula is from "Systematic Differentiation of Warm Pathogen Diseases 溫病條辨", authored by Wu Tang 吳瑭 (1758–1836), in Qing dynasty 清朝 (1636–1912), and published in 1813.

Herbal Ingredients

Fang Ji防己15g, Xing Ren杏仁15g, Hua Shi滑石15g, Lian Qiao連翹9g, Zhi Zi梔子9g, Yi Yi Ren薏苡仁15g, Ban Xia半夏9g, Can Sha蠶沙9g, Chi Xiao Dou赤小豆 9g

Ingredient Explanations

1. Fang Ji防己 promotes urination and reduces edema, especially in the lower body; expels Wind-Dampness and alleviates pain.
2. Xing Ren杏仁 stops cough, calms wheezing, moistens the Intestines.
3. Hua Shi滑石 clears Phlegm Heat and Heat from Urinary Bladder promotes urination, and drains Damp.
4. Lian Qiao連翹 clears Upper Jiao Heat.
5. Zhi Zi梔子 clears Heat, cools the Blood, resolves Damp, reduces swelling, and opens the meridians.
6. Yi Yi Ren薏苡仁 promotes urination, leaches out Dampness, strengthens the Spleen, expels Wind-Dampness, and clears Damp-Heat.

7. Ban Xia半夏 transforms Phlegm.

8. Can Sha蠶沙 expels Wind, drains Damp, and harmonizes the Stomach.

9. Chi Xiao Dou赤小豆 clears Damp Heat, reduces swelling, and promotes urination.

Modifications

- For severe pain, add Jiang Huang 薑黃 and Hai Tong Pi海桐皮.

Formula Actions

- Clears and transforms Heat, opens meridians, and stops pain.

Formula Indication

- Temporomandibular joint disorders due to Wind-Damp-Heat Bi Syndrome.

Cooking Instructions

1) Place the herbs into an herb pot. Traditionally, a clay cooker is used. Gently rinse the herbs once with a strainer.

2) Add 1.6 liters water into the pot.

3) Presoak the herbs for 15 minutes.

4) Turn on the stove to high heat and bring the herbs to a boil. Reduce the heat to medium-low to simmer but with slight bubbling. Turn off the heat when the decoction has reduced to 600 milliliters.

5) Drain the decoction into a container.

6) Divide the decoction into three separate cups.

Dose

- Drink one cup of the decoction after meals three times daily.

5 Liver and Kidney Yin Deficiency肝腎陰虛型

Treatment Principle

Tonify Liver and Kidney Yin, clear Heat, and extinguish Wind.

Herbal Formula

Liu Wei Di Huang Wan六味地黃丸

This formula is from "Key to Therapeutics of Children's Diseases 小兒药证直诀", authored by Qian Yi 钱乙 (1032–1113), a physician of the Song dynasty 北宋 (960–1127), and published in 1119–1125.

Herbal Ingredients

熟地黃24g, 山茱萸12g, 山藥12g, 澤瀉9g, Fu Ling茯苓9g, 牡丹皮9g

Ingredient Explanations

1. Shu Di Huang熟地黃 tonifies Liver and Kidney Yin and benefits Essence and Blood.

2. Shan Zhu Yu山茱萸 tonifies Liver and Kidney, strengthens Bone, benefits Essence and Blood.
3. Shan Yao山藥 tonifies the Kidney and benefits the Yin and the Spleen Qi.
4. Ze Xie澤瀉 drains Damp, especially Damp Heat in Lower Jiao, and clears Kidney deficient Heat.
5. Fu Ling茯苓 drains Damp and tonifies Spleen.
6. Mu Dan Pi牡丹皮 clears Heat, cools the Blood, and moves the Blood.

Modifications

For Liver Blood deficiency, add Dan Gui當歸 and Bai Shao白芍.
7. For frequent urination, remove Ze Xie澤瀉 and add Yi Zhi Ren益智仁 and Fu Pen Zi覆盆子.
8. For pain in the low back and knees, add Niu Xi牛膝, Du Zhong杜仲, and Sang Ji Sheng桑寄生.

Formula Actions
- Nourishes Liver and Kidney Yin.

Formula Indication
- Migraine due to Liver and Kidney Yin deficiency.

Contraindications
- It is contraindicated for patients who suffer from headache caused by Yang deficiency.
- Caution for patients who suffer from diarrhea and indigestion due to Spleen deficiency.

Cooking Instructions
1) Place the herbs into an herb pot. Traditionally, a clay cooker is used. Gently rinse the herbs once with a strainer.
2) Add water into the pot until it just covers the herbs, then add one more cup of water.
3) Presoak the herbs for 15 minutes.
4) Turn on the stove to high heat and bring the herbs to a boil. Then reduce the heat to medium-low to simmer but with slight bubbling. Turn off the heat when the decoction has reduced to one cup.
5) Drain the decoction into a container.
6) Repeat steps 2–5 one more time, and drain each decoction to the same container.
7) Divide the decoction into two separate cups.

Dose
- Drink one cup of the decoction about 15 minutes after meals twice daily.

Acupuncture Treatment

Basic Points

GB2聽會Tinghui, GB12完骨Wangu, ST6頰車Jiache, ST7下關Xiaguan, SJ17翳風 Yifeng

Modifications

1. For Liver Qi stagnation, add LI4合谷Hegu, LIV3太沖Taichong, GB34陽 陵泉Yanglingquan, SJ5外關Waiguan, GB41足臨泣Zulinqi, LIV14期門 Qimen, SI3後溪Houxi, UBI8肝俞Ganshu, UB19膽俞Danshu, and Taiyang 太陽.
2. For Local Qi and Blood stagnation, add LIV3太沖Taichong, LI4合谷Hegu, UB17膈俞Geshu, SP10血海Xuehai, GB34陽陵泉Yanglingquan, LU7列缺 Lieque, SI3後溪Houxi, LIV8曲泉Ququan, and GB21肩井Jianjing.
3. For Wind-Cold Bi Syndrome add GB20風池Fengchi, UB10天柱Tianzhu, UB12風門Fengmen, DU20百會Baihui, GB39懸鐘Xuanzhong, and KD7復 溜Fuliu.

Tui Na Treatments

Tui Na Techniques

Pressing, kneading, One Finger Meditation pushing, pushing, rubbing.

Tui Na Procedures

The patient is in sitting position, the therapist stands in front.

1. Applies kneading techniques on the face to relax the muscles, the force should be light, the speed should be even and slow.
2. Applies One Finger Meditation pushing techniques on ST6頰車Jiache and ST7下關Xiaguan. Figure 2.4.1; Figure 2.4.2

3. Applies pressing and kneading techniques on SJ17翳風Yifeng and LI4合谷 Hegu to release spasm, move Blood, relax Tendon, and stop pain.

The patient is in supine position, the therapist stands near the head facing the patient's legs.

1. Places both thumbs on ST6頰車Jiache while the remaining fingers are on the lower portion of the chin.
2. Applies pressing and kneading techniques with both thumb pads and, at the same time, rotates the chin with the remaining fingers.

The patient is seated, and the therapist stands behind the patient.

1. Places one hand on the effected TMJ and places another hand on the contralateral chin.
2. Asks the patient to open and close the mouth while both hands push toward each other.
3. Applies rubbing techniques on the effected joint until warm.

PATIENT ADVISORY

1. Advise the patient to avoid cold air blowing on the face.
2. Avoid eating hard foods.
3. Advise the patient to avoid extreme jaw opening for three weeks and to support the jaw with a fist under the chin when yawning.
4. Apply heating pad on the affected side of the face.

TEMPOROMANDIBULAR JOINT DISLOCATION顳下頜關節脫位

INTRODUCTION

Temporomandibular joints are the two joints that connect the lower jaw condyle (mandible bone) to the skull socket (temporal bone). Dislocation means the condyle is completely separated from the socket, and the joint cannot return to its normal position by itself. This happens when the mouth is over-opened, such as yawning, laughing, seizures, as well as because of underlying anatomic disorders, such as ligamentous laxity, or from a direct blow to the joint, such from a motor vehicle accident.

ANATOMY

The temporomandibular joints are one of the few synovial joints in the human body with an articular disc, another being the sternoclavicular joint.

WESTERN MEDICINE ETIOLOGY AND PATHOLOGY

Temporal mandibular joint dislocation can occur as a result of atraumatic causes or traumatic factors. Etiologies leading to atraumatic dislocation can include anything that results in prolonged over-opening of the jaw joint that can occur during yawning, laughing, singing, dental procedures, and underlying anatomic causes such as ligamentous laxity.

Most temporal mandibular joint dislocations are due to traumatic factors; the common causes include a direct blow during physical assault, motor vehicle accidents, and falls.

Temporomandibular joint dislocation results when the mandibular condyle loses contact with the fibrocartilaginous articular disc. Dislocations can be defined as subluxation (partially lost contact) or luxation (completely lost contact). Dislocations can be anterior or posterior. Patients who have had one episode of dislocation are predisposed to recurrences. Clinically anterior dislocations are the most common type of dislocation.

Anterior dislocations: The mandibular condyle head is displaced anteriorly from the glenoid fossa to the temporal bone articular eminence and becomes locked in the anterior superior aspect of the eminence that results in the mouth closure. The etiology behind the trismus is the results in stretching of the ligaments and spasm of the temporalis, masseter muscles, and the medial pterygoid. This type of a dislocation occurs when the mouth is closed by the lifting the mandible with the functions of the temporalis and masseter muscles after an extended period of the wide open mouth by the lateral pterygoid muscle.

Anterior temporal mandibular joint dislocations commonly follow extreme opening of the mouth such as in dental surgery.

Posterior dislocations: the mandibular condyle is displaced posteriorly toward the mastoid. The dislocations commonly follow a direct blow to the joint pushing the mandibular condyle posteriorly toward the posterior aspect of the temporal bone behind the ear.

Lateral dislocations: (Saikrishna, Sundar, and Mamata 2015)

Type I (subluxation) – the mandibular condyle is displaced out of the glenoid fossa.

Type II (luxation) – the mandibular condyle passes superiorly and laterally to enter the temporal fossa outside the zygomatic arch.

Type IIA – the mandibular condyle is not hooked above the zygomatic arch.

Type IIB – the mandibular condyle is hooked above the zygomatic arch.

Type IIC – the mandibular condyle lodged inside the fractured zygomatic arch.

Type III (luxation) – complete dislocation without associated fracture of anterior mandible.

Type IIIA – condyle not hooked above the zygomatic arch.

Type IIIB – condyle is hooked above the zygomatic arch.

Type IIIC – condyle is lodged within the zygomatic arch, which is fractured.

The dislocations commonly follow a direct blow laterally to the mandible joint. The trauma usually is associated with mandibular fractures.

Superior dislocations: the mandibular condyle displaces into the middle cranial fossa. The dislocations commonly follow a direct blow to the mouth when it is partially opened.

Ligamentous injury and mandible fracture may occur as a result of the force applied during the reduction (Hillam, Isom 2021).

CHINESE MEDICINE ETIOLOGY AND PATHOLOGY

1. Due to old age and physical weakness, insufficient Qi and Blood, Liver and Kidney weakness, Tendons lack of nourishment from Blood, the joint capsule and ligaments of the temporomandibular joint are loosened, and the stability becomes poor. Therefore, when yawning, laughing, chewing hard objects, or vomiting, the masseter muscles contract and stretch strongly, causing the mouth to open beyond the normal physiological range, and the mandibular condyle slips out of the temporomandibular fossa and loses contact with the fibrocartilaginous articular disc.
2. It can also be Blood stasis that is caused by the impact of external force on the mandible when the mouth is opened.

MANIFESTATIONS

1. Temporomandibular joint dislocation is categorized in three stages: acute dislocation is within two weeks, protracted dislocation refers to a condition that persists for more than one month without reduction, and habitual dislocation is recurrent.
2. The patient may have a history of injury to the face or may have an atraumatic cause right before the episode of dislocation such as yawning or laughing, or the dislocation may happen right after a dental procedure
3. The mouth is open, it is difficult and painful to close.
4. The mandible is locked.
5. The mandible protrudes forward or to one side.
6. Salivation.
7. Difficulty talking.

PHYSICAL EXAMINATION

1. Medical History
   1. General health condition such as old age, cheek color, and any previous fractures or trauma, as well as dental history.
   2. Traumatic history to the TMJ includes a direct blow to the joint, car accident, etc.
   3. Atraumatic history to the TMJ includes yawning and laughing, a dental procedure, or a medical examination procedure, such as endoscopy. Both negative and positive findings should be documented.

4. Ask patients about medication use because haloperidol, phenothiazines, and thiothixene have been associated with an increased risk of jaw dislocation (Hillam, Isom 2021).
5. Painful regions, such as pain in front of the ear tragus.
6. Associated symptoms, such as salivation and difficult communication.
7. History of neurodegenerative or connective tissue disorders may also increase the risk of dislocation; examples include Marfan syndrome, Ehlers–Danlos syndrome, muscular dystrophy, multiple sclerosis, Huntington disease, Parkinson disease, and epilepsy (Hillam, Isom 2021).
8. Nerve injury related symptoms, such as fracture, accompanies damage of the mandibular division of the trigeminal nerve; the injuries with superior dislocation may include facial nerve injury and the eighth cranial nerve, resulting in deafness.

2. Inspection
1. Observe the facial symmetry for jaw deviation, erythema, bruising, and swelling when there is fracture, facial deformation that the mouth is slightly open, and preauricular depression.
2. Ask patients to slowly open and close the mouth to observe the difficulty of movements and the joint crepitus.
3. Bilateral dislocation will have a downward and forward displacement and an open, fixed mandible. In unilateral dislocation, the mandible deviates to one side. In a superior dislocation, a bulge will often be present in the preauricular and temporal areas of the face.

1. Palpation
    1. Examine the preauricular region to note any tenderness and indentation.
    2. Also examine the mandible for symmetry to determine the dislocation that is either unilateral or bilateral. A hollowness on both preauricular regions anterior to both ear's tragus will be palpated in bilateral dislocation; otherwise, a hollowness is palpated on one preauricular region to the ear's tragus if it is a unilateral dislocation.
    3. Palpation may find a tenderness in the preauricular region.
    a. Range of Motion

        1. Stiffness, pain and difficulty when opening/closing the mouth or chewing

1. Special Tests
    1. Radiography such as X-ray or a computed tomography (CT) scan, may be selected for dislocation if there is a concern for fracture or an unclear diagnosis. MRI can be a special test to assess the joint capsule and the surrounding soft tissue conditions. Radiography also is recommended before the reduction procedures to rule out fractures (Hillam, Isom 2021).
    2. Initial laboratory testing may not be necessary for isolated dislocation (Hillam, Isom 2021).

Differential Diagnosis

Mandibular joint dislocation needs to be differentiated from the following conditions.

1. Temporomandibular Joint Disorder

Temporomandibular joint dislocation needs to be differentiated from TMD because the pain can be at the temporomandibular joint, and palpation may reveal tenderness over the location of the TMJ.

4. TMD does not present with jaw deviation.
5. TMD often presents with a clicking noise that prevents mouth opening and closing because of discomfort, even though patients may not like to open or close the mouth, the range of motion of it is still in normal range.
6. The pain in TMD may be released by a warm heating pad and aggravated by rain, wind, and cold temperatures.
7. The disorder doesn't prevent the opening and closing of the mouth.

Treatments

Herbal Treatments

Acute Dislocation Stage

Treatment Principle

Soothe Tendon, move Blood, open the meridians, and stop pain.

Herbal Formula

Fu Yuan Huo Xue Tang復元活血湯

This formula is from "Clarifying the study of medicine醫學發明", authored by Li Dongyuan 李東垣, in Yuan dynasty元朝 (1271–1368), and published in 1315.

Herbal Ingredients

Chai Hu 柴胡9g, Tian Hua Fen 天花粉9g, Dang Gui 當歸9g, Hong Hua 紅花6g, Gan Cao 甘草6g, Chuan San Jia 穿山甲 炮6g, Da Huang 大黄12g, Tao Ren 桃仁9g

Ingredient Explanations

1. Chai Hu 柴胡 harmonizes Shao Yang disorders and relieves stagnation.
2. Tian Hua Fen 天花粉 clears Heat, expels pus, and reduces swelling.
3. Dang Gui 當歸 moves and tonifies the Blood, disperses Cold, and stops pain due to Blood stasis.
4. Hong Hua 紅花 moves the Blood, releases Bruising, opens the meridians, and stops pain.
5. Gan Cao 甘草 releases spasms, alleviates pain, harmonizes the other herbs, and guides the herbs to all 12 meridians.
6. Chuan San Jia 穿山甲 moves the Blood, reduces swelling, and expels Wind Damp.

7. Da Huang 大黃 moves the Blood and disperses Bruising,
8. Tao Ren 桃仁 breaks up stubborn Blood stasis.

Cooking Instructions

9. Place the herbs into an herb pot. Traditionally, a clay cooker is used. Gently rinse the herbs once with a strainer. You may notice some red flowers in the strainer; place them back into the herb pot.
10. Add water to the pot until it just covers the herbs, then add one more cup of water.
11. Presoak the herbs for 15 minutes.
12. Turn on the stove to high heat and bring the herbs to a boil. Reduce the heat to medium-low to a simmer but with slight bubbling. Turn off the heat when the decoction is reduced to one cup.
13. Drain the decoction into a container.
14. Repeat step 2–5 one more time and drain each decoction to the same container.
15. Divide the decoction into two separate cups.

Dose

Drink one cup of the decoction, two times a day.

Formula Action

Moves the Blood, disperses Bruising, soothes the Liver, and opens the meridians.

Formula Indication

Physical trauma, Blood stasis, and severe pain.

Intermediate and Later Dislocation Stage

Treatment Principle: tonifies Liver and Kidney, strengthens Bones and Tendons, regulars Qi and Blood

Herbal Formula

Bu Shen Zhuang Jin Tang補腎壯筋湯
This formula is from "Supplement to Traumatology傷科補要", authored by Qian Xiuchang錢秀昌 (unknown), in Qing dynasty 清朝 (1636–1912), and published in 1808.

Herbal Ingredients

Sheng Di Huang 生地黃15g, Shan Zhu Yu 山茱萸15g, Qing Pi 青皮6g, Bai Shao白芍 10g, Xu Duan 續斷10g, Du Zhong杜仲10g, Dang Gui當歸10g, Fu Ling茯苓10g, Wu Jia Pi 五加皮10g, Niu Xi牛膝6g

Ingredient Explanations

1. Sheng Di Huang 生地 cools the Blood and tonifies Kidney Yin.
2. Shan Zhu Yu山茱萸 tonifies Liver and Kidney Yin and strengthens Kidney Yang.

1. Qing Pi青皮 soothes Liver Qi and breaks up stagnant Qi, dries Dampness, and transforms Phlegm.
2. Bai Shao白芍 nourishes the Blood, preserves Yin, calms Liver Yang, softens Tendons, and alleviates spasm and pain.
1. Xu Duan續斷 nourishes Liver and Kidney, strengthens Bones and Tendons, and alleviates pain.
2. Du Zhong杜仲 tonifies the Kidneys and Liver and strengthens Bones and Tendons.
1. Dang Gui當歸 nourishes the Blood, reduces edema, and stops pain.
2. Fu Ling茯苓 drains Damp and tonifies the Spleen.
1. Wu Jia Pi 10g五加皮 expels Wind, Damp, nourishes and warms Liver and Kidney, and strengthens Bones and tendons.
2. Niu Xi牛膝 expels Wind Dampness, tonifies Liver and Kidneys, and directs herbs downward.

Modifications

1. For strengthening Tendon and Bone, add Gui Ban龜膠 and Gou Qi Zi枸杞.
2. For tonifying Qi, add Dang Shen黨參, Huang Qi黃芪, and Bai Zhu白朮.

Habitual Dislocation Stage

Treatment Principle

Tonify Qi and Blood.

Herbal Formula

Ba Zhen Tang 八珍湯

This formula is from "Categorized Essentials of Repairing the Body 正體類要", authored by Xue Ji 薛己 (1487–1559), in Ming dynasty 明朝 (1368–1644), and published in 1529.

Herbal Ingredients

Ren Shen人參10g, Bai Zhu白朮10g, Fu Ling茯苓10g, Dang Gui當歸10g, Chuan Xiong 川芎10g, Bai Shao白芍10g, Shu Di Huang 熟地黃10g, Zhi Gan Cao炙甘草5g, Sheng Jiang生薑10g, Da Zao大棗3 pieces

Ingredient Explanations

1. Ren Shen人參 tonifies Qi and Yang.
2. Bai Zhu白朮 tonifies the Spleen Qi and dries Dampness.
3. Fu Ling茯苓 drains Damp and tonifies the Spleen.
4. Dang Gui當歸 nourishes the Blood and moves the Blood.
5. Chuan Xiong 川芎 moves the Blood and moves Qi.
6. Bai Shao白芍 nourishes the Blood, soothes Liver, and stops pain.
7. Shu Di Huang 熟地黃 tonifies Liver and Kidney Yin, and benefits Essence and Blood.
8. Zhi Gan Cao炙甘草 tonifies the Spleen, harmonizes the other herbs, and guides the herbs to all 12 meridians.

9. Sheng Jiang生薑 unblocks the pure Yang pathway and harmonizes rebellious Qi.
10. Da Zao大棗 tonifies the Spleen Qi, nourishes the Blood, and moderates and harmonizes the harsh properties of the fragrant herbs.

Modifications

1. For poor appetite, add Shan Yao山藥, Shan Zha山楂, Mai Ya麥芽, and Qian Shi芡實.
2. For headache due to Blood deficiency, add Man Jing Zi蔓荊子 and Gao Ben 藁本.

Formula Actions

1. Tonifies Qi and Blood.

Acupuncture Treatments

Acupuncture Point Formula

Local Points

SJ20角孫Jiaosun, SJ21耳門 Ermen, ST7下關Xiaguan, ST6頰車Jiache, SJ17翳風 Yifeng, SI17天容Tianrong, ST4地倉Dicang

Distal Points

UB18肝俞Ganshu, UB23腎俞Shenshu, LI4合谷Hegu

Tui Na Treatments

Treatment Principle

Relocate dislocation, relax Tendons, open meridians, tonifies Qi and Blood, and stop pain.

Reduction

Reduction techniques include intraoral and extraoral techniques. Intraoral technique is the most common technique. In this book, we introduce the intraoral method because it is a classic, traditional, and effective technique.

Reduction Procedures

The provider is standing facing the patient while the patient is seated upright. Ensure that the patient's mandible is respectively lower than the provider's elbows and the patient's occipital region against the wall, or the head is held still by an assistant.

1. The provider wears gloves and wraps the thumbs in gauze prior to maneuver for protection in case the teeth involuntarily bite.
2. The provider places both thumbs into the patient's open mouth and place them on the external oblique ridge area as far back as possible or about the second and third molar which is the most common site for the reduction method. And place the remaining fingers out of the mouth on the angle of the mandible to elevate the body of the mandible and chin.

3. The thumbs apply pressure to push the mandible downward and then backward to free the condyle from the anterior eminence and push the mandible condyle back into the fossa.
4. Perform the reduction for one mandible joint then another if it is a bilateral dislocation.
5. Post maneuver, reduction can be confirmed if the patient is having less or no pain with jaw movement, both mandible joints are symmetrical, and the hollowness on the preauricular region anterior to the ear tragus has disappeared.

Tui Na Techniques

Pressing, kneading

Tui Na Procedures

1. This procedure is for habitual dislocation to strengthen the Tendons.
2. The patient is seated upright while the therapist standing and facing the patient.
3. Presses with thumb pads on the following acupuncture points: SJ20角孫 Jiaosun, SJ21耳門 Ermen, ST7下關xiaguan, ST6頰車Jiache, SJ17翳風 Yifeng, SI17天容Tianrong, UB18肝俞Ganshu, UB23腎俞Shenshu, and LI4 合谷Hegu.
4. Presses each point for 30 seconds, once a day for two weeks.

PATIENT ADVISORY

1. Right after the reduction, immobilize the joint with a Kling bandage wrap around the chin and the skull for two weeks.
2. Advise the patient to avoid extreme jaw opening for three weeks; support the jaw with a fist under the chin when yawning.
3. Avoid hard biting, maintain soft diet for a week or two.

SINUSITIS 鼻竇炎

INTRODUCTION

Sinuses are the hollow spaces located throughout the body, and there are many types of them. This chapter focuses on the paranasal sinuses which are within the face bones between eyes, behind the cheek, and in the forehead. The functions of the sinuses are to produce mucus to keep the nose moist, heat inhaled air, protect against dust, allergens, and pollutants; they also lighten the weight of the head and serve as a crumple zone to protect vital structures in the event of facial trauma.

Healthy sinuses are filled with air. Sinusitis is an inflammation in the sinuses which leads to accumulation of fluid and becomes a place where the germs can grow and can result an infection and headache.

ANATOMY

There are four paired paranasal sinuses that surround the nasal cavity. They are named following their respective anatomic bone locations.

1. The *frontal sinuses* are in the frontal bone above the orbits. The blood supply is from the anterior and posterior ethmoidal branches of the ophthalmic artery (Henson, Drake, and Edens 2021). The nerve innervation is from the supraorbital nerve which is a branch of the ophthalmic nerve and itself is one of three branches of the trigeminal nerve.
2. The *sphenoidal sinuses* are in the body of the sphenoid behind the eyes. The blood supply is from small branches of the cavernous internal carotid arteries and the pharyngeal branch of the maxillary artery.
3. The largest of the paranasal sinuses are the *maxillary sinuses*. They are located bilaterally under the eyes in the maxilla of the face in the maxillary bone, the back of the semilunar hiatus of the nose. The blood supply is the infraorbital artery and the superior, anterior, and posterior alveolar branches of the maxillary artery. They are innervated by the maxillary nerve via the infra-orbital and alveolar branches.
4. The *ethmoidal sinuses* are between the orbits that house the globes of the eyeballs. The blood supply is also from the anterior and posterior ethmoidal branches of the ophthalmic artery. The nerve supplies are the anterior and posterior ethmoidal branches of the nasociliary nerve which is a branch of the ophthalmic nerve and itself is one of three branches of the trigeminal nerve.

WESTERN MEDICINE ETIOLOGY AND PATHOLOGY

Acute sinusitis is mostly caused by infection, such as common cold. It is almost always viral, and only a small percentage develops secondary bacterial infection. A periapical dental abscess of a maxillary tooth spreads to the overlying sinus resulting acute sinusitis.[15] Chronic sinusitis is caused by a weak immune system in people who are immunocompromised (e.g., diabetes and HIV infection), a bacterial or fungal infection, a blocked airway caused by a nasal allergy from environmental irritants (e.g., airborne pollution and tobacco smoke), an abnormal nose structure (e.g., a deviated septum, nasal polyps, or swelling of the lining of the sinuses).

A sinus inflammation or infection is classified as acute when the duration of symptoms is shorter than four weeks, subacute when the duration is from four weeks to 12 weeks, and chronic if it lasts longer than 12 weeks. If the symptoms of sinusitis are on and off several times a year, it is called recurrent sinusitis.[16,17]

CHINESE MEDICINE ETIOLOGY AND PATHOLOGY

1. Heat in Lung Meridians Type肺經鬱熱型

   According to traditional Chinese medicine, Lungs are the tender visceral organ. Lungs open into the nose, manage the regulation of water passage, and govern the skin and the hair.

Diseases of the nose are closely associated with Lungs. When the body immunity is weak, the body is invaded by Six Yin六淫 also called Six Xie 六邪 (Six Evils, six nefarious factors). In among these six conditions, the most common one is Wind風. When it invades one's body, Lungs are the first to be affected.

1. Lung Qi blocks nefarious Wind-Heat at the exterior: When one's Lungs are able to function, the immune system is able to block the nefarious Wind Heat at the exterior.
2. Heat and Fire reside in the Lungs: Lung Qi fails to block nefarious Wind-Cold or Wind-Heat in the exterior level, Wind induces Heat or Cold to invade and reside in Lungs. When the nefarious Wind-Cold or Wind-Heat resides in Lungs, it blocks Lung Qi resulting in deficiency of circulation. Wind then becomes weak, Cold becomes Heat, Heat becomes Fire, Heat or Fire resides in Lungs. Sinusitis follows the failure of Lung Qi to reach the nose.
3. Lung Qi deficiency: the immune system is unable to clear the nefarious Heat because Lung Qi is deficient.
a. Damp-Heat Accumulation in Spleen and Stomach Type脾胃積熱證

   Dysfunction of Spleen and Stomach due to long-term alcohol drinking and having the excessive diet with rich food, fatty food, sweet food, and strong-tasting food results in Phlegm-Damp accumulation that leads to the accumulation of Damp-Heat to block the Middle Jiao. Evolving over time, causing transformation and transportation failure to function results in pure Yang being unable to rise and turbid Yin unable to sink. When the situation continues to deteriorate, Damp-Heat rises to the sinuses.

   a. Damp-Heat in Liver and Gallbladder Type肝膽濕熱證

   Stress leads to stagnation of Qi, long-term Qi stagnation gives rise to interior Heat or Fire in Gallbladder; the Heat or Fire in the Gallbladder affects the Brain; and the Fluid of the Brain floods the nose along the Gallbladder meridians.

MANIFESTATIONS

In Western Medicine

Acute and chronic sinusitis present similar manifestations, including nasal congestion and obstruction. The associated sinus is often accompanied with facial pain and headaches. Often the pain is more severe in acute sinusitis.

1. Maxillary sinusitis causes pain in the maxillary area, toothache, and frontal headache. The pain can be unilateral or bilateral. And there may be tenderness in the cheek where the sinus located.
2. Frontal sinusitis causes pain around the eye and in the frontal region as well as frontal headache. Tenderness may be felt on the frontal area with erythema of the skin. There is usually fever.

3. Ethmoid sinusitis causes pressure and pain behind and between the eyes and around the nose; a frontal headache often described as splitting, periorbital cellulitis, and tearing. There is tenderness when touching the bridge of the nose.
4. Sphenoid sinusitis causes less well localized pain referred to the frontal, occipital temporal, periorbital, or vertex and can even be vague or occur anywhere in the craniofacial region area (Ruoppi et al. 2000).

In Chinese Medicine

1. Heat accumulation in Lung meridians type肺經鬱熱型
   This is seen in the early and the intermediate stage of acute sinusitis.
   a. Lung Qi blocks Nefarious Wind-Heat at the Exterior
       – Wind Cold type風寒型: this is seen in the early stage of acute sinusitis. The symptoms include large amount of clear discharge, sneezing, fever, headache, cough, no sweat, and aversion to wind and cold temperatures.
       Tongue: pink, thin white coating.
       Pulse: floating and tight.
       – Wind Heat type風熱型: this is seen in the early stage of acute sinusitis. The symptoms include severe congestion, profuse yellow sticky discharge, pain, red and swelling around the entrance of the nose; fever, headache, aversion to wind, sweating, thirsty, and sore throat.
   Tongue: red tongue body, thin yellow coating.
   Pulse: floating and rapid.
   b. Heat and Fire Reside in the Lungs
       This is seen in the intermediate stage of acute sinusitis. The symptoms include enduring congestion but the symptoms are intermittent; the rhinorrhea is yellow, sticky in small amounts, with pressure pain at the associated sinuses; mild aversion to cold temperatures, cough accompanied by head distension, fever, hyposmia, and poor memory.
       Tongue: red tongue body, red tip, thin yellow coating.
       Pulse: floating, wiry, and rapid.
2. Damp-Heat in Spleen and Stomach Type脾胃積熱證
   This is seen in the intermediate and later stage of acute sinusitis. The symptoms include severe nasal congestion with large amounts of sticky, yellow rhinorrhea, hyposmia, and frontal headache; fever, thirst and desire to drink water, halitosis, swelling and red gums, yellow and decreased urination, and dry stool.
   Tongue: red tongue body, yellow coating.
   Pulse: rapid, forceful.

3. Damp-Heat in Liver and Gallbladder Type肝膽濕熱型
   This is seen in the intermediate and later stage of acute sinusitis. The symptoms include enduring congestion with large amount of sticky yellow or green rhinorrhea with a bad odor, hyposmia, poor memory, dry throat,

irritability, tinnitus, abdominal distension, poor appetite, bitter taste and dry throat, sticky saliva in mouth, no desire to drink water, headache and heavy sensation in the head, vertigo, yellow urine, and possibly fever.

Tongue: red tongue body with yellow greasy coating.

Pulse: wiry or slippery and rapid.

PHYSICAL EXAMINATION

1. Medical History
    1. Begin with the duration of symptoms to determine whether it is acute or chronic sinusitis.
    2. General health condition such as AIDS, diabetes, old age, asthma, and allergies or frequent upper respiratory tract infections. Diabetic patients, AIDS patients, and older persons have more compromised immune systems and a greater prevalence of acute sinusitis. Allergies can increase sinus inflammation and can lead to the acute maxillary rhinosinusitis, while asthma patients are more prone to chronic sinusitis.
    3. Patients who suffer from nasal polyps or a deviated nasal septum are also prone to increased risk of both acute and chronic sinusitis.
    4. Dental disease that causes in infection such as periodontal infection and dental abscesses, can precipitate sinusitis.
    5. Travel may increase the risk for sinus blockage, e.g., visiting places where the atmospheric pressure is different, such as in high-altitude country.
    6. Swimming in pools may also be a risk factor. Chlorine can irritate the lining of the nose and the sinuses, resulting in sinusitis.

1. Inspection
    1. Observe and check for nasal polyps, deviated nasal septum, purulent rhinorrhea, swollen turbinate, and erythema of the skin. A speculum or otoscope can be helpful for the examination of the nose.[18]
    2. Fever may present at the beginning of a sinus infection.
    3. Focal facial pain with forward bending.
    a. Palpation
    4. Examine for local sinus tenderness when pressing.

1. Special Tests
    1. Culture of the aspirate from an antral puncture is a test for the diagnosis of acute bacterial sinusitis.

DIFFERENTIAL DIAGNOSIS

2. Sinusitis needs to be differentiated from allergic rhinitis. Rhinitis is also known as hay fever, and it is triggered by a host of outdoor and indoor allergens, such as pollen, dust mites, and pet hair. It can cause similar symptoms

as sinusitis, such as runny nose, nasal congestion, postnasal drip, frequent sneezing, and coughing. There are some symptoms that can distinguish between the two conditions. For instance, itchy nose, sneezing, and watery eyes are more common with allergic rhinitis (distancing oneself from the allergens can improve the allergic symptoms). Clear nasal discharge often coincides with allergic rhinitis, while yellow or green nasal discharge in larger amounts usually indicates sinusitis.[19]

3. Acute sinusitis needs to be differentiated from the common cold and flu because both conditions have the similar symptoms such as nasal congestion with yellow or green nasal discharge. The nasal congestion in common cold usually doesn't last longer than two weeks, and it begins with an occipital headache, fatigue, fever, aversion to wind and cold temperatures, sneezing, sweating, thirst, nasal congestion, nasal discharge color changes from clear to yellow and sticky, from large to small amounts – then vanishes.

TREATMENTS

1. Heat Accumulation in Lung Meridians肺經鬱熱型
   a. Lung Qi Blocks Nefarious Wind-Heat at the Exterior 風熱襲表

Wind Cold Type風寒型

Treatment Principle

Expel Wind, release Cold, and open nasal orifice.

Herbal Treatments

Herbal Formula

Xin Yi San辛夷散

This formula is from the book "Analytic Collection of Medicinal Formulas 醫方集解", authored by Wang Ang汪昂 (1615–1694AD), in Qing dynasty 清朝 (1636–1912), and published in 1682.

Herbal Ingredients

Xin Yi辛夷2.4g, Bai Zhi白芷2.4g, Sheng Ma升麻2.4g, Gao Ben藁本2.4g, Fang Feng防風2.4g, Chuan Xiong川芎2.4g, Xi Xin細辛2.4g, Mu Tong 木通2.4g, Gan Cao甘草2.4g, tea茶葉 2.4g

Ingredient Explanations

4. Xin Yi辛夷 disperses Wind and opens the nasal passages.
5. Bai Zhi白芷 expels Wind, drains Dampness, dispels Cold, and stops pain.
6. Sheng Ma升麻 releases the exterior, clears Heat, raises pure Yang, and lifts sunken Qi.
7. Gao Ben藁本 expels Wind and drains Damp in Tai Yang meridians.
8. Fang Feng防風 expels Wind and drains Damp
9. Chuan Xiong川芎 moves the Blood and moves Qi.

10. Xi Xin細辛 enters Shao Yin meridian to expel Wind, Damp, and Cold.
11. Mu Tong 木通 drains Heat and promotes urination.
12. Gan Cao甘草 harmonizes the ingredients.

Modifications

1. To make the formula stronger for nasal congestion, add Cang Er Zi蒼耳子 and Shi Chang Pu石菖蒲.

Formula Actions

2. Nasal congestion.

Instructions

1. Grind the herbs into fine powder.
2. Place 2 grams of the herbal powder in a cup.
3. Add warm water to fill the cup.

Dose

Drink one cup after meals, 2 times a day.

Wind Heat type風熱型

Treatment Principle

Clear Heat, expel Wind, and open Orifice.

Herbal Treatments

Herbal Formula

Sang Ju Yin 桑菊飲

This formula is from "Systematic Differentiation of Warm Pathogen Diseases 溫病條辨", authored by Wu Tang 吳瑭 (1758–1836), in Qing dynasty 清朝 (1636–1912), and published in 1813.

Herbal Ingredients

Sang Ye桑葉8g, Ju Hua菊花3g, Bo He薄荷2g, Xing Ren杏仁6g, Jie Geng桔梗6g, Lian Qiao連翹5g, Gan Cao甘草2g, Wei Gen葦根6g

Ingredient Explanation

4. Sang Ye桑葉 calms the Liver, clears Heat, cools the Blood.
5. Ju Hua菊花 clears Heat, expels Wind, and calms Liver Yang and Wind.
6. Bo He薄荷 expels Wind Heat, Relieves Liver Qi stagnation, disperses stagnant Heat, and enhances Chai Hu's ability to relieve the Liver.
7. Xing Ren杏仁 stops cough, calms wheezing, and moistens the Intestines.
8. Jie Geng桔梗 opens Lungs, disperses Lung Qi, and expels Phlegm.
9. Lian Qiao連翹 clears Upper Jiao Heat.
10. Gan Cao甘草 harmonizes the ingredients.
11. Wei Gen葦根 clears Heat and generates fluid.

Modifications

To make the formula stronger, add Cang Er Zi San 蒼耳子散.

Formula Actions

12. Expels Wind and clears Heat.
13. Clears Lungs.

Cooking Instructions

1. Place the herbs into an herb pot. Traditionally, a clay cooker is used. Gently rinse the herbs once with a strainer. You may notice some red flowers in the strainer; place them back into the herb pot.
2. Add two cups of water to the pot.
3. Turn on the stove to high heat and bring the herbs to a boil. Reduce the heat to medium-low to simmer but with slight bubbling. Turn off the heat when the decoction is reduced to one cup.
4. Drain the decoction into a container.
5. Repeat steps 2–4 once and drain each decoction to the same container.
6. Divide the decoction into two separate cups.

Dose

Drink one cup each time, two times a day.
    b. Heat and Fire Reside in the Lungs 風邪中人

Treatment Principle

Clear Heat.

Herbal Treatments

Herbal Formula

Cang Er Zi San and Yin Qiao San Jia Jian 蒼耳子散合銀翹散加減

1. Cang Er Zi San 蒼耳子散: This formula is from "A Book of Formulas to Promote Well-Being 嚴氏濟生方", authored by Yan Yonghe 嚴用和 (1200~1268), in Song dynasty 南宋 (1127–1239), and published in 1253.
2. Yin Qiao San 銀翹散: This formula is from "Systematic Differentiation of Warm Pathogen Diseases 溫病條辨", authored by Wu Tang 吳瑭 (1758–1836), in Qing dynasty 清朝 (1636–1912), and published in 1813.

Herbal Ingredients

Can Er Zi蒼耳子10g, Xin Yi辛夷10g, Bai Zhi白芷10g, Bo He薄荷6g, Huang Qin黃芩10g, Jin Yin Hua金銀花15g, Liao Qiao連翹10g, Jing Jie 荊芥10g, Jie Geng桔梗18g, Niu Bang Zi 牛蒡子18g, Tian Hua Fen 天花粉15g, Sang Bai Pi桑白皮10g, She Gan 射幹15g, Chan Tui 蟬蛻15g, Mi Bai Bu 蜜百部15g, Hua Shi 滑石5g, Gua Lou Pi 瓜蔞皮15g, Niu Bang Zi 牛蒡子10g, Ting Li Zi 葶藶子10g, Ban Xia 半夏15g, Mu Hu Die 木蝴蝶2g, Zhi Gan Cao炙甘草 6g

Ingredient Explanations

1. Can Er Zi蒼耳子 disperses Wind, drains Damp, and opens the nasal passages.
2. Xin Yi Hua 辛夷 disperses Wind and opens the nasal passages.
3. Bai Zhi白芷 expels Wind, drains Dampness, dispels Cold, and stops pain.
4. Bo He薄荷 expels Wind Heat, Relieves Liver Qi stagnation, disperses stagnant Heat, and enhances Chai Hu's ability to relieve the Liver.
5. Huang Qin黃芩 clears Heat, dries Damp, cools the Blood, stops bleeding, calms Liver Yang.
6. Jin Yin Hua金銀花 clears Heat, disperses Wind Heat, Clears Damp Heat, and cools the Blood.
7. Liao Qiao連翹 clears Upper Jiao Heat.
8. Jing Jie 荊芥 releases the exterior and disperses Wind.
9. Jie Geng桔梗 opens Lungs, disperses Lung Qi, and expels Phlegm.
10. Niu Bang Zi 牛蒡子 disperses Wind-Heat.
11. Tian Hua Fen 天花粉 relieves toxicity, expels pus, and reduces swelling.
12. Sang Bai Pi桑白皮 clears Lung Heat and stops cough.
13. She Gan 射幹 clears Lung Heat and transforms Phlegm.
14. Chan Tui 蟬蛻 disperses Wind-Heat.
15. Mi Bai Bu 蜜百部 moistens Lungs and stops coughing.
16. Hua Shi 滑石 clears Phlegm Heat and Heat from Urinary Bladder, promotes urination, and drains Damp.
17. Gua Lou Pi 瓜蔞皮 clears and moistens Lungs and transforms Heat Phlegm.
18. Niu Bang Zi 牛蒡子 clears Wind-Heat.
19. Ting Li Zi 葶藶子 sedates Lungs, calms wheezing, and drains Phlegm.
20. Ban Xia 半夏 transforms Phlegm.
21. Mu Hu Die 木蝴蝶 clears Lungs, relieves cough, soothes the Liver, and harmonizes the Stomach.
22. Zhi Gan Cao炙甘草 tonifies the Spleen, harmonizes the other herbs, and guides the herbs to all 12 meridians.

Modifications

1. For sweating and aversion to wind, add Huang Qi黃芪, Fu Xiao Mai浮小麥, and Fang Feng防風.

Formula Actions

2. Clears Heat and expels Wind.
3. Opens nasal orifice and stops headache.

Cooking Instructions

1. Place the herbs into an herb pot. Traditionally, a clay cooker is used. Gently rinse the herbs once with a strainer. You may notice some red flowers in the strainer; place them back into the herb pot.
2. Add water into the pot until it just covers the herbs, then add one more cup of water.

3. Presoak the herbs for 15 minutes.
4. Turn on the stove to high heat and bring the herbs to a boil. Reduce the heat to medium-low to simmer but with slight bubbling. Turn off the heat when the decoction is reduced to 1 cup.
5. Drain the decoction into a container.
6. Repeat steps 2–4 once and drain each decoction to the same container.
7. Divide the decoction into two separate cups.

Dose

Drink one cup each time, two times a day.

2. Damp-Heat in Spleen and Stomach脾胃積熱證

Treatment Principle

Clear Heat in the Spleen and Stomach.

Herbal Treatments

Herbal Formula

Qing Wei Tang 清胃湯

This formula is from "Imperially Commissioned Golden Mirror of the Orthodox Lineage of Medicine醫宗金鑑", authored by the Qianlong emperor 乾隆 (1711–1799), in Qing dynasty 清朝 (1636–1912), and published in 1742.

Herbal Ingredients

Shi Gao石膏12g, Huang Qin黃芩3g, Sheng Di Huang生地3g, Mu Dan Pi丹皮4.5g, Huang Lian黃連3g, Sheng Ma升麻3g

Ingredient Explanations

4. Shi Gao石膏 clears Heat in the Qi Stage and stops thirst.
5. Huang Qin黃芩 clears Heat, dries Damp, cools the Blood, stops bleeding, calms Liver Yang.
6. Sheng Di Huang生地黃 nourishes Yin and clears Heat.
7. Mu Dan Pi丹皮 clears Heat, cools the Blood, and moves the Blood.
8. Huang Lian黃連 clears Heat and drains Damp.
9. Sheng Ma升麻 releases the exterior, clears Heat, raises pure Yang, and lifts sunken Qi.

Modifications

1. To make the formula stronger, add Huang Lian Jie Du Tang 黃連解毒湯.

Formula Actions

2. Clears Stomach Fire.

Cooking Instructions

3. Place the herbs into an herb pot. Traditionally, a clay cooker is used. Gently rinse the herbs once with a strainer. You may notice some red color flowers in the strainer; place them back into the herb pot.

4. Add 400 ml water to the pot
5. Presoak the herbs for 15 minutes.
6. Turn on the stove to high heat and bring the herbs to a boil. Reduce the heat to medium-low and bring to a simmer but with slight bubbling. Turn off the heat when the decoction is reduced to 320 ml.
7. Drain the decoction into a container.
8. Repeat steps 2–5 once and drain each decoction to the same container.
9. Divide the decoction into two separate cups.

Dose

Drink one cup each time after meals, two times a day.
3. Damp-Heat in Liver and Gallbladder type肝膽濕熱型

Treatment Principle

Drain Excess Fire from the Liver and Gallbladder; clear and drain Damp-Heat from the Lower Jiao.

Herbal Treatments

Herbal Formula

Long Dan Xie Gan Tang龍膽瀉肝湯
This formula is from the book "Analytic Collection of Medicinal Formulas醫方集解", authored by Wang Ang汪昂 (1615–1694AD), and published in Qing dynasty清朝 in 1682 AD.

Herbal Ingredients

Long Dan Cao龍膽草12g, Zhi Zi栀子9g, Huang Qin黃芩9g, Chai Hu柴胡6g, Sheng Di Huang生地黃12g, Ze Xie澤瀉9g, Dang Gui當歸5g, Che Qian Zi車前子10g, Mu Tong川木通9g, Gan Cao甘草5g

Ingredient Explanations

10. Long Dan Cao龍膽草 drains Damp-Heat from the Liver and Gallbladder channels.
11. Zhi Zi栀子 clears Heat, cools the Blood, resolves Damp, reduces swelling, and opens the meridians.
12. Huang Qin黃芩 clears Heat, dries Damp, cools the Blood, stops bleeding, and calms Liver Yang.
13. Chai Hu 柴胡 moves Liver Qi and releases stagnation.
14. Sheng Di Huang生地黃 nourishes Yin and clears Heat.
15. Ze Xie澤瀉 drains Damp, especially Damp Heat in Lower Jiao, and clears Kidney deficient Heat.
16. Dang Gui當歸 nourishes the Blood, benefits the Liver, and regulates menstruation.
17. Che Qian Zi車前子 clears Damp Heat and promotes urination.
18. Mu Tong川木通 drains Heat and promotes urination.
19. Gan Cao甘草 harmonizes the herbs, clears Heat, and expectorates phlegm.

Modifications
- For making the formula stronger, add Qi Shou Huo Xiang Wan 奇授藿香丸.

Formula Actions
- Drains excess Fire from the Liver and Gallbladder and drains Damp-Heat from Lower Jiao.

Formula Indication
- Headache due to Liver Fire and Damp Heat in Lower Jiao, such as migraine headache, urethritis, acute glaucoma, and acute cholecystitis.

Contraindications
- Caution for patients who suffer from Cold and Yang deficiency in Spleen and Stomach, due to the Cold herbs in the formula.

Cooking Instructions
1) Place the herbs into an herb pot. Traditionally, a clay cooker is used. Gently rinse the herbs once with a strainer.
2) Add water to the pot until it just covers the herbs, then add one more cup of water.
3) Presoak the herbs for 15 minutes.
4) Turn on the stove to high heat and bring the herbs to a boil. Then reduce the heat to medium-low to simmer but with slight bubbling. Turn off the heat when the decoction is reduced to one cup.
5) Drain the decoction into a container.
6) Repeat steps 2–5 once and drain the decoction to the same container.
7) Divide the decoction into two separate cups.

Dose
- Drink one cup of the decoction after meals twice daily.

Acupuncture Treatments
Acupuncture Point Formula

Local Points
UB2攢竹 Cuanzhu, ST3巨髎 Juliao, LI20迎香Yingxiang

Distal Points
LI4合谷Hegu, UB13 肺俞Feishu, LU9 太淵Taiyuan, GB20風池Fengchi, Yin Tang 印堂

Modifications
1. For Coldness in the Lungs, add moxibustion Therapy on DU23 上星Shangxing.
2. For Lung Yin deficiency, apply even acupuncture technique.

3. For Heat Toxin, apply sedating acupuncture technique on LI4合谷Hegu, ST36 足三里Zusanli.
4. For Damp accumulation underlying Spleen deficiency, add moxibustion Therapy on UB20 脾俞Pishu, and SP9 陰陵泉Yinlingquan.

Auricular Therapy

Internal Nose, Adrenal, Shenmen, Forehead

Tui Na Treatments

Tui Na Techniques

Pressing, kneading, grasping

Tui Na Procedures

Tui Na on Neck

Acupuncture Points: DU14 大椎Dazhui, UB13肺俞Feishu, UB12風門 Fengmen, GB21肩井Jianjing, GB20風池Fengchi, DU16 風府Fengfu

Tui Na Techniques

Grasping, kneading, pressing

Tui Na Procedures

1. The patient is seated upright.
2. The therapist stands behind the patient.
3. Grasp both sides of the neck along the Gallbladder meridians for 2–3 minutes. Figure 2.6.1; Figure 2.6.2; Figure 2.6.3; Figure 2.6.4

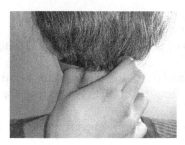
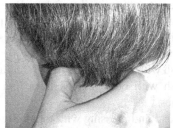
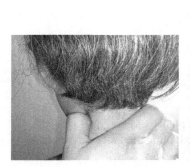
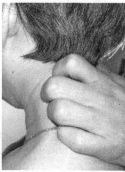

4. Knead and press with the thumb tips on GB20風池Fengchi and DU16 風府 Fengfu until the patient feels sore and distended on the points. Figure 2.6.5

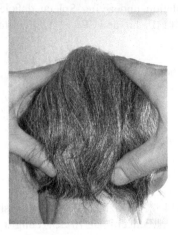

5. Repeat step 3.
6. Grasp GB21肩井Jianjing until the skin is warm.
7. Knead DU14 大椎Dazhui, UB13肺俞Feishu, and UB12風門 Fengmen for 1 minute each point.

Tui Na on the Face

Acupuncture Points: Yin Tang, Tai Yang, LI20迎香Yingxiang
Tui Na Techniques:

1. The patient is seated upright.
2. The therapist stands in front of the patient.
3. Rub with both thumbs on both sides of the nose to LI20迎香Yingxiang for 5 minutes.
4. Push with both thumbs from Yin Tang 印堂to Tai Yang 太陽for 5–6 times.
5. Push with both thumbs from Yin Tang to LI20迎香Yingxiang for 5–6 times.
6. Knead and pressing with thumb tips on LI20迎香Yingxiang and Tai Yang 太陽.
7. Repeat step 3 until warm.

PATIENT ADVISORY

1. Exercise in the outdoors.
2. Avoid nose picking.
3. Avoid common cold.
4. Avoid feeling cold on the face.
5. Avoid working at night, and get plenty of good quality sleep.
6. Avoid cold drinks and deep-fried, pungent, and spicy foods.

TRIGEMINAL NEURALGIA 三叉神經痛

INTRODUCTION

Trigeminal neuralgia is a chronic pain condition that affects the trigeminal nerve, a cranial nerve that carries sensory data from the face to the brain. Trigeminal nerve isn't involved in facial expressions but is involved in facial pain. Patients with this condition can be affected by temperature, the slightest pressure (such as a light wind), cold sensation, touching the face lightly while putting on makeup, brushing the teeth (some sufferers avoid tooth brushing), or even talking.

Trigeminal neuralgia is the most common cause of facial pain. It is not a life-threatening disease, but it is exceptionally painful, and the intensity of the pain can be debilitating.

When trigeminal neuralgia occurs, the pain varies: commonly, the pain presents from excruciating stabbing, as from a knife cutting, to aching, burning, and numb sensations.

If you have facial pain that starts near the top of the ear and splits in three, toward the eye, cheek, and jaw, plus the type and the severity of pain are similar to toothache, but the duration of the pain is much shorter (sometimes only a few seconds), and frequency is sporadic with sudden burning, you likely suffer from trigeminal neuralgia.

ANATOMY

The trigeminal nerve is the fifth, the largest, and the most complex nerve among the 12 pairs of the cranial nerves.

Its primary function is to provide sensory and motor innervation to the contralateral face, which means it senses facial sensations, such as touching, pain, and temperature; it also controls muscles used for chewing of the opposite side of the brain. When the nerves in the divisions sense pain, they send the signal through the Gasserian ganglion to the trigeminal nerve root, then continues going toward the brain stem and inserts into the pons. The signals continue to travel through the trigeminal nerve nucleus and get processed, then send the message to the brain and cerebral cortex, where a conscious perception of facial sensation is generated (Kaufmann, and Patel 2001).

The sensory information from the face crosses over to the contralateral brain hemisphere.

It exits the brain by a large sensory root and a smaller motor root coming out of the pons at its junction with the middle cerebral peduncle. It passes laterally to join the Gasserian ganglion in the Meckel cave of the cranial cavity. It splits into three branches that extend to different areas of the face as outlined below:

- Ophthalmic V1 – innervates and is responsible for sensory of the forehead and skull above the palpebral fissure, eye, and portions of the nasal cavity and the frontal sinus.
- Maxillary V2 – innervates and is responsible for sensory of the area just below the orbits and above the mouth including the cheek, portions of the

nasal cavity, maxillary sinuses, maxillary teeth, palate, and the middle portion of the face and skull above the mouth and below the forehead.

• Mandibular V3 – innervates and is responsible for sensory and motor function of the lower portion of the face and jaw. It is responsible for sensory innervation of pain and temperature including the buccal mucosa, mandibular teeth, temporomandibular joint, the anterior two-thirds of the tongue and the skin below the mouth (below the area of the maxillary nerve). Its motor function is to control all the muscles of mastication.

WESTERN MEDICINE ETIOLOGY AND PATHOLOGY

Trigeminal neuralgia is divided into classical and secondary.

The exact etiology of classical trigeminal neuralgia is still unknown. Commonly, it is believed to be associated with neurovascular compression in the pons. It is caused by the compression of the trigeminal nerve root most frequently by the superior cerebellar artery. The compression usually leads to demyelination and a dysregulation of voltage-gated sodium channel expression in the membrane of nerve fibers.

Secondary trigeminal neuralgia is most likely caused by multiple sclerosis, a space-occupying lesion, such as a tumor, affecting the trigeminal nerve.

CHINESE MEDICINE ETIOLOGY AND PATHOLOGY

Trigeminal neuralgia is classified into acute, intermediate, and chronic conditions. The trigeminal neuralgia evolves from acute condition that is caused by Wind invasion, intermediate condition progressed to accumulation of interior Phlegm Heat, to chronic condition which is caused by Blood stagnation.

1. Wind-Cold Invasion風寒外襲

Wind-Cold invades Yang Ming meridian, due to the character of Wind in Yang, and it ascends to the head and face; while Cold is the evil of Yin, its character is stagnant, it blocks the Blood circulation leading to Qi and Blood occlusion, resulting in pain.

2. Stomach-Fire Uprising胃火上攻

Wind-Heat invasion that evolves to attack Yang Ming Fu organ 陽明腑臟 causes Stomach Fire. Overeating hot food, such as spicy or pungent, sweet foods, or consuming alcohol leads to Heat and Fire accumulation in the Stomach. The Heat or Fire blocks the circulation in the face and head along the meridian, leading to Qi and Blood occlusion and resulting in pain.

3. Liver Fire Uprising肝火上炎

It is mostly due to internal injuries by Seven-Emotions and stagnation of Liver Qi, which turns into Fire. Liver Fire disturbs the cheeks along the Stomach collaterals and develops the disease as well as Wood dryness because of insufficiency of Water due to Kidney Yin deficiency.

4. Phlegm and Blood Stasis Block the Collaterals痰瘀阻絡

Accumulation of turbid Phlegm often due to a long course of disease, as well as abnormal Spleen functions in transformation and transportation by Spleen deficiency, blocks the Blood circulation; or long-term illness affects the collaterals leading to Blood stasis and collaterals occlusion resulting in pain.

MANIFESTATIONS

1. Wind-Cold Invasion風寒外襲

It often occurs or worsens due to cold weather or after a common cold. There is a feeling of tightening of the facial muscles during the pain, which presents paroxysmal short convulsive pain and a preference for hot compresses on the affected face. Patient is not thirsty.
Tongue: coating is thin and white or white and slippery.
Pulse: floating tight or deep slow.

2. Stomach Fire Uprising胃火上攻

The facial pain is severe and paroxysmal, and it is induced by heat. It is accompanied by irritability, swollen gums, halitosis, thirst, dry stool, red and yellow urination, and dull pain in the upper gastric region.
Tongue: red body, the coating is yellow, thick, or greasy.
Pulse: slippery and rapid.

3. Liver Fire Uprising肝火上炎

- The facial pain presents as frequent electric shocks accompanied red face, hypochondriac pain, bitter taste, and dry mouth. The pain is often induced by anger.

Tongue: red tongue body, yellow and greasy coating.
Pulse: deep and wiry.

- If the pain is induced by deficient Fire, the facial pain is usually worsened in the afternoon. The accompanied symptoms often include red cheeks, insomnia, forgetfulness, and convulsion.

Tongue: red with less coating.

Pulse: thin and wiry.

4. Phlegm and Blood Stasis Block the Collaterals痰瘀阻絡

It is an enduring facial pain, but the symptoms can come and go. The pain is intense, sharp, or stabbing.

If it is Phlegm obstruction, the accompanied symptoms include chest fullness, vomiting and salivation, loose stools, and dark complexion.

Tongue: dusky, slippery, sticky coating.

Pulse: deep and slippery pulse.

If it is Blood stasis, the location of pain is fixed and is worse in the afternoon.

Tongue: dusky, or ecchymosis and petechiae are seen.

Pulse: thin and choppy.

PHYSICAL EXAMINATION

- Medical History
  - Obtain the onset history: whether the pain begins while brushing teeth, eating or drinking, speaking, yawning, being exposed to the wind, or cold temperatures, as when washing the face with cold water.
  - Ask about the pain: whether onset is sudden, the pain intensity, the pain quality, and whether numbness in the face preceded the pain.
  - Inquire about the painful location: whether it's on one side of the face, in the forehead, cheek, jaw, teeth, gums, and/or lips. The pain can affect one or more divisions of the trigeminal nerve.
  - The frequency of the episodes: the number of attacks may vary from less than one per day to hundreds per day.
  - The duration of the episodes: the paroxysmal attacks of the facial pain can last from a fraction of a second to two minutes.
  - General health condition: if the pain was preceded by or coincided with a herpes zoster rash in the ipsilateral trigeminal distribution; if the pain occurs after a common cold; does the patient often experience high stress; is the pain preceded by a relevant trauma to the ipsilateral face; did the pain occur after an invasive dental procedure; is the pain evoked by chewing hard foods. For patients who also complain of toothache and pain with brushing teeth, a detailed oral examination should be performed to rule out the cause of secondary trigeminal neuralgia. Patients who have other neurological conditions, like multiple sclerosis, should be asked about other neurological symptoms such as focal weakness, dizziness, vision abnormality, and ataxia. Dizziness, vertigo, hearing loss, and numbness of the face may be accompanying symptoms indicating the condition in Chinese medicine.
- Inspection
  - The physical examination in patients with classical trigeminal neuralgia is generally normal. The examination will focus on the abnormalities that can be the causative factors to the secondary trigeminal neuralgia.

- The inspection in the physical examination should be performed on the head, ears, mouth, teeth, neck, and temporomandibular joint, paying attention to any abnormal findings, such as weakness in any facial muscle.
- Palpation
  - Neurologic examination is often normal in the classical trigeminal neuralgia. Some abnormal findings can be detected in the secondary trigeminal neuralgia, such as absent corneal reflex (Gold, Reich 2017).
- Special Tests
  - MRI may be useful to identify tumors, multiple sclerosis plaques, or blood vessel contact with the trigeminal nerve or any division.

DIFFERENTIAL DIAGNOSIS

- Trigeminal neuralgia needs to be differentiated from atypical facial pain. Both conditions have pain on one side of the face, electric shock-like stabbing pain may occur, and stress may be the factor. The difference is that atypical facial pain usually describes the pain as a burning or aching feeling, and the duration is continuous from several minutes to hours rather than episodic.
- Trigeminal neuralgia needs to be differentiated from temporomandibular joint disorders. Both conditions have pain on one side of the face and the pain is described as high-intensity facial pain. Temporomandibular joint disorders have a clicking, crackling, or crunching sensation in the jaw, the jaw can be locked, and spasm in the jaw muscles can be present. TMD often is accompanied by headache, neck pain near the ear, and difficulty opening or closing the mouth fully.
- Trigeminal neuralgia needs to be differentiated from Bell's palsy. Both conditions have facial symptoms on one side of the face, and the tongue functions (the anterior two-thirds of the tongue) are similar. Bell's palsy is a neurological disease affecting the seventh cranial nerve; it causes paralysis or weakness in the facial muscles, resulting inability to smile on one side of the face. The patient may lose the sense of taste.

TREATMENTS

1. Wind-Cold Invasion風寒外襲

Treatment Principle
Expel Wind, disperse Cold, warm the meridians, and stop pain.

Herbal Treatments
Herbal Formula
Chuan Xiong Cha Diao San 川芎茶調散

This formula is from "Prescriptions of the Bureau of Taiping People's Welfare Pharmacy 太平惠民和劑局方", authored by Imperial Medical Bureau 太醫局, in southern Song dynasty南宋 (1127–1239), and published in 1134.

Herbal Ingredients

Bo He薄荷240g, Fang Feng防風45g, Xi Xin細辛30g, Qiang Huo羌活60g, Bai Zhi 白芷60g, Zhi Gan Cao炙甘草60g, Chuan Xiong川芎120g, Jing Jie荊芥120g

Ingredient Explanations

- Bo He薄荷 expels Wind Heat, relieves Liver Qi stagnation, disperses stagnant Heat, and enhances Chai Hu's ability to relieve the Liver.
- Fang Feng防風 expels Wind and drains Damp.
- Xi Xin細辛 enters Shao Yin meridian to expel Wind, Damp, and Cold.
- Qiang Huo羌活 expels Wind-Cold-Dampness, unblocks painful obstruction, and alleviates pain.
- Bai Zhi白芷 expels Wind, drains Dampness, dispels Cold, and stops pain.
- Zhi Gan Cao炙甘草 tonifies the Spleen, harmonizes the other herbs, and guides the herbs to all 12 meridians.
- Chuan Xiong川芎 moves the Blood and moves Qi.
- Jing Jie荊芥 stops bleeding, dispels Wind, and relieves muscle spasm.

Modifications

- To make the formula stronger, add Sheng Jiang生薑 and Su Ye蘇葉.

Formula Actions

- Expels Wind.
- Stops pain.

Cooking Instructions

- Grind the herbs into granules and mix them.

Dose

- Mix 6 grams of the granules with tea and take it after every meal.

2. Stomach-Fire Uprising胃火上攻

Treatment Principle

Clear Stomach Fire and stop pain.

Herbal Treatments

Herbal Formula

Qing Wei San 清胃散

This formula is from "Secrets of the Orchid Pavilion蘭室秘藏", authored by Li Dongyuan 李東垣, in Yuan dynasty元朝 (1271–1368), and published in 1276.

Herbal Ingredients

Dang Gui當歸3g, Sheng Di Huang生地黃3g, Mu Dan Pi牡丹皮5g, Huang Lian黃連5g, Sheng Ma升麻10g

Ingredient Explanation

- Dang Gui當歸 nourishes the Blood and benefits Liver.
- Sheng Di Huang生地黃 nourishes Yin and clears Heat.
- Mu Dan Pi牡丹皮 clears Heat, cools the Blood, and moves the Blood.
- Huang Lian黃連 clears Heat and drains Damp.
- Sheng Ma升麻 releases the exterior, clears Heat, raises pure Yang, and lifts sunken Qi.

Modifications

- To make the formula stronger, add Xi Xin細辛, Bai Zhi白芷, and Chuan Xiong川芎.
- For halitosis, add Huo Xiang藿香 and Ding Xiang丁香.

Formula Actions

- Clears Stomach Fire.
- Cools the Blood.
- Reduces swelling.

Cooking Instructions

- Grind the herbs into granules and mix them evenly.
- Place the herbs with 1.5 cups of water in a pot; traditionally, a clay cooker is used.
- Turn on the stove to high heat and bring the herbs to a boil. Turn off the heat when the decoction is reduced to 0.7 cup.
- Drain the decoction into a container.

Dose

- Drink the decoction after it cools down 30 minutes after a meal.

3. Liver Fire Uprising肝火上炎

Treatment Principle

Clear Liver Fire and stop pain.

Herbal Treatments

Herbal Formula

Long Dan Xie Gan Tang龍膽瀉肝湯

This formula is from the book "Analytic Collection of Medicinal Formulas醫方集解", authored by Wang Ang汪昂 (1615–1694AD), and published in Qing dynasty 清朝 (1682 AD).

Herbal Ingredients

Long Dan Cao龍膽草12g, Zhi Zi栀子9g, Huang Qin黃芩9g, Chai Hu柴胡6g, Sheng Di Huang生地黃12g, Ze Xie澤瀉9g, Dang Gui當歸5g, Che Qian Zi車前子10g, Mu Tong川木通9g, Gan Cao甘草5g

Ingredient Explanations

Long Dan Cao龍膽草 drains Damp-Heat from the Liver and Gallbladder channels.

Zhi Zi栀子 clears Heat, cools the Blood, resolves Damp, reduces swelling, and opens the meridians.

Huang Qin黃芩 clears Heat, dries Damp, cools the Blood, stops bleeding, and calms Liver Yang.

Chai Hu 柴胡 moves Liver Qi and releases stagnation.

Sheng Di Huang生地黃 nourishes Yin and clears Heat.

Ze Xie澤瀉 drains Damp, especially Damp Heat in Lower Jiao; clears Kidney deficient Heat.

Dang Gui當歸 nourishes the Blood, benefits Liver, and regulates menstruation.

Che Qian Zi車前子 clears Damp Heat and promotes urination.

Mu Tong川木通 drains Heat and promotes urination.

Gan Cao甘草 harmonizes the herbs, clears Heat, and expectorates Phlegm.

Modifications

- For making the formula stronger, add Ju Hua菊花 and Chuan Xiong川芎.

Formula Actions

- Drains excess Fire from the Liver and Gallbladder; drains Damp-Heat from Lower Jiao.

Formula Indication

- Headache due to Liver Fire and Damp Heat in Lower Jiao, such as migraine headache, urethritis, acute glaucoma, and acute cholecystitis.

Contraindications

- Caution for patients who suffer from Cold and Yang deficiency in Spleen and Stomach, due to the Cold herbs in the formula.

Cooking Instructions

1) Place the herbs into an herb pot. Traditionally, a clay cooker is used. Gently rinse the herbs once with a strainer.
2) Add water into the pot until it just covers the herbs, then add one more cup of water.
3) Presoak the herbs for 15 minutes.
4) Turn on the stove to high heat and bring the herbs to a boil. Reduce the heat to medium-low to simmer but with slight bubbling. Turn off the heat when the decoction is reduced to one cup.

5) Drain the decoction into a container.

6) Repeat steps 2–5 once and drain each decoction to the same container.

7) Divide the decoction into two separate cups.

Dose

• Drink one cup of the decoction after meals twice daily.

4. Phlegm and Blood Stasis Block the Collaterals痰瘀阻絡

Treatment Principle

Transform Phlegm, move the Blood stasis, and open the meridians.

Herbal Treatments

Herbal Treatments

Herbal Formula

Xue Yu Zhu Yu Tang血府逐瘀湯

This formula is from "Correction of Errors in Medical Classics醫林改錯", authored by Wang Qingren 王清任 (1768–1831), in Qing dynasty清朝 1636–1912, and published in 1849.

Herbal Ingredients

Dang Gui當歸9g, Sheng Di Huang生地黃9g, Tao Ren 桃仁12g, Hong Hua紅花6g, Zhi Ke枳殼6g, Chi Shao赤芍6g, Chai Hu柴胡3g, Gan Cao 甘草3g, Jie Geng 桔梗 5g, Chuan Xiong川芎5g, Chuan Niu Xi牛膝9g

Ingredient Explanations

- Dang Gui當歸 moves the Blood, disperses Bruising, and stops pain due to Blood stasis.
- Sheng Di Huang生地黃 nourishes Yin and clears Heat.
- Tao Ren 桃仁 moves the Blood, disperses Bruising, and stops pain due to Blood stasis.
- Hong Hua紅花 moves the Blood, disperses Bruising, and stops pain due to Blood stasis.
- Zhi Ke枳殼 directs the actions of the formula to the lower body.
- Chi Shao赤芍 moves the Blood, disperses Bruising, and stops pain.
- Chai Hu柴胡 moves Liver Qi and releases stagnation.
- Gan Cao 甘草 harmonizes the herbs, tonifies Middle Jiao Qi, clears Heat, expectorates phlegm, and stops cough.
- Jie Geng 桔梗 directs the actions of the formula to the upper body.
- Chuan Xiong川芎 moves the Blood, disperses Bruising, and stops pain due to Blood stasis.
- Chuan Niu Xi牛膝 unblocks and promotes movement in the channels and collaterals, and stops spasms and convulsions.

Modifications

- For palpitation and insomnia, add Fu Shen茯神, Wu Wei Zi五味子, and Suan Zao Ren酸棗仁.
- For persistent fever, add Jin Yin Hua金銀花 and Lian Qiao連翹.

Formula Actions

- Moves the Blood, opens the meridians, and stops pain.

Formula Indication

- Headache due to Blood stasis.

Contraindications

- It is contraindicated for patients who suffer from headache that is not caused by Blood stasis.
- It is contraindicated during pregnancy.

Cooking Instructions

1) Place the herbs into an herb pot. Traditionally, a clay cooker is used. Gently rinse the herbs once with a strainer.
2) Add water into the pot until it just covers the herbs, then add one more cup of water.
3) Presoak the herbs for 15 minutes.
4) Turn on the stove to high heat and bring the herbs to a boil. Reduce the heat to medium-low to simmer but with slight bubbling. Turn off the heat when the decoction is reduced to one cup.
5) Drain the decoction into a container.
6) Repeat steps 2–5 once, and drain each decoction to the same container.
7) Divide the decoction into two separate cups.

Dose

- Drink one cup of the decoction after meals twice daily

Acupuncture Treatments

Acupuncture Point Formula
Local Points
GB20風池 Fengchi, SJ17翳風Yifeng, ST7下關Xiaguan

Distal Points
LI4合谷Hegu, LIV3太衝Taichong, LI10手三里Shousanli, ST44內庭Neiting through KD1 涌泉Yongquan

Modifications

- Ophthalmic Division (V1) Trigeminal Neuralgia

Point Selection: Yu Yao鱼腰
    Method: Oblique downward insertion 0.3–0.5 Cun 寸

Other points: Tai Yang 太陽, UB2攢竹Zanzhu, GB14陽白Yangbai, UB3眉衝 Meichong, ST8 頭維Touwei

- Maxillary Division (V2) Trigeminal Neuralgia

Point Selection: GB2 聽會Tinghui
    Method: Perpendicular insertion 1 Cun 寸
    Other points: ST2四白Sibai, LI20迎香Yingxiang

- Mandibular Division (V3) Trigeminal Neuralgia

Point Selection: ST7下關 Xiaguan
    Method: Perpendicular insertion 1.5 Cun 寸
    Other points: ST4地倉Dicang, ST6 頰車Jiache, REN24 承漿Chengjiang

Extra Points

Jia Cheng Jiang Xue夾承漿穴
    Location: 1 Cun 寸 bilateral to REN24 承漿Chengjiang
    Needle direction and depth: 0.1–0.2 Cun 寸, perpendicular insertion. Or 0.3–0.5 Cun 寸, oblique downward insertion.

Auricular Therapy

Liver Yang, Shen Men, Cheek, Forehead, Stomach, Kidney

Tui Na Treatments

Tui Na Techniques
Pressing, kneading, pushing, rubbing

Tui Na Procedures

The patient is in a sitting position; the therapist stands in front of the patient.

- Applies pushing technique with both thumb pads on Yin Tang印堂 to the front hairline for 30 times.
- Applies pressing and kneading techniques with both middle finger pads on ST7下關 Xiaguan for 1 minute Figure 2.7.1

- Applies pressing and kneading techniques with both middle finger pads on ST6頰車 Jiache for 1 minute.
- Applies pressing and kneading techniques with both thumb pads on SJ17翳風Yifeng for 1 minute.
- Applies pressing and kneading techniques with both middle finger pads on GB20風池 Fengchi for 1 minute. Figure 2.7.2

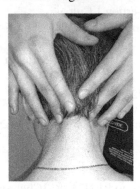

- Applies rubbing technique with both palms on the face for 2 minutes.

PATIENT ADVISORY

Dietary recommendations:

- The food should be soft, easily digested, and light taste.
- If biting hard food is a factor, recommend liquid food
- Avoid spicy, pungent, raw, cold, sweet, and fatty foods.
- Avoid fried food.
- Avoid foods that are cold or cool in nature, if Wind Cold invasion is a factor.
- Use gentle force in brushing teeth, washing face, and rinsing the mouth.
- Be careful when talking, laughing, and coughing.
- Keep the head and face warm.
- Avoid using very hot or very cold water when washing the face.
- Avoid catching cold. Have an appropriate room temperature, and wear warm clothes.
- Be happy, manage the mood, and avoid anger.
- Go to sleep no later than 11 p.m. Ensure that there is plenty of good quality sleep.

BELL'S PALSY 面癱

INTRODUCTION

Sir Charles Bell first described the facial condition in 1829, and the condition is named after him. Bell's palsy is the most common cause of facial paralysis.

Facial paralysis can be bilateral and unilateral, but bilateral facial paralysis is extremely rare (less than 2% of all the facial palsy cases). Bell's palsy accounts for only 23% of bilateral facial paralysis (Pothiawala and Lateef 2012).

Generally, Bell's palsy causes sudden weakness in the facial muscles, with drooping around the mouth, drooling, and inability to fully close the eye.

It affects about 40,000 people in the United States every year. It can affect anyone, any gender, and any age (NINDS 2018).

ANATOMY

The impaired nerve that causes Bell's palsy is the facial nerve. It is the seventh paired cranial nerve.

Cranial nerve VII can be discussed in two parts:

- Intracranial – the course of the nerve within the cranium.
- Extracranial – the course of the nerve outside the cranium which innervates the face, external ear, the anterior two-thirds of the tongue, and the glands of the head and neck.

INTRACRANIAL

Cranial nerve VII arises in the pons of the brain stem at the area posterior to cranial nerve VI (abducens nerve) and anterior to cranial nerve VIII (vestibulocochlear nerve).

The facial nerve comprised of a large motor root and a smaller sensory root.[20] The motor root of the facial nerve originates in the facial nerve nucleus, which receives input from the primary motor cortex and the ophthalmic division of the trigeminal nerve.[21]

The sensory root arises from the intermediate nerve.[22]

Both motor and sensory fibers are fused to comprise the facial nerve and form the geniculate ganglion; then the ganglion gives rise to three nerves:

- *Greater petrosal nerve*: the function of it is taste and secretions[23] from the mucous and lacrimal glands. Damage to this branch may result in reduced tear secretion (Nturibi and Bordoni 2021).
- *Nerve to stapedius muscle* of the middle ear – damage to this branch may result in paralysis of stapedius leading to hypersensitivity to loud noises.[24]
- *Chorda tympani nerve* – this branch carries special sensation from the anterior two-thirds of the tongue via the lingual nerve. Damage to the branch may result in reduced saliva secretion and loss of taste to the anterior two-thirds of the tongue.

Then the facial nerve traverses through the facial canal (also known as the Fallopian canal) located in the petrous portion of the temporal bone then exits the skull at the stylomastoid foramen (Carreiro 2009). The exit is just located posterior to the styloid process of the temporal bone.

Extracranial

After the facial nerve exits the skull, it runs anterior to the outer ear and gives rise to:

- The posterior auricular nerve, which innervates the intrinsic and extrinsic muscles around the outer ear and the occipital part of the occipitofrontalis muscle.
- The motor branches of the facial nerve that innervates the posterior belly of the digastric muscle (some reported cases addressed that the anterior belly of the digastric muscle was innervated by the branches of the facial nerve in addition to the mylohyoid nerve) (Kim and Loukas 2019).
- *Nerve to the stylohyoid muscle* (which is one of four suprahyoid muscles), its function is to elevate and retract the hyoid bone, elongating the floor of the mouth during swallowing.[25]

The facial nerve continues anteriorly and inferiorly into the parotid gland and splits into five motor branches that they are responsible for innervating the muscles of the facial expression.

- Temporal branch – innervates the frontalis, orbicularis oculi and corrugator supercilia.
- Zygomatic branch – innervates the orbicularis oculi.
- Buccal branch – innervates the orbicularis oris, buccinator and zygomaticus.
- Marginal mandibular branch – innervates the depressor labii inferioris, depressor anguli oris and mentalis.
- Cervical branch – innervates the platysma).

Western Medicine Etiology and Pathology

Facial palsy describes any kind of facial muscle paralysis; the origin of palsy includes central and peripheral.

Central facial palsy is the result of stroke that causes facial paralysis. The symptoms of facial palsy will be limited to the lower half of the face; stroke is the upper motor neuron lesion.

Peripheral facial palsy is the result of the facial nerve fiber (cranial nerve VII) being damaged by known or unknown causes.

Clinically, there are several possible known causes resulting in peripheral facial palsy.

- The viruses known to be able to cause viral neuritis are herpes simplex virus, Epstein-Barr virus, varicella-zoster virus, and the chicken pox shingles virus.
- The bacteria known to be able to cause neuritis is the bacterium *Borrelia burgdorferi*, or Lyme disease.

From the initial infection, the body develops immunity that keeps the infection under control. Under certain conditions, such as sleep deprivation that awakens the dormant infection to cause sickness in the nerve, the person's immunity weakens.[26]

When the cause of peripheral facial palsy is unknown or cannot be confirmed, it is an "idiopathic" peripheral facial paralysis, and it is called "Bell's palsy". Even though the exact cause of Bell's palsy is still unknown, clinically, it is believed that the cause can be from cold weather, air-conditioning and wind blowing directly onto the face. Additionally, emotional stress, starvation, and fatigue may be risk factors.[27]

CHINESE MEDICINE ETIOLOGY AND PATHOLOGY

Bell's palsy in Chinese medicine is a condition called "Zhong Feng, Zhong Jing Luo 中風中經絡" which literally translates to "Catching Wind-Wind attack the meridians". According to Chinese medicine, Wind is the factor causing blockages in the facial meridians resulting Bell's palsy.

1. Wind-Cold Invasion 風邪入絡

Wind is a Yang energy, it tends to go upward, and it is characterized by rapid movements and rapid changes. Wind-Cold blocks the circulation of Qi and Blood in the facial upper meridians, resulting in a loss of muscular movement and sensation.

2. Qi and Blood Deficiency氣血兩虛

Wei Qi 衛氣 is the equivalent of the immune system; it circulates between the skin and muscle and protects us from pathogenic environmental factors.

Illness, both physical and mental, fatigue, pregnancy, and emotional stress can weaken Qi and Blood. When Wei Qi Wei Qi 衛氣 is undernourished due to Qi and Blood deficiency, the body is unable to fight against Wind-Cold invasion.

3. Phlegm and Blood Stasis Occlusion痰瘀互阻

Wind-Cold invades meridians with Phlegm and Blood clots, resulting in meridian obstruction.

4. Liver Qi Stagnation 肝氣鬱結

When Liver is unable to regulate the circulation of Qi, Blood, and body fluid, it manifests as depression. When Liver is over-dispersing, it results in agitation. Both under- and over-dispersing can result in dysfunction of regulation.

Insufficient Liver Blood causes rigid Tendons that fail to act nimbly 七八肝氣衰，筋不能動, dry eyes and blurred vision, and Liver Wind stirring. The deficiency of Liver Blood results in facial muscles twisting and failing to move, manifesting as deviation of the mouth and inability to close the eye.

Liver meridian connects with the eyes, continues upward to emerge from the forehead to the end of the vertex of the head. The illness of Liver meridian results in dry eyes and inability to close the eye completely.

5. Liver Wind Stirring 肝風內動

Insufficiency of Liver Blood causes stirring of Liver Wind. Or Liver Wind stirring underlies Liver Yang rising.

MANIFESTATIONS

In Western Medicine

- Rapid onset of mild weakness to total paralysis on one side of the face – occurring within hours to days.
- Facial droop and difficulty making facial expressions, such as closing the eye or smiling.
- An inability to make facial expressions, such as smiling or frowning.
- Drooling.
- Difficulty speaking, eating, and drinking.
- Facial weakness.
- Muscle twitches in the face.
- Pain around the jaw or in or behind the ear on the affected side.
- Increased sensitivity to sound on the affected side.
- Headache.
- Loss of the sense of taste.
- Changes in the amount of tears and saliva the patient produces, dry eye and mouth, irritation of the eye on the involved side.

In Chinese Medicine

1. Wind-Cold Invasion 風邪入絡

Sudden onset of deviation of the eye and mouth, facial paresthesia, headache, stuffy nose, stiff neck, occasional twitching of facial muscles, the eye on the affected side cannot be closed, and facial lines become shallow or flat.

Tongue: thin white coating.

Pulse: superficial.

2. Qi and Blood Deficiency氣血兩虛

Deviation of the eye and mouth, lax facial muscles, forehead lines become shallow or flat, the eye on the affected side cannot be closed, lacrimation, drooling, leakage of food from the mouth while chewing, shortness of breath, and withdrawal. Dizziness, fatigue, poor appetite, palpitation, and blurry vision.

Tongue: pale body with thin white coating.

Pulse: thready and feeble.

3. Phlegm and Blood Stasis Occlusion痰瘀互阻

Deviation of the eye and mouth, numbness on facial muscles, slurred speech, wheezing noise in throat like whistling or squeaking, and salivation.

Tongue: rigid tongue body, white sticky coating.
Pulse: wiry and slippery or wiry and slow.

4. Liver Qi Stagnation 肝氣鬱結

The facial paralysis occurs under the influence of anger, manifested by rigid facial muscles, facial lines become flat, the eye on the affected side cannot be closed, photophobia, lacrimation, swelling and pain at the base of the ear, sighing, feeling of fullness and distress in the chest and rib cage, loss of appetite, occasional grief, and crying.
Tongue: thin white coating.
Pulse: wiry.

5. Liver Wind Stirring 肝風內動

Sudden onset of paralysis, red face, pain behind the ear, numbness on limbs, head feels heavy but the feet feel light, and vertigo.
Tongue: pale body, yellow or less, and dry coating.
Pulse: wiry, rapid, and forceful.

PHYSICAL EXAMINATION

- Medical History
  - General health conditions such as the Shen, generally Bell's palsy patient appear normal. Patients with Bell's palsy may lose the sense of taste or complain of a difference in taste on the anterior two-thirds of the tongue. Patients may also complain of a sensitivity to sound, otalgia, and changes to tearing and salivation.
  - Obtain the history of pregnancy and skull or facial fractures.
  - History of stroke, infection (e.g., ear infection, infection of the skull bone close to the ear, such as mastoiditis; the infection of the parotid gland, such as parotitis; and infection with a herpes virus, such as Ramsay Hunt syndrome), surgery and head trauma, and autoimmune diseases, such as Lyme disease.
  - History of dental procedures because iatrogenic facial palsy may also be as a result of an incorrectly placed dental local anesthetic, such as the inferior alveolar nerve block[28].
  - History of skin cancer is also a concern because it can spread from the skin along the nerve.
  - History of exposure to wind and cold temperature.
  - History of stress and sleep deprivation.
  - History of tumor in the head. The common type that causes facial palsy is a benign growth that occurs inside the skull called a vestibular schwannoma.
- Inspection
  - Observe the region of the face and the ear to rule out swelling. Wrinkling of the forehead when raising the eyebrows is asymmetric or absent on

the affected side. Also ask patients with facial palsy to frown, close
eyes, show teeth, and open mouth (to compare nasolabial folds), and
ask the patient to puff out their cheeks. Patients with Bell's palsy may
present incomplete closure of the eye when attempts to close the eyes
and an inability to puff the cheeks out on the affected side.

- Vital signs in patients with Bell's palsy are usually normal. Patients
  are usually oriented to person, place, and time. Herpes simplex and
  varicella-zoster virus (shingles) may have a fever.
- Skin examination of patients with Bell's palsy is usually normal. It is
  still necessary to differentiate the causative diseases, for example, herpes simplex may present blistering sores in the mouth or on the geni-
  tals and examine bruises for local physical trauma. Some diseases may
  present rashes (e.g., shingles); Ramsay Hunt syndrome may present
  vesicles in the ipsilateral ear, on the hard palate, or on the anterior two-
  thirds of the tongue; shingles may present ulcers at the site of infection
  (e.g., herpes simplex).
- Taste can be tested on the anterior two-thirds of the tongue. This can be
  done by the patient's anterior portion of the tongue touched by a swab
  dipped in a flavored solution.

- Palpation
  - Examine the facial muscle tone; rigid muscle may indicate Liver Qi
    stagnation, and lax muscle indicates Qi and Blood deficiency.
  - Tenderness upon palpation of the ear may be present.[29]
  - Facial tenderness in distribution of facial nerve.[30]
  - Patients with Bell's palsy are usually normal in the examination of
    palpation on neck, back, abdomen, and genitourinary areas. Muscle
    rigidity test may be rigid or lax. Other neuromuscular examinations
    are usually normal, such as Babinski reflex, straight-leg raise test, gait,
    finger-to-nose test, and palm-tapping test.
- Range of Motion
  Patients with Bell's palsy are usually normal in range of motion on limbs
  and neck.
- Special Tests
  - Impaired glabellar reflex: gentle finger percussion of the glabella while
    observing for involuntary blinking with each stimulus; the reflexes may
    be weak or absent because the efferent response of this reflex is carried
    by the facial nerve.
  - Impaired corneal reflex: stand out of the patient's line of vision. Gently
    touch a thin strand of clean cotton ball to the cornea while observing for
    blinking. The orbicularis oculi muscle is innervated by cranial nerve
    VII, contraction of the palpebral portion closes the eyelid. When one
    eye is tested, both should blink. When only the untouched eye blinks,
    the facial nerve palsy is ipsilateral. When the untouched eye doesn't
    blink, a contralateral facial nerve palsy may be the cause.

DIFFERENTIAL DIAGNOSIS

Bell's palsy needs to be differentiated from the following conditions.

*   Stroke-Related Facial Palsy

In facial paralysis due to stroke, the patients can still raise eyebrows, and the eyes can be closed firmly.

*   Ramsay Hunt Syndrome or Shingles-Related Facial Palsy

It typically presents with painful vesicles on the ipsilateral outer ear, on the hard palate, or on the anterior two-thirds of the tongue (Sweeney and Gilden 2001).

*   Lyme Disease

ELISA (enzyme-linked immunosorbent assay) and Western blot are positive.

TREATMENTS

1. Wind-Cold Invasion 風邪入絡

Treatment Principle

Expel Wind, release Exterior, and open the meridians.

Herbal Treatments

Herbal Formula

Qian Zheng San 牽正散

This formula is from "Yang Family Preserved Formulas 楊氏家藏方", authored by Tan Yang 楊倓(子靖) (1120–1185), in southern Song dynasty南宋 (1127–1239), and published in 1178.

Herbal Ingredients

Zhi Bai Fu Zi制白附子1g, Jiang Can殭蠶1g, Quan Xie全蠍1g

Ingredient Explanation

*   Zhi Bai Fu Zi制白附子 dispels Wind, transforms Phlegm, and stops spasms.
*   Jiang Can殭蠶 extinguishes Wind, transforms Phlegm, and stops spasms and convulsions.
*   Quan Xie全蠍 extinguishes Wind, unblocks collaterals; and stops tremors, convulsions, and pain.

Modifications

*   To make the formula stronger, add Wu Gong蜈蚣, Tian Ma天麻, and Di Long地龍to increase the functions of expelling Wind, stopping spasms, and opening the meridians.

Formula Actions
- Expels Wind.
- Dissolves Phlegm.
- Stops spasm.

Cooking Instructions
- Grind the herbs into granules.

Dose
- Drink the granules with warm wine, three times a day.

2. Qi and Blood Deficiency氣血兩虛

Treatment Principle
Tonify Qi and Blood, move the Blood, open the meridians.

Herbal Formula
Gui Pi Tang歸脾湯

This formula is from "Categorized Essentials of Repairing the Body 正體類要", authored by Xue Ji 薛己 (1487–1559), in Ming dynasty 明朝 (1368–1644), and published in 1529.

Herbal Ingredients
Ren Shen人參3g, Zhi Huang Qi 黃耆炙15g, Bai Zhu白朮9g, Fu Ling茯苓9g, Dang Gui當歸9g, Suan Zao Ren酸棗仁炒9g, Long Yan Rou龍眼肉9g, Zhi Yuan Zhi遠志6g, Mu Xiang木香3g, Zhi Gan Cao甘草炙4.5g, Sheng Jiang生薑2 slices, Da Zao大棗3 pieces

Ingredient Explanation
- Ren Shen人參 tonifies Qi and Yang.
- Zhi Huang Qi黃耆炙 tonifies Qi and strengthens the Spleen.
- Bai Zhu白朮 tonifies the Spleen Qi and dries Dampness.
- Fu Ling茯苓 drains Damp and tonifies the Spleen.
- Dang Gui當歸 moves the Blood, disperses Bruising, and stops pain due to Blood stasis.
- Suan Zao Ren酸棗仁 nourishes Heart Yin and tonifies Liver Blood.
- Long Yan Rou龍眼肉 drains Damp-Heat from the Liver and Gallbladder channels.
- Yuan Zhi遠志 expels Phlegm, calms Shen, and reduces abscesses.
- Mu Xiang木香 promotes the movement of Qi and alleviates pain.
- Zhi Gan Cao甘草炙 tonifies the Spleen, harmonizes the other herbs, and guides the herbs to all 12 meridians.
- Sheng Jiang生薑 unblocks the pure Yang pathway and harmonizes rebellious Qi.
- Da Zao紅棗 tonifies the Spleen Qi, nourishes the Blood, and moderates and harmonizes the harsh properties of the fragrant herbs.

Formula Actions
- Nourishes the Blood.
- Tonifies Qi.
- Strengthens the Spleen.

Contraindications
- Contraindicated for those who have bleeding with interior Heat or insomnia.

Modifications
- To make the formula stronger, add Shu Di Huang 熟地and E Jiao阿膠.
- For severe insomnia and forgetfulness, add He Huan Pi 合歡皮, Ye Jiao Teng夜交藤, and Shi Chang Pu 石菖蒲.

Cooking Instructions
1) Place the herbs into an herb pot. Traditionally, a clay cooker is used. Gently rinse the herbs once with a strainer.
2) Add water into the pot until it just covers the herbs, then add one more cup of water.
3) Presoak the herbs for 15 minutes.
4) Turn on the stove to high heat and bring the herbs to a boil. Reduce the heat to medium-low to simmer but with slight bubbling. Turn off the heat when the decoction is reduced to one cup.
5) Drain the decoction into a container.
6) Repeat steps 2-5 once and drain each decoction to the same container.
7) Divide the decoction into two separate cups.

Dose
- Drink one cup of the decoction after meals twice daily.

3. Phlegm and Blood Stasis Occlusion痰瘀互阻

Treatment Principle
Expel Wind, transform Phlegm, tonify the Spleen, and open the meridians.

Herbal Formula
Ban Xie Bai Zhu Tian Ma Tang半夏白朮天麻湯

This formula is from "Medical Insights 醫學心悟", authored by Cheng Guopeng 程國彭 (1662–1735), in Qing dynasty清朝 1636–1912, and published in 1732.

Herbal Ingredients
Zhi Ban Xia 半夏4.5g, Bai Zhu白朮3g, Tian Ma天麻3g, Ju Hong橘紅3g, Fu Ling茯苓3g, Gan Cao甘草1.5g, Sheng Jiang生薑2pieces, Da Zao大棗3g, Man Jing Zi 蔓荊子3g,
水煎服。

Ingredient Explanation
- Zhi Ban Xia半夏 transforms Phlegm.
- Tian Ma天麻 extinguishes Wind, calms the Liver, subdues rising Liver Yang, and alleviates pain.
- Bai Zhu白朮 tonifies the Spleen Qi and dries Dampness.
- Ju Hong橘紅 regulates Qi and transforms Phlegm.
- Bai Zhu白朮 tonifies the Spleen Qi and dries Dampness.
- Fu Ling茯苓 drains Damp and tonifies the Spleen.
- Gan Cao甘草 harmonizes the herbs, tonifies Middle Jiao Qi, clears Heat, expectorates Phlegm, and stops cough.
- Sheng Jiang生薑 unblocks the pure Yang pathway and harmonizes rebellious Qi.
- Da Zao大棗 tonifies the Spleen Qi, nourishes the Blood, and moderates and harmonizes the harsh properties of the fragrant herbs.
- Man Jing Zi 蔓荊子 expels Wind and stops pain.

Formula Actions
- Tonifies Spleen Qi.
- Dissolves Dampness.
- Transforms Phlegm.
- Soothes the Liver.
- Expels Wind.

Contraindications
- Contraindicated for those with interior Heat, due to Kidney and Liver Yin deficiency, or dizziness and vertigo, due to Qi and Blood deficiency.

Modifications
- To make the formula stronger, add Shen Qu神曲, Mai Ya麥芽, Ze Xie澤瀉, Ren Shen人參, Huang Qi黃芪, Gan Jiang乾薑, and Huang Bai黃柏.
- For severe headache, add Bai Ji Li白蒺藜.
- For headache due to Blood deficiency, add Dang Gui當歸 and Chuan Xiong 川芎.
- For severe Qi deficiency, add Ren Shen 人參, Dang Shen黨參, and Huang Qi黃芪.
- For severe vertigo, add Jiang Can殭蠶 and Dan Nan Xing膽南星.
- For arteriosclerosis in the brain, add Gou Teng鉤藤.
- For Liver Yang rising, add Gou Teng鉤藤 and Dai Zhe Shi代赭石.
- For severe dizziness due to Liver excess, add Jiang Can殭蠶, Gou Teng鉤藤, and Dan Nan Xing膽南星.

Cooking Instructions
1) Place the herbs into an herb pot. Traditionally, a clay cooker is used. Gently rinse the herbs once with a strainer.

2) Add water into the pot until it just covers the herbs, then add one more cup of water.
3) Presoak the herbs for 15 minutes.
4) Turn on the stove to high heat and bring the herbs to a boil. Reduce the heat to medium-low to simmer but with slight bubbling. Turn off the heat when the decoction is reduced to one cup.
5) Drain the decoction into a container.
6) Repeat steps 2–5 once and drain each decoction to the same container.
7) Divide the decoction into two separate cups.

Dose

- Drink one cup of the decoction after meals twice daily.

4. Liver Qi Stagnation 肝氣鬱結

Treatment Principle
Move Liver Qi and stop pain.

Herbal Formula
Xiao Yao San逍遙散

This formula is from "Prescriptions of the Bureau of Taiping People's Welfare Pharmacy 太平惠民和劑局", authored by Imperial Medical Bureau 太醫局, in southern Song dynasty南宋 (1127–1239), and published in 1134.

Herbal Ingredients
Chai Hu柴胡10g, Dang Gui當歸10g, Bai Shao白芍10g, Bai Zhu白朮10g, Fu Ling茯苓10g, Zhi Gan Cao甘草5g, Sheng Jiang生薑10g, Bo He 薄荷5g

Ingredient Explanations
- Chai Hu 柴胡 relieves Liver Qi stagnation, pacifies the Liver, relieves the Shao Yang, and reduces fever.
- Dang Gui 當歸 nourishes the Blood, regulates menses, and invigorates and harmonizes the Blood.
- Bai Shao 白芍 calms Liver Yang, alleviates pain, nourishes the Blood, and regulates menstruation.
- Bai Zhu 白朮 tonifies the Spleen and tonifies Qi.
- Fu Ling 茯苓 strengthens the Spleen and harmonizes the Middle Jiao.
- Zhi Gan Cao 甘草 tonifies Spleen Qi, stops pain, and harmonizes the other herbs.
- Bo He 薄荷 relieves Liver Qi stagnation, disperses stagnant Heat, and enhances Chai Hu's ability to relieve the Liver.
- Sheng Jiang 生薑 harmonizes and prevents rebellious Qi and normalizes the flow of Qi at the center.

Modifications

- To make the formula stronger, add Xiang Fu香附, Yu Jin鬱金, and Chen Pi 陳皮.
- For Liver Fire, add Mu Dan Pi丹皮; add Zhi Zi梔子 to clear Heat and cool the Blood.

Formula Actions

- Soothes the Liver.
- Releases stagnation.
- Strengthens the Spleen.
- Harmonizes Liver and Spleen.

Cooking Instructions

3. Grind the herbs into granules.
4. Mix 6 grams in one cup water.
5. Boil the herbs in 0.7 cup water.

Dose

Drink the decoction when it is warm, twice daily.

5. Liver Wind Stirring 肝風內動

Treatment Principle

Soothe the Liver, extinguish Wind, open the meridians, and stop spasm.

Herbal Formula

Tian Ma Gou Teng Yin 天麻鉤藤飲

This formula is in the book "Newly Explained Diagnosis and Treatment of TCM Internal Medicine and Miscellaneous Diseases中醫內科雜病證治新義", authored by Hu Guangci 胡光慈 in 1958.

Herbal Ingredients

Tian Ma天麻6g, Gou Teng鉤藤9g, Chuan Niu Xi牛膝12g, Shi Jue Ming石決明15g, Du Zhong杜仲12g, Huang Qin黃芩6g, Zhi Zi梔子6g, Yi Mu Cao益母草9g, Sang Ji Sheng桑寄生9g, Ye Jiao Teng夜交藤9g, Fu Shen茯神9g

Ingredient Explanations

- Tian Ma天麻 extinguishes Wind, calms the Liver, subdues rising Liver Yang, and alleviates pain.
- Gou Teng鉤藤 expels Wind and releases spasm.
- Chuan Niu Xi川牛膝 unblocks and promotes Blood circulation.
- Yi Mu Cao益母草 moves the Blood, promotes urination, reduces swelling, and clears Heat.
- Shi Jue Ming石決明 drains Liver Fire and expels Wind.
- Du Zhong杜仲 tonifies the Liver and Kidney, strengthens Bone, and benefits Essence and Blood.

- Huang Qin黄芩 clears Heat, dries Damp, cools the Blood, stops bleeding, and calms Liver Yang.
- Zhi Zi栀子 clears Heat, cools the Blood, resolves Damp, reduces swelling, opens the meridians.
- Sang Ji Sheng桑寄生 tonifies the Liver and Kidney, strengthens Bone, and benefits Essence and Blood; expels Wind and drains Damp.
- Ye Jiao Teng夜交藤 nourishes Heart Yin and Blood, calms Shen, and expels Wind and Damp.
- Fu Shen茯神 calms Shen.

Modifications

- To make the formula stronger, add Ju Hua菊花, Xia Ku Cao夏枯草, Ku Ding Cha苦丁茶, and Zhen Zhu Mu珍珠母.
- For severe Liver Heat, add Xia Ku Cao夏枯草 and Long Dan Cao龍膽草.
- For Blood stasis, add Dan Shen丹參 and Chi Shao赤芍.
- For insomnia, add Long Gu龍骨, Mu Li牡蠣, Ye Jiao Teng夜交藤, and Yuan Zhi遠志.

Formula Actions

- Soothes Liver Qi, expels Wind, moves the Blood, clears Heat, and tonifies the Liver and Kidney.

Formula Indication

- Headache, vertigo, insomnia, anger, and bitter taste in the mouth due to Liver Yang uprising.

Contraindications

- Caution for patients who suffer from Blood, Yin, and body Fluid deficiency.
- Stop taking this formula if there is dizziness, chest distention, shortness of breath, nausea, vomiting, or skin itch.
- Do not take Tian Ma overdose.

Cooking Instructions

1) Place the herbs into an herb pot. Traditionally, a clay cooker is used. Gently rinse the herbs once with a strainer.
2) Add water into the pot until it just covers the herbs, then add one more cup of water.
3) Presoak the herbs for 15 minutes.
4) Turn on the stove to high heat and bring the herbs to a boil. Reduce the heat to medium-low to simmer but with slight bubbling. Turn off the heat when the decoction is reduced to one cup.
5) Drain the decoction into a container.
6) Repeat steps 2–5 once and drain each decoction to the same container.
7) Divide the decoction into two separate cups.

Dose

- Drink one cup of the decoction after meals twice daily.

External Formula

Herbal Formula

Jia Wei Qian Zheng San 加味牽正散

Herbal Ingredients

Zhi Bai Fu Zi制白附子3g, Jiang Can殭蠶3g, Quan Xie全蠍3g, Zhi Chuan Wu 制川烏3g, Zhi Cao Wu 制草烏3g, Ban Xia半夏3g, Wei Ling Xian威靈仙3g, Bai Ji白芨 3g, Chen Pi陳皮3g, Ginger juice生姜汁 3g

Ingredient Explanations

- Zhi Bai Fu Zi制白附子 dispels Wind, transforms Phlegm, and stops spasms.
- Jiang Can殭蠶 extinguishes Wind and transforms Phlegm; stops spasms and convulsions.
- Quan Xie全蠍 extinguishes Wind and unblocks the collaterals; stops tremors, convulsions, and pain.
- Zhi Chuan Wu 川烏 expels exterior Wind, drains interior Damp, warms channels, and stops pain.
- Zhi Cao Wu 草烏 expels Wind and disperses Blood and Cold.
- Ban Xia半夏 transforms Phlegm.
- Wei Ling Xian威靈仙 expels Wind Damp and stops pain.
- Bai Ji白芨 astringes leakage of Blood and stops bleeding; reduces swelling and generates flesh.
- Chen Pi陳皮 dries Damp and transforms Phlegm.
- Ginger juice生姜汁 unblocks the pure Yang pathway and harmonizes rebellious Qi.

Direction

Grind the herbs into fine powders and mix with alcohol or vinegar; apply topically three times a day.

Acupuncture Treatments

Acupuncture Point Formula

Local Points

SJ17翳風Yifeng, GB14陽白Yangbai, ST2四白Sibai, ST4地倉Dicang, ST7下關 Xiaguan, LI20迎香 Yingxiang

Distal Points

GB20風池Fengchi, LI4合谷He Gu, LIV3太沖Taichong

Seven Star Needle

Tap the needle on GB14陽白Yangbai, Tai Yang太陽, ST4地倉Dicang, LI4合谷He Gu, LIV3太沖Taichong until red. Treatment frequency is once daily.

Tui Na Treatments
Tui Na Techniques
Pushing

Tui Na Procedures
Pushing on five facial lines

4. Begins from REN24承浆Chengjiang→ST6頰車Jiache→ST7下關xiagu an→ST8頭維Touwei
5. Begins from REN24承浆Chengjiang→ST4地仓Dicang→SI18顴 髎Quan liao→GB1瞳子 髎Tongziliao→Taiyang太 陽→ST8頭維Touwei
6. Begins from contralateral LI20迎 香Yingxiang→DU26人中Renzhong →LI20迎香Yingxiang→ST1承泣Chengqi→GB1瞳子 髎Tongziliao→Tai yang太 陽→ST8頭維Touwei
7. Begins from contralateral ST4地仓Dicang→REN24承浆Chengjiang→ST6 頰車Jiache→SJ17翳風Yifeng → GB20風池Fengchi
8. Begins from contralateral ST1承泣Chengqi→LI20迎香Yingxiang→DU26 人中Renzhong→LI20迎香Yingxiang →SI18顴髎Quanliao→ST7下 關xiag uan→SJ17翳風Yifeng→ends at GB20風池Fengchi

Method: Apply gentle force with the thumb pads pushing in even speed on these five lines orderly, once daily.

PATIENT ADVISORY

- Early treatment is key.
- Avoid catching cold; wearing a mask is helpful.
- Promote blood circulation on the face during the acute stage to improve inflammation and reduce swelling.
- The sooner the treatments in Chinese medicine are received, the better the treatment outcome.
- Avoid drinking alcohol and smoking.
- Stick to a routine in the daily life schedule.
- Exercise regularly.
- Be happy.

NOTES

1. https://www.scientificamerican.com/article/to-every-pathogen-there-is-a-season/.
2. https://www.pasadenahealthcenter.com/blog/family-healthcare/common-winter-illneses -and-their-symptoms/.
3. https://www.greenhillspeds.com/news/well-child-visits/illnesses-and-allergies-to-watch -for-in-the-fall/#:~:text=The%20most%20common%20fall%20illness.
4. https://thejournalofheadacheandpain.biomedcentral.com/track/pdf/10.1186/s10194-020 -01134-1.pdf.

5. "Digastric Muscle." n.d. Physiopedia. https://www.physio-pedia.com/Digastric_Muscle.
6. https://thejournalofheadacheandpain.biomedcentral.com/track/pdf/10.1186/s10194-020
   -01134-1.pdf.
7. https://www.spinalcord.com/types-of-traumatic-brain-injury.
8. https://www.brainfacts.org/ask-an-expert/what-happens-in-the-brain-during-and-after-a
   -concussion.
9. https://www.aans.org/en/Patients/Neurosurgical-Conditions-and-Treatments/
   Concussion.
10. https://www.brainline.org/article/anatomy-brain-trauma-concussion-and-coma.
11. https://www.physio-pedia.com/Head_Impulse_Test.
12. https://pocketdentistry.com/4-mechanics-of-mandibular-movement/.
13. https://howldental.com/tmj-disorder-worse-stress.
14. https://www.physio-pedia.com/Temporomandibular_Disorders?utm_source=physiope-
    dia&utm_medium=related_articles&utm_campaign=ongoing_internal.
15. https://www.merckmanuals.com/professional/ear,-nose,-and-throat-disorders/nose-and
    -paranasal-sinus-disorders/sinusitis.
16. https://my.clevelandclinic.org/health/diseases/17700-chronic-sinusitis.
17. https://www.acponline.org/acp_news/misc/video/sinusitis.pdf.
18. https://www.wikidoc.org/index.php/Rhinosinusitis_physical_examination.
19. https://www.allergytampa.com/2019/01/18/the-difference-between-rhinitis-and
    -sinusitis.
20. https://teachmeanatomy.info/head/cranial-nerves/facial-nerve/.
21. https://www.kenhub.com/en/library/anatomy/facial-nerve.
22. https://en.wikipedia.org/wiki/Facial_nerve.
23. https://www.verywellhealth.com/greater-petrosal-nerve-5111893#:~:text=The%20greater
    %20petrosal%20nerve%20is,dealing%20with%20secretions%20from%20glands.
24. https://radiopaedia.org/articles/nerve-to-stapedius?lang=us#:~:text=The%20nerve%20to
    %20stapedius%20arises,to%20loud%20noises%20(hyperacusis.
25. https://en.wikipedia.org/wiki/Suprahyoid_muscles.
26. https://www.tampabayhearing.com/ear-education/bells-palsy-and-ramsay-hunt
    -syndrome/.
27. https://crystal-touch.nl/what-is-the-difference-between-facial-palsy-and-bells-palsy.
28. https://en.wikipedia.org/wiki/Facial_nerve.
29. https://www.wikidoc.org/index.php/Bell%27s_palsy_physical_examination.
30. https://www.wikidoc.org/index.php/Bell%27s_palsy_physical_examination.

3 Neck Pain

ACUTE CERVICAL FIBROSITIS OR STIFF NECK 落枕

INTRODUCTION

Stiff neck is the most commonly seen neck pain for which the exact cause is not known. Patients usually have pain on one side of the neck, difficulty moving in the morning after waking up, and usually present to medical providers with the head leaning to one side. There is no injury nor over-use experience.

ANATOMY

The muscle that involves a stiff neck condition is the Levator Scapulae. It is a small and slim muscle that begins in the neck and descends to attach to the scapula.

Attachments: it originates from the posterior tubercle of transverse processes of the C1–C4 vertebrae, and the insertion is to the superior part of the medial border of the scapula.

Innervation: dorsal scapular nerve (C5) and cervical nerve (C3, C4).

Blood supply: dorsal scapular artery.

Action: it elevates the scapula and tilts the glenoid cavity inferiorly by rotating the scapula.

WESTERN MEDICINE ETIOLOGY AND PATHOLOGY

Strained levator scapular muscle may be the cause.

Neck strains usually occur when the neck is in an unsupported posture for a long time, such as sleeping in an awkward position or with the head on an inappropriate pillow height, using a cellphone while the neck is slouching, or cradling the phone between the shoulder and the head, especially with prolonged phone use.

CHINESE MEDICINE ETIOLOGY AND PATHOLOGY

Local Qi Stagnation 頸筋受挫型

Due to stress or anxiety as well as an inappropriate sleeping posture, sleeping pillow is too high, too low, or too hard, inappropriately neck and head position on the pillow, or the neck is in an awkward position for a long period of time, the strain tissues establish Qi obstruction in the meridians.

Wind Cold Invasion風寒侵淫型

Attacking of Wind-Cold on the neck during sleeping leads to disturbance of the local circulation of Qi and Blood in the meridians. It occurs mostly in adults.

Liver and Kidney Deficiency肝腎虧虛型

Patients have deficiency of the Liver and Kidney, and the joints and the Tendons lack flexibility. Commonly, patients with cervical spondylosis have this type of Chinese medicine. Wind-Cold attacks the neck when the patients are exhausted from over-working, resulting in Qi and Blood stagnation.

MANIFESTATIONS

In Western Medicine

Stiff neck is an acute neck pain, but the patient with the neck pain may not have an obvious injury history.

The pain intensity can vary. It can be from mild to extremely painful, sharp, and limiting. The pain may radiate to the shoulder or the arm.

Neck pain occurs in the morning upon waking up. It doesn't occur before bed on the night before the episode. Stiffness often is on one side with motor impairment. The neck leans to the effected side. The pain is aggressive if the neck is moved to the healthy side. There is no redness or swelling. Spasm on the levator scapulae muscle may be palpated.

In Chinese Medicine

Local Qi Stagnation頸筋受挫型

Poor sleeping posture, strained soft tissues in the neck, fatigue. Sudden stabbing pain in the neck after waking up, the pain is fixed on one side of the neck.

> Tongue: purple tongue or petechiae marks on the tongue body, thin white coating.
> Pulse: tight.

Wind Cold Invasion風寒侵淫型

The neck pain is severe, the pain radiates on one side, sometimes accompanied by numbness in the neck and shoulders or accompanied by chills, fever, headache, and body ache; sweating comes and goes.

> Tongue: pale body, thin white or slightly yellow coating.
> Pulse: floating, tight, or slow.

Liver and Kidney Deficiency肝腎虧虛型

Weak constitution, reinfluence of external pathogens while the prior neck pain is in endurance condition. The neck pain occurs frequently, the muscles are numb, accompanied by backache and weakness. Five-Center-Heat, heavy pain on the body, cold limbs, aversion to cold, palpitation, shortness of breath.

Tongue: pale body, white coating.
Pulse: thin.

PHYSICAL EXAMINATION

- Begin with obtaining medical history
  - General health condition is such that the neck pain condition is acute or chronic.
  - Whether the patient has a history of extreme stress or physical trauma.
  - History of chronic degenerative disc disease, numbness of hands and fingers, headache (e.g., spondylosis).
  - Also obtain both positive and negative findings of the Liver and Kidney.
- Inspection
  - Vital signs in patient with stiff neck.
  - Check for scars to determine whether the neck pain is related to surgery or trauma.
  - Check for skin color, swelling, and skin temperature. Neck trauma may cause bruising, swelling, and warm temperature. The skin temperature is cooler, and the color has no change if the neck pain is from Wind-Cold invasion.
- Palpation
  - Palpate both the anterior and posterior neck for movable and fixed masses.
  - Palpate the muscle tightness and locate tender spots.
  - Palpate the spinous process and notice if there is any misalignment.
  - The skin temperature on the ipsilateral side is cooler.
- Range of motion
  - Both rotation and the lateral flexion to the contralateral side is limited and painful.

DIFFERENTIAL DIAGNOSIS

- Cervical trauma: there is redness, swelling, heat, and pain at the traumatic region.
- Neurogenic abnormalities: spinal cord tumor or progressive spinal cord diseases can cause acute neck pain. Patients may experience headaches, vomiting and positive neurological signs, speech difficulties, and limb weakness.
- Congenital muscular torticollis: patients have a history of birth trauma. It usually occurs within the first two months of life.
- Postural torticollis: the sleeping infants remains in a position for a prolonged period of time. It usually occurs within the first five months of life.

TREATMENTS

Local Qi Stagnation頸筋受挫型

Treatment Principle
Move Blood, disperse Bruising, regulate Qi, and stop pain.

Herbal Treatments
Herbal Formula

He Ying Zhi Tong Tang 和營止痛湯

This formula is from "Supplement to Traumatology傷科補要", authored by Qian Xiuchang錢秀昌 (unknown) in Qing dynasty 清朝 (1636–1912), and published in 1808.

Herbal Ingredients

Xu Duan 續斷12g, Dang Gui 當歸9g, Chi Shao 赤芍9g, Chuan Xiong川芎 9g, Su Mu蘇木9g, Chen Pi陳皮9g, Tao Ren桃仁 9g, Wu Yao 烏藥9g, Mo Yao沒藥 6g, Ru Xiang乳香 6g, Gan Cao甘草 6g

Ingredient Explanations

- Dang Gui 當歸 nourishes the Blood and moves the Blood.
- Chi Shao 赤芍, Chuan Xiong川芎, Su Mu蘇木, Tao Ren桃仁, Mo Yao沒藥, and Ru Xiang乳香 move the Blood, disperse Bruising, and stop pain.
- Wu Yao 烏藥 and Chen Pi陳皮 move Qi and stop pain.
- Xu Duan 續斷 tonifies Kidney and strengthens Bone.
- Gan Cao甘草 harmonizes the ingredients.

Modifications

- To make the formula stronger for pain, add San Qi 三七, 6 grams.

Formula Actions

- Moves the Blood.
- Breaks up Stagnation.

Wind Cold Invasion風寒侵淫型

1. **Wind-Cold invasion predominantly 風寒偏勝型**

Treatment Principle

Expel Wind, disperse Cold, move the Blood, and stop pain.

Herbal Treatments

Herbal Formula

Ge Gen Tang 葛根湯

This formula is from "Shang Han Lun 傷寒雜病論" authored by Zhang Zhong-Jing 張仲景 (150–219) in eastern Han dynasty 東漢 (25–220), and published in 200–210.

Herbal Ingredients

Ge Gen 葛根12g, Ma Huang麻黃9g, Sheng Jiang生薑9g, Gui Zhi桂枝6g, Bai Shao 芍藥6g, Zhi Gan Cao甘草炙6g, Da Zao大棗12 pieces

Ingredient Explanations

- Ge Gen葛根 relieves Heat, releases muscles, especially of the neck and upper back.
- Ma Huang麻黃 induces sweating, releases the Exterior, and warms and disperses Cold pathogens.
- Sheng Jiang生薑 unblocks the pure Yang pathway and harmonizes rebellious Qi.
- Gui Zhi桂枝 warms the meridians and relieves pain.
- Bai Shao芍藥 nourishes the Blood, soothes the Liver, and stops pain.
- Zhi Gan Cao甘草炙 tonifies the Spleen, harmonizes the other herbs, and guides the herbs to all twelve meridians.
- Da Zao大棗 tonifies the Spleen Qi, nourishes the Blood, and moderates and harmonizes the harsh properties of the fragrant herbs.

Modifications

- To make the formula stronger, add Wei Ling Xian威靈仙, Qin Jiao秦艽, and Ji Xue Teng雞血藤.
- For arm pain or discomfort, add Sang Zhi桑枝 and Gui Zhi桂枝.

Formula Actions

- Promotes sweating and releases Exterior.
- Generates Fluid and opens the meridians.

Cautions

- Patients with Wind Heat need to be cautious when taking this formula.

Pattern Herbs

- Shu Jin Huo Luo Wan 舒筋活絡丸

2. **Wind-Cold invasion predominantly** 風濕偏勝型

Treatment Principle

Expel Wind, disperse Cold, move the Blood, and stop pain.

Herbal Treatments

Herbal Formula

Ge Gen Tang 羌活勝溼湯

This formula is from "Clarifying Doubts about Damage from Internal and External Causes 內外傷辨惑論", authored by Li Dongyuan李東垣 (1180–1251), in Jin dynasty金朝 (1115-1234), and published in 1232.

Herbal Ingredients

Qiang Huo羌活3g, Du Huo獨活3g, Gao Ben藁本1.5g, Fang Feng防風1.5g, Chuan Xiong川芎1.5g, Man Jing Zi蔓荊子0.9, Zhi Gan Cao甘草炙1.5克

Ingredient Explanations

- Qiang Huo羌活 expels Wind-Cold-Dampness, unblocks painful obstruction, and alleviates pain.
- Du Huo獨活 expels Wind, Damp, and Cold in Lower Jiao and joints; removes chronic Bi stagnation in lower limb.
- Gao Ben藁本 expels Wind and drains Damp in the Tai Yang meridians.
- Fang Feng防風 expels Wind and drains Damp.
- Chuan Xiong川芎 moves the Blood, disperses Bruising, and stops pain due to Blood stasis.
- Man Jing Zi蔓荊子 expels Wind and stops pain.
- Zhi Gan Cao甘草炙 tonifies the Spleen, harmonizes the other herbs, and guides the herbs to all twelve meridians.

Modifications

- To make the formula stronger, add Fang Ji防己, Qin Jiao秦艽, Fu Zi附子, and Wu Tou烏頭.
- For headache with neck pain, add Ge Gen葛根 and Chi Shao赤芍.

Formula Actions

- Expels Wind, drains Damp.

Liver and Kidney Deficiency 肝腎虧虛型

Herbal Formula

Du Huo Ji Sheng Tang獨活寄生湯

This formula is from "Essential Formulas Worth a Thousand in Gold to Prepare for Emergencies備急千金要方", authored by Sun Simiao孫思邈 (541–682), in Tang dynasty 唐朝 (618–907), and published in 652.

Herbal Ingredients

Du Huo獨活9g, Sang Ji Sheng桑寄生6g, Du Zhong杜仲6g, Niu Xi牛膝6g, Xi Xing细辛3g, Qin Jiao秦艽6g, Fu Ling茯苓6g, Rou Gui肉桂6g, Fang Feng防風6g, Chuan Xiong川芎6g, Dang Shen党参6g, Zhi Gan Cao甘草3g, Dang Gui當歸6g, Bai Shao白芍6g, Shu Di Huang熟地黃6g

Ingredient Explanations

- Du Huo獨活 expels Wind, Damp, and Cold in Lower Jiao and joints; removes chronic Bi stagnation in lower limb.
- Xi Xing细辛 enters Shao Yin meridian to expel Wind, Damp, and Cold.
- Qin Jiao秦艽 expels Wind, Damp, opens the meridians, and benefits the joints.
- Rou Gui肉桂 warms meridians, expels Cold, and opens the meridians to promote the Blood circulation.
- Fang Feng防風 expels Wind and drains Damp.

- Sang Ji Sheng桑寄生, Du Zhong杜仲, and Niu Xi牛膝 are for deficiency of Qi and Blood in Liver and Kidney and strengthen Bone and Tendons; Sang Ji Sheng桑寄生 can expel Wind, drains Damp; Niu Xi牛膝 moves the Blood and opens the meridians to benefit joints and Tendons.
- Dang Gui當歸, Chuan Xiong川芎, Shu Di Huang熟地黃, and Bai Shao白芍 nourish and harmonize the Blood.
- Dang Shen党参, Fu Ling茯苓, and Gan Cao甘草 tonify Spleen Qi.

Formula Actions

- Dispels Wind, Cold, and Damp.
- Relieves painful obstruction.
- Supplements Qi and Blood.
- Tonifies the Liver and Kidney.

Modifications

- For severe pain, add Zhi Chuan Wu制川烏, Zhi Cao Wu制草烏, Bai Hua She白花蛇, Xu Duan川斷, Gou Ji狗脊, Tu Si Zi菟絲子, Hong Hua紅花, and Di Long地龍.

Contraindications

- Damp Heat, excessive Bi Syndrome.

Acupuncture Treatments

Acupuncture Point Formula

Local points

Ashi point 阿是穴, GB20風池Fengchi, SJ5外關Waiguan

Distal Points

SI3後溪Houxi, GB39懸鐘Xuanzhong, UB60崑崙Kunlun

Extra Points

Luo Zhen落枕穴

The location is on the dorsum of the hand, between the second and third metacarpal bones, about 0.5 cun 寸 posterior to the metacarpophalangeal joint.

Cupping Therapy

Ashi point 阿是穴 on the neck, UB12風門Fengchi, GB21肩井Jianjing, Hua Tuo Jia Ji華佗夾脊 (a half cun 寸 bilateral to DU meridian 督脈from T1 through L5. Classically, there are 17 pairs of points attributed to Hua Tuo華佗, a physician in Han Dynasty漢朝).

Cupping instructions:

1) The patient lies prone with the neck uncovered. The therapist stands beside the affected ankle.

2) Pick up a cotton ball with long medical tweezers.
3) Wet the cotton ball with 90% rubbing alcohol.
4) Hold a glass cup with one hand and hold the tweezers with the other hand.
5) Set the cotton ball on fire.
6) Insert the flaming cotton ball into the glass cup for 3 seconds or until it feels warm. The cup should still be cool enough to handle.
7) As soon as the cotton ball has been removed, place the cup onto the skin over the affected ankle, release the hand, and let go. The glass cup should stay on the skin.
8) Extinguish the flaming cotton ball immediately and drop it into a fire-resistant cup containing water. Cover the cup with a lid.
9) Cupping retaining time should not be longer than 10 minutes.
10) To remove a glass cup, press the skin beside its rim to let air seep in slowly.

Moxibustion Therapy
Points: DU14大椎 Dazhui and SI3後溪Houxi
Duration: 10 to 15 minutes each time, once daily.
Recommendation: rotate the head while applying moxa on SI3後溪Houxi.

Bleeding Therapy
Ashi point 阿是穴 on the neck with cupping
Cupping with bleeding Therapy
Instructions:

1) Prepare a clean work area before each bleeding procedure.
2) The patient lies prone.
3) Choose one or two Ashi points 阿是穴 by pressing the neck with fingers to find the most painful sites.
4) Insert a new lancet into the lancing device. Some lancing devices also have a depth selection dial; select the appropriate depth setting according to the skin thickness and the patient's comfort. The higher the number the deeper insertion; this may draw more blood, but it may cause more pain to the patients when pricking.
5) Wash hands with soap and warm water before performing, and ensure they are thoroughly dry.
6) Put clean medical gloves on both hands.
7) Wipe the skin sites with an alcohol swab on an area bigger than the cupping cup where they are going to be prepared for bleeding.
8) Hold the lancing device against the bleeding site (one or two Ashi points 阿是穴on the neck) and, with gentle pressure, press the release button to prick the skin. Prick several times on different spots to make sure the blood drops are large enough for a cup to suck.
9) Place acupuncture cups to drain about 2 ml of blood.
10) Treatment frequency is once every other day, it is three times a week, then takes a break for two days.

Electro-Acupuncture Therapy
Luo Zhen落枕穴 to SI3後溪Houxi
   2/100 hz for 20 minutes daily
   Auricular Therapy: Neck, Cervical, Shenmen, and Subcortex

Tui Na Treatments
Tui Na Techniques
Pressing, kneading, rolling, pulling, rotating, pushing

Tui Na Procedures
Supine Position
The patient is in a prone position. The Tui Na therapist is standing close to the patient's neck.

1. Apply pressing and kneading with the thumb on the Cervical Superior Blood Point (0.3 Cun 寸 superior to GB20), Cervical Medial Blood Point (1.5 Cun 寸 bilateral to C5), and Cervical Inferior Blood Point.

Sitting Position
The patient is in a sitting position. The Tui Na therapist is standing behind the patient.

1. Apply Dragon Rolling on the affected side for 35 times Figure 3.1.1.

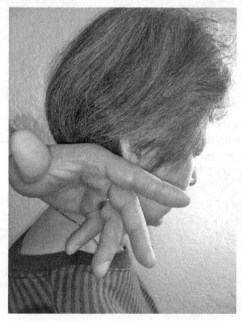

2. Rotate the shoulder 15 times Figure 3.1.2.

3. Grasp both sides of the neck 35 times Figure 3.1.3; Figure 3.1.4; Figure 3.1.5; Figure 3.1.6

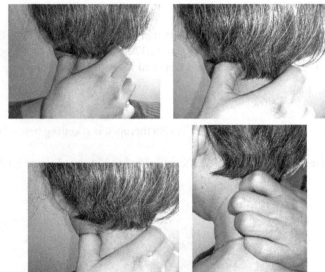

4. Apply Rotating technique on the neck 15 times.
5. Pull the neck with mild force just enough to slightly stretch the neck's soft tissues for 3–5 minutes Figure 3.1.7; Figure 3.1.8.

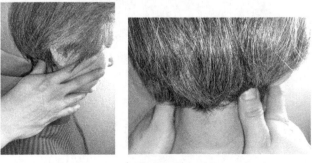

6. Apply pushing the neck for 35 times.

Patient Advisory

- Apply heat to the painful area.
- When working at the computer, set the monitor at eye level and sit up straight. Also, avoid tilting and twisting the head down or to the side while using a computer or laptop.
- If you sit for work, stand up to stretch and move every hour. Put a cup of warm water near your monitor and finish drinking the cup of water within an hour (this will force you to stand up and move because you must empty your Urinary Bladder). I recommend drinking tea with chrysanthemum flowers, as tea is diuretic and both of its cooling functions reduce inflammation. Chrysanthemum flower clears the Liver Fire from over-mental work and brightens the eyes.
- Hold the cellphone at eye level to avoid slouching.
- Avoid cradling the phone between the shoulder and the head.
- Sleep in a supine position. Avoid sleeping prone, as often this will cause the head and neck to twist one way or the other for hours at a time. https://health.clevelandclinic.org/do-you-have-a-stiff-neck-try-these-simple-remedies/
- Avoid jerking the head and neck or other painful activities. Rather, try to move the neck smoothly and slowly. This helps to promote blood circulation and Qi circulation in Chinese medicine.
- Adjust pillow height. You may try sleeping on two soft pillows, this allows you to adjust the height and comfort level by moving either of them. The recommended pillow height is the size of one fist when supine or the length from the shoulder to the neck when lying on the side. The neck should be fully on the pillow; otherwise, the soft tissues of the neck are not able to relax while sleeping. The neck should be on and covered by the pillow. Move the top pillow one third down to cover the neck, this is going to prevent Wind-Cold attacking as it wraps the neck. Patients with neck pain or Yang deficiency are also able to prevent the invasion of Wind-Cold by wearing a scarf or a shower towel around the neck while sleeping.
- Liver controls the Tendons, Kidney controls Bones and joints. To nourish the Liver and Kidney, avoid going to sleep later than 11 p.m. and ensure good quality and sufficient hours of sleep. See a Chinese medicine practitioner for assistance.
- Stress can tense up the soft tissues in the neck. Managing stress can help to treat and prevent neck pain and stiffness. There are a variety of ways to choose from, including exercise (especially hiking in a green environment), meditation, Tai Chi, taking a vacation or a break from the busy schedule (even if it is just a few hours), watching a funny movie, reading fiction, joining a party, eating your favorite food, and drinking alcohol (over drinking hurts the Liver; drink alcohol in moderation).

CERVICAL SPONDYLOSIS 頸椎病

INTRODUCTION

Cervical spondylosis is an age-related neck pain. It is another common neck pain that affects more than 85% of people over the age of 60. Even though it is common, most people experience no symptoms.

It is a condition related to the narrowing of the space needed by the spinal cord and nerve roots to pass through. When symptoms occur, they present pain and stiffness in the neck; if the nerve roots are pinched, the symptoms can manifest in the arms, hands, urinary bladder, intestine, legs, and feet.

ANATOMY

The neck is also called the cervical spine. The human vertebral column has five regions including cervical, thoracic, lumbar, sacral, and coccygeal. The cervical spine is the most important region due to its proximity to the head. It is a well-engineered and strong but flexible structure, housing the bones, vessels, muscles, ligaments, tendons, and spinal cord, which contains the nerves that transmit messages from the brain to the whole body.

The cervical spine lies between the cranium and the thoracic vertebrae. It consists of seven vertebrae (C1–C7), two of which are given unique names: atlas(C1) and axis(C2). These two vertebrae are responsible for spinal rotation, flexion and extension.

- **Atlas (C1)**

Atlas (C1) does not have a body or spinous process, rather it consists of an anterior arch and a posterior arch. It articulates superiorly with the occiput (the atlantooccipital joint) and inferiorly with the axis (the atlantoaxial joint). The atlas (C1) is made up of a thick anterior arch, a thin posterior arch, two prominent lateral masses, and two transverse processes on each side of the atlas which enclose two transverse foramens where the vertebral artery passes in each foramen. The superior facet articulates with the occipital condyles. The inferior facet articulates with the superior facets of the axis.

- **Axis (C2)**

Axis (C2) has a large vertebral body and contains the odontoid process (dens) and two transverse foramens. The anterior surface of odontoid process of axis (C2) articulates with the posterior aspect of the anterior arch of the atlas (C1) and forms the median atlantoaxial joint. The atlantoaxial joint is stabilized by the transverse ligament.

- **C3–C7**

C3–C7 are like the thoracic and lumbar vertebrae. There are cervical intervertebral discs in between each vertebra starting below axis (C2). Each one of C3–C7 consists of two transverse foramens, the openings for the vertebral artery to course through (from C6 to C1). They are on C1–C7, except the thoracic and lumbar vertebrae.

Intervertebral Discs

Intervertebral discs are located between the vertebral bodies of C2–C7. The inside of each cervical intervertebral disc contains a nucleus pulposus, and it is surrounded and protected by the annulus fibrosus. The discs are thicker anteriorly.

There are plates between the adjacent vertebral bodies and the discs. The functions of the plates are to transmit compressive loads throughout a cervical range of motion.

Facet Joints

Except for the very top of the cervical spine, each cervical vertebra has two facet joints on each side.

Neuroforamen

An opening between the facet joints, called a "neuroforamen", is formed on each side of the vertebra when stacked on top of one another. The nerve roots leave the spinal cord through these foramens.

Transverse Foramina

Unlike the thoracic and lumbar vertebrae, the cervical spine has transverse foramina in each vertebra for the vertebral arteries passing through and supplying blood to the brain.

Ligaments

Ligaments secure the joints.

The ligaments that secure the craniocervical junction and the atlantoaxial joints are the external and the internal ligaments.

The external ligaments

- Atlanto-occipital.
- Anterior atlanto-occipital.
- Anterior longitudinal ligaments.

The internal ligaments

- Transverse ligament.
- Accessory ligaments.
- Paired alar ligaments.
- Tectorial membrane.
- Accessory atlantoaxial ligament.

The cervical spine ligaments

- Anterior longitudinal ligament.
- Anterior atlanto-occipital membrane.

- Apical ligament.
- Alar ligaments.
- Cruciate ligament of the atlas.
- Posterior longitudinal ligament.
- Tectorial membrane.
- Ligamentum flavum.
- Posterior atlanto-occipital membrane.
- Ligamentum nuchae.
- Interspinous ligaments.
- Supraspinous ligaments.

Nerves

There are eight cervical spinal nerves (also called nerve roots) that exit the neural foramens above their correspondingly numbered vertebral body from C2–C7. The first spinal nerve exits the space between the occiput and the atlas (C1), and the eighth exits the neuroforamen between C7 and T1.

Etiology and Pathology in Western Medicine

Dehydrated disks occur in the early stage following disk degeneration. Degeneration of the cervical spine has been found on asymptomatic patients age 40 and older (Wang et al. 2019).

Degenerative disc is not a disease, it is a medical condition that produces pain due to the damage and wear and tear of the disc. The degenerative changes mainly affect the vertebral bodies, the neural foramina, and the facet joints. It may pinch the spinal cord or the nerve roots when the space between two adjacent vertebrae narrows.

Years of constant abnormal pressure from daily activities, such as having the neck in a wrong position for a long period, neck injury, and sports activities can dry out and shrink the discs over time and may cause bone-on-bone contact between the vertebrae.

Joint subluxation or herniated disk causes abnormal weight bearing and sometimes can press on the spinal cord and nerve roots.

The abnormal stress on the joint causes the body to produce extra amounts of new bone (bone spurs) in order to compensate for the new weight distribution. These bone spurs can sometimes compress the spinal cord and the nerve roots.

Ligaments can be stiffened with age leading to a less flexible neck.

Risk Factors

1. Repetitive strain injury
   Inappropriate lifestyle while working in front of computers, driving, traveling, intense farm work, etc.
2. Age-related degeneration
   Asymptomatic change begins at age 40 and older, in which the discs between the vertebrae become less spongy and provide a less cushion-like quality. The ligaments get less flexible with age, and the neck becomes stiff.

3. Trauma

Previous injury to the neck from caused by routine job-associated activities may cause neck injury. Similarly, other occupations, like gymnasts, football players, springboard divers, or weightlifters can cause neck injury.

Four Distinct Types of Cervical Spondylosis

Type I. Cervical Radiculopathy

Cervical radiculopathy is often called "pinched nerve". There are eight pairs of cervical nerve roots, C1 to C8. A compression on any root leads to the pain and the loss of sensation along the nerve's pathway into the arm, hands, and fingers.

Pinched nerve can occur as an outcome of pressure from a nearby bone spur, cervical foraminal stenosis, or cervical herniated disc resulting from a ruptured disc from degeneration changes in the bones. It can be also caused by infections in the spine, arthritis, or other injuries that put pressure on the nerve roots.

Type II. Cervical Myelopathy

Cervical myelopathy is often caused by compression of the spinal cord in the cervical spine due to stenosis that often develops from degenerative changes. It also can be caused by arthritis, a tumor or a cervical trauma leading to herniation or fracture, resulting in compression on the spinal cord.

Type III Syndrome

Inappropriate pillow height, poor sleep posture, fatigue, common cold, and cold temperature could produce the pathological changes to the soft tissues underlying cervical degeneration.

Type IV. Cervical Spondylotic Arteriopathy

Cervical spondylotic arteriopathy is often caused by cervical vertebral lesions, thereby leading to compression of the vertebral artery resulting in insufficient blood supply (Liu et al. 2019).

The common causes include hyperplasia and stenosis of the transverse foramen of the cervical vertebrae, compression on the vertebral artery by the hyperplasia, and hypertrophy of superior articular process, with decreased stability of the cervical spine after degeneration.

Type V. Sympathetic Cervical Spondylosis

Sympathetic cervical spondylosis is a part of the presence of proliferative protrusions in the degenerative change of the cervical spine that stimulate the sympathetic nervous system. This condition could produce sympathetic excitation and induce a sympathetic reflex, causing vertigo and hypertension.

MANIFESTATIONS IN WESTERN MEDICINE

Pain is the most reported symptom of cervical spondylosis. It can occur in multiple locations beside the chief complaint of neck pain. It can be intermittent neck pain, shoulder pain, arm pain, finger numbness or tingling and can also manifest in chest and the upper back between the scapular. Beside the bodily pain, Chronic suboccipital headache can be a symptom of cervical spondylosis but often it is understood as the pain on the muscles near the suboccipital region such as obliquus capitis superior muscle and rectus capitis anterior muscle. The headache in patients with cervical spondylosis may radiate to the base of the neck and the top of the head.

In severe conditions, there may be muscle spasms or atrophy, neck movement limitation, and symptoms of the autonomic nervous system disorders such as dizziness, vertigo and fainting, unstable blood pressure, sweating abnormalities, and digestive and vision disorders.

These distinct clinical syndromes can result from cervical spondylosis:

Type I. Cervical Radiculopathy

This type is the easiest one to recognize because its characteristic is neck pain with radiating pain or numbness or tingling from the arm to the fingers. C5-6 disc (C6 nerve root), the C6-7 disc (C7 nerve root) and C7-T1 disc (C8 nerve root) are the most common discs distributing the pain to the fingers. C6 nerve root radiculopathy will present the numbness or tingling sensation on the thumb and index fingers. C7 nerve root radiculopathy will present the numbness or tingling sensation on the middle finger. C8 nerve root radiculopathy will present the numbness or tingling sensation on the pinky.

Patients usually complain of dizziness, soreness, and burning, tingling pain; movement limitation in the neck; and sometime pain that may radiate along the dermatome distribution to the fingers. Muscular atrophy of the upper limbs may be seen in severe condition. Palpation may reveal deviation of the affected spinous process.

Type II. Cervical Myelopathy

As the manifestations are subtle, the diagnosis is difficult. Weakness may start from the upper limbs to legs or the opposite way with reduced or absence of pain, touch, and vibration or position sense. Reflexes are brisk. Muscular tone is increased with rigidity of the extremities.

The patients with cervical myelopathy will often complain of peripheral neuropathy in the limbs, and difficulty with fine movements and control of the fingers, such as fastening and unfastening buttons.

Patients also complain of unsteady gait and increased deep tendon reflexes. They may present in the lower leg, ankle, and foot moving together stiffly as a unit when seated with the legs dangled as loosely as possible over the edge of the examination table and shaking the leg gently back and forth.

There are two commonly used neurological examinations helpful in detecting cervical cord compression or an upper motor neuron disease. Involuntary clawing flexion of the thumb and the other digits when flicking the middle finger nail and an extension of the big toe when stroking the sole of the foot indicate a positive Hoffmann's sign and Babinski's sign. The positive Hoffmann's sign and Babinski's sign may indicate an upper motor neuron and corticospinal tract dysfunction localized to the cervical segments above C5 and C6 of the spinal cord,[1] Even if both Hoffmann's sign and Babinski's sign are positive on the patients with cervical myelopathy, however, up to 3% of the positive Hoffmann's sign were found to have no compression or upper motor neuron disease (Whitney and Munakomi 2021).

Type III Axial Joint Pain

There is no neurologic deficit in this type, hence, pain below the elbow suggests nerve root involvement. This type of cervical condition is characterized by neck pain, and the pain may refer to one or more of the following: medial scapula, chest wall, shoulder area, occipital region and possibly the proximal upper extremity. Headache may present, commonly located at the occipital region occasionally frontal radiation, it is usually associated with cervical muscle spasm. Movement of the neck produces pain, which is relieved by rest and immobilization.

Type IV. Cervical Spondylotic Arteriopathy

Patients with this type present with dizziness, paroxysmal distended headache of the occipital and vertex regions, or it can radiate to the temporal region. Tinnitus, abrupt onset of visual or hearing loss, vomiting, fainting and falling during a head rotation or flexion or extension while turning the head or the body on bed are other symptoms, but patients usually can regain normal activity once standing up.

Type V. Sympathetic Cervical Spondylosis

Symptoms include vertigo that is non-positioning, headache that is aggravated when rotating the head; tinnitus, nausea and vomiting, tachycardia, irregular heartbeat, angina, elevated blood pressure, abnormal sweating, hearing loss, and dysphonia.

Physical Examination

- Cervical spondylosis is a clinical diagnosis that is based upon history, abnormal examination findings, and confirmatory imaging studies.
- Begin with obtaining a medical history such as the timeline of the neck pain, radiating pain onto the arm and fingers, the type of pain, affecting which fingers, and the aggravating factors.
  - There are autonomic nervous system symptoms needed to obtain and differentiate; symptoms include dizziness, vertigo, fainting, unstable blood pressure, abnormal sweating patterns such as alternating between too much sweating and too little sweating; abnormal digestive symptoms, such as dysphagia, loss of appetite, bloating, diarrhea, constipation; abnormal vision, such as photophobia, lacrimation, vision loss, and uneven pupils
  - Commonly, the neck pain in patients with cervical spondylosis is worse when upright, especially after keeping the head in one position for too long time. The pain subsides with bed rest.

- Inspection
  - Observe the pain intensity when having the range of motion of the neck in hyperextension and lateral flexion. These motions manifest the decrease in range of motion, especially the hyperextension motion, and increase the neck pain, nervous system symptoms and the radiating pain onto the upper limb according to the dermatome region.
  - Observe the results when the patient is asked to place and rest the forearm on the top of the head. A few patients may have already discovered that this procedure can relieve radiating arm pain spontaneously (Voorhies 2021).
- Palpation
  - Tenderness may be palpated along the cervical paraspinal muscles, and it is more pronounced along the ipsilateral side of the affected nerve root.
  - Muscle tenderness, hypertonicity and spasm may be palpated along the muscles innervated by the involved nerve root on the ipsilateral shoulder, scapula, and arm.
- Range of motion
  - Abnormal sensation such as numbness and tingling may present on the arm and fingers corresponding to their nerve roots innervation if there is a compression on C5, C6, C7, and C8 as below:
  - (C4–5 disc) C5 nerve root radiculopathy may show the weakness in shoulder abduction and elbow flexion (weakness in the deltoids and biceps muscles).
  - (C5–6 disc) C6 nerve root radiculopathy will present weakness in the biceps muscle (decreased brachioradialis reflex) and wrist extension (weakness in the carpi radialis longus and brevis); weakness in elbow flexion and wrist extension; and numbness or tingling sensation in the thumb and index finger.
  - (C6–7 disc) C7 nerve root radiculopathy will present weakness in the triceps, the brachioradialis, elbow extension, wrist flexion, finger extension and numbness or tingling sensation of the middle finger
  - (C7–T1 disc) C8 nerve root radiculopathy will present weakness in the thumb extension, finger abduction, and gripping (flexion); wrist ulnar deviation as well as the numbness or tingling sensation in the pinky.
- Orthopedic tests
  - Spurling's sign: It is a foraminal compression test. With the patient seated, they extend the neck, flex laterally, and rotate to the side of the pain. The practitioner carefully compresses the head by slight axial loading. A positive Spurling's sign is when the pain arising in the neck radiates in the direction of the corresponding dermatome ipsilaterally.
  - Hoffman sign: It is an involuntary flexion movement of the thumb and/ or index finger when the practitioner flicks the fingernail of the middle finger down.

DIFFERENTIAL DIAGNOSIS

- Type I. Cervical radiculopathy differentiation from carpal tunnel syndrome and brachial plexus neuritis.
  - Carpal tunnel syndrome: The pain may occur at night and often awakens the patient. Shaking the hand helps to ease the pain. Tinel's sign is positive.
  - Brachial plexus neuritis: It is a burning pain in the shoulder and upper arm, it may awaken the patient from sleep. Pain from brachial plexus neuritis involves multiple nerves (C5–T1) while in cervical radiculopathy the pain is usually restricted to one nerve root involvement. In brachial plexus, pain occurs and subsides after a variable period of usually a few weeks (Voorhies 2021) prior to the weakness of the shoulder and upper arm; whereas, the pain, weakness, and sensory loss associated with cervical radiculopathy tend to occur simultaneously (Miller, Pruitt, and McDonald 2000).
- Type II. Cervical Myelopathy differentiation from spinal cord tumors

  Pain and nerve root irritation symptoms; dyskinesia of the ipsilateral limb below the compression plane can sometimes cause difficulty walking and sometimes leading to falls.

  Sensory disturbance may occur on the contralateral limb. Spinal tumors progress at different rates and direction depending on the type of tumor. The progressive sensory loss in the intramedullary tumor is from the upper extremities toward the lower extremity, while the sensory loss in extramedullary tumor begins in the lower extremities and progresses toward the upper extremities. Diagnostic imaging, such as X-ray, CT, MRI, and EMG can be used to confirm a diagnosis. MRI is commonly recommended for the detailed information of the tumor about the location, the condition of the spinal cord, the bones, discs and the soft tissue structures, while tumor histology is able to determine the type of tumor of the suspected tissue that has been excised by biopsy or surgical resection. If the histological type is different from what is usually found in the tumor being examined, it can mean the tumor has spread to that area from some primary site. (Cancer Diagnosis n.d.)
- Type III Axial Joint Pain differentiation from cervical and shoulder issues, such as acute cervical muscle sprain, acute cervical fibrositis (Wind Cold invasion in Chinese medicine), adhesive capsulitis or "Frozen Shoulder", shoulder impingement syndrome, rotator cuff tear, fibromyalgia, and angina pectoris.
  - Acute cervical muscle sprain: there is an injury history prior to neck pain. Localized redness, warm, bruising, and swelling may present.
  - Acute cervical fibrositis: unilateral neck pain occurs in the morning after waking up, localized cold to touch; no sign of bruising, swelling, or redness.

- Wind Cold common cold in Chinese medicine: sneezing, headache, chill, or alternative chill and fever, runny nose, and coughing.
- Adhesive capsulitis or "frozen shoulder": loss of passive and active shoulder range of motion, especially abduction where the higher the abduction the greater the pain, and limited range of motion. Shoulder muscles may be atrophic. It is seen in elderly patients.
- Shoulder impingement syndrome: pain is at the shoulder joint. It is persistent and more painful at night or with overhead lifting. Painful arc test is positive. Shoulder bone spurs may be present.
- Rotator cuff tear: popping or clicking shoulder or sensations when moving the arm. Shoulder pain that is deep inside and worsens at night or when resting the arm. Painful arc test is positive.
- Angina pectoris: the pain is often described as squeezing, pressure, heaviness, tightness, or pain in the chest accompanied by dizziness, fatigue, inability to exercise, lightheadedness, sweating, rapid breathing, or shortness of breath.
- Fibromyalgia: it is widespread pain, morning stiffness, fatigue, headache, irritable bowel, anxiety, and depression.

Etiology and Pathology in Chinese Medicine

1. **Bi Syndrome 痹痛型**

    1.1 Due to aging, Qi and Blood are deficient, Tendons and Bone lack nourishment. Wind-Cold-Damp invades and causes blockages in the meridians leading to Qi and Blood stagnation and resulting in pain.

    Main symptoms: joint stiffness, numbness, and weakness in limbs. Weak or sore pain in shoulders and arms is aggravated with cold, relieved with hot.

    1.2 Degeneration of Bone and Tendons is due to the deficiency of the Liver and Kidney leading to deficient Cold accumulation in Bone and Tendons, resulting in joint stiffness and range of motion limitation.

    Main symptoms: weak arms, difficult movement, weak and lax joints, tinnitus and dizziness, cold arms and hands, and atrophy.

2. **Liver and Kidney Deficiency肝腎虧虛型**

    Cervical spondylosis in Chinese medicine is a condition of bone, joint, and tendon disorders. Liver stores Blood and governs Tendons; Kidney stores Essence and governs Bones. Liver and Kidney are related. When Yin is deficient, Tendons and Bones are undernourished, resulting cervical vertebrae strain and degeneration. Stress, chronic diseases, and sleep deprivation drain Yin energy, and Yin deficiency causes Blood deficiency. Bone and joints lack nourishment when both Yin and Blood are inadequate.

3. **Dizziness Type眩暈型**

    3.1 Liver and Kidney Deficiency 肝陽挾痰型

    Deficiency of Kidney leads to Liver Yang rising; Liver Wind then stirs together with Phlegm upward to the head.

3.2 Spleen and Stomach Impairment 氣血虧虛型

Over thinking and planning leads to Qi and Blood deficiency. Thus, the brain lacks nourishment.

3.3 Phlegm Damp Blockage in Meridian 痰濕中阻型

Due to having too much rich food, Spleen injury when physically exhausted. Or Seven Emotion injury causes disharmony of Spleen and Liver, leading to dysfunction of Transportation and Transformation of Spleen and resulting in dizziness and vertigo. With Damp Phlegm accumulation and blockage in Middle Jiao, pure Yang cannot ascend, and turbid Yin cannot descend.

4. **Wei Syndrome Type痿症型**

Chronic deficiency of the Liver and Kidney leads to the injury of Du meridian, stasis Blood blocks meridians, and Yang Qi fails to ascend, resulting in weakness in the limbs.

5. **Qi and Blood Stagnation 氣滯血瘀型**

The neck in a wrong position for a long time and excessive over-work lead to chronic Tendon injury, resulting in Qi and Blood stagnation. Or Wind-Cold attacks the neck, causing blockages in the meridians; and Qi and Blood fail to flow through, resulting in pain.

6. **Qi and Blood Deficiency 氣血虧虛型**

Over-worrying or over-thinking causes Heart and Spleen injury, leading to insufficiency of Qi and Blood; the Liver and Kidney are insufficient in older age. Bones, joints, and Tendons can be undernourished when there is inadequate Qi and Blood or weak constitution after a physical trauma; Wind, Cold, and Damp attack and invade the meridians, resulting in neck discomfort.

MANIFESTATIONS IN CHINESE MEDICINE AND TREATMENTS

1. **Bi Syndrome痹痛型**

Unilateral or bilateral neck pain, arm pain, tingling fingers, weak fist clenching, difficult finger movements, finger joint stiffness, tender spots in and around the cervical spinous processes, transverse process or near the superiorly inferior corner of the scapula.

Tongue: thin white coating.

Pulse: floating, tight.

Treatment Principle

Move Blood, open the channel, tonify Kidney and Liver.

Herbal Treatments

Herbal Formula

Huo Xue Shu Jin Tang活血舒筋湯

This formula is from "Handouts of Chinese Medicine Traumatology中醫傷科學講義", authored by Shanghai College of Traditional Chinese Medicine上海中醫學院 in 1964.

Herbal Ingredients

Dang Gui Wei當歸尾12g, Chi Shao赤芍12g, Jiang Huang姜黄12 g, Shen Jin Cao伸筋草9 g, Song Jie松節9g, Hai Tong Pi海桐皮9g, Lu Lu Tong路路通9g, Qiang Huo羌活9g, Fang Feng防風9g, Xu Duan續斷12g, Gan Cao甘草5g

Ingredient Explanations

- Dang Gui Wei當歸尾moves the Blood, disperses Bruising, unblocks stagnation, opens the meridians, and stops pain.
- Chi Shao赤芍moves Blood, disperses Bruising, stops pain
- Jiang Huang薑黄moves the Blood, opens the meridians, expels Wind, and reduces swelling.
- Shen Jin Cao伸筋草expels Wind, drains Damp, and relaxes Tendons.
- Song Jie 松節dispels Wind, drains Damp, opens the meridians, and stops pain.
- Hai Tong Pi海桐皮expels Wind, drains Damp, and reduces swelling.
- Lu Lu Tong路路通moves Blood, opens the meridians, expels Wind, and promotes water metabolism.
- Qiang Huo羌活expels Wind-Cold-Dampness, unblocks painful obstruction, and alleviates pain.
- Fang Feng防風expels Wind and drains Damp.
- Xu Duan續斷tonifies Kidney and strengthens Bone.
- Gan Cao甘草harmonizes the herbs, tonifies Middle Jiao Qi, clears Heat, expectorates phlegm, and stops cough.

Modifications

For severe pain, add Ru Xiang乳香 and Mo Yao沒藥

Formula Actions

- Moves the Blood, disperses bruising, soothes Tendons, and opens the meridians.

2. Liver and Kidney Deficiency 肝腎虧虛型

Blurry vision, headache or distended sensation in the head, tinnitus, deafness, pain in the neck, shoulder with limited range of motion, weak legs; dry mouth, irritability, Five-Center-Heat, fatigue, pale face; urgent urination, frequent urination, and urinary incontinence.

Tongue: small size, red body, less coating.
Pulse: wiry, thin or thready, rapid.

Treatment Principle

Nourish the Liver Yin and Blood and tonify Kidney Essence.

Herbal Treatments

Herbal Formula

Zuo Gui Wan左歸丸

This formula is from "The Complete Compendium of Jingyue景岳全書", authored by Zhang Jiebin張介賓(1563–1640AD), in Ming dynasty明朝 (1368–1644), and published in 1624.

Herbal Ingredients

Shu Di Huang熟地黃 24g, Shan Yao山藥12g, Gou Qi Zi枸杞 12g, Shan Zhu Yu山茱萸12g, Chuan Niu Xi牛膝 9g, Tu Si Zi菟絲子12g, Lu Jiao Jiao鹿角膠12g, Gui Ban Jiao龜板膠 12g

Ingredient Explanations

- Shu Di Huang熟地黃 nourishes the Blood; nourishes Liver and Kidney Yin.
- Shan Yao山藥 tonifies Kidney Yin and Spleen Qi.
- Gou Qi Zi枸杞子 nourishes the Blood; tonifies Liver and Kidney Yin.
- Shan Zhu Yu山茱萸 nourishes the Liver and Kidney; strengthens Kidney Yang.
- Chuan Niu Xi牛膝 moves the Blood, expels Wind Dampness, tonifies Liver and Kidney, and directs herbs downward.
- Tu Si Zi 菟絲子 tonifies Kidney Yang, tonifies Kidney and Liver Yin.
- Lu Jiao Jiao鹿角膠 nourishes and tonifies Essence Qi, and Blood.
- Gui Ban Jiao龜板膠 nourishes Kidney Yin and Yang, strengthens bones, and nourishes the Blood.

Formula Actions

- Nourishes Yin.
- Tonifies the Kidneys.
- Supplements Jing.
- Benefits Marrow.

Contraindications

Contraindicated for those with signs of Excess Heat.

Cautions

This formula is not suited for long-term use.

3. Dizziness Type眩暈型

Neck pain, arm pain, tingling fingers, weak fist clenching, tender spots, and weak and numb limbs; vomiting, vertigo, sudden falls, and numb limbs; stabbing pain, limited limbs range of motion, and cold limbs.

Tongue: red body, yellow greasy coating.
Pulse: soggy, rapid.

Treatment Principles
Soothe the Liver, expel Wind.

Herbal Formula:
Tian Ma Gou Teng Yin 天麻鉤藤飲
This formula is in the book "Newly Explained Diagnosis and Treatment of TCM Internal Medicine and Miscellaneous Diseases中醫內科雜病證治新義", authored by Hu Guangci 胡光慈 in 1958.

Herbal Ingredients
Tian Ma天麻6g, Gou Teng鉤藤9g, Chuan Niu Xi牛膝12g, Shi Jue Ming石決明15g, Du Zhong杜仲12g, Huang Qin黃芩6g, Zhi Zi梔子6g, Yi Mu Cao益母草9g, Sang Ji Sheng桑寄生9g, Ye Jiao Teng夜交藤9g, Fu Shen茯神9g

Ingredient Explanations
- Tian Ma天麻extinguishes Wind, calms the Liver, subdues rising Liver Yang, and alleviates pain.
- Gou Teng鉤藤expels Wind and releases spasm.
- Chuan Niu Xi川牛膝unblocks and promotes Blood circulation.
- Yi Mu Cao益母草moves the Blood, promotes urination, reduces swelling, and clears Heat.
- Shi Jue Ming石決明drains Liver Fire and expels Wind .
- Du Zhong杜仲tonifies the Liver and Kidney, strengthens Bones, and benefits Essence and Blood.
- Huang Qin黃芩clears Heat, dries Damp, cools Blood, stops bleeding, and calms Liver Yang.
- Zhi Zi梔子clears Heat, cools the Blood, resolves Damp, reduces swelling, and opens the meridians.
- Sang Ji Sheng桑寄生tonifies the Liver and Kidney, strengthens Bone, and benefits Essence and Blood; expels Wind and drains Damp.
- Ye Jiao Teng夜交藤nourishes Heart Yin and Blood, clams Shen, and expels Wind and Damp.
- Fu Shen茯神calms Shen.

Modifications
- To make to formula stronger, add Ju Hua菊花, Xia Ku Cao夏枯草, Ku Ding Cha苦丁茶, and Zhen Zhu Mu珍珠母.
- For severe Liver Heat, add Xia Ku Cao夏枯草 and Long Dan Cao龍膽草.
- For Blood stasis, add Dan Shen丹參 and Chi Shao赤芍.
- For insomnia, add Long Gu龍骨, Mu Li牡蠣, Ye Jiao Teng夜交藤, and Yuan Zhi遠志.

Formula Actions

- Soothes the Liver Qi, expels Wind, moves the Blood, clears Heat, and tonifies the Liver and Kidney.

Formula Indication

- Headache, vertigo, insomnia, anger, bitter taste in the mouth due to Liver Yang uprising.

Contraindications

- Caution for patients who suffer from Blood, Yin and body Fluid deficiency.
- Stop taking this formula if dizziness, chest distention, shortness of breath, nausea, vomiting, or skin itch are present.
- Do not take Tian Ma overdose.

4. **Wei Syndrome Type**痿症型

Numbness and weakness of arms and hands, tremor and stiffness, unstable walking, muscle atrophy; bowel incontinence, urinary retention.

Tongue: small size, red, little or no coating.
Pulse: wiry, thin, or thin rapid.

Treatment Principles

Move Blood, transform Phlegm, and open the channel.

Herbal Formula

Bu Yang Huan Wu Tang補陽還五湯

This formula is from "Correction of Errors in Medical Classics醫林改錯", authored by Wang Qingren 王清任 (1768–1831), in Qing dynasty 清朝 (1636–1912), and published in 1849.

Herbal Ingredients

Huang Qi黃耆120g, Dang Gui Wei當歸尾6g, Chi Shao 赤芍5g, Di Long地龍3g, Chuan Xiong川芎3g, Tao Ren桃仁3g, Hong Hua 紅花3g

Ingredient Explanations

- Huang Qi黃耆tonifies Qi, strengthens the Spleen, raises the Yang Qi of the Spleen and Stomach, tonifies Wei Qi, stabilizes the exterior, and tonifies the Blood.
- Dang Gui Wei當歸尾invigorates, harmonizes, and nourishes the Blood and alleviates pain.
- Chi Shao 赤芍clears Heat, cools and invigorates the Blood, disperses Blood Stasis, and eliminates pain.

- Di Long地龍unblocks and moves the channels, clears Heat, and releases spasms and convulsions.
- Chuan Xiong川芎moves the Blood, promotes Qi movement, expels Wind, and relieves pain.
- Hong Hua 紅花moves the Blood, dispels Blood Stasis, and alleviates pain.
- Tao Ren桃仁disperses Blood Stasis, moves the Blood circulation, descends Qi, and releases spasms.

Modifications

- For moving Blood, add Niu Xi牛膝.
- For expelling Wind, draining Damp, and tonifying the Liver and Kidney, add Sang Ji Sheng桑寄生 .
- For expelling Wind, draining Damp, and tonifying the Liver and Kidney, add Wu Jia Pi 五加皮.

Pattern Herbs

(1) Tian Ma Wan天麻丸

Ingredients: Tian Mat天麻, Niu Xi牛膝, Du Zhong杜仲, Dang Gui當歸, Qiang Huo 羌活、Du Huo獨活, Sheng Di Huang生地黃
   Functions: Expels Wind, opens channel, moves the Blood, and stops pain.
   Indications: Bi type cervical spondylosis and Liver and Kidney Xu type cervical spondylosis.
   Dose: 5 pills each time, three times daily.
   Cautions: Contraindicated for pregnancy.

(2) Xiao Huo Luo Dan小活絡丹

Ingredients: Zhi Nan Xing制南星 Zhi Chuan Wu制川烏 Di Long地龍 Ru Xiang乳香、Mo Yao沒藥
   Functions: Warms the channel, moves the Blood.
   Indications: Bi type cervical spondylosis, Qi and Blood stagnation type cervical spondylosis.
   Dose: 1 pill, 2 times daily.

(3) Da Huo Luo Dan大活絡丹

Ingredients: Fang Feng防風, Mu Xiang木香, Rou Gui肉桂, Shu Di Huang熟地黃, Ru Xiang乳香, She Xiang麝香, He Shou Wu何首烏, Chen Xiang沉香, Tian Ma天麻, Qiang Huo羌活, Gui Ban龜板, Dang Gui當歸, Xi Xing細辛, Ding Xiang丁香, Gui Zhi桂枝, Hong Shen紅參
   Functions: Moves Qi, expels Wind.
   Indications: Bi type cervical spondylosis, Bi and Blood stagnation cervical spondylosis. Phlegm Damp type cervical spondylosis, Liver and Kidney Xu type cervical spondylosis, and Qi and Blood Xu type cervical spondylosis.

Dose: 1 pill, 2 times daily.

5. Qi and Blood Stagnation 氣滯血瘀型

The duration of the neck pain lasts a long time. Stabbing pain, neck muscle stiffness, headache, insomnia, irritability, and depressive mood.
Tongue: dusky, purple, ecchymoses.
Pulse: wiry, choppy.

Treatment Principles

Move Blood and Qi, stop pain.

Herbal Formula

Shen Tong Zhu Yu Tang身痛逐瘀湯

This formula is from "Correction of Errors in Medical Classics醫林改錯", authored by Wang Qingren 王清任 (1768–1831), in Qing dynasty 清朝 (1636–1912), and published in 1849.

Herbal Ingredients

Dang Gui 當歸9g, Chuan Xiong川芎9g, Tao Ren 桃仁9g, Hong Hua 紅花9g, Mo Yao 沒藥 6g, Wu Ling Zhi 五靈脂6g, Qin Jiao 秦艽9g, Qiang Huo 羌活9g, Xiang Fu 香附3g, Chuan Niu Xi川牛膝9g, Di Long 地龍6g, Gan Cao 甘草6g

Ingredient Explanations

- Dang Gui當歸, Chuan Xiong川芎, Tao Ren桃仁, and Hong Hua紅花 move the Blood, disperse Bruising, and stop pain due to Blood stasis.
- Qin Jiao秦艽and Qiang Huo羌活 expel Wind-Cold-Dampness, unblocks painful obstruction, and alleviates pain.
- Wu Ling Zhi五靈脂, Mo Yao沒藥, and Xiang Fu香附 move Qi and Blood and stops pain due to Qi and Blood stagnation.
- Chuan Niu Xi川牛膝and Di Long地龍 unblock and promote movement in the channels and collaterals; stop spasms and convulsions.
- Gan Cao甘草 harmonizes the herbs.

Modifications

- For Heat symptoms, add Chai Hu柴胡 and Huang Bai黃柏.
- For Qi deficiency, add Dang Shen黨參 and Huang Qi黃芪.
- For pain in lumbar and leg, add Xu Duan續斷, Du Zhong杜仲, and Sang Ji Sheng桑寄生.
- For severe pain, add Quan Xie全蠍 and Wu Gong蜈蚣.
- For Cold symptoms, remove Qin Jiao 秦艽 and add Zhi Chuan Wu制川烏.
- For Cold symptoms and severe pain, add Gui Zhi桂枝and Xi Sin細辛.

- For Wind traveling pain, add Fang Feng防風.
- For radiating pain to the leg, remove Wu Ling Zhi 五靈脂 and add Shen Jin Cao伸筋草.
- For Damp Heat joint edema, add Cang Zhu蒼朮, Huang Bai 黃柏, and Fang Ji防己.
- For stiffness in joints, add Shen Jin Cao伸筋草, Quan Xie全蠍, and Wu Gong蜈蚣.
- For mild joint edema and severe Damp Bi pain, add Fang Ji防己, Yi Yi Ren 薏苡仁, and Cang Zhu蒼朮.

Formula Indications
- Painful obstruction due to Qi and Blood Stagnation.

Contraindications

- Contraindicated during pregnancy.

6. **Qi and Blood Deficiency 氣血虧虛型**

The main manifestation is pain; pain in the neck, shoulders, upper back; limited range of motion and intermittent headache; numbness and tingling in limbs; fatigue, pale face.

Tongue: pale tongue, thin coating.
Pulse: thin, weak.

Treatment Principle
Tonify Qi and Blood.

Herbal Formula
Ba Zhen Tang 八珍湯

This formula is from "Categorized Essentials of Repairing the Body 正體類要", authored by Xue Ji 薛己 (1487–1559), in Ming dynasty 明朝 (1368–1644), and published in 1529.

Herbal Ingredients
Ren Shen人參10g, Bai Zhu白朮10g, Fu Ling茯苓10g, Dang Gui當歸10g, Chuan Xiong 川芎10g, Bai Shao白芍10g, Shu Di Huang 熟地黃10g, Zhi Gan Cao炙甘草5g, Sheng Jiang生薑10g, Da Zao大棗3 pieces

Ingredient Explanations
- Ren Shen人參 tonifies Qi and Yang.
- Bai Zhu白朮 tonifies the Spleen Qi and dries Dampness.
- Fu Ling茯苓 drains Damp and tonifies Spleen.
- Dang Gui當歸 nourishes the Blood and moves the Blood.
- Chuan Xiong 川芎 moves the Blood and moves Qi.

- Bai Shao白芍 nourishes the Blood, soothes the Liver, and stops pain.
- Shu Di Huang 熟地黃 tonifies Liver and Kidney Yin and benefits Essence and Blood.
- Zhi Gan Cao炙甘草 tonifies the Spleen, harmonizes the other herbs, and guides the herbs to all twelve meridians.
- Sheng Jiang生薑 unblocks the pure Yang pathway and harmonizes rebellious Qi.
- Da Zao大棗 tonifies the Spleen Qi, nourishes the Blood, and moderates and harmonizes the harsh properties of the fragrant herbs.

Modifications

For poor appetite, add Shan Yao山藥, Shan Zha山楂, Mai Ya麥芽, and Qian Shi 芡實.

For headache due to Blood deficiency, add Man Jing Zi蔓荊子 and Gao Ben 藁本.

Formula Indications

- Qi and Blood Deficiency.
- Liver and Spleen Deficiency.

Contraindications

- It is contraindicated for those patients with Heat or Excess conditions.

Acupuncture Treatments

Acupuncture point formula:

GB20風池Fengchi, Huo Tuo Jia Ji頸夾脊, UB10 天柱 Tianzhu, GB21肩井 Jianjing, SI3後溪Houxi, LI4合谷Hegu, SJ5外關Waiguan, GB39懸鍾Xuanzhong, Ashi points on the neck阿是穴

Modifications

- For Qi and Blood stagnation, add P6內關Neiguan and UB17膈俞Geshu.
- For Liver and Kidney deficiency, add LIV8曲泉Ququan and KD1湧泉 Yongquan.
- For Qi and Blood deficiency, add ST36足三里Zusanli and SP6三陰交 Sanyinjiao.
- For shoulder pain, add LI15肩髃Jianyu.
- For upper back pain, add SI14肩外俞Jianwaishu and SI11天宗Tianzong.
- For Cold invasion or Yang deficiency, add moxibustion Therapy to the needles.
- For numbness on arms, add electroacupuncture to LI15肩髃Jianyu and P6 內關Neiguan.

Tui Na Treatments

Tui Na Techniques
- Pressing, kneading, rotating, grasping, and lifting.

Tui Na Procedures

The patient is in a sitting position, the therapist stands behind the patient, and performs the following:

1) Applies kneading, pressing, and grasping techniques on the shoulder and both sides of the neck to alleviate the tightness of the muscles and to promote the circulation of Qi and Blood. Figure 3.2.1; Figure 3.2.2; Figure 3.2.3; Figure 3.2.4

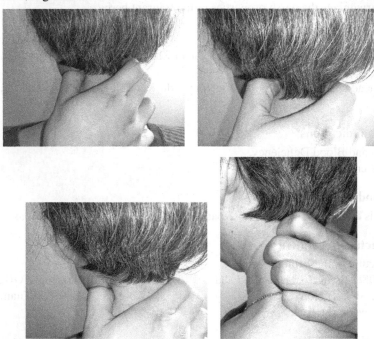

2) Places their left hand on the chin and their right hand on the base of the occiput.
3) Slowly moves the patient's head leaning forward (flexion) then leaning backward (extension) to the maximal range for 3 times. Figure 3.2.5; Figure 3.2.6; Figure 3.2.7

4) Based on the last step, turns the patient's face to the left side 45 degrees and leans the head forward and backward for 3 times; repeat the same procedures on the other side.
5) Places their left elbow at the chin and their right hand on the junction between the skull and the cervical regions. Figure 3.2.8

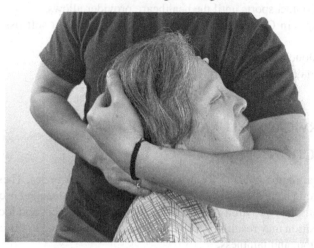

6) Lifts their elbow and right hand to elevate the patient's neck for 1–2 minutes.
7) (Rotating toward the left for the rightward deviation of the spinous process as for an example) With the left elbow still placed at the patient's chin, places the right thumb on right aspect of the deviated vertebra's spinous process or the transverse process, lifts and leans the patient's head forward 15–30 degrees; asks the patient to relax the neck and turn the face toward the left side slowly following the therapist. The therapist pushes the spinous process with the right thumb toward the left when feeling a resistance. The angle in leaning the neck forward depends on the location of the deviated spinous process. For upper cervical spine disease, 15–30 degrees are appropriate; setting the neck at the neutral position is for the middle part of the neck and 30–45 degrees are for the lower cervical spine disease. The manipulative force exerted should be varied in accordance with the patient's age, gender, physical constitution, and severity of the degeneration.

Patient Advisory

- Wear a soft cervical collar or neck brace to limit neck motion and relieve nerve irritation.
- Do appropriate neck exercises help to strengthen the soft tissues.
- Maintain good posture while sitting and walking.
- Keep the neck warm.
- Appropriate massage on the neck helps to promote circulation.
- Avoid long hours working at a computer. Exercise the neck every 1–2 hours.

- Avoid sleeping in prone position. The pillow should not be too high, too low, or too hard. The head and neck should be rested on the pillow. When sleeping on the side, the pillow height should be the length from the shoulder to the neck.
- Avoid wind directly blowing on the head and neck.
- Avoid contact sports until the healthcare provider allows.
- Be happy. In Chinese medicine, Stress can cause tight soft tissues in the neck.
- Don't slouch in a chair or bed.
- Don't make the neck "pop" have spinal manipulations.

UPPER CERVICAL SUBLUXATION 寰樞關節半脫位

INTRODUCTION

Upper cervical region refers to the first and the second neck bones. These two bones are given unique names, the first is called atlas or C1 and the second is called axis or C2. With their close proximity to the brain stem, misalignment and locking into an abnormal position may result in neck pain, headache, disruption to the cerebrospinal fluid circulation, and faintness.

ANATOMY

The cervical spine is lying between the cranium and the thoracic vertebrae. It consists of seven vertebrae (C1–C7), two of which are given unique names: atlas(C1) and axis(C2). These two vertebrae form a joint named the "atlantoaxial joint". It is a pivot joint which is responsible for spinal rotation, flexion, and extension.

Atlas (C1)
- Atlas (C1) does not have a body or spinous process; it consists of an anterior arch and a posterior arch. It articulates superiorly with the occiput (the atlanto-occipital joint) and inferiorly with the axis (the atlantoaxial joint). The atlas (C1) is made up of a thick anterior arch, a thin posterior arch, two prominent lateral masses, and two transverse processes on each side of the atlas which enclose two transverse foramens where the vertebral artery passes in each foramen. The superior facet articulates with the occipital condyles. The inferior facet articulates with the superior facets of the axis.

Axis (C2)
- Axis (C2) has a large vertebral body, and it contains the odontoid process (dens) and two transverse foramens. The anterior surface of odontoid process of axis (C2) articulates with the posterior aspect of the anterior arch

of the atlas (C1) and forms the Median atlantoaxial joint. The atlantoaxial joint is stabilized by the transverse ligament.

Intervertebral Discs

There is no intervertebral disc between C1 and C2; the discs appear in between the vertebrae starting below axis (C2).

Atlantooccipital Joint

The joint between atlas (C1) and the occipital bone is called the atlantooccipital joint (it is also known as the C0–C1 articulation), and it is comprised of a pair of condyloid synovial joints. Figure 3.3 from Shutterstock.com #415445938

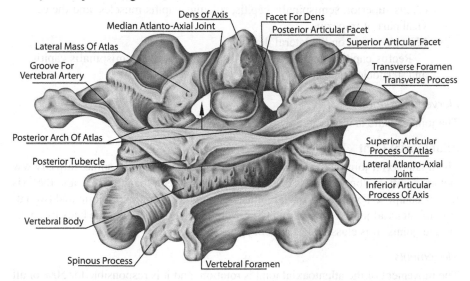

Movements

The joint allows for 10 degrees of flexion, 20 degrees of extension, 8–10 degrees of lateral flexion, and around 5–7 degrees of axial rotation.[2] The 30-degree head movements in flexion and extension is responsible for about 50% of all cervical flexion and extension (Windsor et al. 2017) or about half of the "up and down" mobility which allows us to nod the head to say "yes".

Blood Supplies

The atlantooccipital joint receives its blood supply from an anastomosis formed by the deep cervical, occipital, and vertebral arteries.

Ligaments

The joint is stabilized by two articular capsules, a posterior atlantooccipital ligament and an anterior atlantooccipital ligament.

Muscles
- Muscles acting on the flexion from the upright position:
  Trapezius, splenius capitis, longissimus capitis, semispinalis capitis, rectus capitis posterior major, rectus capitis posterior minor, and obliquus capitis superior.
- Muscles acting on the flexion from the supine position:
  Sternocleidomastoid, longus capitis, and rectus capitis anterior muscles.
- Muscles acting on the extension from the upright position:
  Sternocleidomastoid, longus capitis, and rectus capitis anterior muscles.
- Muscles acting on the Extension from the prone position:
  Rectus capitis posterior major, rectus capitis posterior minor, obliquus capitis superior, semispinalis capitis, splenius capitis muscles, and the cervical part of trapezius.
- Muscles acting on the lateral flexion:
  Rectus capitis lateralis, trapezius, splenius capitis, semispinalis capitis, and sternocleidomastoid.

Nerve Innervation
The joint capsule is innervated by the C1 nerve.

Atlantoaxial Joint
The atlantoaxial joint is the most mobile portion of the cervical spine. https://www .ncbi.nlm.nih.gov/books/NBK563271/. The joint between the atlas (C1) and the axis (C2) is composed of three synovial joints; one median atlantoaxial joint and two lateral atlantoaxial joints. The C1 and C2 vertebrae are connected in the back by a pair of facet joints. It is classified as a pivot joint.

Movements
The movement of the atlantoaxial joint is rotation, and it is responsible for 50% of all cervical rotation. The atlas is together with the head while in rotation. The movement of rotation toward both sides is around the dens of axis. The movement provides about half of the "side to side" mobility which allows us to shake the head to say "no".

Blood Supplies
The atlantoaxial joint receives blood supply from surrounding branches of the deep cervical, occipital, and vertebral arteries.

Ligaments
This joint is secured by four ligaments of which two are the major ligaments including a thick, strong ligament called Cruciform ligament (transverse ligament of the atlas) and the alar ligament; the remaining two are the tectorial membrane and apical ligament of dens.

Muscles
The suboccipital muscle group acts to stabilize the cervical vertebrae and joints. The group comprises four muscles, including the rectus capitis posterior minor, rectus

capitis posterior major, obliquus capitis superior, and obliquus capitis inferior. The last three muscles form the suboccipital triangle to provide fine motor function in movements of the head. https://en.wikipedia.org/wiki/Suboccipital_triangle.

Nerve Innervation

The innervation of the atlantoaxial joint is primarily by the branches of the second cervical spinal nerve. Each cervical vertebra has two nerve roots that arise from the spinal cord. As the nerve roots converge together in the cervical spinal foramen, they are termed cervical spinal nerves.

WESTERN MEDICINE ETIOLOGY AND PATHOLOGY

Even though subluxation was discovered more than 100 years ago (it was first reported by Edred Moss Corner in 1907 (Corner 1907)), the understanding and treatment of the condition is still relatively lagging behind and still in the groping stage.

Because the exact mechanism of the condition is still vague, there are many names of this cervical condition.

The upper cervical disharmony has a number of names: "Upper cervical subluxation", "Atlantoaxial Joint Disorder Syndrome", "Traumatic atlantoaxial rotatory dislocation", "Post-traumatic upper cervical subluxation", "Atlas Subluxation Complex", "Vertebral subluxation (VS)", "Vertebral subluxation complex (VSC)", "Pivot joint injury", "Upper cervical disorder", "Upper cervical instability" and "Silent Killer" because it slowly causes health problems over time, and severe damage to the C1 or C2 can cause paralysis or death.

Atlantoaxial stabilization relies three factors:

- The completion of the atlantoaxial joint's bone structure.
- The functions of the tensile stress resistance of both Cruciform ligament (transverse ligament of the atlas) and the alar ligament.
- The range of movements of the atlantoaxial joint maintaining in the physiological range. The atlantoaxial joint allows 10–15° of flexion/extension, 30° of axial rotation, and only minimal lateral flexion.[3]

Generally, atlantoaxial instability can be classified into three categories:

- Congenital-Dejerine-Sottas Disease (DS) is the most seen with atlantoaxial instability. DS is considered a type 3 Charcot-Marie-Tooth Disease (CMT) that is an inherited peripheral neuropathy condition. It causes atlantoaxial instability due to ligamentous laxity and osseous abnormalities (Lacy, Bajaj, and Gillis 2021).
- Traumatic injury to the head and neck, especially fractures of C1 or C2, can disrupt the structure of the transverse and alar ligaments resulting in atlantoaxial instability. Based on the uniaxial mechanical testing, the alar ligaments had an in vitro strength of 200 N while the transverse ligaments had an in vitro strength of 350 N. Therefore, with the lower strength and

its axial direction of loading, the alar ligament might be prone to injury (Dvorak, Schneider, Saldinger, and Rahn 1988).

- Inflammatory rheumatoid arthritis (RA) is a very common inflammatory condition that affects the craniovertebral junction (CVJ).

It is a disease characterized by a series of symptoms caused by minor dislocations of the atlantoaxial joint and atlantooccipital joint, with surrounding tissue damage, strain, and degenerative changes.

- Instability or subluxation or dislocation of the atlantooccipital joint or the atlantoaxial joint can result from a high-speed impact such as a motor vehicle collision, in which the pathologic mechanism involves hyperflexion of the neck. Whiplash often happens when the head, being unrestrained, strikes the windshield or dashboard. The head is thrown forward, then backward or from side to side, and vice versa. If the transverse ligament is damaged, it loses its function to prevent anterior displacement of the atlas (C1) over axis (C2); hence, it cannot prevent the odontoid process from posterior migration[4]. If the alar ligament is damaged, it loses its function to limit the amount of rotation of the head.

The trauma can lead to either or both atlantooccipital joint and atlantoaxial joint misalignment. The impact is classified in three types: anterior displacement, posterior displacement, or longitudinal distraction. The severity of the injury is classified in three stages: Stage I and II describe none or minimal displacement. Stage III is a severe dislocation and may be associated with injury of the cervical spinal cord; it can be fatal (Grujičić 2021).

- The birthing process can cause misalignment in the spinal column, particularly in the upper neck. The twisting and pulling of a baby's head are common during birth.
- Cervical subluxations in babies can result from falls. The heaviness of the infant's head draws the baby's head down first when falling from a height.
- Congenital defects may lead to upper cervical instability. Atlas hypoplasia occurs when there is an underdeveloped or incompletely developed atlas. The defect condition ranges in severity, as some people may have small defects while others may have a complete lack of the back arch of the atlas. Such instability can easily lead to subluxation when over-turning the neck (Vincent 2019).
- The genetic disorder is commonly associated with Down Syndrome and Ehlers-Danlos Syndrome.[5]
- Holding the head at an awkward angle, leaning the head forward staring at screens and devices too frequently and too long, or sleeping in an odd position may cause tension and straining of one side of the muscles and ligaments. The cumulative effect of these positions leads to misalignment or increased laxity of the joints in the contralateral side.

- Upper cervical spine pyogenic infection involves the occipital condyles, the atlas, and the axis, leading to severe joint instability at C0-C1-C2 (Konbaz et al. 2021).
- In rheumatoid arthritis (RA) and similar arthritic conditions, the progressive deterioration of the joints and vertebrae can lead to cervical instability.
- Infection in the neck region and upper respiratory tract infections can lead to nontraumatic subluxation of the atlantoaxial joint. This condition is called Grisel's syndrome, and it primarily affects children. It is caused by inflammation of the adjacent soft tissues. Other conditions that can cause the subluxation are peritonsillar abscess, retropharyngeal abscess, and post-operative inflammation after certain procedures, such as an adenotonsillectomy. Etiopathogenesis has not been clearly described yet but the cause is thought to be increased muscle contraction due to inflammation. With the head leaning to one side, the dragging force from the spasm muscle increases loosening of the transverse ligament (Bucak 2014).

CHINESE MEDICINE ETIOLOGY AND PATHOLOGY

The etiopathogenesis in upper cervical subluxation in Chinese medicine is considered a condition of "Headache" and "Dizziness".

1. Blockage of Phlegm-Damp in Middle Jiao 痰濕中阻證

The disharmony between the ascending and descending functions of Spleen and Stomach causes Phlegm Damp accumulated in Middle Jiao. Stomach Qi rises to the head and causes dizziness.

2. Liver Yang Rising 肝陽上亢 證
- The patient is usually in excessive Yang energy. With an attack of emotional upset or anger, the stagnation of Qi transforms into Fire, and the Heat drains the Yin fluid. Yin deficiency with Yang rising leads to a stronger imbalance. Qi is a Yang energy, and when Qi is unanchored, it floats upward like the character Wind. Wind disturbs the orifice.
- Deficiency of Kidney leads to Liver Yang rising, Liver Wind combines with Phlegm, the Heat within Yang energy then rises to the head to result in obstruction.

3. Qi and Blood Deficiency氣血兩虛證

Qi and Blood deficiency is caused by insufficiency of resources to produce Blood and Qi, the result of weakness of Spleen and Stomach. The deficiency of Qi and Blood manifests dizziness and vertigo. Such deficiency can occur in both young people, whose diets are irregular and poor, and older people, who may be weak in Spleen and Stomach. Emotional stress and anger can affect the functions of Spleen resulting in Qi and Blood deficiency.

4. Qi Stagnation and Blood Stasis氣滯血瘀證

After the subluxation, the artery compression leads to Blood stasis causing both Qi and Blood stagnation and insufficient blood supply to the brain, resulting in dizziness and vertigo. Even though Blood deficiency is the main presentation in the upper cervical subluxation, the mechanism of the disease is Qi and Blood stagnation.

MANIFESTATIONS

In Western Medicine

Symptoms of upper cervical spine instability are variable, when the displacement of the atlantoaxial joint causes stress to the adjacent C1–C3 cervical nerves, vertebral arteries, and sympathetic nerves, a series of clinical symptoms that can occur one after another.

1. Patients often present with torticollis, where the head leans to one side.
2. With the location of it at the upper cervical region, there are neck pain near the skull and neck stiffness. The pain can be sudden onset with a history of trauma. Patients may or may not have an obvious injury history.
3. Dizziness is a common symptom.
4. Cervicogenic headache is commonly seen at the occipital region. Patients often experience both neck pain and headache together, and the symptoms tend to worsen with movement in the head and neck. Pain can increase during coughing.
5. Other symptoms that patients may experience are nausea and vomiting. Some patients may also present with blurry and/or double vision, pain around the eyes, nasal congestion and allergies; some may have tinnitus, ear infections, and hearing loss, due to misalignment that prevents the ear from draining properly.
6. Symptoms like loss of balance and trouble walking present when the spinal cord is affected (cervical myelopathy).

In Chinese Medicine

1. Blockage of Phlegm-Damp in Middle Jiao 痰濕中阻證

Main symptoms: dizziness, vertigo, headache where the patient cannot tell the exact location (patients often describe the sensation as "the head is wrapped in a damp moist cloth"), chest distention, nausea or vomiting clear sputum, and somnolence.
Tongue: pale body, greasy white coating.
Pulse: soggy, slippery.

2. Liver Yang Rising 肝陽上亢證

Main symptoms: dizziness, vertigo, agitation, and short temper; red face, insomnia with dream disturbances; and dry mouth and bitter taste.

Tongue: red body, yellow coating.
Pulse: wiry.

3. Qi and Blood Deficiency氣血兩虛證

Main symptoms: dizziness, pale face, fatigue, reticent, palpitation, insomnia, poor appetite, and lassitude.
Tongue: pale body, thin white coating.
Pulse: thin, weak.

4. Qi Stagnation and Blood Stasis氣滯血瘀證

Main symptoms: dizziness, stabbing headache; patients often describe the headache location as "not fixed", and the headache is better in daytime and worse at night; patients refuse pressure on the head. Dim-colored face.
Tongue: dusky body with ecchymoses, thin white coating.
Pulse: choppy, deep

PHYSICAL EXAMINATION

- Begin with obtaining medical history, such as the genetic abnormalities of the atlas and axis (e.g., Down syndrome), infections, inflammation, and trauma history including sports injuries. For patients who have had whiplash, inquire about the direction of impact and the symptoms immediately following the accident, such as dizziness, vomiting, headache, double vision and blurry vision, tingling in the arms, fingers and legs; and trouble walking and loss of balance. Trauma can lead to serious abnormalities, including cervical fracture and/or dislocation and/or compression of the spinal cord or nerve roots. The recognition of the severity of the trauma may be noted by how soon the patient reported to the medical facility. When the patients report with symptoms of cervical myelopathy, typically they are taken from the scene of the crash directly to the emergency department, where cervical fracture and/or dislocation and/or compression of the spinal cord or nerve roots can be determined. If the patient visits a general practitioner days later, the trauma is typically less severe, although, atlantoaxial joint, atlantooccipital joint, and surrounding tissue damage and strain are possible (Jensen 2009).
- Inspection

Patients with upper cervical disorders commonly present with postural distortion where the neck and the head leans to one side.

- Palpation

The cervical muscles are rigid. There is a tenderness over the spinous process of the axis with palpation. Palpation may also find that the spinous process of C2 is deviated.

- Range of motion

Upper cervical subluxation results in less cervical rotation range than the normal range which is 45 degrees.

Alar Ligament Test

Active movement: the patient is seated; the examiner palpates the spinous process of C2 and has the patient bend the head to one side. The examiner should feel the coupled movement, in which the spinous process immediately moves toward the opposite side. A positive test for instability is the lack of coupled C2 motion during side bending.

Passive movement: the patient is seated; the examiner palpates, locates, and uses the first 2 fingers to grasp the spinous process of C2. The examiner places a hand on the crown of the patient's head and applies slight compression to facilitate atlantooccipital side bending. Passive side bending then is applied to direct the patient's ear toward the opposite side of the neck. If fixation of the axis is adequate, the normal coupled movement will not be permitted, and side bending will not occur. A positive test for instability is when the side bending occurs while the spinous process of C2 is stabilized.

The Alar ligaments stabilize the odontoid process of C2 with attachments from the lateral aspect of the upper posterior margin of the odontoid process of C2 to the medial surface of the occipital condyles of the skull. Hence, the instability of movement is between C2 and the skull.

- Special tests

Plain radiography can access the C1–C2 articulation. Standard views include open-mouth odontoid and lateral cervical spine radiographs.

The atlantodental interval (ADI) is the horizontal distance between the anterior arch of the atlas and the dens of the axis, used in the diagnosis of atlas and axis injuries. It is the distance between the anterior surface of the dens and the posterior surface of the anterior arch of atlas (C1). The normal values are 3mm or less in male adults, 2.5mm or less in female adults, and 5mm or less in children.[6]

Lateral cervical spine radiograph

On the lateral cervical spine radiograph, atlantoaxial subluxation demonstrates greater than 5 mm of distance between the anterior arch of the atlas and the odontoid (Azar et al. 2021, 1959).

Open mouth (odontoid view) radiograph

On the open-mouth odontoid view, the combined spread of the lateral masses of C1 on C2 should not exceed 6.9 mm; greater than 6.9 mm indicates rupture of the transverse ligament (Leas 2021).

Some nontraumatic conditions that may be associated with increases in the atlantodental interval include Down syndrome, Grisel syndrome, and rheumatoid arthritis. The pathological changes in these conditions are due to laxity of the ligaments and destruction of the articular cartilage

If plain radiograph doesn't produce conclusive results, MRI, CT scans, myelography, angiography, and bone scan can help to reveal a spinal tumor, bone infection, degeneration, rheumatoid arthritis, spinal fracture, torn ligament and to understand other possible genetic abnormalities of the joints.

DIFFERENTIAL DIAGNOSIS

- Occipital headache

The Tai Yang type of headache is located at the occipital region. It is characterized by acute headache, fever, and floating pulse. The headache usually is accompanied by whole body ache, aversion to cold and wind, sneezing, coughing with white sputum, a runny nose with white discharge or a blocked nose, absence of thirst, no desire to drink or may desire a warm drink, and absence of sweat.

- Degenerative intervertebral disc disease of the cervical spine

Symptoms associated with the degenerative discs don't appear on the atlantooccipital joint and atlantoaxial joint since there are no discs between C1 and C2; rather, symptoms may appear between the vertebrae below axis (C2).

Patients with cervical spondylosis manifest neck pain with radiating pain or numbness or tingling in the arm to the fingers.

- Cervical strain, trauma, or fracture

There is a history of injury. The cervical muscles may swell and feel hard or knotted.

- Acute cervical fibrosis

The Alar Ligament Test is negative; radiographs are negative.

TREATMENTS

1. **Blockage of Phlegm-Damp in Middle Jiao** 痰濕中阻證

Treatment Principle
Tonify Qi, raise Pure Yang.

Herbal Treatments

Herbal Formula
He Ying Zhi Tong Tang 益氣聰明湯

Herbal Ingredients

Huang Qi 黃耆15g, Ren Shen人參15g, Ge Gen葛根9g, Man Jing Zi蔓荊子9g, Bai Shao白芍6g, Huang Bai黃柏6g, Sheng Ma升麻4.5g, Zhi Gan Cao甘草灸3g

Ingredient Explanations

- Huang Qi 黃耆tonifies Qi, strengthens the Spleen, raises the Yang Qi of the Spleen and Stomach, tonifies Wei Qi, stabilizes the exterior, and tonifies the Blood.
- Ren Shen人參tonifies Qi and Yang.
- Ge Gen葛根relieves Heat, releases muscles especially of the neck and upper back.
- Man Jing Zi蔓荊子expels Wind and stops pain.
- Bai Shao白芍nourishes the Blood, soothes the Liver, and stops pain.
- Huang Bai黃柏drains Damp-Heat, especially from the Lower Jiao.
- Sheng Ma升麻releases the exterior, clears Heat, raises pure Yang, and lifts sunken Qi.
- Zhi Gan Cao甘草灸 tonifies the Spleen, harmonizes the other herbs, and guides the herbs to all twelve meridians.

Modifications

- To make to formula stronger for draining Phlegm Damp, add Ban Xia半夏，Bai Zhu白朮，and Cang Zhu蒼朮.
- For Kidney Yin deficiency, add Liu Wei Di Huang Wan.
- For Kidney Yang deficiency, add Gui Fu Di Huang Wan.
- For Liver Yang rising and underlying Kidney Yin deficiency, add Qi Ju Di Huang Wan.

Formula Actions

- Tonifies Qi and strengthens Middle Jiao.
- Raises Pure Yang.

Direction

- Drink one cup of the decoction before meals twice daily.

2. Liver Yang Rising 肝陽上亢證

Treatment Principle

Soothes the Liver, suppresses Yang, nourishes Yin, and clears Heat

Herbal Treatments

Herbal Formula

Tian Ma Gou Teng Yin 天麻鉤藤飲

This formula is in the book "Newly Explained Diagnosis and Treatment of TCM Internal Medicine and Miscellaneous Diseases中醫內科雜病證治新義", authored by Hu Guangci 胡光慈 in 1958.

Herbal Ingredients

Tian Ma天麻6g, Gou Teng鉤藤9g, Niu Xi牛膝12g, Shi Jue Ming石決明15g, Du Zhong杜仲12g, Huang Qin黃芩6g, Zhi Zi梔子6g, Yi Mu Cao益母草9g, Sang Ji Sheng桑寄生9g, Ye Jiao Teng夜交藤9g, Fu Shen茯神9g

Ingredient Explanations

- Tian Ma 天麻 extinguishes Wind, calms the Liver, subdues rising Liver Yang, and alleviates pain.
- Gou Teng鉤藤expels Wind, calms the Liver, and releases spasm.
- Niu Xi牛膝 moves the Blood downwards, tonifies the Liver and Kidney; opens the meridians to benefit joints and Tendons.
- Shi Jue Ming石決明drains Liver Fire, expels Wind; assists Tian Ma 天麻 and Gou Teng鉤藤 to enhance extinguishing Wind; and calms the Liver.
- Du Zhong杜仲tonifies the Liver and Kidney, strengthens Bone, and benefits Essence and Blood.
- Huang Qin黃芩clears Heat, dries Damp, cools Blood, stops bleeding, and calms Liver Yang.
- Zhi Zi梔子clears Heat, cools the Blood, calms Liver Yang, resolves Damp, reduces swelling, and opens the meridians.
- Yi Mu Cao益母草moves the Blood, promotes urination, reduces swelling, clears Heat
- Sang Ji Sheng桑寄生tonifies the Liver and Kidney, strengthens Bone, and benefits Essence and Blood; expels Wind and drains Damp.
- Ye Jiao Teng夜交藤nourishes Heart Yin and Blood, calms Shen神, and expels Wind and Damp.
- Fu Shen茯神 calms Shen.

Modifications

- For excessive Liver Fire, add Xia Ku Cao夏枯草 and Long Dan Cao龍膽草.
- For insomnia, add Long Gu龍骨 and Mu Li牡蠣.
- For Blood stasis, add Dan Shen丹參and Chi Shao赤芍.
- For Liver and Kidney Yin deficiency, add Liu Wei Di Huang Wan 六味地黃丸.
- For red face and constipation, add Huang Lian黃連 and Da Huang大黃.

3. Qi and Blood Deficiency氣血兩虛證

Treatment Principles

Tonify Qi and nourish Blood.

Herbal Formula

Gui Pi Tang歸脾湯

This formula is from "Categorized Essentials of Repairing the Body 正體類要" authored by Xue Ji 薛己 (1487-1559) in Ming dynasty 明朝 (1368-1644) and published in 1529

Herbal Ingredients

Ren Shen人參3g, Zhi Huang Qi 黃耆炙15g, Bai Zhu白朮9g, Fu Ling茯苓9g, Dang Gui 當歸9g, Suan Zao Ren酸棗仁炒9g, Long Yan Rou龍眼肉9g, Zhi Yuan Zhi遠志6g, Mu Xiang木香3g, Zhi Gan Cao甘草炙4.5g, Sheng Jiang生薑2 slices Da Zao 大棗3 pieces

Ingredient Explanations

- Ren Shen人參tonifies Qi and Yang.
- Zhi Huang Qi黃耆炙tonifies Qi, strengthens the Spleen.
- Bai Zhu白朮tonifies the Spleen Qi and dries Dampness.
- Fu Ling茯苓drains Damp and tonifies the Spleen.
- Dang Gui當歸moves the Blood, disperses Bruising, and stops pain due to Blood stasis.
- Suan Zao Ren酸棗仁 nourishes Heart Yin and tonifies Liver Blood.
- Long Yan Rou龍眼肉drains Damp-Heat from the Liver and the Gallbladder channels.
- Yuan Zhi遠志expels Phlegm, calms Shen, and reduces abscesses.
- Mu Xiang木香promotes the movement of Qi and alleviates pain.
- Zhi Gan Cao甘草炙tonifies the Spleen, harmonizes the other herbs, and guides the herbs to all twelve meridians.
- Sheng Jiang生薑unblocks the pure Yang pathway and harmonizes rebellious Qi.
- Da Zao紅棗tonifies the Spleen Qi, nourishes the Blood, moderates and harmonizes the harsh properties of the fragrant herbs.

Formula Actions

- Nourishes the Blood.
- Tonifies Qi.
- Strengthens Spleen.

Contraindications

- Contraindicated for those with bleeding or with interior Heat or insomnia.

Modifications

- To make the formula stronger, add Shu Di Huang 熟地and E Jiao阿膠.
- For severe insomnia and memory loss, add He Huan Pi 合歡皮, Ye Jiao Teng夜交藤, Shi Chang Pu 石菖蒲.

4. Qi Stagnation and Blood Stasis氣滯血瘀證

Treatment Principles

Moves Blood and disperses Bruising.

Herbal Formula:

Tong Qiao Huo Xue Tang通竅活血湯

This formula is from "Correction of Errors in Medical Classics醫林改錯", authored by Wang Qingren 王清任 (1768–1831), in Qing dynasty 清朝 (1636–1912), and published in 1849.

Herbal Ingredients

Chi Shao赤芍3g, Chuan Xiong川芎3g, Tao Ren 桃仁9g, Hong Hua紅花9g, She Xiang麝香0.5g, Cong Bai老蔥15g, Sheng Jiang生薑9g, Da Zao紅棗7 pieces

Ingredient Explanations

- Chi Shao赤芍 moves the Blood, disperses Bruising, and stops pain.
- Chuan Xiong川芎 moves the Blood, disperses Bruising, and stops pain due to Blood stasis.
- Tao Ren 桃仁 moves the Blood, disperses Bruising, and stops pain due to Blood stasis.
- Hong Hua紅花 moves the Blood, disperses Bruising, and stops pain due to Blood stasis.
- She Xiang麝香 opens orifices, revives Shen, moves the Blood, opens twelve meridians, and stops pain.
- Cong Bai老蔥 unblocks the pure Yang pathway, disperses Cold, induces sweat, releases the exterior, and relieves toxicity.
- Sheng Jiang生薑 unblocks the pure Yang pathway and harmonizes rebellious Qi.
- Da Zao紅棗 tonifies the Spleen Qi, nourishes the Blood, and moderates and harmonizes the harsh properties of the fragrant herbs.

Formula Actions

- Moves the Blood.
- Opens the orifices.

Contraindications

- Contraindicated during pregnancy.

Modifications

- To make the formula stronger, add Wu Ling Zhi 五靈脂, Man Jing Zi蔓荊子, Tian Ma天麻, Gou Teng鉤藤, and Da Huang大黃.
- For severe headache, add Wu Ling Zhi 五靈脂, Man Jing Zi蔓荊子, Quan Xie全蠍, Wu Gong蜈蚣, and Tu Bie Chong土鱉蟲.

Directions

1) Place the herbs into an herb pot. Traditionally, a clay cooker is used. Gently rinse the herbs once with a strainer.
2) Add 250 ml cooking wine to the pot
3) Turn on the stove to high heat and bring the herbs to a boil. Turn off the heat when the decoction is reduced to 150 ml.

4) Drain the decoction out into a container. Keep only the herbal wine.

5) Pour the herbal wine back into the herb pot and add She Xiang麝香.

6) Bring the herbal wine to a boiling point for 2 times.

7) Drink the herbal wine before bed.

Acupuncture Treatments

Acupuncture Point Formula

Local Points

GB20 風池Fengchi, DU16 風府Fengfu, GB19 N腦空Aokong, DU20百會 Baihui, Huo Tuo Jia Ji 華佗夾脊C2 (select both sides)

Distal Points

SI3 後溪Houxi (select the opposite side, insert the needle direction toward LI4), UB62申脈Shenmai, SJ3中渚Zhongzhu, UB60崑崙Kunlun, 印堂Yintang

Modifications

- For Phlegm-Damp in Middle Jiao, add REN12中脘Zhongwan, ST36足三里 Zusanli, and ST40 豐隆Fengnong.
- For Liver Yang Raising, add KD3 太溪Taixi, KD7復溜Fuliu, LIV2行間 Xingjian, LI4合谷He Gu, and LIV3太沖Taichong.
- For Qi and Blood Deficiency, add DU4命門Mingmen, ST36足三里 Zusanli, and GB39懸鐘Xuanzhong.
- For Qi Stagnation and Blood Stasis, add LI4合谷He Gu, LIV3太沖 Taichong, SP10 Xuehai, and UB17膈俞Geshu.

Tui Na Treatments

Tui Na Techniques

Pressing, kneading, rubbing

Tui Na Procedures

Joint Reduction

Through radiograph diagnosis, ensure that there is no fracture, spinal compression, or nerve lesion.

Joint reduction is effective for this condition.

Preparation:

1. The patient is in a supine position with a pillow under the head; the therapist stands at the head of the table.
2. The therapist applies kneading, pressing, and rubbing techniques to relax the cervical muscles.
3. During the muscle relaxation or the cervical manipulation procedures, if you notice the patient is nervous, ask the patient to inhale and exhale to relax the head.

Manipulation

Atlantoaxial joint subluxation:

1. Set the subluxation toward the left, for example. The therapist places the left thumb over the tip of the transverse process of C2.
2. Rotate the patient's head toward the right.
3. Place the middle 3 fingers of the right hand on the chin, then extend the neck with slight traction of the right hand.
4. Using the right hand, apply a short lift – the line of drive should be toward the right 45° to the therapist's right side – while the left thumb presses the transverse process of C1 toward the right.

Atlantooccipital joint subluxation:

1. Set the subluxation toward the left, for example. The therapist places the left thumb on the tip of the transverse process of C1, which it is located between the mandible and the mastoid process.
2. Rotate the patient's face toward the right side.
3. Place the middle 3 fingers of the right hand on the chin, then extend the neck with a slight traction of the right hand.
4. The right hand applies a short thrust – the line of drive should be toward the therapist's right shoulder – while the left thumb presses the transverse process of C1 toward the right.

Please be aware that state-approved acupuncture scope of practice varies and may not include spinal manipulation. If you are not sure, check with your state board.

Patient Advisory

Traumatic conditions take a longer time to heal. My observation is that 70% of healing is through self-care; only 30% of healing is due to treatment. Here is a list of suggestions for self-care:

1. Immobilize the neck after the reduction.
2. Avoid extreme neck movements for two weeks after the reduction.
3. Patients with vertigo should not rotate the neck to a large angle or too fast.
4. Get plenty of rest when vertigo occurs.
5. Be ready to sleep before 11 p.m. and ensure that the sleep quality is excellent.
6. Supine sleep posture is more suitable for healing a neck injury.
7. Accurate postures help rapid healing and prevent damage to the joints. Avoid keeping the face downward, upward, or sideways for too long. When looking forward at a computer monitor, take a 10–15-minute break every hour. Ensure good sitting posture when driving. Seatbelt is also highly recommended to prevent further neck trauma.
8. Appropriate exercises improve the circulation of Qi and Blood. The exercises that are slow-paced and have gentle neck movements are recommended.

They include walking, yoga, Tai Chi, Qi Gong. Further ligament injury and spinal cord compression can arise or worsen if sussceptible patients are subjected to extreme ranges of motion; hence, avoid exercises that are too aggresive, such as football and basketball.

9. Keep the neck warm when the weather is cold and/or rainy.

NOTES

1. https://www.physio-pedia.com/Hoffmann%27s_Sign.
2. https://www.physio-pedia.com/Atlanto-occipital_joint.
3. https://radiopaedia.org/articles/atlanto-axial-articulation?lang=us.
4. https://www.physio-pedia.com/Transverse_Ligament_of_the_Atlas#:~:text=It%20functions%20to%20prevent%20anterior,preventing%20it%20from%20posterior%20migration.
5. https://ioaregenerative.com/blog/identifying-the-signs-of-cervical-instability.
6. https://radiopaedia.org/articles/atlantodental-interval?lang=us.

4 Chest Pain

INTERCOSTAL NEURALGIA 肋間神經痛

INTRODUCTION

Intercostal neuralgia is a set of symptoms, of which pain in the upper chest is the primary symptom, caused by nerve irritation. The pain can be constant or intermittent. Movements, coughing, sneezing, laughing, and even breathing may worsen the pain. The pain can travel to the shoulder and upper back.

ANATOMY

The ribs are a set of twelve paired bones. Their function is to protect the internal thoracic organs, but they also play an important role in ventilation, as they expand to allow lung inflation.

There are two classifications of ribs:

Typical Ribs

Ribs 3–9 have a head, neck, and body. The head has two articular facets. The neck is between the head and the body, and it has a facet that articulates with the transverse process. The body has a groove to protect an intercostal artery, an intercostal vein, and an intercostal nerve.

Atypical Ribs

Rib 1 is shorter and wider than the other ribs, and it has one facet articulating with the vertebra.

Rib 2 has two articular facets on the head. The serratus anterior muscle originates from the head.

Rib 10 has one facet articulating with the vertebra.

Rib 11 and 12 have no neck, but each of them has one facet articulating with the vertebra.

Anterior Articulation

Only ribs 1–7 attach to the sternum. The ribs 8–10 attach to the costal cartilage superior to them and ribs 11 and 12 are called "floating ribs" as they do not attach to the sternum nor any cartilage.

Posterior Articulation

All twelve pairs articulate with the vertebrae in two joints:

DOI: 10.1201/9781003203018-4

Costotransverse joint is between the tubercle of the rib and the transverse costal facet of the vertebra.

Costovertebral joint is between the head of the rib and the superior costal facet and the inferior costal facet of the vertebra.

Intercostal Nerves

Intercostal nerves are the anterior rami of the thoracic paravertebral nerves from T1 to T11. They have motor and sensory components.

Each intercostal nerve enters the intercostal space, travels forward with the intercostal vessels in the costal groove of each rib. Only the inferior part of the first nerve enters the costal groove, the superior part of it becomes the brachial plexus. The first two intercostal nerves innervate the upper limb in addition to the thoracic innervation. The first six intercostal nerves innervate the intercostal muscles, and the skin near the midline of the chest. The lower intercostal nerves innervate the intercostal muscles, abdominal peritoneum, and skin. The seventh intercostal nerve terminates at the xiphoid process. The tenth intercostal nerve terminates at the umbilicus. The twelfth intercostal nerve is not part of the intercostal nerves; it is known as the subcostal nerve, and it terminates at the abdominal wall and the groin.

Intercostal nerves are important for normal respiration and they are essential to the efficacy of coughing and sneezing.

WESTERN MEDICINE ETIOLOGY AND PATHOLOGY

The causative factors may be known or unknown. Primary intercostal neuralgia is rare, secondary intercostal neuralgia is the most seen clinically. It is commonly caused by injury and inflammation to the intercostal nerves. Direct personal contact sports such as football, wrestling as well as injury from motor vehicle accident are the known causes that can damage these nerves.

Some medical procedures on the thoracic spine, mastectomy, breast surgery, and the procedure to access the lungs and heart, such as thoracotomy, can cause trauma to the intercostal nerve.

A history of shingles or chicken pox may be another cause of the pain. The herpes zoster virus stays in the body after the episodes of shingle or chicken pox, but it can be reactivated and cause shingles again if the immune system weakens.

Other conditions that can cause intercostal neuralgia include degenerative change of the thoracic spine, inflammation of the thoracic joints, malignant or benign tumor, and pregnancy as the body makes room for a growing fetus.

CHINESE MEDICINE ETIOLOGY AND PATHOLOGY

Because the Liver and Gallbladder meridians pass through the thoracic region, intercostal neuralgia mainly involve these two Zang-fu and their meridians.

1. Retention of Blood stasis瘀 血 阻 絡

   Through either the invasion of the external pathogen or due to injury, Blood flow is seized or moving slowly, leading to Blood stasis. Emotional anger can cause Blood stasis because Qi is the commander of Blood, anger causes stagnation of Qi and Qi stagnation leads to Blood stasis (Apichai 2020).

2. Retention of Damp-Heat in the Liver and Gallbladder meridians濕 熱 蘊 結

   External factor: invasion of the external Damp-Heat pathogen attacking the Shao Yang Gallbladder meridians impairs the dispersing functions of the Liver and Gallbladder.

   Internal factor: poor diet with too much sweet, fatty, spicy, and pungent food or too much alcohol impairs the functions of the Spleen and Stomach, leading to the accumulation of Damp-Heat in the Liver and Gallbladder, causing Qi stagnation in the Liver and Gallbladder, resulting in pain in the ribcage.

3. Insufficiency of Liver Yin肝 陰 不 足

   Chronic Kidney deficiency, chronic diseases, overworking, and sleep deprivation can drain Liver Yin. Long-term stagnation of Qi evokes Fire, the Heat from Fire dries the Yin fluid (which is Blood), resulting in insufficiency of both Yin and Blood. Liver's lack of nourishment from Blood results in dryness in the meridians.

MANIFESTATIONS

In Western Medicine

Pain is the main symptom of intercostal neuralgia. The duration of pain can be constant or intermittent.

The sensation of pain is like a band that wraps around the upper back and chest along the distribution of the dermatome. The region of pain often covers from one to several ribs.

The quality of pain can be sharp, burning, stabbing, aching, numbness, tingling, and dullness. It can radiate from the upper back toward the chest along the intercostal space. It can also present as a referred pain into the shoulder blade and upper back if the affected spine is the T1. The manifestations also include involuntary contractions of individual muscles and color changes of skin above the affected area, as well as the loss of sensitivity over the affected dermatome (Fazekas, Doroshenko, and Horn 2021).

The inducing factors that can worsen the pain commonly include movements (e.g., rapid torso turning or jumping), coughing, sneezing, laughing, and even breathing, so that patients often describe that holding the breath helps to prevent pain worsening.

In Chinese Medicine

Manifestations in Chinese medicine

1. Retention of Blood stasis 瘀 血 阻 絡

Stabbing pain in a fixed location, patients refuse direct pressure on the affected area. The pain is worse at night. A mass may be palpable in the painful region.
Tongue: dark purple body with petechiae.
Pulse: choppy and deep.

2. Retention of Damp-Heat in the Liver and Gallbladder meridians 濕 熱 蘊 結

Chest distention, bitter taste in the mouth, dry throat, nausea, vomiting, poor appetite, red eyes; yellow sclera, skin, and urine.
Tongue: yellow, greasy coating.
Pulse: wiry, slippery, rapid.

3. Insufficiency of Liver Yin 肝 陰 不 足

The pain quality is constant and dull. It can be aggravated by fatigue, sleep deprivation or overwork. Dry mouth and throat, red lips, dizziness, vertigo, restlessness in the muscles, insomnia, and irritability are other symptoms.
Tongue: red and small body, little or no coating.
Pulse: wiry, thready, rapid.

PHYSICAL EXAMINATION

- Begin with obtaining a medical history.

When a patient complaining of chest pain presents, obtaining a thorough medical history is helpful to clarify the exact diagnosis. The pain from intercostal neuralgia can be from a condition in the thoracic spine or the ribs; it might also be a referred pain from an organ. Interview questions should include pain location, duration, quality, and when it occurs. In intercostal neuralgia pain, the typical location of pain follows a distribution along the affected dermatome, and the pain radiates from the upper back toward the chest along the ribs like a band. The duration can be short like an electric shock or constant. The quality of pain can be varied from high pain intensity such as sharp, stabbing to dull, numbness, or tingling. The pain can be induced and aggravated by movements such as jumping, quick body turning, laughing, coughing, sneezing, and breathing.

Preexisting conditions are also important to understanding the cause and making the diagnosis. These include history of prior thoracic surgery, pregnancy, benign or malignant tumor in the thoracic spine, herpes zoster infection, history of direct trauma to the ribcage or thoracic spine, history of thoracic disc herniation, and sports activities that might be relevant to intercostal neuralgia.

- Inspection
  - Observe the region to explore the color of the skin, erythema, swelling, and bruising. The individual muscles may have involuntary

contractions. The skin above the affected area may have color changes. Patients with herpes zoster infection may have a vesicular rash in the distribution of the affected dermatome.

- Palpation
  - Palpate or stretching directly over affected nerves may reproduce or aggravate pain.
- Range of motion
  - Range of motion in patients with the condition of intercostal neuralgia is usually not restricted.
- Orthopedic test
  - Schepelmann's sign
    The test will be positive in patients with the condition of intercostal neuralgia when the pain is increased upon lateral lumbo-thoracic bending toward the side of pain. The pain decreases when bending toward the contralateral side.
- Special test
  In western medicine, intercostal nerve blocks may be another effective tool to the diagnosis of intercostal neuralgia.

DIFFERENTIAL DIAGNOSIS

- Pleurisy
  A positive finding for pleuritic pain when performing Schepelmann's sign is that pain increases upon lateral lumbo-thoracic bending toward the side without pain.
- Cardiovascular diseases
  In addition to the chest and upper back pain, a radiating pain into the left arm, accompanied by shortness of breath, sweating, palpitation and sudden confusion, dizziness, or fainting may present. For angina pectoris, sublingual nitroglycerin can relieve sudden angina.
- Pulmonary diseases
  Patients may present with shortness of breath, chronic cough that may produce sputum or blood, unintended weight loss, and/or edema in the legs, ankles, and feet.

TREATMENTS

1. Retention of Blood stasis瘀 血 阻 絡

Treatment principle
Move the Blood, disperse Bruising, and stop pain.

Herbal Formula
Fu Yuan Huo Xue Tang復 元 活 血 湯

This formula is from "Clarifying the Study of Medicine醫 學 發 明", authored by Li Dongyuan 李 東 垣, in Yuan dynasty元 朝 (1271–1368), and published in 1315.

Herbal Ingredients

Chai Hu 柴 胡9g, Tian Hua Fen 天 花 粉9g, Dang Gui 當 歸9g, Hong Hua 紅 花6g, Gan Cao 甘 草6g, Chuan San Jia 穿 山 甲 炮6g, Da Huang 大 黄12g, Tao Ren 桃 仁9g

Ingredient Explanations

- Chai Hu 柴 胡 harmonizes Shao Yang disorders and relieves stagnation.
- Tian Hua Fen 天 花 粉 clears Heat, expels pus, and reduces swelling.
- Dang Gui 當 歸 moves and tonifies the Blood, disperses Cold, and stops pain due to Blood Stasis.
- Hong Hua 紅 花 moves the Blood, releases Bruising, opens the meridians, and stops pain.
- Gan Cao 甘 草 releases spasms, alleviates pain, harmonizes the other herbs, and guides the herbs to all twelve meridians.
- Chuan San Jia 穿 山 甲 moves the Blood, reduces swelling, and expels Wind Damp.
- Da Huang 大 黄 moves the Blood and disperses Bruising
- Tao Ren 桃 仁 breaks up stubborn Blood Stasis.

Cooking Instructions

1. Place the herbs into an herb pot. Traditionally, a clay cooker is used. Gently rinse the herbs once with a strainer. If you notice some red flowers in the strainer, place them back into the herb pot.
2. Add water into the pot until it is just enough to cover the herbs, then add one more cup of water.
3. Presoak the herbs for 15 minutes.
4. Turn on the stove to high heat and bring the herbs to a boil. Then reduce the heat to medium-low to simmer with slight bubbling. Turn off the heat when the decoction is reduced to one cup.
5. Drain the decoction into a container.
6. Repeat step 2–5 one more time and drain each decoction to the same container.
7. Divide the decoction into two separate cups.

Dose

Drink one cup, 2 times a day.

Formula action: moves the Blood, disperses Bruising, soothes the Liver, and opens the meridians.

Formula indication: physical trauma, Blood stasis, and severe pain.

2. Retention of Damp-Heat in the Liver and Gallbladder meridians濕 熱 蘊 結

Treatment Principle

Drain excess Fire from the Liver and Gallbladder, clear and Drain Damp-Heat from the Lower Jiao.

Herbal Formula

Long Dan Xie Gan Tang龍膽瀉肝湯

This formula is from the book "Analytic Collection of Medicinal Formulas 醫方集解", authored by Wang Ang汪昂 (1615–1694AD), published in Qing dynasty清朝 (1682 AD).

Herbal Ingredients

Long Dan Cao龍膽草12g, Zhi Zi栀子9g, Huang Qin黄芩9g, Chai Hu柴胡6g, Sheng Di Huang生地黃12g, Ze Xie澤瀉9g, Dang Gui當歸5g, Che Qian Zi車前子10g, Mu Tong川木通9g, Gan Cao甘草5g

Ingredient Explanations

- Long Dan Cao龍膽草 drains Damp-Heat from the Liver and Gallbladder channels.
- Zhi Zi栀子 clears Heat, cools the Blood, resolves Damp, reduces swelling, and opens the meridians
- Huang Qin黄芩 clears Heat, dries Damp, cools the Blood, stops bleeding, and calms Liver Yang.
- Chai Hu柴胡 moves Liver Qi and releases stagnation.
- Sheng Di Huang生地黃 nourishes Yin and clears Heat.
- Ze Xie澤瀉 drains Damp, especially Damp Heat in Lower Jiao, and clears Kidney deficient Heat.
- Dang Gui當歸 nourishes the Blood and benefits the Liver, and regulates menstruation.
- Che Qian Zi車前子 clears Damp Heat and promotes urination.
- Mu Tong川木通 drains Heat and promotes urination.
- Gan Cao甘草 harmonizes the herbs, clears Heat, and expectorates Phlegm.

Modifications

- To make the formula stronger, add Ju Hua菊花 and Tian Ma天麻.
- For redness, swelling, and pain of the eyes, add Chuan Xiong川芎 and Ju Hua菊花.

Formula Actions

- Drains excess Fire from the Liver and Gallbladder and drains Damp-Heat from Lower Jiao.

Formula Indication

- Headache due to Liver Fire and Damp Heat in Lower Jiao, such as migraine headache, urethritis, acute glaucoma, and acute cholecystitis.

Contraindications

- Caution for patients who suffer from Cold and Yang deficiency in Spleen and Stomach due to the Cold herbs in the formula.

Cooking instructions

1) Place the herbs into an herb pot. Traditionally, a clay cooker is used. Gently rinse the herbs once with a strainer.
2) Add water to the pot until it is just enough to cover the herbs, then add one more cup of water.
3) Presoak the herbs for 15 minutes.
4) Turn on the stove to high heat and bring the herbs to a boil. Then reduce the heat to medium low to simmer but with slight bubbling. Turn off the heat when the decoction is reduced to one cup.
5) Drain the decoction out into a container.
6) Repeat steps 2–5 one more time, and drain each decoction to the same container.
7) Divide the decoctions into two separate cups.

Dose

- Drink one cup of the decoction after meal twice daily.
- Insufficiency of Liver Yin肝 陰 不 足

Treatment Principle

Nourish Yin, soothe the Liver.

Herbal Formula

Yi Guan Jian一 貫 煎

This formula is from "Supplement to 'Classified Case Records of Famous Physicians 續 名 醫 類 案", authored by Wei Zhixiu 魏 之 琇 (1722–1772), in Qing dynasty 清 朝 (1636–1912), and published in 1770.

Herbal Ingredients

Sha Shen沙 參10g, Mai Men Dong麥 冬10g, Dang Gui當 歸10g, Sheng Di Huang生 地 黃30g, Gou Qi Zi枸 杞 子15g, Chuan Lian Zi川 楝 子5g

Ingredient Explanations

- Sha Shen沙 參 nourishes the Stomach and Lung Yin.
- Mai Men Dong麥 冬 moistens the Lungs and nourishes Yin.
- Dang Gui當 歸 nourishes the Blood and benefits the Liver; regulates menstruation.
- Sheng Di Huang生 地 黃 nourishes Yin and clears Heat.
- Gou Qi Zi枸 杞 子 tonifies the Liver and Kidney, strengthens Bone, benefits Essence and the Blood.
- Chuan Lian Zi川 楝 子 moves Liver Qi, removes stagnation, and stops pain.

Modifications

- To make the formula stronger, add Chai Hu Shu Gan San柴 胡 疏 肝 散.
- For bitter taste in mouth, add Huang Lian黃 連, Huang Qin黃 芩, Tian Hua Fen天 花 粉, and Chai Hu柴 胡.
- For insomnia, add Bai Zi Ren柏 子 仁, Suan Zao Ren酸 棗 仁, and Wu Wei Zi五 味 子.
- For thirst and irritability, add Zhi Mu知 母 and Shi Gao石 膏.

Formula Actions

- Nourishes Yin, soothes the Liver.

Formula Indication

- Pain in ribs and abdomen due to deficiency of the Kidney and Liver and Liver Qi stagnation. Dry mouth. Red tongue with little coating, thin and rapid pulse, or weak wiry pulse.

Contraindications

- Contraindicated for patients who suffer from emotional stress in excess condition.
- The formula contains many herbs that are sweet, so patients with weak digestion should be careful.

Cooking instructions

1) Place the herbs into an herb pot. Traditionally, a clay cooker is used. Gently rinse the herbs once with a strainer.
2) Add water into the pot until it is just enough to cover the herbs, then add one more cup of water.
3) Presoak the herbs for 15 minutes.
4) Turn on the stove to high heat and bring the herbs to a boil. Then reduce the heat to medium-low to simmer with slight bubbling. Turn off the heat when the decoction is reduced to one cup.
5) Drain the decoction into a container.
6) Repeat steps 2–5 one more time and drain each decoction to the same container.
7) Divide the decoction into two separate cups.

Dose

- Drink one cup of the decoction after meal twice daily.

Acupuncture Treatments

Acupuncture Point Formula

Basic Points

Hua Tuo Jia Ji 華 陀 夾 脊 穴associated points from T1-T11, UB11大 杼Dazhu, UB12風 門Fengmen, UB13肺 俞Feishu, UB14厥 陰 俞Jueyinshu, UB15心 俞Xinshu,

UB16督 俞Dushu,   UB17膈 俞Geshu,   DU20百 會Baihui,   HT7神 門Shenmen, PC6內 關Neiguan,   PC5間 使Jianshi,   LI4合 谷Hegu,   GB8率 谷Shuaigu, SP6三 陰 交Sanyinjiao, LIV3太 沖Taichong，GB40丘 墟Qiuxu

Modifications

- For Blood stasis, add UB17膈 俞Geshu, SP21大 包Dabao, UBI8肝 俞Ganshu, SP10血 海Xuehai,   GB25京 門Jingmen,   SP6三 陰 交Sanyinjiao,   and LIV2行 間Xingjian.
- For Damp-Heat in the Liver and Gallbladder, add SJ6支 溝Zhigou, LIV14期 門Qimen,   GB34陽 陵 泉Yanglingquan,   GB24日 月Riyue, GB40丘 墟Qiuxu, and LIV3太 沖Taichong.
- For Liver Yin deficiency, add UBI8肝 俞Ganshu, GB20風 池Fengchi, LIV8曲 泉Ququan, SP6三 陰 交Sanyinjiao, and KD3太 溪Taixi.

Extra Points

Sishencong四 神 聰

Auricular Therapy

Shenmen神 門

Cupping Therapy

Adding cupping after the treatment of cutaneous acupuncture can improve the treatment outcome.

Cutaneous Acupuncture Treatments

Tool: plum-blossom needles

Acupuncture points: associated Hua Tuo Jia Ji T1-11華 陀 夾 脊 穴, Ashi points阿 是 穴 along the intercostal space.

Procedures

1. Patient is lying on the contralateral side.
2. After the skin has been wiped with 75% alcohol, tap the plum-blossom needles perpendicularly on the skin at the associated Hua Tuo Jia Ji T1-11華 陀 夾 脊 穴 and the Ashi points阿 是 穴 along the intercostal space with wrist force and then lift immediately. Each point will be tapped repeatedly for 30 times or 3–5 minutes. The tapping force should be gentle enough to cause only a slight pain until the skin gets flushed but is not bleeding. The treatment is once a day.

Cutaneous needles are a special tool composed of a set of seven needles attached to a soft handle like a hammer. The tool is used to stimulate an area or an acupuncture point by either tapping forcefully, to achieve bleeding to promote Blood circulation, or tapping superficially to promote the smooth flow of Qi. It is commonly used in the treatment of acute sprain of soft tissues and neuropathy.

Tui Na Treatments

Tui Na Techniques
Pressing, kneading

Tui Na Procedures

1. The patient is in a supine position, and the therapist stands at the side of the thoracic region next to the patient.
   a. The therapist applies the thumb pads in pressing LU1中 府Zhongfu, LU2雲 門Yunmen, SP21大 包Dabao, RN17膻 中Danzhong, and GB24日 月Riyue for 30 seconds per acupuncture point.
   b. The therapist then applies the palm in kneading and rubbing the anterior ribs including the affected shoulder for 3–5 minutes.
2. The patient is in a sitting position, and the therapist does the following:
   a. Applies the thumb pads in pressing the painful region.
   b. Applies One-Finger-Meditation-Pushing techniques on the ipsilateral thoracic Urinary Bladder meridians for 3–5 minutes.
   c. Repeats the above step on both sides of the Urinary Bladder meridians of the thoracic region.
3. The patient is standing, and the therapist stands behind the patient.
   a. The therapist places both hands on the anterior aspects of the patient's shoulders, by going through the armpits, and asks the patient to relax, breathe naturally, and not hold the breath.
   b. Carries the patient on their back, then shakes the patient upward, downward, leftward, and rightward for 2–5 times, while the patient coughs forcefully during shaking.
   a. figure 4.1.1 from the book "lower body pain"; figure 4.1.2 from the book "lower body pain"; figure 4.1.3 from the book "lower body pain"; figure 4.1.4 from the book "lower body pain"

PATIENT ADVISORY

- Chinese medicine is effective for treating intercostal neuralgia excepts in conditions associated with fractures or tumors.
- Patients are adviced to avoid heavy lifting and overworking.
- Patients also need to get plenty of mental and physical rest.
- A firm bed mattress is more suitable for intercostal neuralgia pain.
- Keep warm, avoid catching cold, especially while working outdoors.

SPRAIN OF THE CHEST AND HYPOCHONDRIUM胸 肋 屏 傷

INTRODUCTION

Chest pain can be caused by an urgent heart condition, such as heart attack or angina pectoris. And although 20% of such emergencies were due to heart conditions, 80% of chest pain complaints were something else, and chest pain from a pulled chest muscle was one of them.[1]

A strained or pulled muscle in the chest may cause a sharp pain. It can happen during a motor vehicle injury or from sports, including direct and indirect contact.

ANATOMY

Sprain of the chest muscles may injure the pectoralis major and the intercostal muscle.

Pectoralis Major

The pectoralis major is a paired muscle and the most superficial on the anterior surface of the thoracic cage.

Origin

It is a large piece of muscle that has three origins.

1) Clavicular head. It is at the anterior surface of the medial half of the clavicle.
2) Sternocostal head. It is at the anterior surface of sternum, the costal cartilages of ribs 1–6.
3) Abdominal head. It is at the aponeurosis of the external oblique muscle.

Insertion

The pectoralis major forms a flat tendon that is about four or five centimeters in diameter then inserts into the lateral lip of the crest of the greater tubercle of humerus.

Action

Its action is different between the clavicular head and the sternocostal head.
    The clavicular head flexes the humerus.
    The sternocostal head extends the humerus back to the anatomical position but no
further toward hyperflexion.
    When both work together, they perform the abduction of the humerus and medial
or internal rotation of the humerus.

Nerve Innervation

There are two nerves innervating, lateral and medial pectoral nerves, that arise from
the brachial plexus. The medial pectoral nerve originates from the medial cord of
the brachial plexus and the lateral pectoral nerve originates from the lateral cord of
the brachial plexus.

Blood Supply

For the blood supply, the pectoralis muscle gets it arterial supply from the pectoral
branches of thoracoacromial artery.

Intercostal Muscle

The intercostal spaces are the spaces between the ribs. There are three muscles from
the front to back.

- The external intercostal muscle

The muscle fibers are oriented inferiorly and medially from the upper rib to the rib
below, where it is about to meet the sternum. It doesn't continue as a muscle, instead
it forms an aponeurosis called the external intercostal membrane.

Origin

The external intercostal muscle originates at the lower border of the upper rib.

Insertion

It inserts into the superior border of the lower rib.

Action

The muscle is important in inspiration. It elevates the ribs, increasing the thoracic volume.

- The internal intercostal muscle

The muscle fibers are oriented in the opposite direction to the external intercostal muscle. They're oriented superiorly and medially. And it doesn't have the aponeurosis which connects it to the sternum directly.

Origin

It originates at the lateral border of the costal groove of the upper rib.

Insertion

It inserts into the superior border of the lower rib.

Action

The muscle is important in expiration. It reduces the thoracic volume by depressing the ribcage; the interchondral part elevates the ribs.

- The innermost intercostal muscle

The muscle fibers are oriented the same way as the internal intercostal muscle. The endothoracic fascia is posterior to this muscle.

Origin

It originates at the medial border of the costal groove upper rib.

Insertion

It inserts into the superior border of the lower rib.

Action

The muscle is important in expiration, which is the same as the internal intercostal muscle.

Nerve Innervation

These three muscles are innervated by the intercostal nerves from T1 to T11.

Blood Supply

Between the internal intercostal muscle and the innermost intercostal muscles, there are the intercostal vein, intercostal artery, and intercostal nerve which lie superiorly to inferiorly in the subcostal groove underneath the rib.

WESTERN MEDICINE ETIOLOGY AND PATHOLOGY

- The traumatic causes are usually from careless use or overuse of the arm. Underestimating the weight of a heavy object before lifting, reaching to look for something in an overhead for too long, or a fall that twists the soft tissues in the chest and hypochondrium are common ways that chest muscles are strained and can lead to chest pain.

- Sport's injury is also a common cause: lifting weights without warming up, increasing the weight too soon, or lifting uneven weights may strain or pull the muscles in the chest. A pectoralis major strain may occur during weightlifting on a bench when the bar is lowered. The immediate need to generate high muscle force to lift may overstretch the pectoralis major, and the muscle subsequently tears.

  Other sports that may cause strain are the activities that require forceful or repetitive motions such as gymnastics, tennis, badminton, and golfing. Any sport that involves upper limbs lifting and torso twisting repeatedly may cause strain, such as rowing and swimming. An intercostal muscle strain or pull is common with activities involving the torso twisting with shoulder movements. They include rowing and lifting while twisting, such as moving large and heavy furniture.

- Severe coughing may cause chest pain. Coughing from a common cold, whooping cough, or bronchitis may injury the muscles or the fascial tissues.
- Poor postures or overuse of the muscles cause muscle fatigue and makes them prone to injury.

CHINESE MEDICINE ETIOLOGY AND PATHOLOGY

The 22nd Difficult Issues of the Book of Difficulty says: "Qi stays and does not move, so Qi first becomes sick; Blood is blocked and does not moisten, so Blood second becomes sick".

《難經·二十二難》曰：氣留而不行者，為氣先病也；血壅而不濡者，為血後病也。

1. Injury of Qi 傷氣型

It is mostly due to holding the breath while heavy lifting, leading to impairment of the Qi Movements and causing stagnation in the meridians.

2. Injury of Blood 傷血型

It is mostly due to an external violent impact leading to Blood leaking outside the meridians and becoming stagnant in the muscles.

MANIFESTATIONS

In Western Medicine

1. Patients may have an obvious injury history; the pain is sudden onset with a history of trauma.
2. The quality of pain can be sharp in the acute state, or dull in the chronic state.

3. Breathing is difficult due to the pain.
4. Bruising may be noticed when the straining is acute.
5. The affected area can be swollen when the injury is acute.

In Chinese Medicine

1. Injury of Qi 傷 氣 型

Distention and dull pain on the thoracic and hypochondrium regions, migratory pain, but pressure pain is not obvious, there may be no tender spots; the pain may be a pulling, shooting pain while breathing and speaking, and patients may not be able to lie down due to shortness of breath.
Tongue: red body, thin white coating.
Pulse: wiry, moderate, or slow.

2. Injury of Blood 傷 血 型

Chest distention. The pain location is fixed, with swelling, and there are obvious tender spots; bruising may appear. There is difficulty in rotating the torso. Deep breath and cough may increase the pain intensity. The pain may be mild on the same day of injury, but the intensity may increase over time, possibly with severe pain a few days later. Vomiting blood and low-grade fever may occur in severe conditions.
Usually there is a history of chest trauma.
Tongue: purple body.
Pulse: choppy.

3. Injury of Both Qi and Blood 氣 血 兩 傷 型

This type of chest muscle injury comprises the previous two types.

4. Chronic Injury 胸 部 陳 傷.

Due to incomplete treatment after the injury, the conditions remain. The pain is dull, constant, and intermittent.
Tongue: pink, with a thin white coating.
Pulse: thin, choppy.

Physical Examination

- Begin with obtaining medical history including chest injury, injury while lifting heavy objects, holding the breath while lifting, or whether the pain is due to a direct attack. Obtain the quality of pain information, Injury of Qi presents dull pain but migratory while Injury of Blood presents stabbing pain, the pain doesn't travel, and the pain may be greater at night while at rest. Breathing, speaking loudly, and coughing aggravate the chest pain.

Also, the information of coughing blood or bloody sputum in Chinese medicine is important to obtain; it indicates that the injury is not just limited to the superficial level, but the trauma has already caused internal injury in the Zang-fu level.

- Inspection

    Observe the effected region to find bruising and swelling. Observation on the traumatic site always requires comparison to the other side. If there is ecchymosis on the tongue, the darker the more chronic. The sublingual veins can also tell the different between both types: slim veins indicate Injury of Qi, while earthworm shapes indicate Injury of Blood.

- Palpation

    Palpation is performed on the chest for the skin temperature, hardness, and tender spots. Acute swelling presents puffiness, chronic swelling present hardness when pressed. Skin temperature may be warm in acute injury for the Injury Blood type in Chinese medicine.

    Pulse taking can also tell the type of injury: wiry pulse indicates Injury of Qi; choppy pulse indicates Injury of Blood.

- Special Test

    Chest X-ray, CT, ultrasound, or MRI can exclude fracture, injury of the lungs, and other injury to the thoracic pleural or the heart.

DIFFERENTIAL DIAGNOSIS

1. Angina pectoris

Beside the pain in the chest, there is a radiating pain to the shoulder, arm, and fingers. It is usually accompanied by shortness of breath, palpitation, cold limbs, and sweating.

2. Pulmonary embolism

Rapid or irregular heartbeat, Cyanosis, coughing bloody sputum, dizziness or lightheadedness, excessive sweating, and fever. Leg pain, swelling, or both, the location is usually in the calf if it is caused by a deep vein thrombosis.

3. Pneumonia

Cough which may produce phlegm, shortness of breath, fever, sweating, lower than normal body temperature, nausea, vomiting, or diarrhea.

4. Pleuritis

Sharp chest pain that may radiate to the shoulder or back. Shortness of breath, fever, cough, unexplained weight loss.

5. Heart attack.

Fainting, sudden dizziness, cold sweat, fatigue, shortness of breath, heartburn or abdominal pain, lightheadedness, nausea, racing pulse, difficulty breathing.

TREATMENTS

1. Injury of Qi 傷氣型

Treatment Principle

Regulates Qi, stops pain

Herbal Treatments

Herbal Formula

Chai Hu Huang Qin Tang柴 胡 疏 肝 散

This formula is from "The Complete Compendium of Jingyue景 岳 全 書", authored by Zhang Jiebin張 介 賓(1111–1117AD), in Ming dynasty明 朝 (1624 AD).

Herbal Ingredients

Chai Hu柴 胡6g, Xiang Fu香 附5 g, Bai Shao白 芍5 g, Chuan Xiong川 芎5 g, Zhi Ke枳 殼5 g, Chen Pi陳 皮6 g, Zhi Gan Cao甘 草2 g

Ingredient Explanations

- Chai Hu 柴 胡 moves Liver Qi and releases stagnation.
- Xiang Fu香 附 moves Liver Qi and the Blood and stops pain due to Qi and Blood stagnation.
- Bai Shao白 芍 nourishes the Blood, soothes the Liver, and stops pain.
- Chuan Xiong川 芎 moves the Blood and moves Qi.
- Zhi Ke枳 殼 regulates Qi and removes blockage.
- Chen Pi陳 皮 regulates Qi and removes blockage.
- Zhi Gan Cao炙 甘 草 tonifies the Spleen, harmonizes the other herbs, and guides the herbs to all twelve meridians.

Modifications

- For better result, add Yu Jin鬱 金.
- For conditions with poor appetite, change Chen Pi陳 皮 to Qing Pi青 皮.
- For Heat symptoms, add Mu Dan Pi牡 丹 皮 and Zhi Zi梔 子.

Formula Indications

- Chest and hypochondriac pain due to Liver Qi stagnation.

Contraindications

- Caution for patients who suffer from headache due to the Yin deficient Heat.

2. Injury of Blood 傷 血型

Treatment Principle

Move the Blood, disperse Bruising, stop pain.

Herbal Formula

Fu Yuan Huo Xue Tang復元活血湯

This formula is from "Clarifying the study of medicine醫學發明", authored by Li Dongyuan 李東垣, in Yuan dynasty元朝 (1271–1368), and published in 1315.

Herbal Ingredients

Chai Hu 柴胡9g, Tian Hua Fen 天花粉9g, Dang Gui 當歸9g, Hong Hua 紅花6g, Gan Cao 甘草6g, Chuan San Jia 炮穿山甲6g, Da Huang 大黃12g, Tao Ren 桃仁9g

Ingredient Explanations

- Chai Hu 柴胡 harmonizes Shao Yang disorders and relieves stagnation.
- Tian Hua Fen 天花粉 clears Heat, expels pus, and reduces swelling.
- Dang Gui 當歸 moves and tonifies the Blood, disperses Cold, and stops pain due to Blood Stasis.
- Hong Hua 紅花 moves the Blood, releases Bruising, opens the meridians, and stops pain.
- Gan Cao 甘草 releases spasms, alleviates pain, harmonizes the other herbs, and guides the herbs to all twelve meridians.
- Chuan San Jia 穿山甲 moves the Blood, reduces swelling, and expels Wind Damp.
- Da Huang 大黃 moves the Blood and disperses Bruising.
- Tao Ren 桃仁 breaks up stubborn Blood Stasis.

Formula Action

Moves the Blood, disperses Bruising, soothes the Liver, and opens the meridians.

Formula Indication

Physical trauma, Blood stasis, and severe pain.

External Herbal Applications

Topical use of the herbal liniments and ointments improves pain management. These include Zheng Gu Shui 正骨水and the products extracted from the leaves of the wintergreen plant.

Acupuncture Treatments

Acupuncture Point Formula

Basic Formula

PC6內關Neiguan, SP4公孫Gongsun, SJ6支溝Zhigou, GB34陽陵泉Yanglingquan, LIV1大敦Dadun. Apply strong sedating techniques on all points for acutestage trauma.

Modifications

1. For Injury of Qi 傷 氣 型, add LIV3太 衝Taichong and UBI8肝 俞Ganshu. Apply even techniques on both points.
2. For Injury of Blood 傷 血 型, add UB17膈 俞Geshu and Ashi acupoints 阿 是 穴.

Bleeding and Cupping with Cutaneous Acupuncture Therapy

Tool: Plum-Blossom Needles

This can promote Blood circulation; it is appropriate for Injury of Blood 傷 血 型. Cupping Therapy is used after bleeding on the affected region.

Acupuncture Points

Ashi points阿 是 穴 on the effected chest pain region.

Procedures

1. Patient position is lying on the contralateral side.
2. After the skin has been wiped with 75% alcohol, tap the plum-blossom needles perpendicularly on the skin of the Ashi points阿 是 穴with wrist force and then lift immediately. Each point will be tapped repeatedly for 30 times or 3–5 minutes. The tapping force should be gentle to cause only slight pain until the skin gets flushed and bleeding. The treatment is once a week.

A cutaneous needle is a special tool composed of a set of seven needles attached to a soft handle like a hammer. The tool is used to stimulate an area or an acupuncture point either tapping forcefully, to achieve bleeding to promote Blood circulation, or tapping superficially to promote the smooth flow of Qi. It is commonly used in the treatment of acute sprain of soft tissues and neuropathy.

Tui Na Treatments

Tui Na Techniques

Pressing, kneading, pushing, plucking, pressing, shaking

Tui Na Procedures

The patient is in a supine position in the dorsal recumbent position; the therapist stands near the top of the head. The assistant stands near the hip and immobilizes the hip joints. The therapist performs the following procedures:

1. Places the hands on both lateral sides of the ribcage and asks the patient to continue deep breathing. Figure 4.2 from Shutterstock #722417578

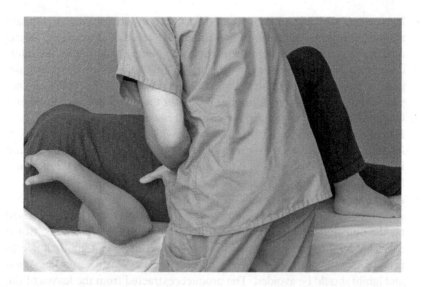

2. Shakes the chest for about 10–15 times: at first, the force should be mild, the shaking range should start with small angles, and the frequency should be slow.
3. Places both hands in the armpits and pull the shoulders upward to stretch the chest. The force should be lower than the patient's pain tolerance.
4. The assistant lifts the patient's arms upward. The therapist applies the index, middle and thumb fingers to arrange and regulate the tissues along the ribs using plucking, pushing, kneading, and pressing techniques.

PATIENT ADVISORY

RICE Protocol. This protocol is effective for acute soft tissue injury.

- **R**est. Avoid any activity that may exacerbate the pain to prevent more blood flowing to the region.
- **I**ce. Apply cold therapy to decrease the blood flow. Wrap a pack of ice in a towel on the affected area for no more than 20 minutes each time; repeat 3–5 times a day. Avoid placing the ice pack directly on the skin.
- Compression. Wrap a compression brace on the torso in order to decrease the blood flow to reduce swelling.
- Elevation. Keep the torso higher than the legs.

MEAT Protocol. In order to reach the rapid healing process while the soft tissue trauma is in the recovery stage, the protocol increases the blood flow to enhance healing.

- Movement. Chronic inflammation from soft tissue injuries precipitate adhesion. It limits the range of motion and decreases the flexibility of the muscles to increase and depress the thoracic volume. Adhesion in the intercostal muscles decreases the ribs elevation while adhesion in the pectoralis major limits the shoulder flexion and abduction movements. "Movement" protocol helps to prevent adhesion, promotes the blood flow or Blood and Qi circulation, and helps the new tissue to grow.
- Exercise. Exercise is a set of movements within the field of kinesiology. Regular and appropriate amount of load of the physical activities can improve muscle strength and boost pain tolerance. It delivers oxygen and nutrients to the damaged tissues to facilitate rapid repairing.
- Analgesia. Some patients may develop pain from the exercise protocol; applying analgesia protocol can provide pain relief. The pain relief is based on the nutrition advice and the use of the topical herbal products. Some everyday food (e.g., turmeric and black pepper) may be a good choice to provide healing. The foods in the category of Fa Wu (e.g., garlic, onion and lamb) should be avoided. The products extracted from the leaves of the wintergreen plant can help to ease the pain.
- Treatment. Appropriate treatment plans help to accelerate the process of healing. Chinese medicine implements treatment plans for each individual patient differently according to their health constitution: for example, patients who suffer from the Injury of Qi 傷氣型 may also manifest Yang陽 deficiency, so moxibustion Therapy is added to the treatment plan. But moxibustion Therapy would not be used for a patient with Yin陰 deficiency condition because it may cause further deficient Fire.

NOTE

1. https://www.health.harvard.edu/heart-health/chest-pain-a-heart-attack-or-something -else

5 Abdominal Pain

GENERAL CONCEPTS OF ABDOMINAL PAIN IN CHINESE MEDICINE 中醫腹痛

INTRODUCTION

Abdominal pain involves a wide range of diseases. It can occur in a variety of internal medicine, surgical, and gynecological diseases. Because there are many Zang-Fu 臟腑 (organs, in Chinese medicine) in the abdomen, and it is the place where there are meridians passing through, therefore, pathological changes, such as the external evil Qi, immoderate diet, emotional upset and the internal change of the Zang-Fu 臟腑 and meridians, such as deficiency of Yang energy can cause abdominal pain.

ANATOMY

The abdomen refers to the anterior trunk from the epigastrium to the pelvic bone. The abdomen contains many Zang-Fu 臟腑, including Spleen, Stomach, Liver, Gallbladder, Kidney, Urinary Bladder, Small Intestine, and Large Intestine in the abdomen; and meridians, including Foot Shao Yin 足少陰meridian, Foot Tai Yin 足太陰meridian, Foot Jue Yin足厥陰 meridian, Foot Shao Yang 足少陽meridians, Foot Yang Ming meridian足陽明, Hand Yang Ming 手陽明meridian, Chong 沖meridians, Ren 任meridians, and Dai 帶meridians. Figure 5.1.1 from shutterstock.com #1005648481

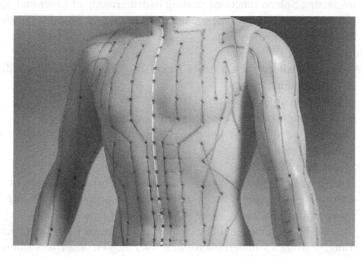

DOI: 10.1201/9781003203018-5

CHINESE MEDICINE ETIOLOGY AND PATHOLOGY

Abdominal pain can be caused by the invasion of exterior pathogens, immoderate diet, emotional upset, physical trauma, insufficiency of Qi and Blood, and deficiency of Yang.

- Invasion of external evil Qi 外邪侵襲.
  - The invasion of the pathogenic Cold 寒邪 and Wind 風邪 to the Middle Jiao 中焦 slows down the flow of Qi and causes blockage in the meridians. It leads to the disturbance of Qi movements and results in Qi and Blood stagnation and Zang-Fu 腑臟 and meridians lack of nourishment. Where there is no free flow, there is pain 不通則痛.
  - The Heat accumulation transforming from the pathogenic Cold when the blockage is chronic or the invasion of the Summer-Heat 署邪 or Damp-Heat 濕熱 can cause blockage in the Intestine leading to the stagnation of Qi resulting pain.
- Food Injury 飲食積滯
  - Immoderate diet and overeating may impair Spleen and Stomach.
  - Extreme diet of fatty, sweet, strong taste, hot spicy, and pungent foods can generate Damp-Heat in the Intestine 小腸實熱.
  - Unhygienic diet or having too much raw and cold foods can cause the accumulation of Cold and Damp internally, leading to impairment of Spleen and Stomach and Qi movements, resulting in Qi stagnation.
- Emotional Injury 情志失和
  - Anger, depression, and emotional upset can disturb the flow of Liver Qi, leading to malfunction of Qi movements.
  - Worry, grief, over-planning, over-thinking, or Stagnation of Liver Qi overacting Spleen functions leading to disharmony of Liver and Spleen can disturb the Qi movements, resulting in Qi and Blood stagnation in Zang-Fu and the meridians.
- Traumatic Injury 跌撲損傷
  Abdominal pain can be a result of Blood stasis blocking meridians due to:
  - Physical trauma.
  - Abdominal surgery.
  - Liver Qi stagnation because Qi is the commander of Blood.
  - Lingering conditions of the diseases in the abdominal Zang-Fu 臟腑 and meridians.
- Cold Occlusion 陽氣虛弱
  - Insufficiency of Spleen Yang can be due to chronic diseases or over-consuming Cold foods, herbs, or medicine, leading to producing Cold internally, and resulting in the inadequacy of Qi and Blood.
  - Chronic Kidney Yang deficiency or chronic diseases can injure Kidney Yang, leading to undernourishment of Zang-Fu 臟腑 meridians and prone to Cold producing internally, resulting in the obstruction of Qi flow by Cold pathogen.

Manifestations

The symptoms can be in a wide range, and it can be varied by its causes. The diagnosis in Chinese medicine doesn't utilize technological equipment, it relies completely on four diagnostic methods to collect the medical history of patients, analyze the etiopathogenesis, and treat the disease. These four methods are Observation, Auscultation, Interrogation, and Palpation.

Quality of Pain

- Cold. The abdominal pain is distending. The intensity is severe. The onset is abrupt. the patients prefer lying in curled position and experience an aversion to cold. The abdominal palpation is firm, and the preference is for a warm pad wrapping the abdomen. It is commonly accompanied by borborygmus, vomiting, diarrhea or loose stool, clear or profuse urine, and absence of thirst. The duration is constant.

Tongue: thin white coating.
  Pulse: tight.
- Heat. The pain often is in the periumbilical region. It is commonly accompanied by fever, constipation, and a preference for cold drinks.

Location of Pain

- Generalized throughout the abdomen. It is often accompanied by severe pain, guarding, rigidity and rebound, or loss of consciousness. It is often due to internal bleeding, infection, and inflammation, and it could be a life-threatening, urgent, and serious condition – the patient should visit the emergency facility.
- Epigastric region. It is often due to the conditions in the Spleen and Stomach, and both Small and Large Intestine.
- Periumbilical region. Pain in this area is from invasion of exterior pathogens, dietary injury, or parasitic infestation.
- Hypogastric region. Pain in this area is from Kidney, Urinary Bladder, and Spleen deficiency.
- Iliac region. Pain in this area is from Liver Meridian (e.g., Liver Qi stagnation) and the conditions in the Large Intestine.

Duration of Pain

1. Acute. The onset is abrupt. It is accompanied by severe pain, and the duration of pain can be from several hours to several days. It is often due to invasion of external pathogens, such as infection or trauma.
2. Chronic. The onset is slow. The pain is constant, low intensity, duration is long, and it can be from several months to several years, often intermittently.

Condition of Pain

- Excessive. It is acute, abrupt. Pressure is not the preference. It is accompanied by abdominal distension and vomiting. Pathogenic Evil Qi invasion is the common cause.
- Deficient. It is slow. The abdominal pain is dull and constant. The duration is long-lasting. There is a preference for pressure.
- Heat. It is acute and abrupt. it is accompanied by thirst, fever, preference of cold drinks, constipation, and foul-smelling stool.
- Cold. The pain is spasmodic and colicky. It is accompanied by borborygmus, worsened by exposure to coldness, and relieved by warmth.

- Qi Stagnation. The abdominal pain is distended with discomfort, and it is accompanied by sighing. The pain fluctuates depending on emotional status, and it can be aggravated by emotional upset, anger, depression and relieved by joy or passing gas. The pain can also migrate. The location is more common in the hypochondrium.
- Blood stasis. The abdominal pain is stabbing, it is in a fixed location. Pressure is not preferred. The pain is worse at night. A mass may be palpable. The tongue is dusky with petechia.
- Overeating or indigestion. The abdominal pain is distended with eructation, and it is alleviated by defecation. The tongue coating is greasy and sticky.

CLASSIFICATIONS AND TREATMENTS

1. Internal Stagnation of Cold 寒 邪 內 阻 證

Abrupt onset of abdominal pain, a desire for warmth, an aversion to cold that is aggravated by cold exposure, absent thirst, preference for sleeping in a curled position, loose stool or constipation, and discharge of a large volume of clear urine.

Tongue: white, or with a white, greasy coating.

Pulse: deep and tight or deep and wiry.

Treatment Principle

Warm the Middle Jiao and disperse Coldness.

Herbal Treatments

Herbal Formula

Zheng Qi Tian Xiang San 正 氣 天 香 散

This formula is from "A Compendium of Medicine 醫學綱目", authored by Lou Ying 楼英 (全善) (1320–1389), in Qing dynasty 清朝 (1636–1912), and published in 1565.

Herbal Ingredients

Xiang Fu 香 附10g, Wu Yao烏 藥10g, Chen Pi陳 皮10g, Zi Su Ye蘇 葉6g, Gan Jiang乾 薑6g

Ingredient Explanations

- Xiang Fu 香附 moves Liver Qi and Blood and stops pain due to Qi and Blood stagnation.
- Wu Yao烏藥 moves Qi and stops pain.
- Chen Pi陳皮 regulates Qi and removes blockage.
- Zi Su Ye蘇葉 releases exterior, disperses Cold, and promotes the functions of Spleen and Stomach Qi.
- Gan Jiang乾薑 warms Middle Jiao, expels interior Cold, and dispels Wind-Dampness.

Modifications

- For better results in relieving Cold stagnation, add Wu Zhu Yu吳茱萸 and Xiao Hui Xiang小茴香.
- For Blood stasis, add Hong Hua紅花 and Tao Ren桃仁.

Formula Indications

- Cold invasion of Middle Jiao 中焦.

Contraindications

- Patients with abdominal pain due to Fire should not use.
- It is contraindicated for Body Fluid deficiency.

2. Deficiency of Middle Jiao and Coldness in Zang-Fu中虛臟寒證

Intermittent dull pain in the abdomen that is aggravated by fatigue and hunger and is alleviated by eating and resting; preference for warmth and pressure, loose stool, listlessness, shortness of breath, and aversion to cold.

Tongue: pale body with white coating.

Pulse: deep and thready.

Treatment Principle

- Warm and tonify the Middle Jiao, disperse Coldness, and relieve spasmodic abdominal pain.

Herbal Treatments

Herbal Formula

Xiao Jian Zhong Tang小建中湯

This formula is from "Shang Han Lun 傷寒雜病論", authored by Zhang Zhong-Jing 張仲景 (150–219), in eastern Han dynasty 東漢 (25–220), and published in 200–210.

Herbal Ingredients

Gui Zhi桂枝去皮9g, Bai Shao白芍18g, Zhi Gan Cao甘草炙9g, Sheng Jiang生薑9g, Da Zao大棗12 pieces, Yi Tang飴糖30g

Ingredient Explanations

- Gui Zhi桂枝 warms the meridians and relieves pain.
- Bai Shao白芍 nourishes the Blood, soothes the Liver, and stops pain.
- Zhi Gan Cao甘草炙 tonifies the Spleen, harmonizes the other herbs, and guides the herbs to all 12 meridians.
- Sheng Jiang生薑 unblocks the pure Yang pathway and harmonizes rebellious Qi.
- Da Zao大棗 tonifies the Spleen Qi, nourishes the Blood, and moderates and harmonizes the harsh properties of the fragrant herbs.
- Yi Tang飴糖 tonifies Middle Jiao Qi, alleviates pain, disperses Cold, and alleviates spasmodic abdominal pain.

Modifications

- For abdominal distension, remove Da Zao大棗 and add Fu Ling茯苓.
- For Qi deficiency, add Huang Qi黃耆 and Ren Shen人參.
- For Blood deficiency, add Dang Gui當歸 and Chuang Xiong川芎.

Formula Indications

- Cold accumulation in Middle Jiao 中焦.
- Spasmodic abdominal pain.

Contraindications

- It is contraindicated for Heat due to Yin deficiency.

3. Qi Stagnation氣滯證

Migratory distending pain over the abdomen and epigastrium, or radiating to the hypochondrium and iliac regions, and relived by eructation but aggravated by change of emotion.

Tongue: thin coating.
Pulse: wiry or wiry and thin.

Treatment Principle

Disperse the stagnated Liver Qi.

Herbal Treatments

Herbal Formula

Chai Hu Huang Qin Tang柴胡疏肝散

This formula is from "The Complete Compendium of Jingyue景岳全書", authored by Zhang Jiebin張介賓(1111–1117AD), in Ming dynasty明代 (1624 AD).

Herbal Ingredients

Chai Hu柴胡6g, Xiang Fu香附5g, Bai Shao白芍5g, Chuan Xiong川芎5g, Zhi Ke枳殼5g, Chen Pi陳皮6g, Zhi Gan Cao甘草2g

Ingredient Explanations

- Chai Hu 柴 胡 moves Liver Qi and releases stagnation.
- Xiang Fu香 附 moves Liver Qi and Blood and stops pain due to Qi and Blood stagnation.
- Bai Shao白 芍 nourishes the Blood, soothes the Liver, and stops pain.
- Chuan Xiong川 芎 moves the Blood and moves Qi.
- Zhi Ke枳 殼 regulates Qi and removes blockage.
- Chen Pi陳 皮 regulates Qi and removes blockage.
- Zhi Gan Cao甘 草 tonifies the Spleen, harmonizes the other herbs, and guides the herbs to all 12 meridians.

Modifications

- For Blood deficiency, add Dang Gui當 歸, Sheng Di Huang生 地, and Shu Di Huang熟 地.
- For Blood stasis, add Dan Shen丹 参.

Formula Actions

- Moves Liver Qi, harmonizes the Blood, and alleviates pain due to Liver Qi stagnation.

Formula Indication

- Headache due to Liver Qi stagnation, Blood stagnation, acid regurgitation, and breast distention.

Contraindications

- Caution for patients who suffer from headache due to the Yin deficient Heat.

4. Blood Stasis 血 瘀 證

Severe localized stabbing pain in the abdomen that can last a long time; there is no preference for pressure.
Tongue: purplish tongue with or without ecchymoses.
Pulse: deep, thready, or choppy.

Herbal Treatments

Herbal Formula

Shao Fu Zhu Yu Tang 少 腹 逐 瘀 湯

This formula is from "Correction of Errors in Medical Classics醫 林 改 錯", authored by Wang Qingren 王 清 任 (1768–1831), in Qing dynasty 清 朝 (1636–1912), and published in 1849).

Herbal Ingredients

Xiao Hui Xiang小 茴 香3g, Gan Jiang乾 薑1g, Yan Hu Suo延 胡 索3g, Mo Yao沒 藥6g, Dang Gui當 歸9g, Chuan Xiong川 芎6g, Rou Gui肉 桂3g, Chi Shao赤 芍6g, Pu Huang蒲 黃9g, Wu Ling Zhi五 靈 脂6g

Ingredient Explanations

1. Xiao Hui Xiang小茴香 expels Cold, regulates Qi, harmonizes the Stomach, and alleviates pain.
2. Gan Jiang乾薑 warms Middle Jiao, expels interior Cold, and dispels Wind-Dampness.
3. Yan Hu Suo延胡索 moves the Blood, promotes Qi circulation, and stops pain.
4. Mo Yao沒藥 moves Qi and Blood and stops pain due to Qi and Blood stagnation.
5. Dang Gui當歸 moves the Blood, disperses Bruising, and stops pain due to Blood stasis.
6. Chuan Xiong川芎 moves the Blood, disperses Bruising, and stops pain due to Blood stasis.
7. Rou Gui肉桂 warms meridians, expels Cold, and opens the meridians to promote Blood circulation.
8. Chi Shao赤芍 moves the Blood, disperses Bruising, and stops pain.
9. Pu Huang蒲黃 stops bleeding and moves the Blood.
10. Wu Ling Zhi五靈脂 moves Qi and Blood and stops pain due to Qi and Blood stagnation.

Modifications

- For Qi deficiency, remove Wu Ling Zhi五靈脂, add Huang Qi黃耆, Dang Sheng黨參, and Bai Zhu白朮.
- For chest and hypochondrium, add Yu Jin鬱金 and Chuan Lian Zi川楝子.

Formula Indications

- Blood stasis in Middle Jiao.

Contraindications

- It is contraindicated during pregnancy.

5. Accumulation of Dampness-Heat濕熱積滯證

Abrupt onset of progressively worsened abdominal pain or intermittent severe distended pain. Accompanying symptoms include aversion to pressure, bitter taste in the mouth, dry mouth, fever, constipation or loose stool, chest and epigastric distension, regurgitation (vomiting), and dark yellow urine.

Tongue: yellow, greasy coating.
Pulse: wiry, rapid.

Treatment Principle

Transform Damp Heat and relieve intestinal stasis by purgation.

Herbal Treatments

Herbal Formula
Tiao Wei Cheng Qi Tang 調胃承氣湯

This formula is from "Shang Han Lun 傷寒雜病論", authored by Zhang Zhong-Jing 張仲景 (150–219), in eastern Han dynasty 東漢 (25–220), and published in 200–210.

Herbal Ingredients

Da Huang大黃12g, Zhi Gan Cao甘草炙6g, Mang Xiao 芒硝10g

Ingredient Explanations

- Da Huang大黃 moves the Blood and disperses Bruising.
- Zhi Gan Cao甘草炙 tonifies the Spleen, harmonizes the other herbs, and guides the herbs to all 12 meridians.
- Mang Xiao 芒硝 clears excess Heat and reduces swelling.

Modifications

- For irritability and abdominal distension, add Zhi Shi枳實 and Hou Pu厚朴.

Formula Indications

- Excessive Heat in Stomach.
- Excessive condition in Yang Ming Fu organ 陽明腑臟.

Contraindications

- It is contraindicated during pregnancy.

6. Retention of Food飲食停滯證

Distended abdominal pain, aversion to pressure, acid eructation, poor appetite, constipation and inclination for diarrhea
Tongue: thick and greasy coating.
Pulse: slippery and full

Treatment Principle

Relieve dyspepsia.

Herbal Treatments

Herbal Formula
Da Cheng Qi Tang 大承氣湯
This formula is from "Shang Han Lun 傷寒雜病論", authored by Zhang Zhong-Jing 張仲景 (150–219), in eastern Han dynasty 東漢 (25–220), and published in 200–210.

Ingredients

Da Huang大黃12g, Zhi Shi枳實12g, Hou Pu厚朴15g, Mang Xiao 芒硝9g

Ingredient Explanations

- Da Huang大 黃 moves the Blood and disperses Bruising.
- Zhi Shi枳 實 breaks up stagnation and transforms Phlegm.
- Hou Pu厚 朴 promotes the movement of Qi in the Middle Jiao, resolves Food Stagnation, descends rebellious Qi, reduces Phlegm, dries Dampness, and calms wheezing.
- Mang Xiao 芒 硝 clears excess Heat and reduces swelling.

Modifications

- For Dry Heat and thirsty, add Shi Gao石 膏, Zhi Mu知 母

Formula Indications

- Excessive Heat in Stomach.
- Excessive condition in Yang Ming Fu organ 陽明腑臟.

Contraindications

- It is contraindicated during pregnancy.
- Use extreme caution if the condition is not interior excessive Heat.

Acupuncture Treatments

Acupuncture Point Formula

Basic Formula

RN12中 脘Zhongwan,   ST36   足 三 里Zusanli,   ST25   天 樞Tianshu, ST39下 巨 虛Xiajuxu, RN4關 元Guanyuan

Modifications

3. Internal stagnation of Cold 寒 邪 內 阻 證, add RN6氣 海Qihai (add moxa on the needle), and ST39下 巨 虛Xiajuxu (sedate the point before adding moxa to the needle), and RN4關 元Guanyuan (add ginger and moxa for warming the needle).
4. Deficiency of Middle Jiao and Coldness in Zang organs中 虛 臟 寒 證, add BL20脾 俞Pishu (add moxa on the needle), ST39下 巨 虛Xiajuxu (sedate the point before adding moxa to the needle), and RN4關 元Guanyuan (add ginger and moxa for warming the needle).
5. Qi stagnation氣 滯 證血 瘀 證(氣 滯 證), add LIV3太 衝Taichong and LIV14期 門Qimen肝 郁 气 滯.
6. Blood stasis氣 滯 血 瘀 證(血 瘀 證), add SP8地 機Diji.
7. Accumulation of Dampness-Heat濕 熱 積 滯 證, add PC6內 關Neiguan, RN6氣 海Qihai, LIV2行 間Xingjian, and RN11建 里Jianli.
8. Retention of food飲 食 停 滯 證, add Lineiting 裡 內 庭 (it is located at the plantar side of ST44內 庭 Neiting).

Auricular Therapy

Large Intestine, Small Intestine, Spleen, Stomach, Shenmen, Sympathetic

Tui Na Treatments

Tui Na Techniques

Pressing, kneading

Tui Na Procedures

The patient is in the prone position, and the therapist stands near the head.

- The therapist pushes both Urinary Bladder meridians with both palms from UB13 肺 俞Feishu to UB25大 腸 俞Dachangshu for 3–5 minutes or until warm.
- The therapist presses and kneads the acupuncture points from UB18肝 俞Ganshu to UB25大 腸 俞Dachangshu for 4–6 minutes.
- Grasps the lumbar for 3–5 minutes.

The patient is in the supine position, and the therapist stands near the head.

- The therapist pushes both hypochondrium with palms until warm or red or for 3–5 minutes.
- Both palms alternately push the REN meridian from RN12中 脘Zhongwan to RN4關 元Guanyuan.
- Both palms rub the abdomen.
- Applies the second to the fourth fingers to press and knead RN12中 脘Zhongwan, ST25天 樞Tianshu, and RN4關 元Guanyuan for 5–8 minutes.
- The therapist stands near the legs and applies the fists to lightly strike ST36足 三 里Zusanli, ST39下 巨 虛Xiajuxu, and ST40豐 隆Fenglong. Figure 5.1.2 from the book "Lower Body Pain" Figure 5.1.3 from the book "Lower Body Pain"

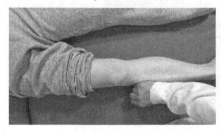 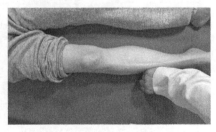

PATIENT ADVISORY

- Avoid cold and raw foods.
- All foods must be fully cooked, including vegetables.
- Avoid fatty, greasy foods.
- Avoid the foods that are too hard for digestion.
- Avoid sleeping on cold floor.
- Avoid extreme emotional changes.
- Eat three meals on time and avoid overeating or starving.

EPIGASTRIC PAIN 胃 脘 痛

INTRODUCTION

Epigastric pain is a pain or discomfort in the region below the ribs but toward the middle upper abdomen. It is usually associated with digestive system disorders. It is a quite common medical condition that it is a result of something harmless, such as inflammation or infection, but sometimes, it is a life-threatening condition, such as an aortic dissection or cancer in the stomach or pancreas.

ANATOMY

For the purposes of study, diagnosis and treatment, the human abdomen is divided into nine different regions using a 3 × 3 grid by anatomists and physicians; the upper three regions are called the "upper quadrants". The left upper quadrant is the *left hypochondriac* region. The central upper quadrant is the *epigastric* region. The right upper quadrant is the *right hypochondriac* region. Figure 5.2.1 from shutterstock.c om #241069693

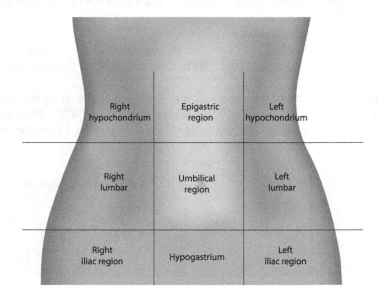

The organs located in the epigastric region are the majority of the stomach, part of the liver, part of the pancreas, part of the duodenum, part of the spleen, and the adrenal glands.

WESTERN MEDICINE ETIOLOGY AND PATHOLOGY

Epigastric pain can be caused by the pathologic changes in the organs located locally in the epigastric quadrant, but it can also be a referring pain from elsewhere.

Pain in Local Organs

The common causes of epigastric pain include:

- *Indigestion*
  - Gastritis.
  - Pancreatitis.
  - Peptic ulcers.
  - Celiac disease.
  - Stomach cancer.
  - Intestinal blockage.
  - Gallstones.
  - Constipation.
  - Diabetes.
  - Thyroid disease.
  - Pregnancy.
- *Inflammation*
  - Gastritis.
  - Esophagitis.
  - Pancreatitis.
  - Gastroenteritis.
  - Gallbladder disorder – gallstones – blocking the opening of the gall-bladder, or the gallbladder itself may be inflamed.
- *Infection*
  - Peptic ulcer.
- Cancers affecting the gallbladder, pancreas, or stomach may result in epigastric pain.
- Lactose intolerance.
- Anatomic changes.
  - Hiatal hernia.
  - Gastroesophageal reflux disease (GERD).
- Pregnancy: it increases abdominal pressure, and it slows digestion due to hormonal changes.
- Overeating: it causes the stomach to expand beyond its normal capacity, puts pressure on the organs around the stomach, and causes epigastric pain.

Radiating Pain to the Epigastrium

- Angina.
- Myocardial infarction.
- Aortic aneurysm.[1]

CHINESE MEDICINE ETIOLOGY AND PATHOLOGY

Excessive Conditions

1. *Acute Invasion of Cold in the Stomach*寒 邪 客 胃

Cold invasion is due to chronic Yang deficiency with underlying weak constitution and malfunction of the Spleen and Stomach.

a. This often is the result of prolonged illness, which impairs Spleen and Stomach Yang.
b. Irregular diet also can damage the Spleen and Stomach Yang: excessive consumption of cold foods, icy beverages, raw foods, and Cold herbs. Spleen and Stomach like warm and dry. The transport and transformation functions of Spleen and Stomach is also influenced by the warming and nourishing functions of Kidney Yang. On the other hand, cold and raw foods are considered Cold and Damp, which the internal environment will start taking away from the Spleen's energy leading to Qi deficiency and later to Yang deficiency when the Cold and Damp impair both Spleen and Kidney.
c. Irregular mealtimes is also a cause of deficiency of Spleen and Stomach. Breakfast is the most important meal in among the three daily meals, providing Qi and Blood for both the mental and physical activities of the day. Breakfast is best in the Stomach time, which is from 7 a.m. to 9 a.m. Dinner should be between 5 p.m. and 7 p.m.; it should not be at 7–9 p.m. because it is the weakest time for digestion, which will impair Spleen and Stomach.

2. *Chronic Retention of Fluids Due to Cold*胃 虛 水 停

There are excess and deficiency conditions.

Excess
Qi from the interior reaches the exterior to assist the body fighting external pathogen invading Tai Yang meridians. Qi in the Middle Jiao becomes weak.

The Yang Qi in Middle Jiao is not strong enough to digest too much water or cold foods, and it results in water retention in the Stomach. Because the Stomach has water, the patient does not desire to drink. If this type of water retention doesn't affect the Qi Movements of Urinary Bladder, the urinary is normal.

Deficiency
Due to water retention in Middle Jiao, water blocks the Movements of Middle Jiao Yang Qi, and there is cold in the extremities.

3. *Retention of Phlegm-Damp*痰 飲 停 胃

Phlegm-Damp stays in the Middle Jiao because of malfunctions of transportation and transformation due to the chronic Spleen Qi deficiency. The Qi Movement of Stomach doesn't move downward, leading to vomiting frothy sputum. Phlegm-Damp that stays in the Middle Jiao disturbs the upward direction of the Qi Movement of Spleen, leading to Pure Yang inability to rise; this results in Phlegm-Damp misting the orifice.

4. *Retention of Food*飲食停滯

Irregular diet is the cause: eating too much or starving damages Spleen and Stomach Qi, resulting in stagnation of Qi in Stomach.

5. *Liver Qi Overacting on the Stomach*肝氣犯胃

Stomach is easily affected by excessive thinking and worry. Excessive mental activity over a long time causes Stomach Qi deficiency. Anger and frustration affect the Liver and cause Stagnation of Liver Qi. Liver then invades Spleen and Stomach. These can cause stagnation of Qi in the Stomach, resulting in Stomach pain.

6. *Heat in the Stomach*肝胃鬱熱

Liver Qi overacts on Stomach, and long-term stagnation evokes Fire.

7. *Phlegm/Heat in the Stomach*脾胃濕熱中阻

Damp-Heat in Middle Jiao can come from several sources:

- d. Overeating foods that are pungent, hot spicy, fatty, sweet, and strong tasting. Taking too much or for too long-term of warm herbs can also cause internal Heat. The Heat burns the Intestine meridians, causing hematochezia.
- e. Invasion of Six Evil Qi to Stomach causing Heat that blocks meridians in the Middle Jiao, resulting in malfunctions of Spleen.

8. *Blood Stagnation*瘀血停胃

Qi is the commander of Blood, long-term Qi stagnation blocks the Blood circulation, leading to Blood stasis.

Deficient Conditions

1. *Stomach/Spleen Qi Deficiency*脾胃虛寒

Malfunctions of transportation and transformation due to the chronic Spleen Qi deficiency. The Qi Movements of Stomach doesn't move downward leading to vomiting frothy sputum. Phlegm-Damp that stays in the Middle Jiao disturbs the upward direction of the Qi Movements of Spleen leading to Pure Yang inability to rise; this results in Phlegm-Damp misting the orifice.

2. Stomach Yin Deficiency胃陰虧虛

Yin insufficiency is due to chronic diseases draining the Body Fluid. Stomach Yin insufficiency can be caused by:

- Long-term Heat in Stomach dries Yin Fluid.
- Long-term Heat in Liver drains Stomach Fluid.
- Long-term Heat in Heart injures Stomach Yin.

Manifestations

Excessive Conditions

1. *Acute Invasion of Cold in the Stomach*寒邪客胃

Signs and symptoms:
Sudden pain, usually severe, spasmodic, some distention that is worse with cold and pressure; nausea, vomiting, aversion to cold, preference for a warm drink, but no desire to drink; pain is relieved with the application of heat.
Tongue: pale tongue with a moist white coat.
Pulse: tight and wiry.

2. *Chronic Retention of Fluids Due to Cold*胃虚水停

Signs and symptoms:
The condition can demonstrate in both excess and deficiency, and it usually occurs after excessively drinking water during Tai Yang meridian disease.
Sensation of borborygmus in the stomach, nausea, intermittent pain relieved by the application of heat.
Patient has a lack of thirst and appetite, normal urination; cold hands and feet.
Tongue: greasy, white coat.
Pulse: tight, deep or wiry, deep.

3. *Retention of Phlegm-Damp* 痰飲停胃

Signs and symptoms:
Abdominal distention, fullness, nausea, poor appetite, possibly frothy vomiting and loose stools; alternating thirst with no desire to drink; vomiting after drinking.
Turbid Phlegm in Middle Jiao blocks the Pure Yang from rising to the head, leading to symptoms of heavy headedness, fuzzy head, and palpitation.
Tongue: pale, white greasy coating.
Pulse: wiry, slippery.

4. *Retention of Food*飲食停滯

Signs and symptoms:
There is a history of overeating of food. It can be chronic, and it is associated with emotional factors; the pain may radiate to hypochondrium.
Persistent fullness and pain in the epigastric area that is constant or colic-like; it gets worse with pressure and eating; intolerance for pressure; nausea, foul belching,

and vomiting of undigested food; pain is reduced after vomiting; no desire to eat; constipation and/or flatulence.

Tongue: thick, greasy coat.

Pulse: full, slippery.

5. *Liver Qi Overacting on the Stomach*肝氣犯胃

Signs and symptoms:

Intermittent pain directly related to emotional disturbances, pain is at the epigastric region, and it is worse with pressure; fullness, distention of the epigastrium and/or hypochondrium, nausea or vomiting, belching, sighing, heartburn, acid regurgitation, flatulence脹 氣, and poor appetite; plum pit throat, dysmenorrhea, depression, alternating constipation and diarrhea.

The Liver can overact on the Stomach, Spleen, or both.

Tongue: thin white or thin yellow coating.

Pulse: wiry.

6. *Heat in the Stomach*肝胃鬱熱

Signs and symptoms:

Pain is severe, urgent, and colicky, worse with pressure, burning sensation in the epigastric area, fullness and tightness of the abdomen, thirst and desire to drink, with preference for cold drink/dislikes hot drink, heartburn and acid reflux or may vomit bitter-tasting water or blood, voracious appetite, irritability, short temper, red face, red eyes, rough breath, constipation, and dark urine.

Tongue: red with a yellow coating.

Pulse: full, rapid or wiry, rapid.

7. *Phlegm/Heat in the Stomach*脾胃濕熱中阻

Signs and symptoms:

Distention, fullness, nausea, vomiting is possible, dry sticky feeling in the mouth with bitter taste, thirst but no desire to drink, heavy body, fatigue, poor appetite, diarrhea with mucus/blood along with a strong odor, burning anus, urgent feeling but constipated and feeling of incomplete movement, and yellow urine.

Stomach signs – epigastric pain, signs of heaviness, and bad breath.

Tongue: red, thick, yellow, greasy coating.

Pulse: slippery, possibly rapid.

8. *Blood Stagnation in the Stomach*瘀血停胃

Signs and symptoms:

Strong pain that is fixed and/or stabbing, pain is worse with pressure and eating, palpable masses are possible, vomiting of a coffee-like substance (indicates blood in the Stomach that has congealed and dried), black stools possible, dark face, and purple lips.

Tongue: purple body, ecchymoses.
Pulse: wiry, choppy.

Deficient Conditions

1. *Stomach/Spleen Qi Deficiency*脾胃虛寒

Signs and symptoms:
   Pain is chronic and dull in nature, it is constant with alternating severity, it is worsened when hungry or exposed to cold, pain is better with the application of pressure and/or heat or after eating, fatigue after eating, nausea or vomiting with clear liquid, no appetite, loose stools or bloody stools (if it is severe), clear and profuse urination, cold limbs and aversion to cold, vomiting blood (if it is severe), listless complexion, slim body shape, possible lower abdominal distention, loose stools, and prolapse of anus.
   Tongue: pale, puffy, with teeth marks; thin, white, and slippery coating.
   Pulse: deep, slow or weak, soggy.

2. *Stomach Yin Deficiency*胃陰虧虛

Signs and symptoms:
   Continuous dull pain, pain is worse after eating, difficulty swallowing, feeling emptiness in the stomach or no appetite even when there is hunger; dry heaves, belching, regurgitation when it is worse, dry mouth; small masses possible in the Stomach; constipation, and emaciation.
   Yin Deficiency signs – malar flush, Five-Center-Heat and night sweats.
   Tongue: dry, red or mirrored, peeled in Stomach area, or coating may not be rooted on the tongue.
   Pulse: thin, wiry, rapid.

PHYSICAL EXAMINATION

- Begin with obtaining medical history, dietary and emotional states. First is to rule out the diseases that may radiate to the epigastric region, such as the heart, lung, liver, and pancreatic conditions. Second is to differentiate the epigastric symptoms from intercostal neuralgia.
  - Interrogate the medical history of the stomach, such as the location of the pain, whether it is located in the epigastric region and radiates to the hypochondriac region; the length of the history of the epigastric pain and whether chronic or urgent; the quality of the pain, whether it is sharp, dull, stabbing, colicky, and tight pain; water consumption and whether the patient is thirsty and desire to drink water. Obtain the history of the duration of pain: is it constant, intermittent, or on and off; the gastrointestinal conditions, such as burning anus, vomiting blood, frothy sputum, and diarrhea with foul feces or constipation, or with

bloody stool; and the state of the patient's energy, such as whether the pain is worse with fatigue or after overworking.

- For dietary history. Ask the patients to describe the foods and drinks in as much detail as possible. If the epigastric pain occurs after eating too big a meal or too fast, causing food retention in the stomach, which upon passing stool or vomiting improves the pain. Undercooked foods, fatty, sweet, cold temperature, raw and foods that are not fresh may cause pain. The Chinese herbal temperatures are necessary knowledge: Hot herbs cause Yin deficiency; Cold herbs cause Yang deficiency. Yang deficiency results in a fuzzy head, cold limbs, and palpitation.
- Obtain the time of the pain and what improves the pain: whether the epigastric pain occurs before or after eating; is the pain worse at night (Blood stasis epigastric pain is more pain at night); what relieves the pain, such as a heating pad on the belly, rest, or drinking warm water. Hot temperatures usually temporarily relieve the pain if it is due to Cold invasion.
- Emotional state. Patients usually will not let us know the pain is associated with the stress or anger, it is very valuable in Chinese medicine. Ask the patient if the pain occurred during a bout of anger or stress and whether the pain was dissipated when the emotional issues were resolved.
- Inspection

   Observe the patient to notice belching and the smell of it, and pay attention to sighing. Also observe the tongue, whether it is red, pale, large, or small; whether there are ecchymosis marks on the tongue; check the coating whether it is thick, greasy, thin, peeled, or no coating at all.
- Palpation

   Palpation on the abdomen tells the temperature, hardness, and pain location. Also pay attention to the pressure to the epigastric region, whether the patients like or dislike pressure.

   Take the pulse to learn the type: whether it is deep, slow, rapid, thin, thready, tight, slippery, soggy, or wiry.

DIFFERENTIAL DIAGNOSIS

1. Angina Pectoris

   Beside the pain on the chest, there may be a radiating pain to the shoulder, arm, and fingers. It usually accompanied by shortness of breath, palpitation, cold limbs, and sweating.
2. Aortic Rupture

   There is a tearing abdominal pain and loss of consciousness, low blood pressure, and fast heart rate.

HERBAL TREATMENTS

Excessive Conditions

1. Acute invasion of Cold in the Stomach寒 邪 客 胃

Treatment Principle

Soothe the Liver, regulate to descend the Stomach Qi; warm Middle Jiao and stop pain.

Herbal Treatments

Herbal Formula

Liang Fu Wan 良 附 丸

This formula is from "A Collection of Fine Formulas from Various Source 良 方 集 腋", authored by Xie Yuanqing 謝 元 慶 (1798–1860), in Qing dynasty 清 朝 (1636–1912), and published in 1842.

Herbal Ingredients

Gao Liang Jiang良 薑3g, Xiang Fu香 附12g

Ingredient Explanations

- Gao Liang Jiang良 薑 warms Middle Jiao, disperses Cold, stops vomiting, and alleviates pain.
- Xiang Fu香 附 moves Liver Qi and Blood and stops pain due to Qi and Blood stagnation.

Modifications

- For Qi stagnation, add Qing Pi青 皮 and Mu Xiang木 香.

Formula Indications

- Stomach Cold and Liver Qi stagnation.

Contraindications

- It is contraindicated for abdominal pain due to Fire in Middle Jiao.
- It is contraindicated for Body Fluid deficiency.
- It is contraindicated during pregnancy.

2. Chronic retention of fluids due to Cold胃 虛 水 停.

Treatment Principle

Warm yang, resolve retained fluid, and promote diuresis.

Herbal Treatments

Herbal Formula

Fu Ling Gan Cao Tang 茯 苓 甘 草 湯

This formula is from "Shang Han Lun 傷 寒 雜 病 論", authored by Zhang Zhong-Jing 張 仲 景 (150–219), in eastern Han dynasty 東 漢 (25–220), and published in 200–210.

Ingredients

Fu Ling茯 苓6g, Gui Zhi桂 枝6g, Gan Cao甘 草3g, Sheng Jiang生 薑9g

Ingredient Explanations
- Fu Ling茯苓 drains Damp and tonifies Spleen.
- Gui Zhi桂枝 warms the meridians and relieves pain.
- Gan Cao甘草 harmonizes the herbs, tonifies Middle Jiao Qi, clears Heat, expectorates phlegm, and stops cough.
- Sheng Jiang生薑 unblocks the pure Yang pathway and harmonizes rebellious Qi.

Modifications
- For loose stool, add Yi Yi Ren薏苡仁.
- For pulsation in lower abdomen, add Da Zao大枣.

Formula Indications
- Retained fluid, palpitation, and faintness.

Contraindications
- Patients with abdominal pain due to Fire should not use.
- It is contraindicated for Body Fluid deficiency.

3. Retention of Phlegm-Damp 痰飲停胃

Treatment Principle
- Warm the Middle Jiao and disperse Coldness.

Herbal Treatments

Herbal Formula
Zheng Qi Tian Xiang San 小半夏湯合苓桂术甘湯

This formula is from "Shang Han Lun 傷寒雜病論", authored by Zhang Zhong-Jing 張仲景 (150–219), in eastern Han dynasty 東漢 (25–220), and published in 200–210.

Herbal Ingredients
Zhi Ban Xia半夏18g, Sheng Jiang生薑15g, Fu Ling茯苓12g, Gui Zhi桂枝9g, Bai Zhu白术6g, Zhi Gan Cao甘草炙6g

Ingredient Explanations
- Zhi Ban Xia半夏 transforms phlegm.
- Sheng Jiang生薑 unblocks the pure Yang pathway and harmonizes rebellious Qi.
- Fu Ling茯苓 drains Damp and tonifies Spleen.
- Gui Zhi桂枝 warms the meridians and relieves pain.
- Bai Zhu白术 tonifies the Spleen Qi and dries Dampness.
- Zhi Gan Cao甘草炙 tonifies the Spleen, harmonizes the other herbs, and guides the herbs to all 12 meridians.

Modifications
- For Spleen Qi deficiency, add Dang Shen黨參 and Huang Qi黃耆.
- For the condition with Cold, add Fu Zi附子.

Formula Indications
- Congested Fluids in the epigastrium.

Contraindications
- It is contraindicated for Liver Yang rising.
- It is contraindicated for those with congested Fluids due to Damp-Heat.

4. Retention of food飲食傷胃 (飲食停滯)

Treatment Principle
Relieve dyspepsia.

Herbal Treatments
Herbal Formula
Da Cheng Qi Tang 大承氣湯
   This formula is from "Shang Han Lun 傷寒雜病論", authored by Zhang Zhong-Jing 張仲景 (150–219), in eastern Han dynasty 東漢 (25–220), and published in 200–210.

Herbal Ingredients
Da Huang大黃12g, Zhi Shi枳實12g, Hou Pu厚朴15g

Ingredient Explanations
- Da Huang大黃 moves the Blood and disperses Bruising.

Zhi Shi枳實 breaks up stagnation and transforms Phlegm.
- Hou Pu厚朴 promotes the movement of Qi in the Middle Jiao, resolves Food Stagnation, descends Rebellious Qi, reduces Phlegm, dries Dampness, and calms wheezing.

Modifications
- For better treatment result in opening Stomach, add Mu Xiang木香 and Xiang Fu香附.

Formula Indications
- Excessive condition in Yang Ming Fu.
- Mild Heat in Stomach.
- Dry Heat in Intestine.
- Early stage of dysentery.

Contraindications

- It is contraindicated during pregnancy.
- Use extreme caution if the condition is weak, tonic herbs may need to be added.

5. Liver Qi overacting on Stomach肝氣犯胃

Treatment Principle

Disperse Liver, regulate Qi, harmonize Stomach, promote digestion, and alleviate pain.

Herbal Treatments

Herbal Formula

Chai Hu Huang Qin Tang柴胡疏肝散

This formula is from "The Complete Compendium of Jingyue景岳全書", authored by Zhang Jiebin張介賓(1111–1117AD), in Ming dynasty明代 (1624 AD).

Herbal Ingredients

Chai Hu柴胡6g, Xiang Fu香附5g, Bai Shao白芍5g, Chuan Xiong川芎5g, Zhi Ke枳殼5g, Chen Pi陳皮6g, Zhi Gan Cao甘草2g

Ingredient Explanations

- Chai Hu 柴胡 moves Liver Qi and releases stagnation.
- Xiang Fu香附 moves Liver Qi and Blood and stops pain due to Qi and Blood stagnation.
- Bai Shao白芍 nourishes the Blood, soothes Liver, and stops pain.
- Chuan Xiong川芎 moves the Blood and moves Qi.
- Zhi Ke枳殼 regulates Qi and removes blockage.
- Chen Pi陳皮 regulates Qi and removes blockage.
- Zhi Gan Cao甘草 tonifies the Spleen, harmonizes the other herbs, and guides the herbs to all 12 meridians.

Modifications

- For Blood deficiency, add Dang Gui當歸, Sheng Di Huang生地, and Shu Di Huang熟地.
- For Blood stasis, add Dan Shen丹參.

Formula Actions

- Moves Liver Qi, harmonizes Blood, and alleviates pain due to Liver Qi stagnation.

Formula Indication

- Headache due to Liver Qi stagnation, Blood stagnation, acid regurgitation, and breast distention.

Contraindications

- Caution for patients who suffer from headache due to the Yin deficient Heat.

6. Heat in the Stomach肝胃鬱熱.

Treatment Principle

Clear stagnant Heat in Stomach and stop epigastric pain.

Herbal Treatments

Herbal Formula

Qing Re Jie Yu Tang 清熱解鬱湯

This formula is from "Restoration of Health from the Myriad Diseases 萬病回春", authored by Gong Tingxian 龔廷賢 (1522~1619), in Ming dynasty 明朝 (1368–1644), and published in 1587.

Ingredients

Zhi Zi梔子6g, Zhi Ke枳殼3g, Xi Qiong 西芎3g, Huang Lian黃連3g, Xiang Fu 香附10g, Gan Jiang乾薑1.5g, Chen Pi 陳皮1.5g, Gan Cao甘草0.9g, Cang Zhu蒼朮2.1g

Ingredient Explanations

- Zhi Zi梔子 clears Heat, cools the Blood, resolves Damp, reduces swelling, and opens the meridians.
- Zhi Ke枳殼 regulates Qi and removes blockage.
- Xi Qiong 西芎 assists Pure Yang to open orifice, moisturizes the dryness of Liver and tonifies its deficiency; moves and disperses Blood and stops pain.
- Huang Lian黃連 clears Heat and drains Damp.
- Xiang Fu 香附 moves Liver Qi and Blood and stops pain due to Qi and Blood stagnation.
- Gan Jiang乾薑 warms Middle Jiao, expels interior Cold, dispels Wind-Dampness.
- Chen Pi 陳皮 regulates Qi and removes blockage.
- Gan Cao甘草 harmonizes the herbs, tonifies Middle Jiao Qi, clears Heat, expectorates phlegm, and stops cough.
- Cang Zhu蒼朮 strongly dries Dampness, tonifies the Spleen, and clears Dampness from the Lower Jiao.

Formula Indications

- Stomach pain due to stress.

Cautions

- For better results in assisting Pure Yang, add Sheng Jiang生薑3 slices.
- Avoid eating for half a day after taking this formula.

7. Phlegm/Heat in the Stomach脾胃濕熱中阻

Treatment Principle
Warm the Middle Jiao and
   disperse Coldness.

Herbal Treatments
Herbal Formula
Zheng Qi Tian Xiang San 地榆散
   This formula is from "Prescriptions of the Bureau of Taiping People's Welfare
Pharmacy 太平惠民和劑局方", authored by Imperial Medical Bureau 太醫局,
in southern Song dynasty南宋 (1127–1239), and published in 1134.

Herbal Ingredients
Di Yu地榆30g, Huang Lian黃連30g, Xi Jiao犀角30g, Xi Gen茜根30g, Huang
Qin黃芩30g, Zhi Zi Ren梔子仁15g 上藥為散。每服12克，用水150毫升,
入薤白5寸，煎至90毫升,

Ingredient Explanations
- Di Yu地榆 cools the Blood, clears Toxic Heat, and stops bleeding.
- Huang Lian黃連 clears Heat and drains Damp.
- Xi Jiao犀角 clears Excessive Heat from the Ying營 and Xue 血stages,
  relieves Fire toxicity熱毒, cools the Blood, clears skin blotches涼血化斑,
  and stops bleeding.
- Qian Cao Gen茜根 cools the Blood and stops bleeding.
- Huang Qin黃芩 clears Heat, dries Damp, cools the Blood, stops bleeding,
  calms Liver Yang.
- Zhi Zi Ren梔子仁 clears Heat in San Jiao, cools the Blood, resolves Damp,
  reduces swelling, opens the meridians, and promotes bowel movements.

Modifications
- For better result in cooling the Blood and stopping bleeding, add Huai Hua
  San.

Formula Indications
- Cools the Blood and stops bleeding and pain.

Contraindications
- The formula is Cold, so it should not be used for a long time; it can irritate
  the stomach lining.
- Those who are Yin deficient should take caution.

8. Blood Stagnation in the Stomach瘀血停胃

Treatment Principle

- Move the Blood.
- Disperse Bruising.
- Open the meridians.

Herbal Treatments

Herbal Formula

Ge Xia Zhu Yu Tang 膈下逐瘀湯

This formula is from "Correction of Errors in Medical Classics醫林改錯", authored by Wang Qingren 王清任 (1768–1831), in Qing dynasty 清朝 (1636–1912), and published in 1849).

Herbal Ingredients

Wu Ling Zhi五靈脂6g, Dang Gui當歸9g, Chuan Xiong 川芎6g, Tao Ren桃仁9g, Mu Dan Pi牡丹皮6g, Chi Shao赤芍6g, Wu Yao烏藥6g, Yan Hu Suo延胡索3g, Gan Cao甘草9g, Xiang Fu香附5g, Hong Hua紅花9g, Zhi Ke枳殼5g

Ingredient Explanations

- Wu Ling Zhi五靈脂 moves Qi and Blood and stops pain due to Qi and Blood stagnation.
- Dang Gui當歸 moves the Blood, disperses Bruising, and stops pain due to Blood stasis.
- Chuan Xiong 川芎 moves the Blood, disperses Bruising, and stops pain due to Blood stasis.
- Tao Ren桃仁 moves the Blood, disperses Bruising, and stops pain due to Blood stasis.
- Mu Dan Pi牡丹皮 clears Heat, cools the Blood, and moves the Blood.
- Chi Shao赤芍 moves the Blood, disperses Bruising, and stops pain.
- Wu Yao烏藥 moves Qi and stops pain.
- Yan Hu Suo延胡索 moves the Blood, promotes Qi circulation, and stops pain.
- Gan Cao甘草 harmonizes the herbs, tonifies Middle Jiao Qi, clears Heat, expectorates phlegm, and stops cough.
- Xiang Fu香附 moves Liver Qi and Blood and stops pain due to Qi and Blood stagnation.
- Hong Hua紅花 moves the Blood, disperses Bruising, and stops pain due to Blood stasis.
- Zhi Ke枳殼 regulates Qi and removes blockage.

Modifications

- For Blood deficiency, add Bai Shao 白芍 and Dan Shen丹參.

Formula Indications

- Blood stasis due to long-term Liver Qi stagnation.

Contraindications

- Patients with weak constitution without Blood stasis should not use.
- It is contraindicated during pregnancy.

Deficient Conditions

9. Stomach /Spleen Qi Deficiency脾 胃 虛 寒

Herbal Treatments

Herbal Formula

Huang Qi Jian Zhong Tang 黃 耆 建 中 湯

This formula is from "Shang Han Lun 傷 寒 雜 病 論", authored by Zhang Zhong-Jing 張 仲 景 (150–219), in eastern Han dynasty 東 漢 (25–220), and published in 200–210.

Herbal Ingredients

Huang Qi黃 耆4.5g, Gui Zhi桂 枝9g, Bai Shao白 芍18g, Gan Cao甘 草 炙9g, Sheng Jiang生 薑9g, Da Zao大 棗12 pieces, Yi Tang飴 糖30g

Ingredient Explanations

- Huang Qi黃 耆 tonifies Qi, strengthens the Spleen, raises the Yang Qi of the Spleen and Stomach, tonifies Wei Qi, stabilizes the exterior, and tonifies the Blood.
- Gui Zhi桂 枝 warms the meridians and relieves pain.
- Bai Shao白 芍 nourishes the Blood, soothes Liver, and stops pain.
- Gan Cao甘 草 harmonizes the herbs, tonifies Middle Jiao Qi, clears Heat, expectorates phlegm, and stops cough.
- Sheng Jiang生 薑 unblocks the pure Yang pathway and harmonizes rebellious Qi.
- Da Zao大 棗 tonifies the Spleen Qi, nourishes the Blood, moderates and harmonizes the harsh properties of the fragrant herbs.
- Yi Tang飴 糖 tonifies Middle Jiao Qi, alleviates pain, disperses Cold, and alleviates spasmodic abdominal pain.

Modifications

- For Blood deficiency, add Dang Gui 當 歸, Dan Shen丹 參, and Chi Shao赤 芍.
- For extremely Cold, add Fu Zi附 子.
- For abdominal distension, add Hou Pu厚 朴, Sha Ren砂 仁, and Mu Xiang木 香.

Formula Indications

- Warms Middle Jiao.
- Alleviates pain.

Contraindications
- It is contraindicated for those with Heat from Yin deficiency.

10. Stomach Yin Deficiency胃 陰 虧 虛

Treatment Principle
Nourish Yin and promote digestion.

Herbal Formula
Yi Guan Jian一 貫 煎

This formula is from "Supplement to 'Classified Case Records of Famous Physicians 續 名 醫 類 案", authored by Wei Zhixiu 魏 之 琇 (1722–1772), in Qing dynasty 清 朝 (1636–1912), and published in 1770.

Herbal Ingredients
Sha Shen沙 參10g, Mai Men Dong麥 冬10g, Dang Gui當 歸10g, Sheng Di Huang生 地 黃30g, Gou Qi Zi枸 杞 子15g, Chuan Lian Zi川 楝 子5g

Ingredient Explanations
- Sha Shen沙 參 nourishes Stomach and Lung Yin.
- Mai Men Dong麥 冬 moistens Lungs and nourishes Yin.
- Dang Gui當 歸 nourishes the Blood, benefits Liver, and regulates menstruation.
- Sheng Di Huang生 地 黃 nourishes Yin and clears Heat.
- Gou Qi Zi枸 杞 子 tonifies Liver and Kidney, strengthens Bone, and benefits Essence and Blood.
- Chuan Lian Zi川 楝 子 moves Liver Qi, removes stagnation, and stops pain.

Modifications
- For making the formula stronger, add Mai Men Dong Tang麥 門 冬 湯.

For insomnia, add Bai Zi Ren柏 子 仁, Suan Zao Ren酸 棗 仁 Wu Wei Zi五 味 子.
- For thirst and irritability, add Zhi Mu知 母 and Shi Gao石 膏.

Formula Actions
- Nourishes Yin.

Formula Indications
- Pain in ribs and abdomen due to deficiency of Kidney and Liver and Liver Qi stagnation. Dry mouth, red tongue with less coating, and thin and rapid or weak and wiry pulse.

Contraindications

- Contraindicated for patients who suffer from emotional stress in excess condition.

The formula contains many herbs that are sweet; patients who are weak in digestion should be cautious.

ACUPUNCTURE TREATMENTS

Acupuncture Point Formula

BASIC POINTS

PC6內 關Neiguan, RN12中 脘Zhongwan, ST36足 三 里Zusanli

Excessive Conditions

1. *Acute Invasion of Cold in the Stomach*寒 邪 客 胃

Moxa on Basic Points after applying reducing techniques on all three of these points. Apply warming needle techniques on RN4關 元Guanyuan and RN6氣 海Qihai.

2. Chronic retention of fluids due to Cold胃 虛 水 停 型

Basic Points, plus UB20脾 俞Pishu, BL21胃 俞Weishu, and LIV13章 門Zhangmen. Apply tonifying techniques on all points.

3. Retention of Phlegm-Damp 痰 飲 停 胃 型

Basic Points, plus ST25天 樞 Tianshu and ST37上 巨 虛Shangjuxu.

4. Retention of food飲 食 傷 胃 (飲 食 停 滯)

Basic Points, plus ST25天 樞 Tianshu, SP15 大 橫Daheng, LI4合 谷Hegu, and LIV3太 衝Taichong.

5. Liver Qi overacting on the Stomach肝 氣 犯 胃

Basic Points, plus LI4合 谷Hegu and LIV3太 衝Taichong.

6. Heat in the Stomach肝 胃 鬱 熱

Basic Points, plus GB34陽 陵 泉Yanglingquan, LIV2行 間Xingjian, and LIV3太 衝Taichong.

7. Phlegm/Heat in the Stomach脾 胃 濕 熱 中 阻

Basic   Points,   plus   GB34陽 陵 泉Yanglingquan,   ST40豐 隆Fenglong,
UB20脾 俞Pishu,   UB25大 腸 俞Dachangshu,   ST44內 庭   Neiting,   and
SJ5外 關Waiguan. Bleeding on SP1隱 白Yinbai.

8. Blood Stagnation in the Stomach瘀 血 停 胃

Basic Points, plus SP4公 孫Gongsun, LI4合 谷Hegu; for chronic condition, add
SP10血 海Xuehai; for acute condition, add ST34梁 丘Liangqiu.

Deficient Conditions

1. Stomach /Spleen Qi Deficiency脾 胃 虛 寒 （ 脾 胃 虛 弱 ）

Basic Points, plus DU20百 會Baihui; warming needle on UB20脾 俞Pishu and
BL21胃 俞Weishu, RN4關 元Guanyuan, and RN6氣 海Qihai; apply ginger moxibustion Therapy on RN8 神 闕Shenque.

2. Stomach Yin Deficiency胃 陰 虧 虛

Basic Points, plus SP3太 白Taibai, LIV8曲 泉Ququan, KD3太 溪Taixi, and
RN3中 極Zhongji.

Cupping Therapy

Cupping is good for both Cold types. Apply the large size cups on RN12中 脘Zhongwan,
UB20脾 俞Pishu for 10 minutes after needling.

Auricular Therapy

Stomach, Spleen, Liver, Sympathetic, Shenmen, Subcortex, Sanjiao, and Abdomen.
Select 3–5 points each visit; needle retention time is 20–30 minutes.

Apply strong sedating stimulation if the epigastric pain is severe, otherwise the
needles can be replaced by ear seeds, teach patients to press each point 20 times each
time of pressing, 3 times a day.

Tui Na Treatments

Tui Na Techniques

Pressing, kneading, chafing, rubbing, one finger meditation pushing, percussion,
grasping

Tui Na Procedures

For acute epigastric pain due to the deficient conditions or foods injures Stomach

1. The patient is in the prone position, the therapist stands beside the thoracic
   region.

2. Places a thumb pad on DU9至 陽Zhiyang and applies pressing and kneading techniques.
3. Presses the point when the patient is exhaling.
4. Kneads the point when the patient is inhaling.
5. The time spending on the point is about 3–5 minutes.
6. Applies cupping Therapy and moxibustion Therapy on the point afterward.

For spasmodic epigastric pain on acute condition:

1. The patient is in the supine position, the therapist stands beside the thigh region
2. Applies grasping, pressing and kneading techniques on ST34梁 丘Liangqiu and SP10血 海Xuehai, this step can be applied on both legs

For chronic epigastric pain

1. The patient is in the supine position, the therapist stands beside the chest region.
2. Places the second to the fifth fingers on REN12中 脘Zhongwan. Figure 5.2.2 from shutterstock.com #2092749745

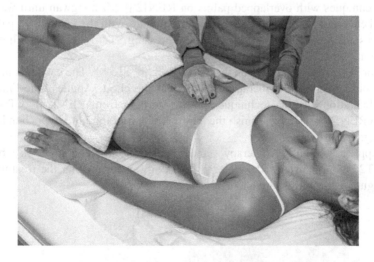

3. Presses the point when the patient is exhaling.
4. Kneads the point when the patient is inhaling.
5. The time spending on the point is about 3–5 minutes.
6. Applies percussion techniques with the fingertips on the epigastric region between REN14巨 闕Juque and RN8神 闕Shenque until the skin turns red.
7. For "Chronic retention of fluids due to Cold胃 虛 水 停 型", places all four fingertips on REN12中 脘Zhongwan, presses the point when the patient is exhaling and vibrates the point for 5–10 times.

8. Applies kneading techniques on the palmer side of the fingers from the 2nd to the 5th on REN12中 脘Zhongwan.

9. Applies pressing and grasping technique with both hands on both ST25天 樞Tianshu. Presses the point when exhaling, grasping the point when inhaling. This step promotes peristalsis for 3–5 minutes or until warm.

10. Overlaps the hands on REN8神 闕Shenque and kneads the point in clock-wise direction for 3–5 minutes.

11. For "Acute invasion of Cold in the Stomach寒 邪 客 胃". Applies chafing techniques on the spine between DU6脊 中Jizhonguntil and DU8筋 縮Jinsuo until warm, places the palm on the chafing area until it cools down, then repeat for about 8–10 times.

12. For "Stomach /Spleen Qi Deficiency脾胃虛寒（脾胃虛弱）". The therapist stands near the head, applies pushing techniques from DU14大 椎 DaiZhui to DU4命 門Ming Men until the skin turns red or for 3–5 times.

13. For "Blood Stagnation瘀 血 停 胃". Applies pinching–pushing techniques on the inner Urinary Bladder lines from the thoracic region to the lumber region until the skin turns red or for 3–5 times.

14. For "Retention of food飲 食 傷 胃 (飲 食 停 滯)". The patient is in the supine position, the therapist stands near the abdomen. Applies kneading techniques with overlapped palms on REN12中 脘Zhongwan until warm. Then applies pressing and vibrating techniques with the overlapped palms on same point. Finish the treatment by pressing ST36足 三 里Zusanli with thumb tips.

15. For "Liver Qi overacting on the Stomach肝 氣 犯 胃 ". The patient is in the supine position, the therapist stands near the chest. Applies One-Finger-Meditation-Pushing techniques on the sternum from REN22天 突Tiantu to REN14巨 闕Juque. Finish the treatment by rubbing both hypochondriac region until warm.

16. Applies cupping Therapy and moxibustion Therapy on both ST25天 樞Tianshu and REN12中 脘Zhongwan for closing the treatment. figure 5.2.3 from shutterstock.com # 1055759024

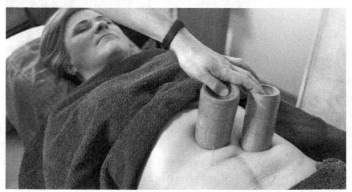

PATIENT ADVISORY

- For patients with Yin 陰deficiency, it is recommended to have meals on time, small amount of food but more frequent times because the epigastric pain occurs when they are hungry. For life style recommendation, they should not stay up late at night, get plenty of sleep to nourish Yin陰.
- For patients with Yang 陽deficiency, it is recommended to have the food only when it is warm or right after cooking, avoid any cold or icy foods, also avoid raw foods because raw in nature is also Cold in Chinese medicine. They also need to know the temperature of food in the nature, some foods are cold or cool in the nature, such as water melon, mung beans, duck, and many melons. Increase ginger in the diet because it is warm in nature. For life style recommendation, regular exercise helps to nourish Yang陽. Should let the stomach rest 2 hours before bed, avoid lying supine after eating and avoid overeating.
- For patients with stress, it is recommended to eat meals on time, avoid having emotional upset before meals. Entertainments during eating meals helps to ease the stomach pain. For life style recommendation, regular exercise helps to ease the stress.

HYPOCHONDRIAC PAIN IN CHINESE MEDICINE中醫脅痛

INTRODUCTION

Hypochondriac Pain refers to the condition of a disorder characterized by pain over the unilateral or bilateral sides of the hypochondriac region. In Chinese medicine, it is commonly associated with the emotional upset, such as stress. It is one of the most treated medical conditions with Chinese medicine.

ANATOMY

For the purposes of study, diagnosis and treatment, the human abdomen is divided into 9 different regions using a 3x3 grid by anatomists and physicians, the upper 3 regions are the upper quadrant. The central upper quadrant is the epigastric region. The right upper section is the right upper quadrant or also called the right hypochondriac region. The left upper section is the left upper quadrant called the left hypochondriac region.

Both hypochondriac regions are located on the lateral sides of the abdominal wall, they are inferior to the ribcage, the middle between them is the epigastrium.

Right Upper Quadrant

The primary structures in the right hypochondriac region include the right portion of the liver, the gallbladder, biliary system, portions of the right kidney, and parts of the small intestine, and ascending and transverse colon.

Left Upper Quadrant

The left hypochondriac region contains part of the spleen, portions of the left kidney, part of the stomach, the pancreas, small intestine and transverse and descending colon. Figure 5.3.1 from Shutterstock.com #1758940178

ABDOMINAL REGIONS

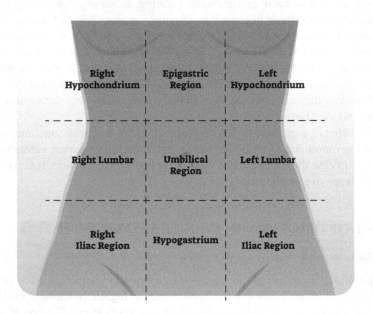

Right Hypochondrium	Epigastric Region	Left Hypochondrium
Right Lumbar	Umbilical Region	Left Lumbar
Right Iliac Region	Hypogastrium	Left Iliac Region

WESTERN MEDICINE ETIOLOGY AND PATHOLOGY

Hypochondriac pain in western medicine is primary caused by the organs located in the sections.

RIGHT UPPER QUADRANT

- The pathology of liver, biliary, gallbladder and lung cause the right upper quadrant pain.
- The liver organ may have the pathology include hepatitis, liver cancer, hepatic abscess and cirrhosis.
- The biliary may cause pain in the right upper quadrant include cholangitis.
- The gallbladder pathology may include gallstones.
- The lung right lower lobe pneumonia can cause a referred pain onto the right upper quadrant pain.[2]
- Pleural effusion in the right lung can cause pain onto the hypochondrium.
- The pneumothorax on the right lung can cause pain onto the right hypochondrium.
- Pulmonary embolism.

Left Upper Quadrant

- The pathology of pancreas, stomach and spleen can cause the left upper quadrant pain.
- The stomach pathology can cause the left upper quadrant pain.
- The spleen pathology (e.g., splenic rupture, splenic abscess, splenic infarct) can cause the left upper quadrant pain.
- The lung left lower lobe pneumonia can cause a referred pain onto the left upper quadrant pain.
- The pleural effusion in the left lung can cause pain onto the left hypochondrium.
- The pneumothorax on the left lung can cause pain onto the left hypochondrium.
- Pulmonary embolism.

CHINESE MEDICINE ETIOLOGY AND PATHOLOGY

Classic text The Yellow Emperor's Inner Classic黃帝內經written in 2000 years ago explained the relationship between Liver and Gallbladder and hypochondrium.

- Ling Shu chapter 20 Five Evils said: When evil energy is in the Liver, both hypochondrium will be painful《黃帝內經·靈樞》五邪篇第二十曰：「邪在肝、則兩脅中痛」。
- Su Wen chapter 22 The Relation Between Energies of Five Viscera and the Four Seasons said: In the Liver disease, there will be pain over the hypochondrium。
- 《黃帝內經·素問》藏氣法時論篇第二十二曰：「肝病者,兩脅下痛」。
- Su Wen chapter 63 Contralateral Pricking said: When the evil energy invades the collateral branch of the Foot Shaoyang Channel, it will cause one to have syndromes of hypochondriac pain, difficult respiration。《黃帝內經·素問》繆刺論曰：「邪客於足少陽之絡，令人脅痛，不得息」。

The causes of pain in hypochondrium are classified into excessive and deficient conditions:

Excessive Condition

1. Liver Qi Stagnation 肝鬱脅痛

University of Edinburgh published a research result in 2015 in the journal Gastroenterology, the study suggested that suffering from anxiety or depression could carry an increased risk of death from liver disease. Researchers at the University of Edinburgh did the investigation on more than 165,000 people over a ten-year period, the outcome of the study showed that individuals who scored highly for symptoms of psychological distress were more likely to later die from liver disease than those with lower scores, Dr Tom Russ of the University of Edinburgh's Centre for Clinical Brain Sciences led the research.[3]

In the classic text books "The Yellow Emperor's Inner Classic黃帝內經", there were numerous statements explaining the relationship between the emotion and the injury of the Liver:

- Su Wen chapter 73 The Source of Disease said: When a person's Qi, perhaps owing to hate or anger, ascends counterflow and fails to descend, this damages the Liver.
  《黃帝內經・素問》本病論篇第七十三曰：「人或恚怒，氣逆上而不下，即傷肝也」。
- Su Wen chapter 39 The Pathogen of Pain said: Anger results in Qi counterflow. Severe anger results in retching of Blood。
  《素問・舉痛論》曰：怒則氣逆，甚則嘔血。
- Ling Shu chapter 4 The Visceral Diseases Caused by Evil Energy said: If fury causes Qi to ascend without descend, the energy will be Stagnant under the hypochondrium and injures Liver。
  《靈樞經・邪氣藏府病形篇》又曰：若有所大怒，氣上而不下，積於脅下，則傷肝。

Based on the above statements, emotional and mental disharmony, such as mental depression, violent anger, frustration, and resentment can damage the liver organ and the Liver Zang-fu 中醫肝臟. The Liver Zang-fu 中醫肝臟 can also become injured as a result of excessive thought, anxiety, and fear. The term Zang-Fu 臟腑 refers to the internal organs in Chinese medicine. The term Zang 臟 refers to the 5 Yin 陰 organs (Liver 肝, Heart 心, Spleen 脾, Lung 肺, and Kidney 腎) while the term Fu 腑 refers to the 6 Yang 陽 organs (Gallbladder 胆, Small Intestine 小肠, Stomach 胃, Large Intestine 大肠, Urinary Bladder 膀胱, and Sanjiao 三焦). Each Yin Zang 臟 organ is paired with a Yang Fu 腑 organ (Liver-Gallbladder, Heart-Small Intestine, Spleen-Stomach, Lung-Large Intestine, and Kidney-Urinary Bladder). Zang-Fu 臟腑 are connected to the 12 meridians, each Yin Zang 臟 organ connects to a Yin meridian 陰經 while each Yang fu 腑 organ connects to a Yang meridian 陽經.

2. Qi stagnation and Blood stasis血瘀脅痛

The Qi and Blood can be stagnant due to physical or mental trauma.

1) Mental trauma: This is an internal cause of pain. Liver is associated with emotional stress and Wind, and Liver governs Tendons. Emotional stress changes the quality of tendons, making them vulnerable.

Liver is associated with Wind, and Wind is associated with movements; its attack is fast and is often recognizable by signs of pain.

2) Physical trauma: This is an external cause of pain. Qi stagnation may also be the result of a physical trauma. When trauma occurs, bodily injury causes Blood stasis in the meridians, it blocks the flow of Qi, then, Qi flow is impeded. The trauma leads to both Qi stagnation and Blood stasis simultaneously. As a result of Qi and Blood stagnation in

the meridians, pain occurs, this is according to a classic quote: "If there is no free flow, there is pain不 通 則 痛."

3. Dampness and Heat in Liver and Gallbladder濕 熱 脅 痛
   - External Damp-Heat pathogen invading the (e.g., viral infection)
   - Overeating in wrong diets (e.g., spicy, greasy, sweet foods)
   - Dysfunction of Spleen in transportation

   The above three factors cause Damp-Heat invading Liver and Gallbladder leading to the impairment of dispersion and dredging resulting hypochondriac pain.

Deficiency Condition:

1. Liver Yin Xu肝 陰 虛

   - Kidney Yin deficiency
   - Long-term Liver Qi stagnation evoking Liver Fire, Heat from Fire excretes Body Fluids and Yin
   - Chronic diseases with weakness
   - Overworking

These factors consume Essence and Blood leading to the insufficiency of Liver Yin causing undernourishment of meridians and collaterals resulting pain.

2. Fluid Retention in Hypochondrium水 飲 內 停
   - Middle Jiao Yang deficiency
   - Invasion of the external Cold and Damp pathogen
   - Impairment of wrong diet

These factors cause hypofunction of dispersing and descending of Lung Qi, malfunction of transportation and transformation of Spleen Qi, and poor dominating water metabolism of Kidney. The impairments of these three Zang-Fu cause obstructions in meridians leading to Fluid retention in chest and hypochondrium resulting pain.

3. Evil Pathogen in Shao Yang Meridian邪 在 少 陽
   - Wind Cold invasion directly attack Shao Yang meridians少 陽 經
   - Wind Cold invades Tai Yang meridian 太 陽 經then attacks Shao Yang meridians少 陽 經

Gallbladder Shao Yang meridians少 陽 經 distribute lateral aspects of the body where includes the hypochondrium. When the pathogenic Wind Cold invade Shao Yang meridians少 陽 經, Gallbladder Shao Yang少 陽 losses its normal functions in controlling the opening of Tai Yang 太 陽and closing of Yang Ming陽 明, the excessive opening of Tai Yang太 陽causes cold sensation, the excessive opening of Yang Ming 陽 明causes heat sensation, these cause the alternating fever and chill symptoms.

If Gallbladder Shao Yang meridians少 陽 經 is attacked by the pathogenic Wind Cold, the hypochondrium is painful accompanied by the main symptom of it which is alternating fever and chill.

MANIFESTATIONS:

Excessive Condition

1. Liver Qi Stagnation 肝鬱脅痛

Hypochondriac pain is migratory, distending. The pain may radiate to the chest, upper back, shoulder, and arm.

The pain is aggravated by emotional stress and anger. It is accompanied by depressive moods, sighing, chest distension, shortness of breath. Breast tenderness, irregular menstruation and dysmenorrhea in female patients.

Tongue: thin coating.

Pulse: wiry

2. Qi stagnation and Blood stasis 血瘀脅痛

Hypochondriac pain is localized stabbing pain, the pain is more at night. Patients do not prefer hand pressed pressure. A mass may be palpable near the painful hypochondrium.

Tongue: purple body, petechiae spots

Pulse: irregular pulse

3. Dampness and Heat in Liver and Gallbladder 濕熱脅痛

Hypochondriac pain is distending, the pain may radiate to the shoulder and upper back. Patients do not prefer hand pressed pressure. It is accompanied by bitter taste in the mouth, chest oppression, nausea, vomiting, red eye or yellow sclera, skin and urine

Tongue: yellow greasy coating.

Pulse: wiry, slippery, rapid

Deficiency Condition:

1. Liver Yin Xu 肝陰虛

Hypochondriac pain is burning, dull and constant, it is aggravated by fatigue or overworking, patients usually prefer hand pressing. It is accompanied by dry mouth and throat, irritability, dizziness and vertigo or blurry vision.

Tongue: red, little coating.

Pulse: wiry, thin and rapid

2. Fluid Retention in Hypochondrium 水飲內停

Hypochondriac pain is distending, it is aggravated by breathing, cough and torso turning. It is accompanied by shortness of breath.

Tongue: white coating.

Pulse: deep, wiry or deep slippery

3. Evil Pathogen in Shao Yang Meridian邪 在 少 陽

Hypochondriac pain is fullness and distending, it is mainly located at the costal region. It is accompanied by fever and chill alternating, bitter taste in the mouth, dry throat, vertigo, hearing loss, poor appetite, irritability, nausea.
Tongue: white slippery coating.
Pulse: wiry

PHYSICAL EXAMINATION:

Begin with obtaining medical history. the questions should include dietary (e.g., sweet, fatty foods), the location of pain (e.g., more pain on the left of right), the type of pain (e.g., burning, dull, constant, stabbing or a mass is palpatable), the time of pain (e.g., more pain at night time, or after work or when fatigued or pain is aggravated by emotional changes, cough, and torso turning). There are many other questions that may differentiate the type of hypochondriac pain include alternating fever and chill, nausea, vomiting, copious cough with clear sputum, shortness of breath, palpitation. The injury history may indicate Blood stasis.

DIFFERENTIAL DIAGNOSIS:

Hypochondriac pain needs to be differentiated with stomach pain, intercostal neuralgia, cardiac disease.

Stomach pain is located in the epigastric region. The pain is usually associated with the meal time.

Intercostal neuralgia is that the pain is like a band-wrap sensation along the ribs from the thoracic toward the chest. Schepelmann's sign is positive.

Cardiovascular disease is commonly located in the left upper chest and upper back, the pain may radiate onto the left arm. The duration of pain can be short as only several minutes. It is accompanied by shortness of breath, sweating, and palpitation and may present as sudden confusion, dizziness, or faintness. For angina pectoris, sublingual nitroglycerin can relieve sudden angina.

TREATMENTS

1. Liver Qi Stagnation 肝 鬱 脅 痛

TREATMENT PRINCIPLE

Sooth Liver and regulate Qi flow.

Herbal Treatments

Chai Hu Shu Gan Tang柴 胡 疏 肝 散

This formula is from "The Complete Compendium of Jingyue景 岳 全 書", authored by Zhang Jiebin張 介 賓(1111–1117AD), in Ming dynasty明 代 (1624 AD).

Herbal Ingredients

Chai Hu柴 胡6g, Xiang Fu香 附5g, Bai Shao白 芍5g, Chuan Xiong川 芎5g, Zhi Ke枳 殼5g, Chen Pi陳 皮6g, Zhi Gan Cao甘 草2g

Ingredient Explanations

- Chai Hu 柴 胡 moves Liver Qi and releases stagnation.
- Xiang Fu香 附 moves Liver Qi and Blood and stops pain due to Qi and Blood stagnation.
- Bai Shao白 芍 nourishes the Blood, soothes Liver, and stops pain.
- Chuan Xiong川 芎 moves the Blood and moves Qi.
- Zhi Ke枳 殼 regulates Qi and removes blockage.
- Chen Pi陳 皮 regulates Qi and removes blockage.
- Zhi Gan Cao甘 草 tonifies the Spleen, harmonizes the other herbs, and guides the herbs to all 12 meridians.

Modifications

- For best results, add Yu Jin鬱 金.
- For conditions with poor appetite, change Chen Pi陳 皮 to Qing Pi青 皮.
- For Heat symptoms, add Mu Dan Pi牡 丹 皮 and Zhi Zi梔 子.

Formula Actions

- Moves Liver Qi, harmonizes Blood and alleviates pain due to Liver Qi stagnation.

Formula Indications

- Hypochondriac pain due to Liver Qi stagnation.

Contraindications

- Caution for patients who suffer from headache due to Yin deficient Heat.

2. Qi stagnation and Blood stasis血 瘀 脅 痛

Treatment Principle
Move Blood, disperse Bruising, and stop pain.

Herbal Formula

Fu Yuan Huo Xue Tang復 元 活 血 湯

This formula is from "Clarifying the study of medicine醫 學 發 明", authored by Li Dongyuan 李 東 垣, in Yuan dynasty元 朝 (1271–1368), and published in 1315.

Formula Ingredients

Chai Hu 柴 胡9g, Tian Hua Fen 天 花 粉9g, Dang Gui 當 歸9g, Hong Hua 紅 花6g, Gan Cao 甘 草6g, Chuan San Jia 穿 山 甲 炮6g, Da Huang 大 黄12g, Tao Ren 桃 仁9g

Ingredient Explanations

- Chai Hu 柴 胡 harmonizes Shao Yang disorders and relieves stagnation.
- Tian Hua Fen 天 花 粉 clears Heat, expels pus, and reduces swelling.
- Dang Gui 當 歸 moves and tonifies the Blood, disperses Cold, and stops pain due to Blood Stasis.
- Hong Hua 紅 花 moves the Blood, releases Bruising, opens the meridians, and stops pain.
- Gan Cao 甘 草 releases spasms, alleviates pain, harmonizes the other herbs, and guides the herbs to all 12 meridians.
- Chuan San Jia 穿 山 甲 moves the Blood, reduces swelling, and expels Wind Damp.
- Da Huang 大 黄 moves the Blood and disperses Bruising.
- Tao Ren 桃 仁 breaks up stubborn Blood Stasis.

Formula action: moves the Blood, disperses Bruising, soothes Liver, and opens the meridians.

Formula indication: physical trauma, Blood stasis, and severe pain.

3. Dampness and Heat in Liver and Gallbladder濕 熱 脅 痛

Treatment Principle

Drain Excess Fire from the Liver and Gallbladder and clear and drain Damp-Heat from the lower Jiao.

Herbal Formula

Long Dan Xie Gan Tang龍 膽 瀉 肝 湯

This formula is from the book "Analytic Collection of Medicinal Formulas 醫 方 集 解", authored by Wang Ang汪 昂 (1615–1694AD), and published in Qing dynasty清 朝 (1682AD).

Herbal Ingredients

Long Dan Cao龍 膽 草12g, Zhi Zi栀 子9g, Huang Qin黃 芩9g, Chai Hu柴 胡6g, Sheng Di Huang生 地 黃12g, Ze Xie澤 瀉9g, Dang Gui當 歸5g, Che Qian Zi車 前 子10g, Mu Tong川 木 通9g, Gan Cao甘 草5g

Ingredient Explanations

1. Long Dan Cao龍 膽 草drains Damp-Heat from the Liver and Gallbladder channels.
2. Zhi Zi栀 子clears Heat, cools the Blood, resolves Damp, reduces swelling, and opens the meridians.
3. Huang Qin黃 芩clears Heat, dries Damp, cools the Blood, stops bleeding, and calms Liver Yang.
4. Chai Hu柴 胡 moves Liver Qi and releases stagnation.
5. Sheng Di Huang生 地 黃 nourishes Yin and clears Heat.
6. Ze Xie澤 瀉 drains Damp, especially Damp Heat in Lower Jiao; clears Kidney deficient Heat.

7. Dang Gui當歸 nourishes the Blood, benefits the Liver, and regulates menstruation.
8. Che Qian Zi車前子 clears Damp Heat and promotes urination.
9. Mu Tong川木通 drains Heat and promotes urination.
10. Gan Cao甘草 harmonizes the herbs, clears Heat, and expectorates Phlegm.

Modifications
- For making the formula stronger, add Ju Hua菊花 and Tian Ma天麻.
- For redness, swelling, and pain of the eyes, add Chuan Xiong川芎 and Ju Hua菊花.

Formula Actions
- Drains excess Fire from the Liver and Gallbladder and drains Damp-Heat from Lower Jiao.

Formula Indication
- Headache due to Liver Fire and Damp Heat in Lower Jiao, such as migraine headache, urethritis, acute glaucoma, and acute cholecystitis.

Contraindications
- Caution for patients who suffer from Cold and Yang deficiency in Spleen and Stomach due to the Cold herbs in the formula.

Deficiency Condition
1. Liver Yin Xu肝陰虛

Treatment Principle
Nourish Yin and soothe the Liver 養陰柔肝.

Herbal Formula
Yi Guan Jian一貫煎

This formula is from "Supplement to 'Classified Case Records of Famous Physicians 續名醫類案'", authored by Wei Zhixiu 魏之琇 (1722–1772), in Qing dynasty 清朝 (1636–1912), and published in 1770.

Herbal Ingredients
Sha Shen沙參10g, Mai Men Dong麥冬10g, Dang Gui當歸10g, Sheng Di Huang生地黃30g, Gou Qi Zi枸杞子15g, Chuan Lian Zi川楝子5g

Ingredient Explanations
- Sha Shen沙參 nourishes Stomach and Lung Yin.
- Mai Men Dong麥冬 moistens Lungs, nourishes Yin.
- Dang Gui當歸 nourishes the Blood, benefits Liver, and regulates menstruation.
- Sheng Di Huang生地黃 nourishes Yin and clears Heat.

- Gou Qi Zi枸 杞 子 tonifies the Liver and Kidney, strengthens Bone, and benefits Essence and Blood.
- Chuan Lian Zi川 楝 子 moves Liver Qi, removes stagnation, and stops pain.

Modifications
- For making the formula stronger, add Chai Hu Shu Gan San柴 胡 疏 肝 散.
- For bitter taste in the mouth, add Huang Lian黃 連, Huang Qin黃 芩, Tian Hua Fen天 花 粉, and Chai Hu柴 胡.
- For insomnia, add Bai Zi Ren柏 子 仁, Suan Zao Ren酸 棗 仁, and Wu Wei Zi五 味 子.
- For thirst and irritability, add Zhi Mu知 母 and Shi Gao石 膏.

Formula Actions
- Nourishes Yin, soothes the Liver.

Formula Indication
- Pain in ribs and abdomen due to deficiency of Kidney and Liver and Liver Qi stagnation; dry mouth, red tongue with less coating; thin, rapid pulse or weak, wiry pulse.

Contraindications
- Contraindicated for patients who suffer from emotional stress in excess condition.
- The formula contains many herbs that are sweet; patients with weak digestion should take caution.

2. Fluid Retention in Hypochondrium水 飲 內 停

Treatment Principle
Relax and prevent adhesion of the tendons, move Blood, and stop pain.

Herbal Treatments
Herbal Formula
Ting Li Da Zao Xie Fei Tang葶 藶 大 棗 瀉 肺 湯
This formula is from "Jin Gui Yao Lue 金 匱 要 略", authored by Zhang Zhong-Jing 張 仲 景 (150–219), in eastern Han dynasty 東 漢 (25–220), and published in 200–210.

Herbal Ingredients
Ting Li葶 藶15g, Da Zao大 棗12 pieces

Ingredient Explanations
- Ting Li Zi 葶 藶 子 sedates Lungs, calms wheezing, and drains Phlegm.
- Da Zao大 棗 tonifies the Spleen Qi, nourishes the Blood, and moderates and harmonizes the harsh properties of the fragrant herbs.

Modifications

- To make to formula stronger, add Dan Shen丹參 and Yu Jin鬱金 for greater pain relief.
- For water retention with difficulty breathing, add Gan Sui甘遂, 0.5–1g.

Formula Actions

- Drains Lungs.
- Stops cough.

3. Evil Pathogen in Shao Yang Meridian邪在少陽

Treatment Principle

Relax and prevent adhesion of the tendons, move Blood, and stop pain.

Herbal Treatments

Herbal Formula

Xiao Chai Hu Tang 小柴胡湯

This formula is from "Shang Han Lun 傷寒雜病論", authored by Zhang Zhong-Jing 張仲景 (150–219), in eastern Han dynasty 東漢 (25–220), and published in 200–210.

Herbal Ingredients

Chai Hu 柴胡 24g, Huang Qin黃芩9g, Ren Shen人參9g, Zhi Ban Xia半夏9g, Gan Cao甘草9g, Sheng Jiang生薑9g, Da Zao大棗 12 pieces

Ingredient Explanations

- Chai Hu 柴胡 moves Liver Qi and releases stagnation.
- Huang Qin黃芩 cools the Blood and dries Dampness.
- Ren Shen人參 tonifies the Spleen and Stomach Qi, tonifies Lung Qi, generates Body Fluids, and stops thirst.
- Zhi Ban Xia半夏 dries Dampness and transforms Phlegm.
- Gan Cao甘草 harmonizes the ingredients.
- Sheng Jiang生薑 warms the Lungs, releases the Exterior, induces perspiration, and disperses Cold.
- Da Zao大棗 tonifies the Spleen and Stomach and augments Qi.

Modifications

- For irritability without vomiting, remove Zhi Ban Xia半夏 and Ren Shen人參 and add Gua Lou Shi栝蔞實.
- For thirst, remove Zhi Ban Xia半夏 and add Gua Lou Gen栝蔞根.
- For mass palpable in hypochondriac region, remove Da Zao大棗 and add Mu Li牡蠣.

Formula Actions

- Harmonizes and releases Shao Yang Stage disorders.
- Harmonizes the Liver and Spleen.

Acupuncture Treatments

Acupuncture Point Formula

Local Points

LIV14 期 門Qimen, SJ6支 溝Zhigou, ST3巨 髎Juliao, RN12中 脘Zhongwan, RN17膻 中Danzhong, LIV3太 沖Taichong, UBI8肝 俞Ganshu, UB19膽 俞Danshu, PC6內 關Neiguan

Modifications
- For Liver Qi Stagnation, add GB34陽 陵 泉Yanglingquan, GB40丘 墟Qiuxu; perform the same technique on LIV14期 門Qimen and UBI8肝 俞Ganshu; sedate LIV3太 沖Taichong.
- For Blood Stasis, add and sedate LIV2行 間Xingjian, SP6三 陰 交Sanyinjiao, and UB17膈 俞Geshu; use cupping Therapy on the affected hypochondriac region after treatment with Seven–Star needle.
- For Pathogenic in Shao Yang少 陽, add and sedate DU14大 椎DaiZhui; add and perform the same technique on SJ3中 渚ZhongZhu, GB41足 臨 泣Zulinqi, and SJ5外 關Waiguan.
- For Fluid Retention, add and sedate on LU5尺 澤Chize and LU7列 缺Lieque; add RN22天 突Tiantu and make the Qi sensation to the chest then remove the needle; add and perform the same technique on ST36足 三 里Zusanli and ST40豐 隆Fenglong.
- For Liver Yin insufficiency, add and perform the same technique on UBI8肝 俞Ganshu; add GB20風 池Fengchi, LIV8曲 泉Ququan, SP6三 陰 交Sanyinjiao, and KD3太 溪Taixi.
- For Damp Heat in Liver and Gallbladder, add SP6三 陰 交Sanyinjiao and GB34陽 陵 泉Yanglingquan.

Auricular Therapy

Chest, Shenmen, Sympathetic, Occiput, Lung

Electro-Acupuncture Therapy

SJ6支 溝Zhigou to GB34陽 陵 泉Yanglingquan with 2/100 hz for 20 minutes daily.

Tui Na Treatments

Tui Na Techniques

Pushing, kneading, pressing, and rubbing

Tui Na Procedures
1. The patient lies prone on a treatment table, and the therapist stands near the head.
2. The therapist kneads and presses with the thumb pads on UB13肺 俞Feishu, UB17膈 俞 Geshu, UB18 肝 俞Ganshu, UB20脾 俞 Pishu, and UB23

Shenshu腎俞 for 2–5 minutes. Figure 5.3.2 from Shutterstock.com #383165062

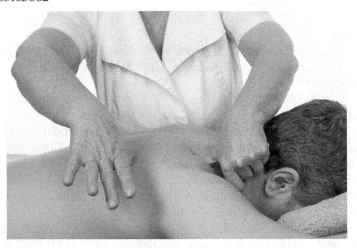

3. The therapist overlaps their hands and places them on UB13肺俞Feishu, then pushes both hands toward UB25 大腸俞Dachangshu. Repeat on both meridians until the skin turns red or warm.
4. Repeats step 3. The patient takes a deep breath, and the therapist pushes the meridians with intermittent pushing techniques while the patient exhales.
5. The patient lies supine on the treatment table laterally with both knees bent and a pillow under the head. The therapist lifts the patient's arm upward toward the head, pushes and kneads with the whole palm from the armpit to the hypochondrium until warm. Repeat this step on the other side.
6. The patient sits on the edge of the treatment table, the therapist stands behind, rubs both hypochondrium with palms until warm.

PATIENT ADVISORY

- Patients are recommended to keep the diet light; avoid fatty, sweet, hot spicy, and pungent foods.
- Ensure enough good quality sleep.
- Be happy, enjoy life. Avoid excessive emotional changes.

IRRITABLE BOWEL SYNDROME 腸易激綜合症

INTRODUCTION

Irritable bowel syndrome (IBS) is a bothersome gastrointestinal disorder, with abdominal pain and discomfort It distracts people, as they may not be able to work, attend social events, or travel even short distances.

IBS is a chronic lower gastrointestinal tract disorder. Recent research has found that it is a neurologic bowel disorder, and emotional upset plays an important role.

Fortunately, IBS does not lead to more serious disease, as it is not an inflammatory, infectious, or malignant disease

ANATOMY

The gastrointestinal tract is where humans and animals digest food to extract and absorb nutrients at the same time as it expels waste as feces. The human gastrointestinal tract consists of the upper and lower gastrointestinal tracts.

The upper gastrointestinal tract is from the mouth to the duodenum.

The lower gastrointestinal tract includes most of the small intestine and all the large intestine. Part of the duodenum, jejunum, and ileum are included in the small intestine in humans, while the large intestine includes the cecum, ascending, transverse, descending colons, sigmoid colon, rectum, and anal canal.

The symptoms of IBS are in the lower gastrointestinal tract. There are five parts as follows:

- Cecum. Its shape is like a pouch. It is the origin of the large intestine. The appendix extends from the base of the cecum. The Ileum, the final and longest segment of the small intestine, ends at the ileocecal valve which empties into the cecum of the colon. The large intestine includes three parts: ascending, transverse, and descending colons.
- Ascending colon rises from the cecum in the right abdomen.
- Transverse colon extends from the ascending colon from the right side of the abdomen crossing the abdominal cavity toward the spleen to the left side of the abdomen.
- Descending colon extends from the transverse colon then turns medially and inferiorly to form the sigmoid colon.
- Rectum extends from the sigmoid colon, then it continues as the anal canal, terminating at the anus.

The gastrointestinal tract is innervated by intrinsic neurons of the enteric nervous system and by the axons of extrinsic sympathetic, parasympathetic, and visceral afferent neurons.

The enteric nervous system is known our "second brain". Stress, depression, and anxiety are associated with it. It may trigger big emotional shifts. Irritation in the gastrointestinal system may send signals to the central nervous system that triggers mood changes. The first and the second brains talk to each other[4]: everyday, emotional well-being may rely on messages from the brain below to the brain above[5]; therapies that help one may help the other – soothing the second brain heals the stress, depression, and anxiety from the first one that is located within the skull.

WESTERN MEDICINE ETIOLOGY AND PATHOLOGY

The cause of irritable bowel syndrome is unknown; however, emotional factors, such as stress, lust, greed, pride, depression, anger, and anxiety may precipitate or aggravate the symptoms. These negative emotions can be from psychosocial or psychological behaviors.[6]

CHINESE MEDICINE ETIOLOGY AND PATHOLOGY

1. Diarrhea Predominant IBS
   a. *Liver Qi Stagnation Overacting on Spleen*肝氣乘脾證

      The preexisting deficiency of Spleen can set the stage for the stagnation of Liver to overact. Long-term excessive mental work leads to Liver Qi Stagnation. Liver overacting on Spleen impairs the functions of Spleen: it disturbs the ascending Pure Yang of Spleen, leading to be descending the water to the gastrointestinal tracts causing loose stool or diarrhea.

   b. *Spleen Qi Deficiency*脾 虛濕盛證

      The deficiency of Spleen is due to over-consumption of fatty, sweet, pungent, and spicy foods. Also, consuming foods that are not clean or are excessive, raw, cold, sweet, or fried, causes the functions of Spleen to become impaired. Spleen dominates, raising Pure Yang, prefers Dryness, dislike Dampness, transforms foods into nutrients, and transports the nutrients to the meridians and Zang-Fu and wastes to the Large Intestines. Due to Spleen Qi deficiency, the functions of transformation and transportation and ascending direction of Spleen become weak, leading to water accumulating in the gastrointestinal tracts.

   c. *Yang Deficiency of Spleen and Kidney*脾肾陽虛證

      Long-term Spleen-Stomach deficiency and Liver overacts on Spleen. With long-term Spleen Yang deficiency, Kidney Yang is undernourished, leading to insufficiency of Ming Men Fire命 門 火. Deficiency of both Spleen and Kidney Yang results in Dampness accumulation.

   d. *Large Intestine Damp Heat*大腸濕熱證
      - External Damp Heat foods, which are the fatty, sweet, shrimp, crabs, fish, or the raw and cold foods, and the strong taste foods attach Yang Ming Meridians.
      - Under-using air conditioning, perspiration is poor, and Heat is blocked interiorly, causes Damp Heat descending to the colon, and resulting in diarrhea.

2. Constipation Predominant IBS
   a. *Liver Qi Stagnation*肝鬱氣滯證

      Stress and depression will affect the dispersing function of Liver, which will result in stagnation in the meridians. Gastrointestinal tracts lack the distribution function of Liver.

   b. *Dryness and Heat in Large Intestine*大腸燥熱證

      Long-term deficiency of Yin dries the body fluid, resulting Dryness in the colon.

   c. *Interior Cold-Heat Complication*寒熱夾雜證

Chronic IBS that was not treated correctly leads to the complication of both Cold and Heat retention in the gastrointestinal tracts. This is the condition due to wrong transferring between Yin-Yang 陰陽. Spleen is a Yin Zang 臟, while Stomach is a Yang Fu 腑. Spleen dominates raising Pure Yang, prefers Dryness, dislikes Dampness, and transports nutrients and wastes. Stomach prefers Dampness, dislikes Dryness, and its function is to drain the turbid Damp downward. Both Spleen and Stomach stay in Middle Jiao; with their opposite conditions and functions, Middle Jiao manages the body to control the balance of Yin-Yang 陰陽, and there is no illness when both Yin 陰 and Yang 陽 are able to connect and transfer to each other. Middle Jiao is the gastrointestinal tracts in western medicine.

Interior Cold. The deficiency of Spleen Qi leads to malfunction of Spleen in transferring and transporting nutrients and wastes. The malfunction results in Phlegm Damp and water retention in the meridians. They are the Yin evil 陰邪 which injures the Yang Qi of Spleen, and deficiency of Spleen Yang 脾 陽虛, causes retention of interior Cold.

Interior Heat. Chronic Dryness in Stomach causes impairment of the Stomach descending function. The impairment results in stagnation of foods. The stagnation transforms to interior Heat.

MANIFESTATIONS

In Western Medicine

Abdominal pain may be associated with either diarrhea, constipation, or both, alternately. The pain can be bloating, cramping or discomfort, and distention. Pain can be recurrent in the abdomen. Other symptoms include borborygmus, indigestion, passing excessive amounts of gas, or an urgent need to defecate; symptoms may be alleviated by defecation, anxiety, depression, discomfort, loss of appetite, and nausea.

In Chinese Medicine

1. *Diarrhea Predominant IBS*
   a. Liver Qi Stagnation Overacting on Spleen肝 氣 乘 脾 證

   Diarrhea after abdominal pain, pain is alleviated after passing stool, fatigue, borborygmus, abdominal distention, occurring with or aggravating with emotional upsets. Accompanied by hypochondriac pain, bitter taste in mouth, and irritability.

   Tongue: red, pink, or dusky body, and thin, white coating.
   Pulse: wiry or wiry and thin.
   b. Spleen Qi Deficiency脾 虛 濕 盛 證

   Loose stool or diarrhea after meals, aggravated by drinking cold water or eating cold food; loss of appetite, borborygmus, sweating, fatigue, and reluctance to talk, even without aphasia.

   Tongue: pale body with teeth marks and white coating.
   Pulse: thin, weak.

    c. Yang Deficiency of Spleen and Kidney脾腎陽虛證

Diarrhea in the morning, cold sensation in the abdomen, pain is relieved
    with the application of heat, there may have some undigested food in
    the stool, aversion to cold, cold limbs, fatigue.

Tongue: pale and big body, with teeth marks and white slippery coating.

Pulse: deep, thin.

    d. Large Intestine Damp Heat大腸濕熱證

Diarrhea right after abdominal pain, it is urgent or impeded defecation.
    Sticky and greasy stools, fetid stools, abdominal distention, dry mouth,
    sticky feeling in the mouth, thirsty with no desire to drink and a burn-
    ing sensation in the anus.

Tongue: red body and greasy, yellow coating.

Pulse: slippery or slippery and rapid.

  2. ***Constipation Predominant IBS***

    a. Liver Qi Stagnation肝鬱氣滯證

Impeded defecation with abdominal pain, dry stool that is difficult to
    pass, aggravated by depression or emotional irritation, bloating, and
    eructation.

Tongue: pale or dusky body with thin, white coating.

Pulse: wiry.

    b. Dryness and Heat in Large Intestine大腸燥熱證

Defecation with abdominal pain, dry stool, abdominal distention aggra-
    vated by pressure; dry mouth with bad smell.

Tongue: red body and thin, yellow, sticky coating.

Pulse: thin or thin and rapid.

    c. Interior Cold-Heat Complication寒熱夾雜證

Defecation with abdominal pain, preference for pressure, diarrhea alter-
    nately with constipation, borborygmus; dry and bitter taste in mouth,
    sensation of heavy anus, and impeded defecation.

Tongue: dusky with sticky, white coating.

Pulse: thin, wiry or slippery, wiry.

PHYSICAL EXAMINATION

  1. Begin with obtaining a medical history.

    Because IBS doesn't rely on lab, orthopedic, or neurological tests or
colonoscopy, when a patient presents with symptoms of irritable bowel syndrome, obtaining a thorough medical history is helpful in the diagnosis and
classification in Chinese medicine.

    • Obtain the history of the emotions, such as depression, stress, nausea,
      and anxiety, hypochondriac pain, and irritability.

    • Obtain information regarding the abdominal pain, including disten-
      tion, colic, watery and impeded defecation, and amount of gas; whether
      the abdominal pain is aggravated with emotional upset or drinking or

eating cold food; whether pain is relieved with heat application or pressure. Before palpation, ask the patient to point to the painful areas.

- Obtain the history of the defecation: for example, whether the diarrhea alternates with constipation; whether the abdominal pain is relieved after defecation; whether the defecation is in the morning, or if there is diarrhea after a meal. Obtain the type of the defecation, such as foul smell and undigested food in the stool.

2. Inspection
   - Patients generally appear to be healthy. Patients with chronic major depressive disorder may present the appearance as listless, frowning, or dejected face.
   - Observe the tongue, such as the body color, coating thickness and color, and the dryness.

3. Auscultation

   Listen to the abdomen with a stethoscope for bowel sounds. Begin in the right lower quadrant, and move in sequence up to the right upper quadrant, left upper quadrant, and finally the left lower quadrant.

4. Percussion

   Percussion can help to assess the fluid in the abdomen, tapping the abdomen to locate the tenderness or pain.

5. Palpation

   Abdominal palpation is valuable in diagnosing IBS and ruling out other abdominal diseases. Palpation needs to include the guarding and rigidity of the abdominal muscle; either it is voluntary or involuntary, pain associated with pressure, and the type of the pulse.

   Avoid percussing and palpating the abdomen if the patient has had an organ transplant or if an abdominal aortic aneurysm or appendicitis is suspected.

DIFFERENTIAL DIAGNOSIS

Inflammatory Bowel Disease

Since IBS is a digestive disorder that affects the large intestine, it has similar symptoms to inflammatory bowel disease (IBD) conditions. IBD is a chronic inflammation of the digestive tract with symptoms that are different than IBS: the symptoms include bloody or black stools, loss of appetite, weight loss, and fever.

Colorectal Cancer

With the change of the defecation, and abdominal discomfort, IBS has similar symptoms to colorectal cancer:

- Rectal bleeding, dark stool, or blood in the stool.
- A change in the size of the stool: it may be thin like a pencil.
- Unexplained weight loss.

TREATMENTS

1. Diarrhea Predominant IBS

1. Liver Qi Stagnation Overacting on Spleen肝 氣 乘 脾 證

Treatment Principle
Regulate and harmonize Liver and Spleen.

Herbal Treatments
Herbal Formula
Tong Xie Yao Fang 痛 瀉 要 方
   This formula is from "The Complete Compendium of Jingyue景 岳 全 書", authored by Zhang Jiebin張 介 賓(1563–1640AD), in Ming dynasty明 朝 (1368–1644), and published in 1624.

Herbal Ingredients
Bai Zhu白 朮18g, Bai Shao白 芍12g, Chen Pi陳 皮9g, Fang Feng防 風6g

Ingredient Explanations
- Bai Zhu白 朮 tonifies the Spleen Qi and dries Dampness.
- Bai Shao白 芍 nourishes the Blood, soothes the Liver, and stops pain.
- Chen Pi陳 皮 regulates Qi and removes blockage.
- Fang Feng防 風 expels Wind and drains Damp.

Modifications
- For chronic diarrhea, add Sheng Ma升 麻.
- If the tongue coating is yellow and greasy, add Huang Lian黃 連.

Formula Indications
- Painful diarrhea due to Liver Qi Stagnation overacting Spleen.
- Wiry pulse in the left Guan關 and moderate pulse in the right Guan關.

Contraindications
- It is contraindicated for patients with diarrhea due to Food Stagnation.

2. Spleen Qi Deficiency脾 虛 濕 盛 證

Treatment Principle
Strengthen Spleen and transform Damp.

Herbal Treatments
Herbal Formula
Shen Lin Bai Zhu San 參 苓 白 朮 散

This formula is from "Prescriptions of the Bureau of Taiping People's Welfare Pharmacy 太 平 惠 民 和 劑 局 方", authored by Imperial Medical Bureau 太 醫 局, in southern Song dynasty南 宋 (1127–1239), and published in 1134.

Herbal Ingredients

Bai Bian Dou扁 豆9g, Ren Shen人 參12g, Bai Zhu白 朮12g, Fu Ling茯 苓12g, Gan Cao甘 草12g, Shan Yao山 藥12g, Lian Zi蓮 子6g, Yi Yi Ren薏 苡 仁6g, Jie Geng桔 梗6g, Sha Ren砂 仁6g, Da Zao大 棗3 pieces

Ingredient Explanations

- Bai Bian Dou扁 豆 strengthens Spleen and resolves Dampness.
- Ren Shen人 參 tonifies Qi and Yang.
- Bai Zhu白 朮 tonifies the Spleen Qi and dries Dampness.
- Fu Ling茯 苓 drains Damp and tonifies Spleen.
- Gan Cao甘 草 harmonizes the herbs, tonifies Middle Jiao Qi, clears Heat, and expectorates phlegm.
- Shan Yao山 藥 tonifies Kidneys and Benefits the Yin and the Spleen Qi.
- Lian Zi蓮 子 tonifies the Spleen and astringes diarrhea.
- Yi Yi Ren薏 苡 仁 promotes urination, leaches out Dampness, strengthens the Spleen, expels Wind-Dampness, and clears Damp-Heat.
- Jie Geng桔 梗 opens Lungs, disperses Lung Qi, and expels Phlegm.
- Sha Ren砂 仁 aromatically transforms Dampness, moves Qi, and stops diarrhea.
- Da Zao大 棗tonifies the Spleen Qi, nourishes the Blood, and moderates and harmonizes the harsh properties of the fragrant herbs.

Modifications

- For Damp-Heat diarrhea, add Huang Lian黃 連 and Mu Xiang木 香.
- For excessive Damp, add Huo Xiang 藿 香, Hou Po厚 朴, and Cang Zhu蒼 朮.
- For poor appetite, add Shan Zha山 楂 and Ji Nei Jin雞 內 金.

Formula Indications

- Dampness due to Spleen Qi deficiency.

Contraindications

- It is contraindicated for patients with Excess syndrome.
- Use with caution during pregnancy.

3. Yang Deficiency of Spleen and Kidney脾 肾 陽 虛 證

Treatment Principle

Warm Spleen and Kidney.

Herbal Treatments

Herbal Formula

Fu Zi Li Zhong Tang 附子理中湯

This formula is from "Treatise on Diseases, Patterns, and Formulas Related to theUnification of the Three Etiologies三因極一病証方論", authored by Chen Yan 陳言 (1131–1189), in Song dynasty 南宋 (1127–1239), and published in 1174.

Herbal Ingredients

Ren Shen人參5g, Zhi Fu Zi炮附子5g, Gan Jiang炮乾薑5g, Zhi Gan Cao炙甘草5g, Bai Zhu 白朮5g

Ingredient Explanations

- Ren Shen人參 tonifies Qi and Yang.
- Zhi Fu Zi炮附子 warms Ming Men Fire.
- Gan Jiang乾薑 warms Middle Jiao, expels interior Cold, and dispels Wind-Dampness.
- Zhi Gan Cao炙甘草 tonifies the Spleen, harmonizes the other herbs, and guides the herbs to all 12 meridians.
- Bai Zhu白朮 tonifies the Spleen Qi and dries Dampness.

Modifications

- For abdominal fullness and distention, add Yi Zhi Ren益智仁.
- For Damp accumulation, add 茯苓Fu Ling and 澤瀉Ze Xie.
- For poor appetite, borborygmus, and undigested food in the stools, add Shan Zha山楂, 莱菔子Lai Fu Zi, and 麥芽 Mai Ya.

Formula Indications

- Spleen Yang deficiency and Spleen and Kidney Yang deficiency.

Contraindications

- It is contraindicated for patients with external invasion with fever.
- It is contraindicated for patients during pregnancy.

4. Large Intestine Damp Heat大腸濕熱證

Treatment Principle

Clear Heat and drain Damp.

Herbal Treatments

Herbal Formula

Ge Gen Huang Qin Huang Liang Tang 葛根黃芩黃連湯

This formula is from "Shang Han Lun 傷寒雜病論", authored by Zhang Zhong-Jing 張仲景 (150–219), in eastern Han dynasty 東漢 (25–220), and published in 200–210.

Herbal Ingredients

Ge Gen葛 根15g, Huang Qin 黃 芩9g, Huang Lian黃 連9g, Gan Cao甘 草6g

Ingredient Explanations

- Ge Gen葛 根 relieves Heat and releases muscles, especially of the neck and upper back.
- Huang Qin 黃 芩 clears Heat, dries Damp, cools the Blood, stops bleeding, and calms Liver Yang.
- Huang Lian黃 連 clears Heat and drains Damp.
- Gan Cao甘 草 harmonizes the herbs, tonifies Middle Jiao Qi, clears Heat, expectorates phlegm, and stops cough.

Modifications

- For abdominal pain, add Mu Xiang木 香 and Bai Shao白 芍.
- For excess Heat, add Ju Hua銀 花.

Formula Indications

- Lower Jiao Intestinal Damp-Heat.

Contraindications

- It is contraindicated for patients with dysentery or diarrhea with no fever.

Constipation Predominant IBS

1. Liver Qi stagnation肝 鬱 氣 滯 證

Treatment Principle

Clear Heat and drain Damp.

Herbal Treatments

Herbal Formula

Liu Mo Liang Tang 六 磨 湯

This formula is from "Effective Formulas from Generations of Physicians世 醫 得 效 方", authored by Wei Yilin 危 亦 林 (1277–1347), in Yuan dynasty元 朝 (1271–1368), and published in 1345).

Herbal Ingredients

Bing Lang檳 榔3g, Chen Xiang 沉 香3g, Mu Xiang木 香3g, Wu Yao烏 藥3g, Da Huang大 黃3g, Zhi Shi枳 殼3g

Ingredient Explanations

- Bing Lang檳 榔 regulates Qi, reduces accumulations, drains downward, and unblocks the bowels.
- Chen Xiang 沉 香 promotes Qi Movement for descending Qi, alleviates pain, warms the Middle Jiao, dispels Cold, and warms the Kidney.

- Mu Xiang木 香 promotes the movement of Qi and alleviates pain.
- Wu Yao烏 藥 moves Qi and stops pain.
- Da Huang大 黃 moves the Blood and disperses Bruising.
- Zhi Shi枳 殼 regulates Qi and removes blockage.

Modifications
- For abdominal pain, add Chen Pi陳 皮 and Chuan Xiong川 芎.
- For excess Heat that causes constipation, add Zhi Zi栀 子, Huang Qin黃 芩, and Mu Dan Pi牡 丹 皮.

Formula Indications
Liver and Spleen Qi Stagnation causing constipation.

Contraindications
- It is contraindicated for patients with Qi Sinking.
- Do not take long-term.

2. Dryness and Heat in Large Intestine大 腸 燥 熱 證

Treatment Principle
Clear Heat and moisten Intestine.

Herbal Treatments
Herbal Formula
Ma Zi Ren Wan 麻 子 仁 丸
This formula is from "Shang Han Lun 傷 寒 雜 病 論", authored by Zhang Zhong-Jing 張 仲 景 (150–219), in eastern Han dynasty 東 漢 (25–220), and published in 200–210.

Herbal Ingredients
It is a pattern herb that each pill is made to the size of about 0.5 cm.
Dose: Take 10 pills each time, 3 times per day.
Huo Ma Ren麻 子 仁, Bai Shao白 芍, Zhi Shi枳 實, Da Huang大 黃, Hou Po厚 朴, Xing Ren杏 仁

Ingredient Explanations
- Huo Ma Ren麻 子 仁 moistens the Intestines and unblocks the bowels.
- Bai Shao白 芍 nourishes the Blood, soothes the Liver, and stops pain.
- Zhi Shi枳 實 breaks up stagnation and transforms Phlegm.
- Da Huang大 黃 moves the Blood and disperses Bruising.
- Hou Po厚 朴 promotes the movement of Qi in the Middle Jiao, resolves Food Stagnation, descends Rebellious Qi, reduces Phlegm, and dries Dampness.
- Xing Ren杏 仁 stops cough, calms wheezing, and moistens Intestines.

Modifications

- For constipation due to Blood deficiency, add Dang Gui當 歸 and Tao Ren桃 仁.
- For severe constipation, add Mang Xiao芒硝.
- For constipation due to Heat injuring Body Fluid, add Sheng Di Huang生 地, Gua Lou Ren栝蔞仁 , Xuan Shen玄參, Bai Zi Zen柏 子 仁, and Shi Hu石 斛.

Formula Indications

- Insufficient Fluid in Intestine.
- Heat in Intestine.

Contraindications

- It is contraindicated for pregnancy.
- Do not take long-term for patients with Body Fluid insufficiency and Blood deficiency.

3. Interior Cold-Heat Complication寒 熱 夾 雜 證

Treatment Principle

Balance Yin-Yang and regulate Cold-Heat.

Herbal Treatments

Herbal Formula

Wu Mei Wan 乌 梅 丸

This formula is from "Shang Han Lun 傷 寒 雜 病 論", authored by Zhang Zhong-Jing 張 仲 景 (150–219), in eastern Han dynasty 東 漢 (25–220), and published in 200–210.

Herbal Ingredients

Wu Mei烏 梅 肉15g, Ren Shen人 參3g, Gui Zhi桂 枝3g, Xi Xin細 辛3g, Huang Lian黃 連8g, Dang Gui當 歸2g, Chuan Jiao 川 椒2g, Huang Bai黃 柏3g, Zhi Fu Zi附 子3g, Gan Jiang乾 薑5g

Ingredient Explanations

- Wu Mei烏 梅 肉 eliminates roundworms and alleviates pain.
- Ren Shen人 參 tonifies Qi and Yang.
- Gui Zhi桂 枝 warms the meridians and relieves pain.
- Xi Xin細 辛 enters Shao Yin meridian to expel Wind, Damp, and Cold.
- Huang Lian黃 連 clears Heat and drains Damp.
- Dang Gui當 歸 moves the Blood, disperses Bruising, and stops pain due to Blood stasis.
- Chuan Jiao 川 椒 warms Middle Jiao, disperses Cold, dispels Damp, and stops pain.

- Huang Bai黃 柏 drains Damp-Heat, especially from the Lower Jiao.
- Zhi Fu Zi制 附 子 warms Ming Men Fire.
- Gan Jiang乾 薑 warms Middle Jiao, expels interior Cold, and dispels Wind-Dampness.

Modifications

- For constipation, add Bing Lang檳 榔 and Zhi Shi枳 實.
- For severe abdominal pain, add Chuan Lian Zi川 楝 子 and Mu Xiang木 香.
- For IBS with abdominal pain, add Ying Su Ke罌 粟 殼 and Yan Hu Suo延 胡 索.

Formula Indications

- Jue Yin 厥 陰Stage Cold and Heat.
- Chronic diarrhea or dysentery.
- Collapse from roundworms.

Contraindications

- It is contraindicated for patients with Damp-Heat dysentery.
- It is contraindicated for patients with the condition of kidney disease.
- Use with caution for infant, children, and during pregnancy.
- Do not take long-term.

Acupuncture Treatments

Acupuncture Point Formula

Basic Formula

ST36足 三 里Zusanli, ST25天 樞Tianshu, SP6三 陰 交Sanyinjiao

Modifications

- For Liver Qi Stagnation Overacting on Spleen肝 氣 乘 脾 證, UBI8肝 俞Ganshu, UB20脾 俞Pishu, and LIV3太 衝Taichong.
- For Spleen Qi Deficiency脾 虛 濕 盛 證, add UB20脾 俞Pishu, UB21胃 俞Weishu, LIV13章 門Zhangmen, and REN12中 脘Zhongwan.
- For Yang Deficiency of Spleen and Kidney脾 腎 陽 虛 證, add UB20脾 俞Pishu, UB23腎 俞Shenshu, UB25大 腸 俞Dachangshu, DU4命 門Mingmen, LIV13章 門Zhangmen, and REN4關 元Guanyuan with moxibustion Therapy on all these points.
- For Large Intestine Damp Heat大 腸 濕 熱 證, add ST37上 巨 虛Shangjuxu.
- For Liver Qi stagnation肝 鬱 氣 滯 證, add UBI8肝 俞Ganshu and LIV2行 間Xingjian.
- For Dryness and Heat in Large Intestine大 腸 燥 熱 證, add LI4合 谷Hegu and LI11曲 池Quchi.
- For Interior Cold-Heat Complication寒 熱 夾 雜 證, add moxibustion Therapy, as stated later.

- For constipation, add UB25大腸俞Dachangshu, SJ6支溝Zhigou, and ST40豐隆Fenglong. Reducing techniques may be applied for excessive condition, while tonifying techniques may be utilized for deficient condition, and moxibustion Therapy could be added for Cold condition.

Moxibustion Therapy

Acupuncture Point: REN8神闕Shenque

Materials: a slice of raw fresh ginger about 2–3 mm thick and 5–10 cones of loose moxa

1. The patient is in the supine position.
2. Punch 2–3 holes in ginger, and place it on the acupuncture point.
3. Place 1 cone of moxa on top of ginger and ignite it.
4. Replace the moxa cone when each piece has no more smoke.
5. Use 5–7 cones or until the acupuncture is hot or red. Figure 5.4.1 from shutterstock.com #250347505

Tui Na Treatments

Tui Na Techniques

Pressing, kneading, rolling, rubbing, stretching

Tui Na Procedures

The patient is in a sitting position, and the therapist stands behind.

1. First, the therapist locates the spinous processes that are painful and deviated between T9 and L4.
2. Applies rubbing techniques on the sacrum region and both sides of the spine between T9 and L4 for about 6–10 minutes.

3. Presses and kneads the painful and deviated spinous processes until sore and distended.

4. The patient raises the left arm and places their left hand on the neck, the therapist places their left hand on the top of the patient's hand by putting the left forearm under the patient's armpit. Figure 5.4.2 from shutterstock.com #735583471

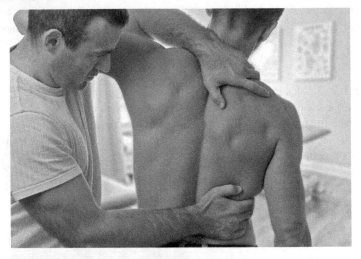

5. Places the right thumb over the painful spinous process.

6. Uses the left hand to lean the patient's trunk forward and to the right. At the end of this movement, the right thumb presses the spinous process leftward. Figure 5.4.3 from shutterstock.com #1255674847

The patient is in the supine position, and the therapist stands near the lumbar.

1. Asks the patient to breathe using the abdomen.
2. Rubs the abdomen counterclockwise for 10 minutes or until warm.
3. Pays attention on REN6氣 海Qihai and REN12中 脘Zhongwan while rubbing the abdomen. Figure 5.4.4 from shutterstock.com #624148640

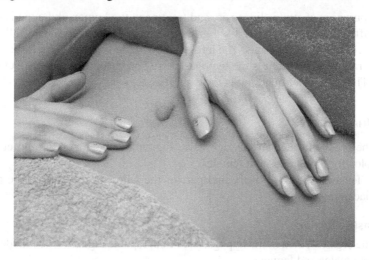

PATIENT ADVISORY

- Avoid foods that are cold, raw, and fatty.
- Avoid foods that may cause dry stool, such as crackers.
- Avoid dairy products if it causes diarrhea.
- Recommend foods that are light-tasting and easy to digest, such as congee and noodle.
- For patients with constipation, recommend foods that are high in fiber and drinking plenty of fluids.
- Recommend appropriate exercise: it strengthens immunity, reduces stress.
- Recommend going to sleep not later than 11 p.m. and getting up at about 7 a.m. for exercise. This balances Yin-Yang 陰陽.

LOWER ABDOMINAL PAIN 下 腹 痛

INTRODUCTION

The cause of pain in the lower abdomen can be complex, as it can be related to the digestive tract, urinary tract, both male and female reproductive organs, and blood vessels. The symptoms can be wide-ranging, such as inflammation, infection, shingles, appendicitis, bowel obstruction, indigestion, gas in intestine, bowel perforation, kidney stones, testicular torsion or injury, menstrual cramps, endometriosis, ovarian torsion (twisting of an ovary), rupture of an ovarian cyst, ectopic pregnancy, or pelvic inflammatory disease.

Lower abdominal pain can be a life-threatening medical condition that needs immediate medical care. All medical providers should be aware of that, make an initial diagnosis in a timely manner, and refer the patients to the urgent care unit if necessary.

ANATOMY

For the purposes of study, diagnosis, and treatment, the human abdomen is divided into 9 different regions using a 3 × 3 grid by anatomists and physicians, all 3 lower regions are the lower quadrants. The left lower quadrant is called the left iliac region. The central lower quadrant is the hypogastric region, also known as the suprapubic region. The right lower quadrant is the right iliac region.

RIGHT ILIAC REGION

- The organs located in the right iliac region are the appendix and cecum colon for both males and females.
- In females, the right ovary and part of the fallopian tube are in the right iliac region.

Hypogastric Region

- The organ located in the hypogastric or suprapubic region is the bladder for both males and females.
- In females, the uterus is also in the suprapubic region.

Left Iliac Region

- The sigmoid colon is in the left iliac region.
- In females, the left ovary and fallopian tube are also in the left iliac region. Figure 5.5 from Shutterstock #1758940178

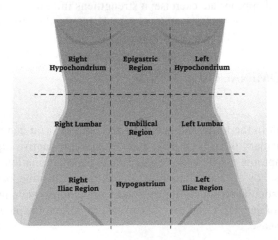

ABDOMINAL REGIONS

WESTERN MEDICINE ETIOLOGY AND PATHOLOGY

Finding the cause of lower abdominal pain can be tricky; the location of the pain can be a physical indicator that lets the medical providers know something is wrong with the organs in the region. Whereas, some diseases may manifest the pain in multiple regions.

MULTIPLE REGIONS

- Irritable bowel syndrome.
- Inflammatory bowel disease.
- Colon cancer.
- Indigestion.
- Trapped gas.
- Constipation.
- Muscle strain.
- Muscle spasm.
- Gastroenteritis.
- Gastritis.

Left Iliac Region

- Constipation.
- Diverticulitis.
- Ischemic colitis.
- Left ovary conditions, e.g., ovarian cysts.
- Ovarian torsion.

Right Iliac Region

- Appendicitis.
- Right ovary conditions, e.g., ovarian cysts, ovarian torsion.

Suprapubic Region

- Gynecologic conditions, e.g., uterine fibroids and cancer.
  - Ruptured ovarian cyst.
  - Ovarian torsion (twisted ovary).
  - Ectopic pregnancies.
  - And some women experience abdominal pain during pregnancy.
- Urological conditions, e.g., urinary retention, urinary tract infection (UTI), bladder stones, bladder cancer, bladder prolapse, prostate cancer, and interstitial cystitis (IC).

Referred Pain

Lower abdominal pain may not be always from the pathologic change from the organs in the regions of the lower abdomen – it can be a referred pain. When the pain is referred to the lower abdomen, it is usually referred from the pelvis or hip.

In Males

- Testicular pathology, e.g., testicular tumors, testicular torsion, orchitis, and epididymitis.

Inguinal hernia.

In Females

- Ovary pathology, e.g., ovarian cysts.
- Ovarian torsion.

Others

Hip fractures, osteoarthritis, and musculoskeletal injuries in the hip or pelvis.

Chinese Medicine Etiology and Pathology

Lower abdomen in Chinese is also divided into 3 parts:

- Both left and right iliac regions are called Shao Fu (少 腹).
- The suprapubic region is called Xiao Fu (小 腹).
    1. The pathology of Shao Fu (少 腹) is related to Liver meridians disorders.
    2. The pathology of Xiao Fu (小 腹) is related to:
        - Internal medical disorders including pathology of the Kidney, Urinary Bladder, and Small Intestine.
        - The deficiency of Womb Palace 胞 宮, Ren 任, and Chong衝 meridians, which are mostly seen in gynecological conditions.
        - Womb Palace 胞 宮 is a gynecological system associated with pregnancy in Chinese medicine; it includes the organs of the uterus, fallopian tubes, and ovaries in western medicine.
        - Ren Meridian任 脈is known as the "Sea of Yin陰 脈 之 海". It manages the Yin 陰 and Blood in the Yin meridian 陰經, to regulate menstrual cycles, and nourishes the fetus.
        - Chong Meridian衝 脈is known as the "Sea of 12 Meridians十 二 經 脈 之 海" and "Sea of Blood血 海". It is closely associated with the reproductive system. Regular menstruations rely on the functions of Chong衝 脈meridian and Ren meridian任 脈. It originates in Bao Zhong 胞 中 where it is the "Root of Life生 命 之 根" "Root of Ren Meridian任 脈, Du Meridian督 脈, Chong Meridian衝 脈, Dai Meridian帶 脈 and Kidney Meridian". Bao Zhong 胞 中regulates Yin-Yang陰 陽, Qi 氣 and Blood, it plays the role in assisting Womb Palace 胞 宮. The location of Bao Zhong 胞 中 includes Dan Tian 丹田, Lower Jiao 下 焦, Liver, Gallbladder, Kidney, and Urinary Bladder.

MANIFESTATIONS

In Western Medicine

- It is uncomfortable, and it may be associated with anxiety. If the pain is severe, it may be a symptom of inflammation, appendicitis, bowel obstruction, or bowel perforation.

- If there is cramping, it may be associated with trapped gas in the intestine, indigestion, and menstrual pain or endometriosis in women. It may be a symptom resulting from diverticulitis; although it is commonly located in the left lower abdomen, it can be anywhere in the abdomen.
- Lower abdominal pain accompanied by abnormal vaginal bleeding may result from gynecologic cancer (e.g., cervical, ovarian, uterine, vaginal, and vulvar cancers). These cancers are more common in older women than in younger women. Abnormal vaginal bleeding includes extremely heavy bleeding during the period, heavy bleeding that lasts more than seven days, unusual bleeding between periods, and vaginal bleeding after menopause. Abnormal vaginal bleeding is not always caused by gynecologic cancer; many other causes can lead to uterine bleeding – hormone replacement therapy is one of them.
- Severe pain in women may result from menstrual cramps, ectopic pregnancy, ovary disorders (e.g., ovarian torsion, ovarian cyst, or ovarian rupture), and pelvic inflammatory diseases (e.g., gonorrhea or chlamydia infections).
- Severe pain in men may result from testicular torsion, testicular injury, infection, or incarcerated hernia.
- Severe pain that starts suddenly and comes in waves may result from kidney stones. The pain can be a sharp, cramping pain in the lower abdomen, back, and groin; it comes and goes. Male patients may feel pain at the tip of the penis.

In Chinese Medicine

Suprapubic Region Disorders

1. Damp-Heat in Urinary Bladder膀胱濕熱小腹痛

    Lower abdominal fullness, less urine than usual, dark colored urine, hematuria, burning pain when urinating, urinary tract obstruction when the condition is worse, pulling pain in the suprapubic region, thirst, and constipation are the symptoms.

    Tongue: red body, yellow coating.

    Pulse: rapid or thin rapid.

2. Obstruction in Bladder膀胱阻滯小腹痛

    This is the condition caused by the blockage from Damp-Heat in Urinary Bladder膀胱濕熱小腹痛.

    Stagnation of Qi氣滯阻滯: lower abdomen distended pain, aggravated after urination, chest distension, with hypochondriac pain. Wiry pulse.

    Blood stasis瘀血阻滯: lower abdominal severe pain and hematuria.

    Stone obstruction砂石阻滯: lower abdominal severe pain that radiates to the groin and hematuria. The pain is less if there is sand or stone visible in the urine.

3. Deficiency and Cold in Lower Jiao下焦虛寒小腹痛

    Intermittent lower abdominal dull pain, cold abdomen (or icy cold when it is worsened), it is not warm, even in summer. It is aggravated by coldness and alleviated by warmth. Cold limb, pale lips but red in the mouth, with profuse clear urine. Pale tongue body with white coating, and deep and thin pulse.

Both Left and Right Iliac Region Disorders

1. Cold Stagnation in Liver Meridian寒滯肝脈少腹痛

   Pulling pain in the iliac region and the testis, it is a distended pain that is aggravated by cold and relieved with the application of heat. Patients often present with a pale face and cold limbs.

   Tongue: white slippery coating.

   Pulse: deep, wiry, or slow.

2. Liver Qi Stagnation 肝氣鬱結少腹痛

   Intermittent pain radiates to the groin and the testis, and it is aggravated by emotional upset. Pressing on the iliac regions and the umbilical region exacerbates the pain. It is accompanied by hypochondriac pain, chest fullness, diarrhea, irritability, and short temper.

   Tongue: thin white coating.

   Pulse: wiry or deep.

3. Damp-Heat in Large Intestine大腸濕熱少腹痛

   Pain in the iliac region, rectal tenesmus, diarrhea with mucus and blood, thirst and desire to drink.

   Tongue: red body, greasy, yellow coating.

   Pulse: slippery and rapid.

4. Deficiency and Cold in Lower Jiao下焦虛寒少腹痛

   Constant dull pain, which is more in the left iliac abdomen, pale face, fatigue, lassitude, cold limbs, aversion to cold, vomiting, and diarrhea are the symptoms.

   Tongue: thin, white coating.

   Pulse: wiry, slow.

PHYSICAL EXAMINATION

When examining the abdomen, the quadrants are uncovered. The patient is asked to dress in a gown and underwear, if they wish. Because the patient is undressed, keep the room as warm as possible. The lighting should be adequate.

The patient lies supine on an exam table with the hands placed at the sides, and the head rests on a pillow without being flexed; if it is flexed, the abdominal musculature becomes tensed. The knees should be bent to relax the abdominal muscles; the therapist stands on the patient's right side.

- Medical History
  - The location of the lower abdominal pain is the first information to obtain, as the initial diagnosis can be based on the pain either at the suprapubic region or any side of the iliac regions. Abdominal pain can migrate from one to another place, for instance, appendicitis initially presents the pain in the periumbilical region then migrates to the right lower quadrant. Liver or gallbladder conditions may cause right shoulder pain, while spleen condition may cause left shoulder pain.

- The quality of the lower abdominal pain is also important information that assists in the diagnosis. The pain can be colicky, tight, waving, sharp, stabbing, or dull. Accompanying symptoms, such as fever, nausea, vomiting, loose stool, diarrhea, pain whether is alleviated by passing stool, feeling cold, aversion to cold, cold limbs and whether the pain is aggravated by cold also need to be ascertained. The quality of pain can transform from one to another, for example, small bowel obstruction can progress from colicky pain to distended pain.
- The time and duration of the lower abdominal pain. The questions include the when the first episode of pain occurred, whether the onset was an acute or gradual, the duration of each episode, whether it occurs during the day or night, and if the time of the pain is always same, such as before or after meal, before bed time or in the morning after bed, constant or intermittent. For example, small bowel obstruction typically progresses from an intermittent pain to a constant pain. For instance, in cases of a ruptured abdominal aortic aneurysm, aortic dissection, a perforated stomach ulcer, or torsion of the ovary or testis, the onset of abdominal pain is usually acute. Conversely, the onset of abdominal pain from Crohn's disease is usually gradual.
- The accompanied symptoms. Vomiting may occur in almost any abdominal condition. For example, small bowel obstruction and ovarian or testis torsion are usually accompanied with vomiting. In liver condition, the patient may vomit blood or "coffee ground emesis". Patients with prior abdominal aortic aneurysm repair may vomit blood. Diarrhea occurs in early small bowel obstruction and appendicitis. Melena (black, tarry stools) suggests upper GI bleeding, while hematochezia suggests a lower GI bleeding. The urge to defecate in a patient with acute abdominal pain suggests a serious disease, including a ruptured aneurysm in an older patient or ruptured ectopic pregnancy in women. Abdominal pain that is accompanied by cough and dyspnea suggests a non-abdominal cause of abdominal pain, rather this type of abdominal pain could be an accompanied symptom from a cardiopulmonary condition. Syncope may suggest the origin of disease is in the chest (e.g., pulmonary embolism or dissection) or abdomen (e.g., acute abdominal aneurysm or ectopic pregnancy).
- Self-care. Before meeting with the medical provider, the patient may find that the application of a heating pad on the lower abdomen alleviates pain, topical use of pain relievers (such as wintergreen peppermint oil) helps to decrease the intensity (while an ice pack may aggravate the pain), and using oral pain medication or receiving other treatments may or may not help.
- Inspection
  - Even when the pain is only in the lower abdomen, observation should begin with the entire abdomen.
  - Pay attention to the size of the abdomen. Is it flat? Distended? Enlarged? If it is bigger than normal, is it symmetrical? If it is not, is there any

distinct protrusion? Sometimes, protrusion or unsymmetrical enlarge-
ment of the abdomen may not be easily perceived; a better way to ascer-
tain is to look up toward the patient's head from their feet (Goldberg
2018).

- Pay attention to surgical scars or other skin abnormalities (Goldberg
  2018), such as distended veins, skin discoloration (e.g., Grey Turner's
  sign correlates with severe acute necrotizing pancreatitis, in which
  ecchymosis is observed and the color of the skin may be green, yellow, or purple; or there is discoloration of the flacks, depending on
  the degree of red blood cell breakdown in the abdominal wall tissues)
  (Goldberg 2018); and signs of herpes zoster blisters or caput medusa (a
  cluster of swollen veins around the navel that branch out from a central
  point, which often relates to liver disease).
- Pay attention to the patient's movement (Goldberg 2018). Patients with
  kidney stones have a hard time to finding a comfortable position while
  lying on the exam table, so they often keep moving. Contrary to this,
  patients with appendicitis are afraid to move, as moving will cause more
  peritoneal irritation and pain, so, they often stay still on the exam table.
- Observation should always note the appearance of the patient, a "chron-
  ically ill appearance" or an "acutely ill appearance". Also pay attention
  to: where the hands are pressing on the abdomen, the patient's position,
  spontaneous movements, respiratory pattern, and facial expression.
- Abdominal pain with abnormal vital signs should alert the practitioner
  to a condition with a serious underlying cause. High fever may be the
  result of infection, low blood pressure may indicate internal bleeding,
  tachypnea may be caused by a metabolic acidosis, such as diabetic keto-
  acidosis, and tachycardia may indicate an imbalance of electrolytes, a
  high fever, or high or low blood pressure and anemia. The above conditions can be life-threatening.
- Auscultation
  - Listening to the abdomen a stethoscope can detect pathologic condi-
    tions of the intestines, vessels, and organs.
  - Listening to the abdomen before touching is significant, as percussion
    of the abdomen may disturb the intestine and artificially alter the activ-
    ity and bowel sounds.
  - Before placing the stethoscope onto the abdomen, ensure that the dia-
    phragm of the stethoscope is warm (the easiest way to warm up the
    diaphragm of the stethoscope is by placing it on the hands or rubbing
    the diaphragm of the stethoscope against the front of the shirt).
  - The abdomen is divided into 4 quadrants: right upper quadrant, left
    upper quadrant, right lower quadrant, and left lower quadrant.
  - It does not matter which quadrant to start with and how long each quad-
    rant should be listened to. Clinically, listening for bowel sounds in the
    abdomen is preferred, beginning with the small bowel by placing the
    diaphragm of stethoscope on to the right side of the umbilicus. Then

proceed to all four quadrants. Listen for at least 5–10 seconds in each quadrant.

- Normal bowel sounds consist of low-pitched gurgling sounds that occur 4–5 times per minute.
- Very high-pitched gurgling sounds may indicate there is a mechanical bowel obstruction.
- Excessive bowel sounds that are more frequent than 10 times per minute may indicate partial obstruction of the bowel or diarrhea.
- An absence of bowel sounds for greater than 3–5 minutes may indicate there is no peristalsis, indicating paralytic ileus or generalized peritonitis.
- In addition to bowel sounds, listening to the abdomen with a stethoscope can also determine the presence of vascular murmurs called bruits. A bruit is an abnormal blowing sound from blood flowing through a narrowed artery.
- Abdominal bruits are associated with aortic, renal, or mesenteric stenosis and can be accessed in these areas:
  - Aorta: located above the umbilicus at the intersection of the lines between the lowest point of both ribcages and the front midline. A murmur is a whooshing sound as blood rushes through an abdominal aortic aneurysm or abdominal aortic stenosis.
  - Renal arteries: located at the intersection of the lines between the lowest point of the ribcage and the lateral edge of the rectus muscle bilaterally. A high-pitched sound is a murmur which is caused by turbulent blood flow through a vessel narrowed by atherosclerosis.
  - Iliac arteries: located at the intersection of the lines between the two anterior superior iliac spines and the lateral edge of the rectus muscle bilaterally in the iliac regions. A systolic murmur may indicate iliac artery stenosis.
  - Splenic artery: right hypochondrium for hepatic artery and left hypochondrium for splenic artery.
- Percussion

The sounds from percussion may indicate the contents in the abdomen, and they can be helpful in determining the cause of the abdominal distention. To perform percussion, place the left hand firmly against the abdominal wall, strike with the distal interphalangeal joint of the left middle finger for 2–3 times; and with the tip of the right middle finger, lift the right middle finger immediately by floppy wrist action.

There are two basic sounds to identify from the percussion procedure:
1. Drum-like sounds indicate the percussion is over air-filled structures.
2. Dull sounds indicate the percussion is over a solid structure or fluid. Solid structure can be an organ, such as liver; fluid can be ascites.
3. Pain produced by percussion indicates inflammation, such as peritonitis. It is positive if there is percussion pain of the liver, spleen, kidney, and urinary bladder.

• Palpation

Examine each quadrant separately in a counterclockwise circle from surface structures to the deeper organs. Begin in the left lower quadrant and examine all nine quadrants. Pay attention to the tension of the abdominal wall: whether there is tenderness, rebound pain, guarding, hepatosplenomegaly, fluid waves, and/or masses, such as the abdominal aorta. Tenderness in the right lower abdomen with a positive test of McBurney's sign may suggest the cause of the abdominal pain associated with appendicitis.

A rigid, tense abdominal wall with involuntary guarding is highly indicative of peritonitis.

A tender pulse sensation and a mass near the navel may also be a key distinguishing feature of an acute abdominal aortic aneurysm.

To identify the abdominal guarding, the patient flexes both knees and hips to further relax the abdomen. The tone decreasing with inspiration is voluntary guarding. If the abdominal wall tension is rigid throughout the respiratory cycle, it is true guarding (Macaluso and McNamara 2012).

The fluid wave test or fluid thrill test can be positive for ascites when there is a large amount of free fluid in the peritoneal cavity. Ask an assistant to press the umbilicus firmly with the ulnar surface of a hand, with the fingertips pointing toward the patient's toes. The examiner places a hand on the latera aspect of the patient's abdomen between the costal margin and the ilium in the anterior axillary line. The examiner then taps sharply with the fingertips on the opposite side of the patient's abdomen and feels the wave impulse transmitted through the fluid by the hand placed on the other side of the fingertip tapping. Rotate the sides by tapping sharply with the hands previously placed and feel the waves by the hand previously tapped and vice versa. The test is negative if no impulse is felt.[7]

Differential Diagnosis

Many illnesses can cause pain in the abdomen, too many to be listed here. However, the most critical conditions that need immediate medical attention and that providers need to consider referring to the emergency department are listed here.

Abdominal Aortic Aneurysm

Aortic aneurysm can happen in thoracically and abdominally. Abdominal aortic aneurysm occurs along the part of the aorta that passes through the abdomen.

A ruptured aortic aneurysm can become life-threatening if it is undetected, ignored, or left untreated. Abdominal aortic aneurysms often grow slowly without noticeable symptoms, making it difficult to detect. It causes internal bleeding once it ruptures and causes sudden death. There are some symptoms that may be early signs. There is an abdominal pain that is sudden and severe, deep and constant in the area above or beside the navel, syncope, and the urge to defecate in older patient (https://www.ncbi.nlm.nih.gov/pmc/articles/PMC3468117/). A pulsing sensation and a mass near the navel might also be a key distinguishing feature of an acute abdominal aortic aneurysm. The normal size of the abdominal aorta is about 1.5–2.5 cm. When the abdominal aorta is abnormally enlarged and exceeds 50% of the normal diameter, it is considered an abdominal aortic aneurysm.[8]

Ectopic Pregnancy

Ectopic pregnancy is a life-threatening, medical emergency condition; it is not a pregnancy that can carry to term. The egg has implanted in the fallopian tube, and the tube bursts as the fertilized egg develops, leading to severe internal bleeding and can result in maternal death.[9]

There are some symptoms that can help to detect ectopic pregnancy. The early symptoms can be like typical pregnancy. Additionally, pain in the lower abdomen may also present in the pelvis and lower back, as well as syncope. Vaginal bleeding during pregnancy is a key signal to visit the physician. The urge to defecate may be an accompanied symptom indicating a ruptured ectopic pregnancy (Macaluso and McNamara 2012).

If the fallopian tube bursts, the pain in the lower abdomen, pelvis and lower back could become intense and sharp; fainting, low blood pressure, rectal pressure, and shoulder pain are also symptoms.

Ovarian Torsion

Ovarian torsion is another medical emergency. If it is ignored or untreated, it can cut off blood supply to the ovary and result in the loss of an ovary.[10]

Ovarian torsion can cause intermittent sudden and unbearably intense pelvic and lower abdominal pain, nausea, and vomiting. The sudden pain is often preceded by occasional cramps for several days or weeks. Some patients may experience fever, vaginal bleeding, discharge, and dysuria. Ovarian torsion often occurs in one ovary, the lower abdominal pain hence is in either the left or right side. Palpation may find that the ipsilateral abdomen exhibits tenderness and guarding.

Testicular Torsion

Testicular torsion usually requires emergency surgery. If it is ignored or untreated, the blood flow to the testis may be cut off, and the testis must be removed if it is badly damaged.

Manifestations include intermittent sudden severe pain in the scrotum, a testis is at an unusual angle or position, abdominal pain, nausea and vomiting, fever, and frequent urination.[11]

Acute Appendicitis

The abdominal pain initially begins in periumbilical regions, then it migrates to the right lower quadrant a few hours later once the inflammatory process is detected by the somatic sensors of the parietal peritoneum. McBurney's point is positive (Macaluso and McNamara 2012).

TREATMENTS

Suprapubic Region Disorders

1. Damp-Heat in Urinary Bladder膀 胱 濕 熱 小 腹 痛

Treatment Principle

Clear Damp-Heat in Lower Jiao and promote urination.

Herbal Treatments

Herbal Formula

Ba Zheng San 八 正 散

This formula is from "Prescriptions of the Bureau of Taiping People's Welfare Pharmacy太 平 惠 民 和 劑 局 方", authored by Imperial Medical Bureau 太 醫 局, in southern Song dynasty南 宋 (1127–1239), and published in 1134.

Herbal Ingredients

Mu Tong木 通10g, Qu Mai瞿 麥10g, Che Qian Zi車 前 子10g, Bian Xu萹 蓄10g, Hua Shi滑 石10g, Zhi Gan Cao甘 草 炙10g, Zhi Da Huang大 黃10g, Zhi Zi 栀 子10g, Deng Xin Cao燈 心 草2g

Ingredient Explanations

- Mu Tong木 通 drains Heat and promotes urination.
- Qu Mai瞿 麥 promotes urination, drains Damp-Heat from the Bladder, and unblocks painful urinary dysfunction.
- Che Qian Zi車 前 子 clears Damp Heat and promotes urination.
- Bian Xu萹 蓄 promotes urination, drains Damp-Heat from the Bladder, and eliminates painful urination.
- Hua Shi滑 石 clears Phlegm Heat and Heat from Urinary Bladder, promotes urination, and drains Damp.
- Zhi Gan Cao甘 草 炙 tonifies the Spleen, harmonizes the other herbs, and guides the herbs to all 12 meridians.
- Zhi Da Huang大 黃 drains Damp-Heat and clears Heat through the stool.
- Zhi Zi 栀 子 clears Heat, cools the Blood, resolves Damp, reduces swelling, and opens the meridians.
- Deng Xin Cao燈 心 草 promotes urination, clears Heat from the Heart channel by directing Fire downward, and calms the Spirit.

Modifications

- For blood in urine, add Xiao Ji小 薊, Sheng Di生 地, Han Lian Cao旱 蓮 草, and Bai Mao Gen白 茅 根.
- For painful burning urination, add Huang Lian黃 連 and Sheng Di生 地.

Formula Indications

- Damp-Heat in Urinary Bladder, Blood Lin, and Heat in Heart meridian transferring to Small Intestine meridian.

Contraindications
- It is contraindicated for pregnancy.
- It is contraindicated for Lin Syndrome 淋證 underlying deficiency.
- It is contraindicated for older patients with weakness.

2. Obstruction in Bladder膀 胱 阻 滯 小 腹 痛

Stagnation of Qi氣 滯 阻 滯

Treatment Principle
Dredge the Liver and promote Qi circulation.

Herbal Treatments
Herbal Formula
Chen Xiang San 沉 香 散

This formula is from "Treatise on Diseases, Patterns, and Formulas Related to the Unification of the Three Etiologies三 因 極 一 病 証 方 論", authored by Chen Yan 陳 言 (1131–1189), in Song dynasty 南 宋 (1127–1239), and published in 1174.

Herbal Ingredients
Chen Xiang沉 香 15g, Shi Wei石 韋 15g, Hua Shi滑 石15g, Wang Bu Liu Xing王 不 留 行15g, Dang Gui炒 當 歸15g, Dong Kui Zi 炒 葵 子 23g, Bai Shao白 芍23g, Zhi Gan Cao炙 甘 草 7.5g, Chen Pi橘 皮7.5g

Administrations
Grind the herbs to a fine powder, dissolve 6g in water, and take the herbs before a meal.

Ingredient Explanations
- Chen Xiang沉 香 promotes Qi Movement for descending Qi, alleviates pain, warms the Middle Jiao, dispels Cold, and warms Kidney.
- Shi Wei石 韋 clears Damp-Heat, cools the Blood, stops bleeding, and benefits Essence.
- Hua Shi滑 石 clears Phlegm Heat and Heat from Urinary Bladder, promotes urination, and drains Damp.
- Wang Bu Liu Xing王 不 留 行 moves the Blood, promotes lactation, reduces swelling, and drains abscesses.
- Dang Gui當 歸 moves the Blood, disperses Bruising, and stops pain due to Blood stasis.
- Dong Kui Zi 冬 葵 子 promotes urination and unblocks Lin.
- Bai Shao白 芍 nourishes the Blood, soothes the Liver, and stops pain.
- Zhi Gan Cao甘 草(炙) tonifies the Spleen, harmonizes the other herbs, and guides the herbs to all 12 meridians.
- Chen Pi橘 皮 dries Damp and transforms Phlegm.

Modifications
- For chest fullness, add Qing Pi青皮, Wu Yao烏藥, and Xiao Hui Xiang小茴香.
- For Blood stasis, add Hong Hua紅花, Chi Shao赤芍, and Chuan Niu Xi川牛膝.

Formula Indications
- Liver Qi Stagnation.

Blood Stasis瘀血阻滯

Treatment Principle
Unblock urination, move the Blood, and clear Heat.

Herbal Treatments
Herbal Formula
Tao He Cheng Qi Tang 桃核承氣湯
   This formula is from "Shang Han Lun 傷寒雜病論", authored by Zhang Zhong-Jing 張仲景 (150–219), in eastern Han dynasty 東漢 (25–220), and published in 200–210.

Formula Ingredients
Tao Ren桃仁12g, Da Huang大黃12g, Gui Zhi桂枝6g, Zhi Gan Cao甘草6g, Mang Xiao芒硝6g

Ingredient Explanations
- Tao Ren桃仁 moves the Blood, disperses Bruising, and stops pain due to Blood stasis.
- Da Huang大黃 moves the Blood and disperses Bruising.
- Gui Zhi桂枝 warms the meridians and relieves pain.
- Zhi Gan Cao炙甘草 tonifies the Spleen, harmonizes the other herbs, and guides the herbs to all 12 meridians.
- Mang Xiao芒硝 clears excess Heat and reduces swelling.

Modifications
- For Qi stagnation, add Xiang Fu 香附, Wu Yao烏藥, and Qing Pi青皮.
- For urination difficulty, add Che Qian Zi車前子, Fu Ling茯苓, and Mu Tong木通.

Formula Indications
- Blood stasis with Heat in Lower Jiao.

Contraindications
- It is contraindicated for pregnancy.

Stone Obstruction砂石阻滯

Treatment Principle
Clear Heat, drain Dampness, free and promote urination, and eliminate stones.

Herbal Treatments
Herbal Formula
Pai Shi Tang 排石湯
   This formula is from "Famous Classic and Contemporary Formulas古今名方", authored by Jie Faliang 解發良, and published in 2001.

Formula Ingredients
Chai Hu柴胡15g, Huang Qin黃芩15g, Yu Jin 鬱金15g, Zhi Ke枳殼15g, Jiang Huang薑黃15g, Qing Pi青皮15g, Da Huang大黃15g, Bai Shao白芍 各15g, Shan Zha山楂10g, Chuan Lian Zi川楝子12g, Jin Qian Cao金錢草30g

Ingredient Explanations
- Chai Hu柴胡 moves Liver Qi and releases stagnation.
- Huang Qin黃芩 clears Heat, dries Damp, cools the Blood, stops bleeding, and calms Liver Yang.
- Yu Jin鬱金 moves the Blood, relieves pain, soothes the Liver, moves Qi, cools the Blood, and reduces jaundice.
- Zhi Ke枳殼 regulates Qi and removes blockage.
- Jiang Huang薑黃 moves the Blood, opens the meridians, expels Wind, and reduces swelling.
- Qing Pi青皮 disperses Liver Qi, drains Damp, and transforms Phlegm.
- Da Huang大黃 moves the Blood and disperses Bruising.
- Bai Shao白芍 nourishes the Blood, soothes the Liver, and stops pain.
- Shan Zha山楂 promotes digestion, transforms food stagnation, moves Qi, and disperses Blood.
- Chuan Lian Zi川楝子 moves Liver Qi, removes stagnation, and stops pain.
- Jin Qian Cao金錢草 promotes urination, unblocks Lin, clears Heat, disperses Bruising, and reduces swelling.

Modifications
- For severe abdominal pain, add Yan Hu Suo延胡索.
- For vomiting, add Zhu Ru竹茹 and Ban Xia半夏.
- For fever, add Jin Yin Hua金銀花, Pu Gong Ying蒲公英, and Lian Qiao連翹.
- For jaundice, add Yin Chen Hao茵陳, Zhi Zi梔子, and Long Dan Cao龍膽草.

Formula Indications
- Painful urination due to Stone Lin 石淋.

Contraindications
- The formula should not be used long-term.
- Take caution for patients with kidney failure.

3. Deficiency and Cold in Lower Jiao下 焦 虛 寒 小 腹 痛

Treatment principle
Tonify Kidney Yang for Yang deficiency; tonify Kidney Yin for Yin deficiency.

Herbal Formula for Kidney Yang deficiency
You Gui Wan 右 歸 丸

This formula is from "The Complete Compendium of Jingyue景 岳 全 書", authored by Zhang Jiebin張 介 賓(1563–1640AD), in Ming dynasty明 朝 (1368–1644), and published in 1624.

Herbal Ingredients
Shu Di Huang 熟 地 黃240g, Shan Yao 山 藥120g, Gou Qi Zi 枸 杞 子120g, Tu Si Zi 菟 絲 子120g, Du Zhong 杜 仲120g, Lu Jiao Jiao 鹿 角 膠120g, Shan Zhu Yu 山 茱 萸90g, Dang Gui 當 歸90g, Zhi Fu Zi 附 子60g, Rou Gui 肉 桂60g

Ingredient Explanations
- Shu Di Huang 熟 地 黃 nourishes the Blood and nourishes Liver and Kidney Yin.
- Shan Yao山 藥 tonifies Kidneys and benefits the Yin and the Spleen.
- Qi Gou Qi Zi枸 杞 子 nourishes the Blood and tonifies the Liver and Kidney Yin.
- Tu Si Zi菟 絲 子 tonifies Kidney Yang, tonifies Kidney and Liver Yin.
- Du Zhong杜 仲 tonifies the Kidney and Liver, strengthens the Sinews and Bones, and lowers blood pressure.
- Lu Jiao Jiao鹿 角 膠 tonifies the Kidney and nourishes and tonifies Essence Qi, and Blood.
- Shan Zhu Yu山 茱 萸 nourishes Liver and Kidney Yin and strengthens Kidney Yang.
- Dang Gui當 歸moves and tonifies the Blood, disperses Cold, and stops pain due to Blood Stasis.
- Zhi Fu Zi附 子 warms Ming Men Fire.
- Rou Gui肉 桂 warms Kidney Yang and Ming Men Fire and assists the generation of Qi and Blood.

Modifications
- For Qi deficiency, add Ren Shen人 參 , Bai Zhu白 朮 , Fu Ling茯 苓 , and Huang Qi黃 耆.
- For 5 a.m. diarrhea, add Wu Wei Zi五味子 and Rou Dou Kou肉豆蔻.
- For Cold in Spleen and Stomach, add Gan Jiang乾薑 and Wu Zhu Yu吳茱萸.
- For impotence, add Bai Ji Tian巴 戟 天 , Rou Cong Rong肉蓯蓉 , and Hai Gou Shen海狗腎.

Formula Indications
- Ming Men Fire deficiency.

Contraindications
It is contraindicated for patients who suffer from deficient Fire underlying Kidney Yin deficiency.

Both Left and Right Iliac Region Disorders
1. Cold Stagnation in Liver Meridian寒滯肝脈少腹痛

Treatment Principle
Warm Middle Jiao and Liver meridian.

Herbal Treatments
Herbal Formula
Si Ni Tang 四逆湯

This formula is from "Shang Han Lun 傷寒雜病論", authored by Zhang Zhong-Jing 張仲景 (150–219), in eastern Han dynasty 東漢 (25–220), and published in 200–210.

Herbal Ingredients
Zhi Gan Cao甘草炙6g, Gan Jiang乾薑4.5g, Zhi Fu Zi附子10g

Ingredient Explanations
- Zhi Gan Cao甘草炙 tonifies the Spleen, harmonizes the other herbs, and guides the herbs to all 12 meridians.
- Gan Jiang乾薑 warms Middle Jiao, expels interior Cold, and dispels Wind-Dampness.
- Zhi Fu Zi附子 warms Ming Men Fire.

Modifications
- For abdominal pain due to Cold, add Gui Zhi桂枝 and Bai Shao白芍.
- For cold limbs, add Xi Xin細辛 and Gui Zhi桂枝.

Formula Indications
- Cold in Shao Yin 少陰 and Kidney Yang deficiency with internal Cold.

Contraindications
- Patients with abdominal pain due to Fire should not use.
- It is contraindicated for patient's constitution with True Heat with False Cold.

2. Liver Qi Stagnation 肝氣鬱結少腹痛

Treatment Principle
Disperses Liver, regulates Qi, harmonizes Blood, and alleviates pain.

Herbal Treatments

Herbal Formula

Chai Hu Shu Gan Tang柴 胡 疏 肝 散

This formula is from "The Complete Compendium of Jingyue景 岳 全 書", authored by Zhang Jiebin張 介 賓(1111–1117AD), in Ming dynasty明 代 (1624 AD).

Herbal Ingredients

Chai Hu柴 胡6g, Xiang Fu香 附5g, Bai Shao白 芍5g, Chuan Xiong川 芎5g, Zhi Ke枳 殼5g, Chen Pi陳 皮6g, Zhi Gan Cao甘 草2g

Ingredient Explanations

- Chai Hu 柴 胡 moves Liver Qi and releases stagnation.
- Xiang Fu香 附 moves Liver Qi and Blood and stops pain due to Qi and Blood stagnation.
- Bai Shao白 芍 nourishes the Blood, soothes the Liver, and stops pain.
- Chuan Xiong川 芎 moves the Blood and moves Qi.
- Zhi Ke枳 殼 regulates Qi and removes blockage.
- Chen Pi陳 皮 regulates Qi and removes blockage.
- Zhi Gan Cao甘 草 tonifies the Spleen, harmonizes the other herbs, and guides the herbs to all 12 meridians.

Modifications

- For Blood deficiency, add Dang Gui當 歸, Sheng Di Huang生 地, and Shu Di Huang熟 地.
- For Blood stasis, add Dan Shen丹 參.

Formula Indication

- Abdominal pain due to Liver Qi stagnation and Blood stagnation.

Contraindications

- Caution for patients who suffer from Yin deficient Heat.

3. Damp-Heat in Large Intestine大 腸 濕 熱 少 腹 痛

Treatment Principle

Clears Heat, detoxicates and cools the Blood, and stops diarrhea.

Herbal Treatments

Herbal Formula

Bai Tou Weng Tang 白 頭 翁 湯

This formula is from "Shang Han Lun 傷 寒 雜 病 論", authored by Zhang Zhong-Jing 張 仲 景 (150–219), in eastern Han dynasty 東 漢 (25–220), and published in 200–210.

Herbal Ingredients

Bai Tou Weng白 頭 翁15g, Huang Bai黃 柏12g, Huang Lian 黃 連6g, Qin Pi 秦 皮12g

Ingredient Explanations

- Bai Tou Weng白 頭 翁 clears Heat, detoxicates, cools the Blood, and astringes diarrhea.
- Huang Bai黃 柏 drains Damp-Heat, especially from the Lower Jiao.
- Huang Lian 黃 連 clears Heat and drains Damp.
- Qin Pi 秦 皮 astringes diarrhea, drains Damp-Heat, and clears Fire from Liver Qi stagnation.

Modifications

- For borborygmus, add Mu Xiang木 香, Bin Lang檳 榔, and Zhi Ke枳 殼.

Formula Indications

- Damp Heat in Intestine.

Contraindications

- Do not use long-term.
- It is contraindicated for Spleen and Stomach Yang deficiency.

4. Deficiency and Cold in Lower Jiao下 焦 虛 寒 少 腹 痛

Treatment Principle

Warm Middle Jiao, tonify the deficiency, descend rebellious Qi, and stop vomiting.

Herbal Treatments

Herbal Formula

Wu Zhu Yu Tang吳 茱 萸 湯

This formula is from "Shang Han Lun傷 寒 雜 病 論", authored by Zhang Zhong-Jing張 仲 景in 200–210AD, in eastern Han dynasty東 漢 25–220AD.

Herbal Ingredients

Wu Zhu Yu吳 茱 萸9g, Ren Shen人 蔘9g, Sheng Jiang生 薑18g, Da Zao大 棗4 pieces

Ingredient Explanations

- Wu Zhu Yu吳 茱 萸 warms Middle Jiao, disperses Cold, expels Damp Cold, and alleviates pain.
- Ren Shen人 蔘 tonifies Qi and Yang.
- Sheng Jiang生 薑 unblocks the pure Yang pathway and harmonizes rebellious Qi.
- Da Zao大 棗 tonifies the Spleen Qi, nourishes the Blood, and moderates and harmonizes the harsh properties of the fragrant herbs.

Modifications

- For abdominal pain, add Bai Shao白芍.
- For persistent fever, add Chai Hu柴胡, Huang Qin黃芩, and Fang Feng防風.
- For extreme Cold, add Fu Zi附子, Gan Jiang乾薑, and Xi Xin細辛.

Formula Indications

Jue Yin headache, cold hands and feet, Cold attacking Middle Jiao, and Jue Yin Cold in Liver.

Contraindications

- It is contraindicated for patients who suffer from vomiting or acid reflux due to Heat in Stomach or Yin deficiency.
- It is contraindicated for patients who suffer from headache due to Liver Yang uprising.

Acupuncture Treatments

Hypogastrium Quadrant

Acupuncture Point Formula

Basic FormulaST36足三里Zusanli, ST25天樞Tianshu, RN12中脘Zhongwan, ST37上巨虛Shangjuxu, GB24日月Riyue, and 阿是穴 Ashi acupoints on lower abdomen.

Modifications

- If the medical providers in western medicine decide that the type of appendicitis doesn't require surgery, needling Lanwei 闌尾穴 (1 Cun 寸 superior to ST37上巨虛Shangjuxu) may be able to alleviate the pain.

  To prevent rupture, a CT scan is recommended to detect the type of appendicitis. A rupture of the appendix spreads infection throughout the abdomen, leading to peritonitis, which can be life-threatening.
- For Deficiency and Cold in Lower Jiao下焦虛寒小腹痛, add UB26關元俞Guanyuanshu.
- For Damp-Heat in Urinary Bladder膀胱濕熱小腹痛, add SP6 三陰交Sanyinjiao.
- For Stagnation of Qi氣滯阻滯 in Obstruction in Bladder膀胱阻滯小腹痛, add LIV3 太衝Taichong.
- For Blood stasis瘀血阻滯 in Obstruction in Bladder膀胱阻滯小腹痛, add SP10 血海Xuehai.
- For Stone obstruction砂石阻滯in Obstruction in Bladder膀胱阻滯小腹痛, add REN3中極Zhongji.

Both Left and Right Iliac Region Disorders

Acupuncture Point Formula

Basic Formula

ST36足三里Zusanli, ST25天樞Tianshu, RN6氣海Qihai, RN4關元Guanyuan, and 阿是穴Ashi acupoints on lower abdomen.

Modifications

- For Cold Stagnation in Liver Meridian寒滯肝脈少腹痛, add LIV5 蠡溝Ligou.
- For Deficiency of Spleen and Stomach, add moxibustion Therapy to ST36足三里Zusanli together with add UB21胃俞Weishu, UB20脾俞Pishu, and REN13上脘Shangwan, with both tonifying needling and moxibustion Therapy, plus needling REN17膻中Danzhong.
- For diarrhea underlying Spleen Qi Deficiency, add REN6氣海Qihai.
- For insomnia and irritability, add SP6三陰交Sanyinjiao and HT7神門Shenmen.
- For Liver overacting on Spleen and Stomach, add ST34梁丘Liangqiu, SP4公孫Gongsun, PC6內關Neiguan, and REN17膻中Danzhong.
- For Deficiency and Cold in Lower Jiao下焦虛寒少腹痛, add UB26關元俞Guanyuanshu.
- For Liver Qi Stagnation 肝氣鬱結少腹痛, add LIV3太衝Taichong, GB34陽陵泉Yanglingquan, PC6內關Neiguan, HT7神門Shenmen, UB17膈俞Geshu, and UBI8肝俞Ganshu.
- For Damp-Heat in Large Intestine大腸濕熱少腹痛, add ST44內庭Neiting.

Electro-Acupuncture Therapy

ST25天樞Tianshu ~ ST36足三里Zusanli
  Intensity: 2
  Frequency: 18 times/minute
  Duration: 30–60 minutes daily

Tui Na Treatments

Tui Na Techniques
Rubbing, kneading, chafing, manipulating

Tui Na Procedures

1. Patient is in the supine position, and the therapist rubs with whole palm for RN12中脘Zhongwan, ST25天樞Tianshu, RN4關元Guanyuan, and REN6氣海Qihai for 8–10 minutes. 1619897770

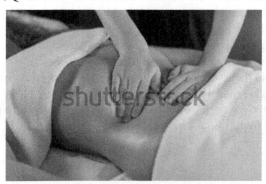

2. Applies the heel of one hand to knead RN12中 脘Zhongwan, RN4關 元Guanyuan, and REN6氣 海Qihai for 8–10 minutes.

3. The patient is in the prone position, the therapist applies a heel of hand to knead the area between UBI8肝 俞Ganshu and UB26關 元 俞Guanyuanshu for 3–5 times.

4. Applies chafing techniques on both the Urinary Bladder meridians and the sacrum region until warm.

5. Applies manipulation techniques to correct the misaligned vertebrae that are painful.

Auricular Therapy

Points: Stomach, Spleen, Sympathetic, Subcortex, Shenmen 神門, Liver, Large Intestine, and Small Intestine

Method: select 2–4 points each time for needling

Needling retention time: 30 minutes

Frequency: once daily

PATIENT ADVISORY

- Self-Managing Emotion
- Keep practicing appropriate exercise.
- Eat easy digested foods and avoid the foods that are strong or stimulating.
- Avoid foods that are cold, raw, fatty, very hard, and/or very dry.
- Avoid sleeping in a cold environment.
- Avoid long-hours of working under the hot sun.
- Do not go running right after a meal.
- Lower abdominal pain can be a serious medical condition; seek immediate medical care or call the emergency line (police line 911) if there is any below symptom
  - Sudden onset of severe sharp pain in the lower abdomen or pelvic region.
  - Pain in scrotum.
  - Rigidity of abdominal muscle.
  - Rapid heart rate.
  - Lower abdominal pain accompanied by vomiting.
  - Vaginal bleeding during pregnancy.
  - Vomiting blood, rectal bleeding, or bloody stool or blood in urine.
  - Persistent lower abdominal pain that causes concern.
  - Change in the level of consciousness or alertness and/or fainting.
  - Fever higher than 103° Fahrenheit (39.4° C).
  - Inability to pass urine even though there is an urgency to pass urine.
  - Inability to pass bowel movements.

DYSMENORRHEA 痛 經

Introduction

Dysmenorrhea is a medical term for painful menstrual periods; it can be severe and accompanied by frequent cramps during the period.

Anatomy

The dysmenorrhea-related female reproductive anatomy includes external and internal structures.

The external structures of the female reproductive system are to protect the internal genital organs from infectious organisms and to enable sperm to enter the body. The structures include labia majora, labia minora, Bartholin's glands, and clitoris.

The internal structures of the female reproductive system are to produce egg cells and host the fetus. The related structures include the uterus and fallopian tubes.

The Uterus

The uterus is connected distally to the vagina and laterally to the fallopian tubes. It includes three parts:

- Fundus: it is at the top part of the uterus above the entry point of the fallopian tubes.
- Body: it is the middle part.
- Cervix: it is the lower part of uterus, and it connects with the vagina.

The uterus's histology is divided into two structures

The fundus and the body include:

- Perimetrium: it is a double layered membrane that is continuous with the abdominal peritoneum.
- Myometrium: it is a smooth muscle layer.
- Endometrium: it is a mucous membrane in the innermost of the uterus.

The Uterus's Ligaments

- Broad ligament: it is the peritoneum attaching to the sides of the uterus to the pelvis. It is a mesentery for the uterus and its function is to keep the uterus in position.
- Round ligament: it is the extended gubernaculum from the uterine horns to the labia majora via the inguinal canal. Its function is to keep the uterus in the anteverted position.
- Ovarian ligament: it is to connect the ovaries to the uterus.
- Cardinal ligament: it is the extended ligament form the cervix to the lateral pelvic walls. Its function is to provide support to the uterus.

- Uterosacral ligament: it extends form the cervix to the sacrum. Its function is to provide support to the uterus.

The Uterus's Vascular Supply and Lymphatics

- The uterine artery and the uterine veins are the vascular supply to the uterus.
- The iliac, sacral, aortic, and inguinal lymph nodes are the lymphatic drainage of the uterus.

The Uterus's Nerve Innervation

- The uterovaginal plexus is the sympathetic nerve fibers which comprise the anterior and intermediate parts of the inferior hypogastric plexus, which enters the spinal cord via T10–T12 and L1 nerve fibers.
- Pelvic splanchnic is the parasympathetic fiber of the uterus.
- The uterovaginal plexus innervates the cervix.

The Fallopian (Uterine) Tubes

There is one bilateral to the uterus. They are the muscular tubes comprising four parts lying in the upper border of the broad ligament.

- Fimbriae: it is at the far end of the tube to capture the ovum from the surface of the ovary.
- Infundibulum: it is near the ovary connecting to the fimbriae.
- Ampulla: it is in the middle part of the tube where fertilization usually occurs.
- Isthmus: it is the narrow part connecting the ampulla to the uterine cavity.

The Fallopian Tubes' Vascular Supply and Lymphatics

The uterine and ovarian arteries are the arterial supply. The uterine and ovarian veins are the venous supply. The iliac, sacral, and aortic lymph nodes are the lymphatic drainage.

The Fallopian Tubes' Nerve Innervation

The ovarian and uterine (pelvic) plexuses provide both sympathetic and parasympathetic innervation. The sensory afferent fibers are from T11–L1.

The Cervix

Anatomically and histologically, the cervix is considered a separate anatomical structure. It is made up mostly of connective tissue and muscle. The cervix is composed of the endocervical canal and the ectocervix.

- The endocervical canal is the inner part of the cervix, and it connects to the body of the uterus.
- The ectocervix is the outer part of the cervix, and it connects to the vagina.

The functions of the cervix are to make and release mucus. The mucus becomes thick for most of the menstrual cycle, and during pregnancy, it stops sperm from

entering the uterus and protects the uterus and the upper female reproductive struc
tures from harmful bacteria. The mucus becomes thin during ovulation, when the
mature egg is released each month, and the thin mucus allows sperm to pass through
the cervix into the uterus.

The Cervix's Vascular Supply and Lymphatics

- The uterine artery and the uterine veins are the vascular supply to the
  cervix.
- The iliac, sacral, aortic, and inguinal lymph nodes are the lymphatic drainage of the cervix.

The Cervix's Nerve Innervation

The uterovaginal plexus innervates the cervix.

WESTERN MEDICINE ETIOLOGY AND PATHOLOGY

- In a normal menstrual cycle, the follicle-stimulating hormone promotes the
  growth of ovarian follicles in the ovary, and the hormone estrogen rises
  in the follicular phase. This causes the endometrium to grow and thicken,
  which takes place from days 6 to 14 in a 28-day menstrual cycle.
- The increase of luteinizing hormone triggers the release of the egg – this
  event is called ovulation. This takes place at around day 14 in a 28-day
  menstrual cycle.
- The egg travels from the ovary through the fallopian tubes toward the uterus.
  At the same time, the hormone progesterone prepares the endometrium for
  pregnancy. If the egg becomes fertilized by a sperm, it attaches itself to the
  uterine wall. If the egg is not fertilized the pregnancy does not occur, and
  the estrogen and progesterone levels drop. Then the endometrium is shed
  out through the vagina, marking the beginning of the menstrual period. The
  first day of bleeding is day one in a 28-day menstrual cycle, and it typically
  lasts from day 1 to day 5.

Lower abdominal pain associated with menstruation is called dysmenorrhea which
has two types of dysmenorrhea: Primary and Secondary

Primary Dysmenorrhea

It is the painful menstrual period that is recurrent and is not caused by other medical
conditions.

Uterine contractions are the cause of pain, and pain associated with primary
dysmenorrhea is caused by hypersecretion of uterine prostaglandins. Women with
dysmenorrhea have higher levels of prostaglandins, especially during the first two
days of menses. Prostaglandin production is controlled by progesterone; when progesterone levels drop (when the egg is not fertilized and there is no pregnancy),
prostaglandin levels increase.

Prostaglandin triggers the uterus muscle to contract. The contraction constricts
the blood supply to the endometrium, and it also blocks the delivery of oxygen to the

tissue of the endometrium. When the tissue lacks blood and oxygen, it breaks down and dies. The contraction squeezes and expels the dead endometrium tissue through the cervix while it is shedding.

Secondary Dysmenorrhea

It is the painful menstrual period that is caused by a medical condition or an infection.

- Endometriosis: the normal endometrium is in the inside of the uterus. Endometriosis is the condition of the endometrium growing outside of the uterus. Because it bleeds during the period, it can cause swelling, scarring, and pain in the lower abdomen.[12]
- Pelvic inflammatory disease (PID): an infection caused by bacteria (e.g., chlamydia and gonorrhea) that spreads from the vagina to the uterus and can spread to fallopian tubes or ovaries. PID can cause pain in the upper and lower abdomen or pain during urination and sex.
- Adenomyosis: a condition where the endometrial tissue grows into the muscular wall of the uterus. The tissue continues to behave the same as the lining cells of the uterus – thickening, breaking down, and bleeding – but because they are trapped in a muscle layer, they form little pockets of blood within the uterine muscle wall that can cause the uterus to get much bigger than it should be – doubling or tripling in size.[13]
- Fibroids: these are benign muscular tumors that grow in the wall of the uterus, and they are almost always benign. Fibroids range in size from small, like an apple seed, to as big as a grapefruit.[14]
- Ectopic pregnancy: this is a life-threatening medical emergency, and it is not a pregnancy that can carry to term. The egg has implanted in the fallopian tube, and the tube bursts as the fertilized egg develops, leading to severe internal bleeding and can result in maternal death.

CHINESE MEDICINE ETIOLOGY AND PATHOLOGY

Excessive Types

1. *Qi and Blood Stagnation* 氣滯血瘀

It is due to chronic stress and anger. An extended period of anger, stress, worry, or emotional frustration leads to Liver Qi stagnation. Qi is the commander of Blood, stagnation of Qi disturbs the circulation of Blood, causing Blood stasis in the uterus and obstruction in both Chong 衝 脈and Ren任 脈meridians, resulting in dysmenorrhea.

2. *Cold-Damp Obstruction*寒濕凝滯

When Blood contacts Cold, it loses thermal energy, it undergoes freezing, changing state from a liquid to a solid. Pain occurs when Blood circulation is sluggish.

There are two possible causes:

- Long-term exposure to wet, damp environments, such as living in a basement, can lead to Cold and Damp accumulation in the body, resulting in injury to the body Yang Qi.
- Prior to menstruation, swimming, exposure to rainy climates, or excessive consumption of cold and raw foods (such as ice cream, salads, ice water, water melon) leads to Cold and Damp accumulation in the uterus.

3. *Damp-Heat Obstruction* 濕熱鬱結

Poor feminine hygiene can lead to many issues, such as fungal infections and urinary tract infections. The term "infection" in western medicine is "Damp-Heat invasion" in Chinese medicine. The invasion of Damp-Heat stays in uterus, struggles with the Blood in the uterus, and lodges in Chong 衝 脈 and Ren 任 脈 meridians, resulting in dysmenorrhea. Another cause is that Damp-Heat develops from consuming excessive amounts of rich, fatty, spicy, and sweet foods during menstruation or after pregnancy, leading to an accumulation of Damp-Heat in the uterus. These two factors lead to impedance of Qi and Blood flow, resulting in stagnation of Qi and Blood.

Deficient Types

1. *Qi and Blood Deficiency* 氣血虛弱

This refers to Qi and Blood deficiency in general or after a severe disease, in which the deficiency leads to insufficiency of Blood in Chong 衝 脈 and Ren 任 脈 meridians.

2. *Liver and Kidney Deficiency* 肝腎虧損

Constitutional weakness in both Liver and Kidney and multiple childbirths in women lead to depletion of both Liver and Kidney. The uterus and both Chong 衝 脈 and Ren 任 脈 meridians become more insufficient by the loss of Blood during menstruation.

MANIFESTATIONS

In Western Medicine
Primary Dysmenorrhea
The pain usually occurs one or two days prior to the period, although bleeding may begin prior to menstrual pain. The pain is defined as cramping pain in the lower abdomen ranging from mild to severe, lasting one to two days when the flow of menstruation becomes heaviest. The pain usually is cramping, sharp, intermittent spasms. The location of pain is usually at the center in the suprapubic area, but it may also radiate to the lower back or the backs of the legs. It may be accompanied by nausea and vomiting, fatigue, and diarrhea.

Primary dysmenorrhea usually presents during adolescence within the first few years of menarche.

Secondary Dysmenorrhea

It is the result of specific pelvic pathology. Endometriosis is the most common cause of secondary dysmenorrhea.

Endometriosis has symptoms that are similar to primary dysmenorrhea, such as lower abdominal cramping, pain radiating to the lower back and legs, painful sex, and painful bowel movements. However, endometriosis may cause infertility, and many patients with endometriosis may have no symptoms. Laparoscopy with biopsy can confirm the diagnosis.

Ectopic Pregnancy

There are some symptoms that can help to detect ectopic pregnancy. Although ectopic pregnancy can start out just like typical pregnancy, pain in the lower abdomen, pelvis, lower back, and syncope can be early symptoms. If vaginal bleeding occurs during pregnancy it is a key signal to visit the physician. The urge to defecate may be an accompanied symptom, indicating a ruptured ectopic pregnancy (Macaluso and McNamara 2012).

If the fallopian tube bursts, the pain in the lower abdomen, pelvis and lower back may become intense and sharp, with fainting, low blood pressure, rectal pressure, and shoulder pain as other symptoms.

In Chinese Medicine

Excessive Types

1. Qi and Blood Stagnation 氣 滯 血 瘀

Symptoms of Qi and Blood Stagnation present as abdominal pain before or during menstruation. The pain is distending or stabbing, and the abdomen cannot tolerate pressure. Emotional stress, depression, and irritability, premenstrual tension, irregular menstruation, and scanty menstruation are accompanied symptoms. The menstrual blood color is dark with blood clots, the abdominal pain is relieved after passing clots, the period is light, the chest feels full, and the breasts are tender.

Tongue: purple, ecchymoses, and white, greasy coating.

Pulse: deep, wiry or deep, choppy.

Qi Stagnation – more distention than pain, abdominal bloating, chest fullness, breast tenderness, irritability, and short temper.

Tongue: normal color or purple body, thin, white coating.

Pulse: wiry.

Blood Stasis – more stabbing pain than distention, pain in hypogastric region is worsened with pressure, a dragging feeling in the lower abdomen before bleeding, blood clots, the pain is relieved after passing clots.

Tongue: purple body with ecchymoses on the edges, thin, white coating.
Pulse: deep, choppy or deep and wiry.

2. Cold-Damp Obstruction寒濕凝滯

Cold sensation and cramping or colicky pain in hypogastric toward the end of or after the menstrual flow, no tolerance for pressure, pain is relieved by using applications of heat, and scanty or profuse menstruation. The menstrual blood color is bright red or pink with dark clots (the clots are smaller than the ones in Blood Stasis), sore lower back, aversion to cold, and loose stool.
Tongue: pink on the edges, white, sticky coating, with teeth marks.
Pulse: deep, tight.

3. Damp-Heat Obstruction 濕熱鬱結

Burning pain in the hypogastric region prior to bleeding, no tolerance for pressure, distention in the lumbar and sacrum regions, intermittent low-grade fever. Some patients may suffer from intermittent hypogastric pain between periods, and it worsens with menstruation. Thick and sticky dark red menstrual blood, clotted, viscid and strong, malodorous vaginal discharge; and scanty, dark urine.
Tongue: red body, yellow coating.
Pulse; wiry, rapid or slippery, rapid.

Deficient Types

1. *Qi and Blood Deficiency*氣血虛弱

Dull pain after menstruation in hypogastric region, it is relieved with pressure, a dragging feeling in the lower abdomen after bleeding. Lower back sore and weak, pale face, scanty menstruation, pale red color blood. Pale complexion, dizziness, loose stools.
Tongue: pale body, teeth marks, thin coating.
Pulse: thin.

2. Liver and Kidney Deficiency肝腎虧損

Thin, scanty menstruation, with pale, red blood. Dull hypogastric pain that is relieved by pressure. Sore lower back and knees, exhaustion, dizziness, tinnitus, and blurred vision when Kidney Qi deficiency is prominent.
Tongue: pale body and thin or no coating.
Pulse: deep, thin.

PHYSICAL EXAMINATION

The physical examination for primary dysmenorrhea can be completely normal. The diagnosis can be made based on the history, and conditions such as the timing, location, pressure preference, character of the pain, and the temperature can affect the outcome of the pain. The physical examination should focus on excluding secondary dysmenorrhea, such as a tumor. With a typical history and a lack of abnormal findings on the routine pelvic examination, further diagnostic evaluation is not needed, but a urine test should be ordered to rule out pregnancy or infection.

- A detailed sexual history such as previous sexually transmitted diseases (STDs), number of sexual partners, and unprotected sex events may lead to the diagnosis of pelvic inflammatory disease.
- The dysmenorrhea that does not respond to a course of NSAID (e.g., ibuprofen, naproxen) therapy may rule out the primary dysmenorrhea, further diagnostic evaluation is needed to consider the second dysmenorrhea (Coco 1999).
- The time of onset can be helpful to understand the types of dysmenorrhea. Pain prior to or during the bleeding usually indicates the excessive types. Pain after the menstruation is often the deficient types.
- Whether the pain is alleviated by pressure or not is also helpful in identifying the types of the dysmenorrhea. If pressure on the lower abdomen is tolerated by the patient and provides some degree of pain relief, it indicates the deficient types. If the patient cannot tolerate pressure on the lower abdomen and it worsens the pain, it is the excessive types.
- If application of heating pad helps to ease some degree of pain, it usually indicates the Cold or Yang Deficiency types of dysmenorrhea.
- If the pain is alleviated after passing blood clots, it indicates Blood Stasis.
- If the pain is stabbing, it is Blood Stasis.
- If the pain is distension, it is Qi Stagnation.
- If the pain is burning, it is Blood Heat or Yin Deficiency.
- If the pain is cramping or colicky, it is Cold-Damp Obstruction.
- If there is a dragging down sensation before menstruation, it is Blood Stasis.
- If there is a dragging down sensation after menstruation, it is Qi and Blood Deficiency.
- If the pain is located at the hypogastric region before bleeding, it is Blood Stasis or Cold-Damp Obstruction.
- If the pain is located at the hypogastric region after bleeding, it is Qi and Blood Deficiency.
- If the pain is located at the hypogastric region during or after bleeding, pain in the lower back or sacrum, sore or weak knees, it is Kidney Qi Deficiency.
- If the pain is located at the iliac regions, it is Liver Qi Stagnation.
- If the color of the menstrual blood is dark and clotted, it is Blood Stasis.
- If the color of the menstrual blood is light red with small blood clots, it is Cold-Damp Obstruction.
- If the bleeding is heavy, fresh red color, big-size dark blood clots, and the period is one week or more early, with irritability, dry stool, red tongue, red lips, strong pulse, and it has been consistent longer than three months, it is Blood Heat.

- If the bleeding is heavy, the period is one or more weeks early or there is one short bleeding in between two cycles, and there is no sign of Heat, weak pulse, low blood pressure, dry stool, or weak digestion, it is Blood Deficiency or both Qi and Blood Deficiency.
- If the bleeding is shorter than three days, and the period is one or more weeks early, with dry stool, insomnia, skinny body size, dry skin, dry vagina, painful sex, hot palms, small tongue size and thin tongue body, and red tongue tip, it is Yin Deficiency.
- If the period is at least one week late (longer than 35 days) for over three months (or cycles), the body may be obese, it is Kidney Essence Deficiency.
- If the period is at least one week late (longer than 35 days) for over three months (or cycles), the body may be obese, with acne on skin, dry stool, or constipation, it is Damp Heat type stagnation.
- If the period is at least one week late, scanty bleeding, skinny body size, poor appetite, low blood pressure, red tongue tip, it is Yin Deficiency.
- If the period is at least one week late, with loose stool, poor appetite, but the body size is normal, it is Qi Deficiency.
- If the period is at least one week late or amenorrhea, and there is no other symptom, this type of late period is not due to Qi Deficiency, rather it is due to malnutrition from fasting or limiting the amount and types of food to get slim.
- If the period is at least one week late, with heavy bleeding and profuse amounts of dark blood clots, it is Blood Stasis。
  - Palpation
- Rectovaginal examination for endometriosis may reveal a fixed or retroverted uterus, reduced uterine mobility, adnexal masses, or uterosacral nodularity.
- Pelvic examination for PID may have abnormal findings including consisting of cervical motion tenderness, uterine tenderness, and/or adnexal tenderness.
- Pelvic examination for adenomyosis may have abnormal findings, including enlarged, tender, boggy uterus.

DIFFERENTIAL DIAGNOSIS

Differentiation is necessary to distinguish some commonly seen secondary dysmenorrhea from primary dysmenorrhea.

Endometriosis

The common symptoms besides lower abdominal pain and dysmenorrhea include dyspareunia and infertility in young women. It doesn't respond to NSAID treatment.

Treatments

Treatment with Chinese medicine is effective for primary dysmenorrhea. The treatments introduced here are only for primary dysmenorrhea. Treatments for the secondary dysmenorrhea will be discussed separately accordingly to the diseases.

Excessive Types

1. Qi and Blood Stagnation 氣滯血瘀

Treatment Principle

Regulate Qi, move the Blood, disperse Bruising, and stop pain.

Herbal Treatments

Herbal Formula

Ge Xia Zhu Yu Tang 膈下逐瘀湯

This formula is from "Correction of Errors in Medical Classics醫林改錯", authored by Wang Qingren 王清任 (1768–1831), in Qing dynasty 清朝 (1636–1912), and published in 1849.

Herbal Ingredients

Chao Wu Ling Zhi靈 脂6g, Dang Gui當 歸9g, Chuan Xiong川 芎6g, Tao Ren桃 仁9g, Mu Dan Pi丹 皮6g, Chi Shao赤 芍6g, Wu Yao烏 藥6g, Yan Hu Suo元 胡3g, Gan Cao甘 草9g, Xiang Fu香 附4.5g, Hong Hua紅 花9g, Zhi Ke枳 殼4.5g

Ingredient Explanations

- Chao Wu Ling Zhi炒 靈脂 moves Qi and Blood, stops bleeding, has a unique ability to treat continuous uterine bleeding, and stops pain due to Qi and Blood stagnation.
- Dang Gui當 歸 nourishes the Blood and benefits Liver and regulates menstruation.
- Chuan Xiong川 芎 moves the Blood, disperses Bruising, and stops pain due to Blood stasis.
- Tao Ren桃 仁 moves the Blood, disperses Bruising, and stops pain due to Blood stasis.
- Mu Dan Pi丹 皮 clears Heat, cools the Blood, and moves the Blood.
- Chi Shao赤 芍 moves the Blood, disperses Bruising, and stops pain.
- Wu Yao烏 藥 moves Qi and stops pain.
- Yan Hu Suo元 胡 moves the Blood, promotes Qi circulation, and stops pain.
- Gan Cao甘 草 harmonizes the herbs, tonifies Middle Jiao Qi, clears Heat, expectorates phlegm, and stops cough.
- Xiang Fu香 附 moves Liver Qi and Blood and stops pain due to Qi and Blood stagnation.
- Hong Hua紅 花 moves the Blood, disperses Bruising, and stops pain due to Blood stasis.
- Zhi Ke枳 殼 regulates Qi and removes blockage.

Modifications

- For Qi deficiency, add Dang Shen黨 參.
- For dry stool, add Da Huang大 黃.
- For Blood deficiency, add Bai Shao白 芍 and Dan Shen丹 參.
- For dysmenorrhea, add Yi Mu Cao益 母 草 and Dan Shen丹 參.

Formula Indications

- Liver Qi Stagnation with Blood stasis in Lower Jiao.

Contraindications

- Contraindicated for patients with weak constitution without stasis of Blood.
- It is contraindicated for pregnancy.

2. Cold-Damp Obstruction寒濕凝滯

Treatment Principle

Disperse Cold, drain Damp, move the Blood, and alleviate pain.

Herbal Treatments

Herbal Formula

Wu Zhu Yu Tang吳茱萸湯

This formula is from "Shang Han Lun傷寒雜病論", authored by Zhang Zhong-Jing張仲景in 200–210AD, in eastern Han dynasty東漢 25–220 AD.

Herbal Ingredients

Wu Zhu Yu吳茱萸9g, Ren Shen人蔘9g, Sheng Jiang生薑18g, Da Zao大棗4 pieces

Ingredient Explanations

- Wu Zhu Yu吳茱萸 warms Middle Jiao, disperses Cold, expels Damp Cold, and alleviates pain.
- Ren Shen人蔘 tonifies Qi and Yang.
- Sheng Jiang生薑 unblocks the pure Yang pathway and harmonizes rebellious Qi.
- Da Zao大棗 tonifies the Spleen Qi, nourishes the Blood, and moderates and harmonizes the harsh properties of the fragrant herbs.

Modifications

- For abdominal pain, add Shao Yao Gan Cao Tang芍藥甘草湯.
- For extreme Cold, add Fu Zi附子, Gan Jiang乾薑, and Xi Xin細辛.

Formula Indications

Jue Yin headache, cold hands and feet, Cold attacking Middle Jiao, and Jue Yin Cold in the Liver.

Contraindications

• It is contraindicated for patients who suffer from vomiting or acid reflux due to Heat in Stomach or Yin deficiency.

- It is contraindicated for patients who suffer from headache due to Liver Yang uprising.

3. Damp-Heat Obstruction 濕熱鬱結

Treatment Principle

Soothe Liver, release stagnation, clear Heat, and cool the Blood.

Herbal Treatments

Herbal Formula

Dan Zhi Xiao Yao San 丹梔逍遙散

This formula is from "Prescriptions of the Bureau of Taiping People's Welfare Pharmacy 太平惠民和劑局方", authored by Imperial Medical Bureau 太醫局, in southern Song dynasty南宋 (1127–1239), and published in 1134.

Herbal Ingredients

Chai Hu柴胡10g, Dang Gui當歸10g, Bai Shao白芍10g, Bai Zhu白朮10g, Fu Ling茯苓10g, Zhi Gan Cao甘草5g, Wei Jiang煨薑10g, Bo He 薄荷5g, Mu Dan Pi 牡丹皮6g, Zhi Zi 梔子6g

Ingredient Explanations

- Chai Hu 柴胡 relieves Liver Qi Stagnation, pacifies the Liver, relieves the Shao Yang, and reduces fever.
- Dang Gui 當歸 nourishes the Blood, regulates menses, and invigorates and harmonizes the Blood.
- Bai Shao 白芍 calms Liver Yang, alleviates pain, nourishes the Blood, and regulates menstruation.
- Bai Zhu 白朮 tonifies the Spleen and tonifies Qi.
- Fu Ling 茯苓 strengthens the Spleen and harmonizes the Middle Jiao.
- Zhi Gan Cao 甘草 tonifies Spleen Qi, stops pain, and harmonizes the other herbs.
- Bo He 薄荷 relieves Liver Qi Stagnation, disperses stagnant Heat, and enhances Chai Hu's ability to relieve the Liver.
- Wei Jiang煨薑 harmonizes and prevents rebellious Qi and normalizes the flow of Qi at the center. 入脾胃，重在溫胃止呕。
- Mu Dan Pi 牡丹皮 clears Heat, cools the Blood, and moves the Blood.
- Zhi Zi 梔子 clears Heat, cools the Blood, resolves Damp, reduces swelling, and opens the meridians.

Modifications

- For painful urination, add Che Qian Zi車前子.
- For excessive Heat, add Sheng Di Huang生地 and Di Gu Pi地骨皮.

Formula Indications

- Liver Qi Stagnation turning to Heat, with underlying Blood and Spleen Qi deficiency.

Contraindications

- Contraindicated for weak constitution with Cold.
- It is contraindicated for pregnancy.

Deficient Types

1. Qi and Blood Deficiency氣血虛弱

Treatment Principle
Tonify Qi and Blood, regulate menstruation, and stop pain.

Herbal Treatments
Herbal Formula
Shen Yu Tang聖愈湯
This formula is from "Imperially Commissioned Golden Mirror of the Orthodox Lineage of Medicine醫宗金鑑", authored by the Qianlong emperor 乾隆 (1711–1799), in Qing dynasty 清朝 (1636–1912), and published in 1742.

Herbal Ingredients
Chuan Xiong川芎2.5g, Shu Di Huang熟地黃5g, Bai Shao白芍5g, Ren Shen人參5g, Huang Qi黃耆5g, Dang Gui當歸2.5g

Ingredient Explanations
- Chuan Xiong川芎 moves the Blood and moves Qi.
- Shu Di Huang熟地黃 tonifies the Liver and Kidney, strengthens Bone, and benefits Essence and Blood.
- Bai Shao白芍 nourishes the Blood, soothes the Liver, and stops pain.
- Ren Shen人參 tonifies Qi and Yang.
- Huang Qi黃耆 tonifies Qi, strengthens the Spleen, raises the Yang Qi of the Spleen and Stomach, tonifies Wei Qi, stabilizes the exterior, and tonifies the Blood.
- Dang Gui當歸 nourishes the Blood, benefits the Liver, and regulates menstruation.

Modifications
- For excessive Cold, add Rou Gui肉桂 and Zhi Fu Zi附子.
- For excessive Heat, add Huang Qin黃芩 and Huang Lian黃連.
- For promoting digestion, add Chen Pi陳皮.

Formula Indications
- Qi and Blood Deficiency with lower abdominal aching.

2. Liver and Kidney Deficiency肝腎虧損

Treatment Principle
Nourish the Yin, soothes the Liver.

Herbal Formula
Yi Guan Jian一貫煎

This formula is from "Supplement to 'Classified Case Records of Famous Physicians 續 名 醫 類 案", authored by Wei Zhixiu 魏 之 琇 (1722–1772), in Qing dynasty 清 朝 (1636–1912), and published in 1770.

Herbal Ingredients

Sha Shen沙 參10g, Mai Men Dong麥 冬10g, Dang Gui當 歸10g, Sheng Di Huang生 地 黃30g, Gou Qi Zi枸 杞 子15g, Chuan Lian Zi川 棟 子5g

Ingredient Explanations

- Sha Shen沙 參 nourishes Stomach and Lung Yin.
- Mai Men Dong麥 冬 moistens Lungs and nourishes the Yin.
- Dang Gui當歸 nourishes the Blood, benefits Liver, and regulates menstruation.
- Sheng Di Huang生 地 黃 nourishes Yin and clears Heat.
- Gou Qi Zi枸 杞 子 tonifies the Liver and Kidney, strengthens Bone, and benefits Essence and Blood.
- Chuan Lian Zi川 棟 子 moves Liver Qi, removes stagnation, and stops pain.

Modifications

- For making the formula stronger, add Shao Yao Gan Cao Tang芍 藥 甘 草 湯.
- For bitter taste in mouth, add Huang Lian黃 連, Huang Qin黃 芩, Tian Hua Fen天 花 粉, and Chai Hu柴 胡.
- For insomnia, add Bai Zi Ren柏 子 仁, Suan Zao Ren酸 棗 仁 and Wu Wei Zi五 味 子.
- For thirst and irritability, add Zhi Mu知 母 and Shi Gao石 膏.
- For weak legs and knees, add Niu Xi牛 膝, Yi Yi Ren薏 苡 仁, and Mu Gua木 瓜.

Formula Actions

- Nourish Yin, and soothes the Liver.

Formula Indication

- Pain in ribs and abdomen due to deficiency of Kidney and Liver, Liver Qi stagnation; dry mouth; red tongue with less coating, thin and rapid pulse, or weak wiry pulse.

Contraindications

- Contraindicated for patients who suffer from emotional stress in excess condition.
- The formula contains many herbs that are sweet; patients who are weak in digestion should take caution.

Acupuncture Treatments

Acupuncture Point Formula
Basic Formula

SP6三 陰 交Sanyinjiao, REN3中 極Zhongji, REN4 關 元Guanyuan, ST36足 三 里Zusanli

Modifications

Excessive Types

- For Qi and Blood Stagnation 氣 滯 血 瘀, add LIV3太 衝Taichong, REN6氣 海Qihai, ST25天 樞Tianshu, and GB34陽 陵 泉Yanglingquan.
- For Cold-Damp Obstruction寒 濕 凝 滯, add SP8地 機Diji, REN6氣 海Qihai, UB32次 髎Ciliao, DU4命 門Mingmen, and add moxibustion Therapy to the needles.
- For Damp-Heat Obstruction 濕 熱 鬱 結, add LIV2行 間Xingjian, SP9陰 陵 泉Yinlingquan, and ST40豐 隆Fenglong.

Deficient Types

- For Qi and Blood Deficiency氣 血 虛 弱, add REN6氣 海Qihai, SP10血 海Xuehai, KD3太 溪Taixi, and GB39懸 鐘Xuanzhong.
- For Liver and Kidney Deficiency肝 腎 虧 損, add UB23腎 俞Shenshu, DU4命 門Mingmen, REN6氣 海Qihai, KD3太 溪Taixi, and GB39懸 鐘Xuanzhong.

Auricular Therapy

Uterus, Sympathetic, Kidney, Liver, Endocrine, Shen Men

Moxibustion Therapy

- Mix and heat up salt, fine white onion, and fine ginger.
- Put them on the navel while they are warm.
- Heat the ingredients and reapply on the navel for a total of five times.
- Repeat the application five times as a set, 2–3 sets daily.

Tui Na Treatments

Tui Na Techniques

Pressing, kneading, rubbing, pushing, grasping, striking, percussing

Tui Na Procedures

1. The patient is in the prone position with both legs straightened, and the therapist stands on the patient's left side.
   1. Apply thumb pressing and kneading on Shi Qi Zhui Xia十 七 椎 下 (it is an extra acupuncture point located below the spinous process of the fifth lumbar vertebra). The kneading and pressing is cooperated with the patient's breathing, while the patient is exhaling, the pressure on the point is more, then release the pressure while inhaling. The direction of the pressure is vertical. The kneading is in a clockwise circle.
   2. Apply rubbing on "Ba Liao" acupuncture points八 髎 穴 (UB31–UB34) until warm.

2. The patient is in the supine position with both legs straightened, and the therapist stands on the patient's right side.
    1. Applies pushing with thumb on the midline from the sternum on the chest to the navel. The force is light.
    2. Overlaps the left hand on the top of the right hand and puts the right palm on the navel.
    3. Kneads the navel with the right palm in clockwise circles.
    4. Kneads with a thumb tip on one side of the lower abdomen while the middle fingertip kneads on the other side, the acupuncture point in the kneading is Zigongxue 子宫穴 (it is an extra acupuncture point located three Cun 寸 lateral to REN3中 极Zhongji) for 1–3 minutes.
    5. Grasps with the thumb and the middle fingertips on Zigongxue 子宫穴.
    6. Grasps and kneads with the thumb and the middle fingertips on SP10血 海Xuehai and ST34梁 丘Liangqiu.
    7. Presses with both thumb tips on SP9陰 陵 泉Yinlingquan, SP6三 陰 交Sanyinjiao, and LIV3太 衝Taichong for 1–3 minutes each point.
    8. Rubs the medial aspect of the lower leg between the knee and the medial malleolus until the skin is red or warm.
    9. Strikes with the palmar side of an empty fist on the medial aspect of the lower leg between the knee and the medial malleolus until the skin is red or warm.
    10. Percusses with the fingertips on the medial aspect of the lower leg between the knee and the medial malleolus until the skin is red or warm.

The patient is in sitting position, and the therapist stands behind the patient.
    1. Rubs both iliac regions with both palms for 1–3 minutes.

PATIENT ADVISORY

1. During menstruation, avoid overworking.

ALWAYS KEEP THE ABDOMEN, LOWER BACK, AND BOTH FEET WARM.

3. Avoid taking cold shower and avoid swimming in cold water.
4. Applying a heating pad over the lower abdomen may ease the pain.
5. Before menstruation, avoid catching cold.
6. Avoid consuming cold, raw foods.
7. Eat three meals on time and eat the food while it is still warm.
8. Lack of sleep or going to bed later than midnight can be a cause of dysmenorrhea; always go to sleep before 11 p.m. and get eight hours of sleep each night.

9. Appropriate exercise relaxes the mind and promotes the circulation of Qi and Blood flows, which can ease the pain.
10. Be happy and do pleasurable activities to ease the stress.

Acupuncture treatments are effective for primary dysmenorrhea, the time to begin the treatment is best 3–5 days daily prior to the bleeding until the end of the menstruation. The treatment duration usually is 2–3 cycles.

NOTES

1. https://www.healthgrades.com/right-care/symptoms-and-conditions/epigastric-pain.
2. https://www.ezmedlearning.com/blog/upper-abdominal-pain-causes-anatomy.
3. https://www.sciencedaily.com/releases/2015/05/150519105856.htm.
4. https://www.hopkinsmedicine.org/health/wellness-and-prevention/the-brain-gut-connection.
5. https://www.scientificamerican.com/article/gut-second-brain/?print=true.
6. https://www.differencebetween.com/difference-between-psychosocial-and-vs-psychological/.
7. https://www.ebmconsult.com/articles/physical-exam-fluid-wave-ascites.
8. https://kknews.cc/zh-my/health/6rr24qv.html.
9. https://my.clevelandclinic.org/health/diseases/9687-ectopic-pregnancy.
10. https://www.healthline.com/health/womens-health/ovarian-torsion.
11. https://www.mayoclinic.org/diseases-conditions/testicular-torsion/symptoms-causes/syc-20378270.
12. https://my.clevelandclinic.org/health/diseases/4148-dysmenorrhea.
13. https://www.jeanhailes.org.au/health-a-z/vulva-vagina-ovaries-uterus/adenomyosis.
14. https://www.womenshealth.gov/a-z-topics/uterine-fibroids#:~:text=What%20are%20fibroids%3F-,Fibroids%20are%20muscular%20tumors%20that%20grow%20in%20the%20wall%20of,of%20them%20in%20the%20uterus.

6 Upper Back Pain

INFLAMMATION OF THE TRAPEZIUS MUSCLE 斜方肌筋膜炎

INTRODUCTION

- Inflammation of the trapezius muscle can literally be a pain in the neck, the shoulder, and the upper back and can be particularly disruptive to the patient's daily life.
- Approximately 20% of the adult population lives with severe chronic pain and 10–20% among these have the specific concern of a painful trapezius muscle (Gerdle et al. 2014).
- The upper trapezius muscle has been implicated among those muscles that effect neck and shoulder pain.

ANATOMY

The trapezius is a large superficial fan-shaped muscle located on the upper back that extends from the external protuberance of the occipital bone to the thoracic region on the posterior aspect of the trunk and attaches to the clavicle and scapula.

The function of the trapezius is to stabilize and move the scapula such as moving the scapula together, shrugging the shoulders up toward the ears and rotating the scapula to allow the arm to reach overhead.

It consists of three parts with different actions:

- Superior fibers of trapezius – elevate and upwardly rotate the scapula and move the neck (Ourieff, Scheckel, and Agarwal 2021).
- Middle fibers of trapezius – adduct the scapula and stabilize the shoulder during arm movement (Ourieff, Scheckel, and Agarwal 2021).
- Inferior fibers of trapezius – depress and stabilize the scapula and upwardly rotate the scapula together with the superior fibers.

Clinically, the area manifesting pain involved with trapezius muscle is commonly seen in the superior fibers of trapezius.[1]

WESTERN MEDICINE ETIOLOGY AND PATHOLOGY

There are several possible causes of trapezius pain, including:

- Overuse: pain in the trapezius often develops due to overuse. Repetitive and prolonged period of upper body activities, especially involving the shoulders, can put stress on the muscle. These activities may include lifting heavy objects or participating in specific sports, such as swimming.

DOI: 10.1201/9781003203018-6

- Emotional stress: it is common for people to tense the muscles of the shoulder and neck when they are under emotional stress. This excess tension can lead to muscle soreness over time.
- Poor posture: prolonged poor posture can place added stress on the trapezius. For example, hunching over a desk or computer keyboard for many hours can result in a shortened and tight trapezius muscle.
- Trauma: injuries to the trapezius, for example, a muscle or tendon tear from applying too much force on the muscle, such as a violent twist or collision or a bad fall, can result in immediate pain.

CHINESE MEDICINE ETIOLOGY AND PATHOLOGY

- Liver Qi Stagnation 肝氣鬱結: stress causes tight and spasmodic tendons; because Liver rules the Tendons and Sinews in traditional Chinese medicine, Liver is the Zang-fu (organ, in Chinese medicine) most immediately influenced when one is under stress or anxious.
- Qi and Blood stagnation 氣滯血瘀: many factors can cause poor blood flow to the trapezius muscle. Working on a computer for an extended period of time with protracted shoulders and gazing a monitor and the still posture of the neck, arms, and shoulders may cause tight muscles that don't allow blood vessels to flow and bring QI and Blood to nourish the muscles. Carrying a heavy backpack puts extra tension on the trapezius muscles; and whiplash from a motor vehicle accident may cause a sore trapezius muscle from the head getting snapped backward. Work or sports activities that involve repetitive overhead movement or sustained positioning can lead to QI and Blood stagnation predisposing the trapezius to muscle strain.

MANIFESTATIONS

- Headaches, occipital headache, or feeling heaviness of the head.
- Unilateral or bilateral neck pain.
- Shoulder is painful or tight.
- Jaw is painful.
- Stiff neck after a prolonged period of inactivity, such as gazing a monitor.
- Tight upper back.
- Burning pain in the shoulder and arm.
- Tingling/numbness down the arm and/or into the hand.
- Low mood.
- Anxiety.

PHYSICAL EXAMINATION

- Begin with obtaining a medical history.
  - General health condition, including whether the patient is under emotional stress.

- Injury history, such as motor vehicle collision or a fall that over-stretched the trapezius muscle.
- Careers that involve lifting heavy objects, such as construction workers, landscapers, nurses or nurses' aids, furniture movers, office workers who hunch over a computer keyboard for prolonged periods of time; or dentists and surgeons who require prolonged standing, stooping, bending, and awkward body positioning.
- Sports activities that involve repetitive shoulder movement, such as swimming, golfing, and quarterbacking.
- Inspection
  1. Patients with inflammation of the trapezius muscles tend to be in a low mood and depressed.
- Palpation

  1. Tenderness of the upper trapezius area.

1. Range of motion
   Muscle strength test for upper fiber of trapezius is positive.
       Testing procedures:
- The patient is in a sitting position, and the therapist stands behind the patient.
- Ask the patient to shrug the shoulder that is being tested with the arms slightly abducted and the head and neck flexed laterally to the tested side with rotation of the head slightly away from the tested side.
- Press the patient's shoulder downward, push the head away from the tested side, and ask the patient to resist the movements.
- Another way to test is to ask the patient to resist abduction and ipsilateral lateral flexion of the head. The tested arm is abducted at 90 degrees, and the head leans toward the arm to resist the therapist's hands, one of which is placed on the patient's elbow, and the other is placed at the patient's temporal region.
1. Special tests
- An X-ray cannot reveal the muscle damage, but it can help to determine whether the pain is due to a bone fracture or bone spur.
- MRI can reveal the condition of the muscle, and it is able to identify whether there is a complete muscle tear or just a strain.

DIFFERENTIAL DIAGNOSIS
- Shoulder bursitis:
  - May have a history of injury, and the shoulder may be red and swollen.
  - May limit the shoulder's range of motion.
  - The pain location is on the top of shoulder.

- There is more pain when the arm is abducted to the level of shoulder or above the shoulder.

- Cervical radiculopathy
  1. History of neck pain.
  2. Dermatomal distribution on the hand and finger from C6–C8.
- Dislocation of SC, AC, and shoulder joint.
  - History of injury.
  - SC and AC joints are "floating", or the clavicle is out of the joint.
  - The shoulder joint is square that the humeral head falls out of the glenoid fossa.

TREATMENT

Treatment Principle
- Move Liver Qi.
- Move Qi and Blood.

Herbal Medicine
Mild Stage

Herbal Formula
Gui Zhi Tang桂枝湯

Herbal Ingredients
Gui Zhi 桂枝9g, Zhi Ke枳壳9g, Chen Pi陳皮9g, Hong Hua紅花9g, Xiang Fu香附 9g, Sheng Di生地9g, Dang Gui Wei當歸尾9g, Yan Hu Suo 延胡索9g, Fang Feng 防風9g, Chi Shao 赤芍9g, Du Huo 獨活9g

Ingredient Explanations
1. Gui Zhi 桂枝 warms the meridians and releases the Exterior.
2. Zhi Ke枳壳 removes stagnant food, promotes Qi movement, and reduces distention and pressure.
3. Chen Pi陳皮 regulates Middle Jiao Qi, dries Dampness, and transforms Phlegm.
4. Hong Hua紅花 moves the Blood, releases Bruising, opens the meridians, and stops pain.
5. Xiang Fu香附 spreads and regulates Liver Qi and alleviates pain.
6. Sheng Di Huang生地 clears Heat for Heat symptoms due to Yin Deficiency, cools the Blood, nourishes Yin and generates fluids, and tonifies Kidney Yin.
7. Dang Gui Wei 當歸尾 moves and harmonizes the Blood.
8. Yan Hu Suo 延胡索 moves the Blood and stops pain.
9. Fang Feng 防風 releases the exterior, expels External Wind, and expels Wind-Dampness.

10. Chi Shao 赤芍 moves the Blood, reduces swelling, and stops pain.
11. Du Huo 獨活 expels Wind Damp and stops pain.

Action: moves Qi and Blood, opens the channel, and stops pain.
 Contraindication: pregnancy.
 Moderate Stage
 Herbal Formula
 Fu Yuan Huo Xue Tang復元活血湯

Formula Ingredients

Da Huang大黃12g, Chai Hu柴胡9g, Tao Ren桃仁9g, Hong Hua紅花6g, Chuan Shan Jia穿山甲6g, Hua Fen花粉 9g, Dang Gui當歸9g, Gan Cao甘草6g

Ingredient Explanations

1. Da Huang大黃 moves the Blood and disperses Bruising.
2. Chai Hu柴胡 releases Liver Qi, raises Yang Qi, and disperses Wind Heat and Phlegm.
3. Tao Ren桃仁 breaks up stubborn Blood stasis.
4. Hong Hua紅花 moves the Blood, releases Bruising, opens the meridians, and stops pain.
5. Chuan Shan Jia穿山甲 moves the Blood, reduces swelling, and expels Wind and Damp.
6. Hua Fen花粉 clears Heat and generates Fluids.
7. Dang Gui當歸 moves and tonifies the Blood, disperses Cold, and stops pain due to Blood stasis.
8. Gan Cao甘草 releases spasms, alleviates pain, harmonizes the other herbs, and guides the herbs to all twelve meridians.

Action: moves the Blood, stops pain.
 Contraindication: pregnancy.
 Severe Stage

Herbal Formula

Jia Jian Bu Jin Wan加減補筋丸

Formula Ingredients

Dang Gui當歸 30g, Shu Di熟地黃60g, Bai Shao白芍60g, Hong Hua紅花30g, Ru Xiang乳香30g, Mo Yao沒藥9g, Ding Xiang丁香15g, Fu Ling茯苓30g, Gu Sui Bu骨碎補30g, Chen Pi陳皮60g

Ingredient Explanations

- Dang Gui當歸 moves and tonifies the Blood, disperses Cold, and stops pain due to Blood stasis.
- Shu Di Huang熟地黃 nourishes the Blood and nourishes Liver and Kidney Yin.

- Bai Shao白 芍 nourishes the Blood, preserves Yin, calms Liver Yang and Liver Wind, softens Tendons, and alleviates spasm and pain.
- Hong Hua紅 花 moves the Blood, releases Bruising, opens the meridians, and stops pain.
- Ru Xiang乳 香 moves the Blood and stops pain.
- Mo Yao没 藥 moves the Blood, reduces swelling, and stops pain.
- Ding Xiang丁 香 warms Middle Jiao and relieves pain.
- Fu Ling茯 苓 drains Damp and tonifies Spleen.
- Gu Sui Bu骨 碎 補 tonifies Kidney Yang, strengthens Tendons and Bones, and moves the Blood, especially due to trauma.
- Chen Pi陳 皮 dries Damp and transforms Phlegm.

Method: mix the herbs with honey; make into pills, 9g each.
  Dose: 1 pill with warm water, three times daily.
  Action: trapezius muscle pain and spasm, shoulder range of motion limitation.
  Contraindication: pregnancy.

Acupuncture Treatments
Local acupuncture point formula
  GB20風 池Fengchi, GB21肩 井Jianjing, UB15心 俞Xinshu, UB17膈 俞Geshu

Distal Acupuncture Point Formula
UB23腎 俞Shenshu,   SI3後 溪Houxi,   UB62申 脈Shenmai,   LI4合 谷Hegu,
LIV3太 沖Taichong, LIV8曲 泉Ququan, SP10血 海Xuehai

Cupping Therapy
GB21肩 井Jianjing, UB17膈 俞Geshu

Moxibustion Therapy
Warming needling techniques on GB21肩 井Jianjing, UB23腎 俞Shenshu

Bleeding Therapy
UB40委 中Weizhong
  Point selected is UB40委 中Weizhong.

Procedures
- Prior to the procedure, tools should be prepared and ready to be used at the work area.
- Set up a clean field with paper towels.
- Place these tools in the clean field along with dry sterile cotton balls, medical alcohol wipes or cotton swabs soaked in alcohol, sterile gloves, a biohazard sharp container, a biohazard trash container, goggles and face mask, and paper towels. A medical lancet (a traditional three-edge needle) or the seven-star needle (a bleeding tool) is selected for bleeding UB40委 中Weizhong in this case.

- The patient lies prone with the legs extended, UB40委 中Weizhong is facing upward.
- The room should be well lit.
- After the therapist washes and dries their hands thoroughly, they should put on sterile gloves.
- Wipe the acupuncture point UB40委 中Weizhong with a medical alcohol wipe or a cotton swab soaked in alcohol, moving from the center of the acupuncture point to the periphery.
- Place the medical lancet at a 90-degree angle to the skin.
- In one quick move, prick the needle into the acupuncture point about 0.1 Cun 寸 deep and allow a few drops of blood to discharge.
- If more drops of blood are needed, absorb the blood with a sterile cotton ball.
- Lastly, the point is pressed with a sterile cotton ball until the bleeding ceases.
- Dispose of the tools in the proper containers.
- Wash hands thoroughly.

Electro-Acupuncture Therapy
GB21肩 井Jianjing ~SI3 後 溪Houxi in the frequency of 2/100 hz, 20 minutes daily

Tui Na Treatments
Tui Na Techniques
Pushing, grasping, striking

Tui Na Procedures
- Patient lies prone with both arms placed at the side of the torso and both legs are extended; the therapist stands near the patient's head.
- Apply pushing techniques with both palms on the Urinary Bladder meridians from the top of the upper back to the lower back for 5–10 times Figure 6.1.1; Figure 6.1.2; Figure 6.1.3.

- Apply pushing techniques with the heels of hands on top of the shoulder where the Gallbladder meridians are located. Pushing direction is from the spine toward the shoulder joints until warm or about 5–10 times.
    - Patient sits straight up; the therapist stands behind.
    - Apply grasping techniques on both GB21肩 井Jianjing about 10 times.
    - Apply striking techniques with ulnar side of the hands for 1 minute Figure 6.1.4; Figure 6.1.5; Figure 6.1.6; Figure 6.1.7

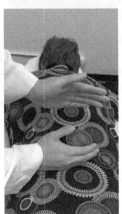

PATIENT ADVISORY

- Recommend patients to apply heating pad on both shoulder trapezius muscles, once or twice daily for 20 minutes each time.
- Food should be simple and light with plenty of green vegetables.
- Recommend routine upper body exercise, such as Tai Chi Chuan and swimming.

RHOMBOID MUSCLE PAIN 菱 形 肌 勞 損

INTRODUCTION

Rhomboid muscle connects the scapula to the spine. The scapula (the shoulder blades) become unstable when the rhomboid muscle is weak, and pain at the upper back occurs when raising the arm.

ANATOMY

There are two rhomboid muscles, they are major and minor.

Rhomboid major originates from the spinous processes of T2–T5, and it attaches to the medial border of the scapula. It is innervated by the dorsal scapular nerve. Its functions are to retract and rotate the scapula.[2]

Rhomboid minor originates from the nuchal ligament and the spinous process of C7–T1, and it attaches to the medial border of the scapula. It is innervated by the dorsal scapular nerve. Its functions are to retract and rotate the scapula.

WESTERN MEDICINE ETIOLOGY AND PATHOLOGY

Rhomboid muscle can be strained by putting too much weight on the muscle. Common causes include carrying a heavy backpack, especially carrying it over only one shoulder; lifting heavy objects; incorrect posture while sleeping (e.g., side lying) that put pressure on the muscle; sitting for extended periods with hunched shoulders, head forward, and a rounded mid-back. These are most seen in patients who do computer and desk work. Repetitive motions of the shoulder (painters), raising the arm over the head for extended periods, and certain strength training in sports such as serving a tennis ball, golfing, and rowing are other activities that cause rhomboid muscle strain.

The muscle in the condition of chronic strain can lead to adhesion between tissues. It is like scar tissue that has built up after an untreated injury. Scar tissue or adhesion can restrict the movement of the muscles between the scapula and the spine. The pain occurs when there is movement of the upper back and the arm.[3]

Emotional stress can build up the muscle tension, which mostly occurs in the neck and upper back. Over time when the tension becomes chronic, the muscle under tension develops painful knots that lead to muscle pain between the scapula and the upper back spine.

CHINESE MEDICINE ETIOLOGY AND PATHOLOGY

1. Invasion of Wind-Damp-Cold風 寒 濕 邪 型

Externally, Wind, an external pathogenic factor of mysterious nature, is thought to carry the pathogenic influences into the body and contribute to the variable nature of the pain.

Internally, a deficiency of body constitution under the post-traumatic circumstance, poor Blood circulation of the rhomboid muscle, and exposure to the wet, wind, and cooler environment, such as sitting or sleeping on the cold lawn during a camping trip causes Bi Syndrome. Bi Syndrome manifest as pain, soreness, aches, numbness, or heaviness of muscles, sinews, and joints, and/or swelling and burning pain. In general, the causative factors of Bi Syndrome are due to the blockage on the meridians and collateral and vessel invasion by Wind, Cold, Damp, and Heat.

2. Deficiency of Spleen and Kidney脾腎虛弱型

- Irregular diet can damage the Spleen and Stomach Yang, for instance, excessive consumption of cold foods, icy beverages, raw foods, and Cold herbs. Spleen and Stomach prefer warm and dry. The transport and transformation functions of Spleen and Stomach is also influenced by warming and nourishing functions of Kidney Yang. On the other hand, cold and raw foods are considered Cold and Damp which the internal environment will start taking away from the Spleen's energy, leading to Qi deficiency and later to Yang deficiency when the Cold and Damp impair both Spleen and Kidney.
- Irregular mealtime is also a cause to deficiency of Spleen and Stomach. Breakfast is the most important among the three daily meals, providing Qi and Blood for both the mental and physical activities of the day. Breakfast is best in the Stomach time, which is from 7 a.m. to 9 a.m. Dinner should be between 5 p.m. and 7 p.m. It should not be at 7 p.m. to 9 p.m. because that is the weakest time for digestion. Eating dinner at this time will impair Spleen and Stomach.
- Muscles and tendons rely on Blood nourishment. The source of Blood in the body relies on the functions of Transformation and Transportation of Spleen. Spleen Qi Deficiency impairs both functions. Rhomboid muscle pain occurs when it lacks Blood nourishment.
- the accumulation of Phlegm, Damp, and Blood stasis will occur under the deficiency of Spleen and Kidney. Phlegm and Damp block the meridians, resulting in rhomboid muscle pain.

3. Stagnation of Liver Qi肝氣鬱結型

Liver stores the Blood and governs the Tendons. This is the internal cause of pain. Liver is associated with emotional stress, and Liver governs Tendons. Emotional stress changes the quality of Tendons, making them vulnerable. Wind-Cold invasion is the external cause of pain. Attacking of Wind-Cold on the rhomboid muscle leads to the disturbance of local circulation of Qi and Blood in the meridians.

MANIFESTATIONS

In Western Medicine

The symptoms of a rhomboid muscle issue include aching pain, stiffness, tightness, and tenderness in the upper back region; it is commonly a unilateral, pinching, shooting pain and can radiate to the arm along the C5 and C6 dermatomes. Tingling, numbness, and pain with clicking, popping, or grinding noise; pain or difficulty when moving the shoulder; muscle knots in the muscle; and painful breathing, swelling, and atrophy of the rhomboids may be evident.

In Chinese Medicine

1. Invasion of Wind-Damp-Cold風寒濕邪型

With acute onset, the patient may feel heavy, like being wrapped, and experience stiff joints, pale color in the affected upper back, cool temperature when touched, aggravated pain when exposed to cold, pain relieved when applying heat. Wondering pain presents when it is more Wind pathogen; numbness on the skin presents when it is more Damp pathogen.

Tongue: pale body, white sticky coating.

Pulse: wiry, tight.

2. Deficiency of Spleen and Kidney脾腎虛弱型

Cold limbs, cold and painful back, pale face, poor appetite, loose stool, sore and weak lumber, and knees.

Tongue: pale tongue body, teeth marks, white coating.

Pulse: deep, thin.

3. Stagnation of Liver Qi肝氣鬱結型

Dull pain and numbness in the shoulder and upper back, spasm and trembling, muscle atrophy, Five-Center-Heat, emotional upset, or a bitter taste in the mouth.

Tongue: pale, pink, ecchymosis on the edges, white coating.

Pulse: wiry, tight.

PHYSICAL EXAMINATION

- Begin with obtaining a medical history.
  - Sleeping position. Rhomboid muscle pain can occur after sleeping, and the pain can be from side sleeping, which puts pressure on the muscle. Inquiry on the sleep posture is important, as pain occurs when sleeping on the affected side.
  - Occupation and sports. Any activity that may be the cause of the muscle sprain should be investigated.
  - Location of the pain, which can radiate to the arm along the C5 and C6 dermatomes (Martin, and Fish 2008).
  - Heart history. When the rhomboid muscle pain is in the left upper back, inquire about the medical history of the heart and symptoms of cardiac angina, such as left chest pain and pain radiating into the left arm, and the duration of the pain.

- Posture. The history of sitting posture, lifting objects, reaching (cabinets) above the head, and sitting or sleeping on the floor, especially on the lawn, will need to be obtained.
- Trauma history. Direct injury to the dorsal scapular nerve, such as anterior shoulder dislocation, overuse/overhead athletics (e.g., volleyball, tennis serving) can be the cause of dorsal scapular nerve entrapment, leading to rhomboid palsy (Farrell and Kiel 2021).
- Environmental factors. Pain can be aggravated by changes in the weather, especially during cold and rainy days.
- Dietary history is also helpful in determining the constitution of Spleen.
- Inspection
  - Identify the rhomboid muscle condition by observing the skin color and looking for swelling.
  - Identify the atrophic condition of the muscle.
- Palpation
  - Locate the painful muscle, make sure that it is the one located near the medial aspect of the scapula.
  - Palpate the muscle to determine whether the affected one is the minor or the major rhomboid muscle.
  - Examine the muscle to check for knots and painful spots. If it is cold, it is Cold invasion; if it is hot, it is due to trauma; if it is puffy, it is swollen.
- Range of motion
  - Check the ipsilateral shoulder movements for abduction, lifting, flexion, hyperextension, inward rotation, and outward rotation. If there is any limitation, find out if it is due to the shoulder condition or because the patient is afraid of moving the rhomboid muscle.
  - Check the shoulder movement during shrugging motion; pain in the rhomboid muscle may increase.
- Orthopedic tests
  Scapular Adduction and Downward Rotation Test
  - The patient is in a prone position with hand facing forward and the elbow bent near the body, the head is rotated to the opposite side.
  - Position the rhomboid muscle and add resistance pressure to the medial scapula in the direction of scapula abduction and upward rotation.
  - Ask the patient to move the elbow backward (adduction).
  - Place a hand on the patient's elbow to resist the arm (abduction).
  - It is positive if the scapula doesn't push the hand that is placed against the medial border.

DIFFERENTIAL DIAGNOSIS

- Pleurisy causes sharp chest pain that worsens during breathing, and pain also may refer to the shoulders or the back. But pleurisy usually presents with fever. A Blood test or chest radiograph can aid in diagnosis.

- Costovertebral joint sprains have similar manifestations: the pain refers along the rib to the chest, the scapula, and sometimes the upper arms. Chest x-ray in severe trauma may reveal subluxation or dislocation of the joints.

TREATMENTS

1. Invasion of Wind-Damp-Cold風寒濕邪型

Treatment Principle

Expel Wind, disperse Cold, drain Damp, and open the meridians.

Herbal Formula

Juan Bi Tang蠲痹湯

This formula is from "A Book of Formulas to Promote Well-Being 嚴氏濟生方", authored by Yan Yonghe 嚴用和 (1200~1268), in Song dynasty 南宋 (1127–1239), and published in 1253.

Herbal Ingredients

Dang Gui當歸9g, Chi Shao赤芍9g, Jiang Huang薑黃9g, Huang Qi黃耆9g, Qiang Huo羌活9g, Gan Cao甘草3g, Sheng Jiang生薑15g, Da Zao大棗3 pieces

Ingredient Explanations

- Dang Gui當歸 nourishes the Blood, benefits the Liver, and regulates menstruation.
- Chi Shao赤芍 moves the Blood, disperses Bruising, and stops pain.
- Jiang Huang薑黃 moves the Blood, opens the meridians, expels Wind, and reduces swelling.
- Huang Qi黃耆 tonifies Qi, strengthens the Spleen, raises the Yang Qi of the Spleen and Stomach, tonifies Wei Qi, stabilizes the exterior, and tonifies the Blood.
- Qiang Huo羌活 expels Wind-Cold-Dampness, unblocks painful obstruction, and alleviates pain.
- Gan Cao甘草 harmonizes the herbs, tonifies Middle Jiao Qi, clears Heat, expectorates Phlegm, and stops coughing.
- Sheng Jiang生薑 unblocks the pure Yang pathway and harmonizes rebellious Qi.
- Da Zao大棗 tonifies the Spleen Qi, nourishes the Blood, and moderates and harmonizes the harsh properties of the fragrant herbs.

Modifications

- When Wind and Cold are more prominent, remove Dang Gui當歸 and Huang Qi黃耆 and add Ma Huang 麻黃, Xi Xin细辛, and Bai Zhi白芷.
- When Damp is more prominent, add Cang Zhu蒼朮 and Yi Yi Ren薏苡仁.

Formula Indication
- Wind-Damp-Cold Bi Syndrome.

Formula Contraindication
- It is contraindicated for patients with Damp-Heat in the body.

2. Deficiency of Spleen and Kidney 脾 腎 虛 弱 型

Treatment Principle
Warm Yang, disperse Cold, disperse Bruising, and stop pain.

Herbal Treatments
Herbal Formula
Si Shen Wan 四 神 丸
   This formula is from "Standards for Diagnosis and Treatment 證 治 準 繩", authored by Wang Kentang 王 肯 堂 (1549–1613), in Ming dynasty 明 朝 (1368–1644), and published in 1602.

Herbal Ingredients
It is a pattern herb which contains the following:
   Bu Gu Zhi補 骨 脂120g, Wu Zhu Yu吳 茱 萸30g, Rou Dou Kou肉 豆 蔻60g, Wu Wei Zi五 味 子60g, Sheng Jiang生 薑240g, Da Zao大 棗100 pieces

Administrations
Take 10 g with warm salt water, 3 times a day.

Ingredient Explanation
- Bu Gu Zhi補 骨 脂 tonifies Liver and Kidney, strengthens Bone, and benefits Essence and Blood.
- Wu Zhu Yu吳 茱 萸 warms the Middle Jiao, disperses Cold, stops vomiting, dries Dampness, and expels Damp-Cold.
- Rou Dou Kou肉 豆 蔻 astringes the Intestines, warms the Middle Jiao, and stops diarrhea.
- Wu Wei Zi五 味 子 replenishes Qi, nourishes Kidney, calms Shen, and stops diarrhea.
- Sheng Jiang生 薑 unblocks the pure Yang pathway and harmonizes rebellious Qi.
- Da Zao大 棗 tonifies the Spleen Qi, nourishes the Blood, and moderates and harmonizes the harsh properties of the fragrant herbs.

Formula Indication
- Spleen and Kidney Yang Deficiency

Formula Contraindication
- It is contraindicated for patients with food stagnation in Stomach and Intestine.

- While taking the formula, avoid consuming cold, raw foods.

3. Stagnation of Liver Qi 肝 氣 鬱 結 型

Treatment Principle
Disperse Liver, regulate Qi, harmonize the Blood, and alleviate pain.

Herbal Treatments
Herbal Formula
Chai Hu Shu Gan Tang柴 胡 疏 肝 散
This formula is from "The Complete Compendium of Jingyue景 岳 全 書",
authored by Zhang Jiebin張 介 賓(1111–1117AD), in Ming dynasty明 代 (1624 AD).

Herbal Ingredients
Chai Hu柴 胡6g, Xiang Fu香 附5g, Bai Shao白 芍5g, Chuan Xiong川 芎5g, Zhi
Ke枳 殼5g, Chen Pi陳 皮6g, Zhi Gan Cao甘 草2g

Ingredient Explanations
- Chai Hu 柴 胡 moves Liver Qi and releases stagnation.
- Xiang Fu香 附 moves Liver Qi and Blood and stops pain due to Qi and
  Blood stagnation.
- Bai Shao白 芍 nourishes the Blood, soothes the Liver, and stops pain.
- Chuan Xiong川 芎 moves the Blood and moves Qi.
- Zhi Ke枳 殼 regulates Qi and removes blockage.
- Chen Pi陳 皮 regulates Qi and removes blockage.
- Zhi Gan Cao甘 草 tonifies the Spleen, harmonizes the other herbs, and
  guides the herbs to all twelve meridians.

Modifications
For Blood deficiency, add Dang Gui當 歸, Sheng Di Huang生 地, and Shu Di
Huang熟 地.
For Blood stasis, add Dan Shen丹 參.

Formula Actions
Moves Liver Qi, harmonizes the Blood, and alleviates pain due to Liver Qi stagnation.

Formula Indication
Liver Qi stagnation, Liver Qi overacts on the Stomach.

Contraindications
Caution is advised for patients who suffer from headache due to Yin deficient Heat.

Acupuncture Treatments
Acupuncture Point Formula
Basic Points

SI11天 宗Tianzong, GB21肩 井Jianjing, GB20風 池Fengchi, SI14肩 外 俞Jian-waishu, SI15肩 中 俞Jianzhongshu, UB10天 柱Tianzhu, SI3後 溪Houxi (Tonifying), UB40委 中WeiZhong (reducing), SJ5外 關Waiguan, LI4合 谷Hegu, GB34陽 陵 泉Yanglingquan, UB60崑 崙Kunlun, Ashi acupoints阿 是 穴

Administrations
Select 3–5 points for each treatment.
   Frequency: once a day.

Auricular Therapy
Shoulder, Thoracic Vertebra, Shenmen, Endocrine, Adrenal Gland

Cupping Therapy
Ashi acupoints 阿 是 穴after Cutaneous needles treatment. Figure 6.2 from shutterstock.com # 1573639324.

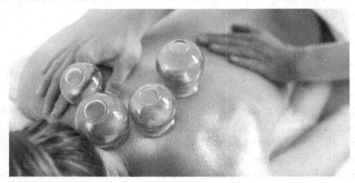

Moxibustion Therapy
Same points as the acupuncture needling points.

Bleeding Therapy
Ashi acupoints 阿 是 穴with Cutaneous needles before cupping Therapy.

Tui Na Treatments
Tui Na Techniques
Pressing, kneading, rubbing, rolling

Tui Na Procedures
The patient is in a sitting position; the therapist stands behind.

   1. Place the patient's hands on his/her knees.
   2. Put a hand on the patient's forehead and use the other hand to apply grasping techniques on the patient's neck and trapezius muscles for about 3 minutes.

3. Place the patient's ipsilateral hand on the contralateral shoulder.
4. Locate and mark the Ashi acupoints阿 是 穴.
5. Apply the radius aspect of the thumb to knead and press the Ashi acupoints阿 是 穴. Begins with slow, gentle pressure and gradually increase the speed and force.
6. Place the patient's hands on his/her knees.
7. Apply kneading and rubbing techniques with the hand's hypothenar aspect on the knots and the Ashi acupoints 阿 是 穴for 4–5 times.
8. Place a hand on the front of the affected shoulder, apply rolling techniques with the other hand while pushing the shoulder backwards for 6–7 times.

PATIENT ADVISORY

1. Avoid sitting or sleeping on the lawn or a cold floor.
2. Change to dry cloths if they are wet from rain or sweating during sports.
3. Wear appropriate clothing according to the weather.
4. Correct poor posture in sitting and sleeping. Avoid sleeping on the side. The appropriate sleeping posture is lying flat on the back, one pillow under the head and neck and another pillow under the knees. Note: This position is not appropriate for pregnant women because it decreases Blood circulation to the Heart and the baby. When pregnant, side sleeping on the unaffected side is appropriate.
5. During acute onset, avoid playing overhead sports and avoid reaching high above the head.
6. Apply heating pad for Cold conditions.

COSTOVERTEBRAL JOINT SPRAIN 胸 壁 扭 挫 傷

INTRODUCTION

Costovertebral joint sprain is damage or tearing of connective tissue (such as ligaments, cartilage, and joint capsules) of one of the upper back's costovertebral joints.

ANATOMY

The spine comprises of many vertebrae. Each vertebra articulates with the vertebra above and below it via two types of joints: (1) the facet joints on either side of the spine and (2) the intervertebral discs centrally.

Left and right superior articular facets articulate with the vertebra above, and the inferior articular facets articulate with the vertebra below. Vertebral bodies indirectly articulate with each other via the intervertebral discs.[4]

These joints are designed to bear body weight and enable some gliding motions between the vertebrae. Each vertebra in the thoracic spine also articulates to each rib via the transverse process of the costovertebral joint on either side of the spine.[5]

WESTERN MEDICINE ETIOLOGY AND PATHOLOGY

2. Costovertebral joint sprains typically occur during excessive weight lifting, repetitive forward or backward arching movements, or sideways bending. The injury may also occur traumatically by repetitive torso twisting. This is more likely due to lack of warming up the relevant muscles or from overestimating the ability to carry out the extensive strenuous activities.
3. The injury may also occur due to being in poor posture for prolonged periods of time such as sitting slouched (particularly) while using a computer, working at a desk, or driving; or sleeping in the fetal position.
4. Costovertebral joint sprains may also occur due to repetitive or prolonged work with the arms in front of the body, such as during house work, gardening, or other manual work particularly in a poor posture.
5. In some cases, excessive coughing, sneezing, or laughing may contribute to the development of the injury.[6]

CHINESE MEDICINE ETIOLOGY AND PATHOLOGY

Qi and Blood stagnation 氣滯血瘀

A direct blow to the costovertebral joint causes blockages to the flow of the local Qi and Blood in the meridians. Bodily injury causes Blood stasis in the meridians, then Qi flow is impeded. As a result of Qi and Blood stagnation in the meridians, pain occurs as the saying goes, "If there is no free flow, there is pain 不通則痛".

Maintaining joints in a wrong position for a long time, such as sitting in a poor posture or doing repetitive or prolonged work with the arms in front of the body, results in Qi and Blood stagnation.

MANIFESTATIONS

The pain is a sudden and immediate onset in the upper back and ribs during the causative activity. It may also be common for patients to experience stabbing pain and stiffness on the next morning after the provocative activity a day prior.

Symptoms are typically felt on one side of the spine and ribcage and muscle spasm is experienced around the affected costovertebral joint. Pain refers along the rib to the chest, scapula occasionally, and sometimes the upper arms.

Symptoms are exacerbated during coughing, sneezing, laughing, and deep breathing so that patients are reluctant to take deep breaths. Pain may be worsened with physical activity involving lifting; pain is also aggravated by axial rotating or flexing. And sometimes the pain restricts movement in the upper back.

PHYSICAL EXAMINATION

- Begin with obtaining a medical history
  - Patients generally are in good health; however, certain diseases that manifest coughing and sneezing may cause a sprain in the rhomboid

muscle. Careers that may make one vulnerable to the rhomboid injury include construction, landscaping, nursing, office work, surgery, and dentistry, among many others. Whatever the career, poor posture, repetitive motion and overuse, and prolonged positioning can make one vulnerable to rhomboid muscle injury.

- Inspection
  - The location of pain is unilateral.
  - Pain upon deep inspiration, coughing, sneezing, and even laughing.
- Palpation
  - Patients with a costovertebral joint sprain typically experience tenderness on palpation at the costotransverse joint and rib angle, but patients usually are not able to locate the exact tender spots.
- Range of motion
  - Thoracolumbar rotating, flexing, lifting, and ipsilateral flexion may induce pain.
  - The costotransverse and costovertebral joint movements are restricted.[7]
- Special tests
  - Usually, the chest x-ray is normal. However, in severe trauma, these joints can subluxate or dislocate; then it will be shown on a chest x-ray.[8] X-ray may also reveal fracture, pneumothorax, or hemothorax.

DIFFERENTIAL DIAGNOSIS

Pleurisy causes sharp chest pain that worsens during breathing, and the pain also may refer to the shoulders or back. Pleurisy usually presents with fever. A Blood test or chest radiograph can aid in diagnosis.

TREATMENT

Treatment Principle
Regulate Tendons, open the channels, and move Qi and Blood.

Herbal Treatments
Qi stagnation sprain
   Symptoms: pain is not fixed; it can be aggravated by deep breathing and coughing accompanied by chest distention.
   Pulse: wiry.

Herbal Formula
Jia Wei Jin Ling Zi San 加味金铃散

Formula Ingredients
Chuan Lian Zi 川楝子 (金玲子) 9g, Yan Hu Suo 延胡索 9g, Chai Hu 柴胡9g, Xiang Fu 香附9g, Zhi Ke 枳壳6g

Ingredient Explanations

- Chuan Lian Zi 川楝子 moves Liver Qi and stops pain.
- Yan Hu Suo 延胡索 moves the Blood and stops pain.
- Chai Hu 柴胡 releases Liver Qi, raises Yang Qi, and disperses Wind Heat and Phlegm.
- Xiang Fu 香附 spreads and regulates Liver Qi and alleviates pain.
- Zhi Ke 枳壳 removes stagnant food, promotes Qi movement, and reduces distention and pressure.

Blood Stasis Sprain

Symptoms: pain is fixed and resists pressure; it can be aggravated by deep breathing and coughing; accompanied with chest distention.

Tongue: purple and dim.

Pulse: thin and choppy.

Herbal Formula

Sheng Jin Shu Gan Tang勝金梳肝湯

Formula Ingredients

Chai Hu 柴胡9g, Dang Gui 當歸9g, Di Bie Cong 地鱉蟲4.5g, Tian Hua Fen 天花粉9g, Tao Ren 桃仁9g, Hong Hua 紅花9g, Da Huang 大黃9g, Chi Shao 赤芍9g, Bai Shao 白芍9g, Gan Cao 甘草9g, Che Qian Zi 車前子15g, Jiang Xiang 降香3g

Ingredient Explanations

- Chai Hu 柴胡 releases Liver Qi, raises Yang Qi, disperses Wind Heat and Phlegm.
- Dang Gui 當歸 moves and tonifies the Blood, disperses Cold, and stops pain due to Blood stasis.
- Di Bie Cong 地鱉蟲 breaks up stubborn Blood stasis.
- Tian Hua Fen 天花粉 drains Heat, generates Fluids, transforms Phlegm, expels pus, and reduces swelling.
- Tao Ren 桃仁 breaks up stubborn Blood stasis.
- Hong Hua 紅花 moves the Blood, releases Bruising, opens the meridians, and stops pain.
- Da Huang 大黃 moves the Blood and disperses Bruising.
- Chi Shao 赤芍 moves the Blood, reduces swelling, and stops pain.
- Bai Shao 白芍 nourishes the Blood, preserves Yin, calms Liver Yang and Liver Wind, softens tendons, and alleviates spasm and pain.
- Gan Cao 甘草 releases spasms, alleviates pain, harmonizes the other herbs, and guides the herbs to all twelve meridians.
- Che Qian Zi 車前子 clears Damp Heat and promotes urination.
- Jiang Xiang 降香 moves the Blood, stops bleeding, and alleviates pain.

Acupuncture Treatments Procedures

1. The patient is in a supine position. Insert acupuncture needles into LIV13章 門Zhangmen, LIV14期 門Qimen, SP21大 包Dabao, REN17膻 中Danzhong, and GB24 日 月Riyue.
2. Turn the patient over to be in a prone position, find the Ashi points along the UB meridians on the affected area, and insert acupuncture needles into about 2–3 Ashi points.
3. Apply cupping Therapy on the Ashi points immediately after withdrawing the needles, or apply Seven-Star bleeding Therapy on the Ashi points after the needles have been removed and before the cups have been put on to draw Blood, to promote Qi and Blood circulation.

These acupuncture procedures promote the flow of Qi and Blood to stop pain.

Cupping Therapy

Ashi points along the UB meridians.

Moxibustion Therapy

Stick moxa on UB17膈 俞Geshu.

Bleeding Therapy

UB40委 中Weizhong

The selected point is UB40委 中Weizhong.

Procedures

1) Prior to the procedure, tools should be prepared and ready be used at the work area.
2) Set up a clean field with paper towels.
3) Place the tools in the clean field along with dry sterile cotton balls, medical alcohol wipes or cotton swabs soaked in alcohol, sterile gloves, a biohazard sharp container, a biohazard trash container, goggles and face mask, and paper towels. A medical lancet (a traditional three-edge needle) or the seven-star needle (a bleeding tool), is selected for bleeding UB40委 中Weizhong in this case.
4) The patient lies prone with the legs extended, UB40委 中Weizhong is facing upward.
5) The room should be well lit.
6) After the therapist washing hands thoroughly, and wipe the hands with a single-use napkin. Put on sterile gloves.
7) Wipe the acupuncture point UB40委 中Weizhong with a medical alcohol wipe or a cotton swab soaked in alcohol. Move from the center of the acupuncture point to the periphery.
8) Place the medical lancet at 90 degrees angle to the skin.

9) In one quick move, prick the needle into the acupuncture point just about 0.1 Cun 寸 deep and allow a few drops of blood to discharge.
10) If more drops of blood are needed, absorb the blood with a sterile cotton ball.
11) Lastly, the point is pressed with a sterile cotton ball until the bleeding ceases.
12) Dispose of the tools in the proper containers.
13) Wash hands thoroughly.

Electro-Acupuncture Therapy
UB17膈 俞Geshu ~SI3後 溪Houxi 2/100 hz, 20 minutes daily.

Tui Na Treatments
Tui Na Techniques
Pressing, kneading, striking, pushing, rubbing

Tui Na Procedures
1. The patient is in a prone position; the therapist stands next to the lumbar region.
2. Applies thumb pad pressing techniques to press and knead the following acupuncture points to promote Qi circulation for the purpose of pain management, the acupuncture points are SI11天 宗Tianzong, UB13肺 俞Feishu, UB15心 俞Xinshu, UB17膈 俞Geshu, UB18肝 俞Ganshu, UB20脾 俞Pishu
3. Applies palm kneading, pushing and rubbing techniques on the upper back tender regions. The therapist stands near the patient's head facing the patient, the patient is still on prone. The therapist places both palms on bilateral Urinary Bladder meridians and apply pressure to perform kneading, pushing and rubbing from the acupuncture point UB13 肺 俞Feishu to UB20脾 俞Pishu. Repeat several times until the upper back is warm. Figure 6.3.1. Figure 6.3.2. Figure 6.3.3.

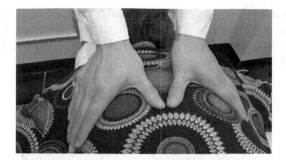

4. With the patient in a sitting position, the therapist stands at the affected side. For instance, the pain is on the right upper back, so the therapist stands at the right side facing the patient.
5. Place the right forearm at the patient's right armpit to lift the patient's right shoulder joint upward vertically, and utilizes the left hand to regulate the right upper back's painful muscles especially the spasm muscles and the trigger points. Figure 6.3.4. Figure 6.3.5. Figure 6.3.6. Figure 6.3.7.

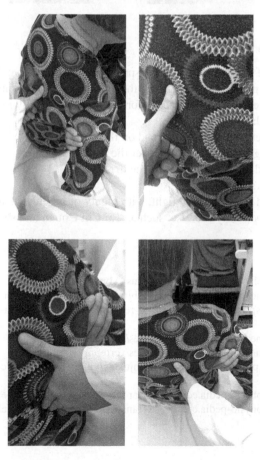

6. Ask the patient to repeatedly inhale deeply and cough; meanwhile, apply percussion with the heel of the left hand on the patient's right upper back where the trigger points located for about 2-3 times. Figure 6.3.8. Figure 6.3.9.

PATIENT ADVISORY

1. The onset of Costovertebral Joints sprain may be able to partially release by holding the posture and managing the breath to relax the muscles.
2. Warm up the traumatic regions before exercise especially in the cold, rainy and windy days.
3. Advise patients to apply warm pad on the painful area
4. Avoid lifting object for 1 week after the first treatment
5. Recommend patients to sleep on the firm bed
6. Keep the torso warm
7. Food should be simple and light, with plenty of green vegetables.
8. Routine upper body exercise, such as Tai Chi Chuan and swimming.

NOTES

1. https://www.physio-pedia.com/Trapezius_Myalgia.
2. https://teachmeanatomy.info/upper-limb/muscles/shoulder/extrinsic/.
3. https://excelsportspt.com/top-three-reasons-for-muscle-pain-between-shoulder-blades/.
4. https://teachmeanatomy.info/back/bones/vertebral-column/.
5. https://teachmeanatomy.info/back/bones/vertebral-column/.
6. https://www.physioadvisor.com.au/injuries/upper-back-chest/costovertebral-joint-sprain/.
7. https://www.physio-pedia.com/Costotransverse_Disorders#cite_ref-:4_17-0.
8. https://www.physio-pedia.com/Costotransverse_Disorders#cite_ref-:4_17-0.

7 Shoulder Pain

GENERAL CONCEPT OF SHOULDER PAIN IN CHINESE MEDICINE 中醫肩痛

INTRODUCTION

Shoulder pain is the pain occurring in the shoulder joints and the surrounding soft tissue, including pain from the bursae, the scapula region, and shoulder pain that refers to the arm.

Shoulder pain, whether it is caused by frozen shoulder, shoulder impingement, supraspinatus tendonitis, or neck conditions that refer pain to the shoulder region, is the medical condition associated with the Tai Yang meridians太 陽 經.

CHINESE MEDICINE MECHANISM OF DISEASE

Shoulder pain in Chinese medicine is called "Shoulder Bi Syndrome肩 痹" and "Shoulder Withering Syndrome萎 證". It is part of the Bi Syndrome 痹 證and Wei Syndrome萎 證.

1. Bi Syndrome 痹 證

Bi Syndrome is the main cause of chronic pain resulting from Qi and Blood Stagnation.

It is associated with the conditions of Qi and Blood Stagnation due to the invasion of Wind, Cold, Damp, and Heat.

Bi Syndrome痹 證 is caused by three factors: deficiency虛, external pathogens邪, and stasis of Blood 瘀.

Deficiency虛: the deficiency of Liver and Kidney is primary; deficiency of Blood and Qi is secondary.

- Due to the anatomic structures of the shoulder region consisting of bone and tendons, Liver governs Tendons 筋 (tendons, nerves, vessels, and ligaments) and Kidney governs Bones (bones, joints, cartilages, and fasciae). Chronic deficiency impairs the meridians leading to insufficiency of the Qi and Blood circulating in the meridians to nourish Tendons and Bones, resulting in the limitation of joint movement. When Liver and Kidney are deficient in the shoulder region, the Bone and ligaments lack nourishment and become tight, spasmodic, and stiff. Deficiency is the internal cause.
- The Liver function is to store Blood. Due to aging, the Blood returning to Liver is weak and there's less of it; Liver Blood deficiency may delay the self-healing of trauma in the shoulder.

DOI: 10.1201/9781003203018-7

External Pathogens邪: Wind風, Cold寒, and Dampness濕 are the pathogens; when they invade the meridians, they cause blockage. Even Wind is one of the pathogens, although it cannot cause blockage by itself – it is a pathogen inducing the other two to cause a blockage in the deficient area, such as the shoulder region.

In Chapter 43 of "Discourse on Blockage痺 論 篇 of Su Wen (素 問 *Plain Questions*)", a part of The Yellow Emperor's Canon of Internal Medicine (黃 帝 內 經, BC475–221) stated that

> Huang Di asked: "How does a blockage emerge?"
> 黃 帝 問 曰 : 痺 之 安 生 ?
> Qi Bo responded: "When the three Qi Wind, Cold, and Dampness arrive together, they merge and cause a block".
> 岐 伯 對 曰 : 風 寒 濕 三 氣 雜 至 合 而 為 痺 也 。

Stasis of Blood 瘀: There are two conditions under this category

- Unsmooth Blood flow in meridians.

Stress, anger, depression, anxiety, and any other emotional upset can cause stagnation of Qi in the meridians. Because Qi is the commander of Blood, the stagnation of Qi leads to stasis of Blood. When Qi and Blood are stagnant, the soft tissues over the shoulder region lack nourishment. This is the internal cause of the shoulder pain.

- Blood flowing outside meridians.

When the shoulder is under improper use, overuse, strain, or trauma, inflammation occurs in the bursae leading to the synovial fluid multiplying, and microscopic tears occur in tendons, ligaments, and muscles in the anatomic structures, resulting in pain, stiffness, and reduced range of motion. The color of the Blood may or may not be visible from the skin, but acute inflammation of the subacromial bursitis may present red skin accompanied by warm temperature. The appearance of the fluid from aspiration of the bursa may be bloody.

Other causes of chronic pain from Qi and Blood Stagnation can be due to physical trauma.

1. Extended physical labor causing the body Qi 氣, Blood 血, Yin 陰, Yang 陽, and the body fluid 津液 to be excessively consumed. The invasion of the exterior pathogen of Wind, Cold, and Damp block the meridians when the body is deficient.
2. The shoulder is pressed due to habitual side-sleeping position, leading to poor circulation of Qi and Blood. Cold invades and stays in the shoulder meridians resulting in Qi and Blood stagnation.

In Chapter 34 of "Maladjustments逆 調 論 篇 of Su Wen (素 問 *Plain Questions*)", a part of The Yellow Emperor's Canon of Internal Medicine (黃 帝 內 經, BC475–221), the Yellow Emperor and his physician Qi Bo岐 伯 have a discussion:

Huang Di: "When a person's muscle is numb and remains numb although his muscle touches the cotton clothes, what disease is this?"

帝曰：人之肉苛者，雖近亦絮，猶尚苛也，是謂何疾？

Qi Bo: "The Essence Qi is depleted; the Guard Qi is replete. When the Essence Qi is depleted, the skin and muscles become numb. When the Guard Qi is depleted, the limbs loss of function. When both the Essence Qi and the Guard Qi are depleted, numbness and debility occur.

岐伯曰：榮氣虛，衛氣實也，榮氣虛則不仁，衛氣虛則不用，榮衛俱虛，則不仁且不用。

1. Wei Syndrome萎 證

Wei syndrome is the medical condition involving motor impairment to the limbs, due to muscular weakness or atrophy. It is a group of symptoms characterized by muscular flaccidity, weakness, or atrophy of the extremities with motor impairment or paralysis.

Chinese medicine believes that Chronic deficiency of Liver and Kidney leads to the injury of Du meridian 督脈, stasis Blood blocks meridians, Yang Qi fails to ascend, resulting in weakness of the limbs.

CHINESE MEDICINE ETIOLOGY AND PATHOLOGY

Wind Cold Invasion Shoulder Pain风 寒 肩 痛

Invasion of Wind while sweating or sleeping at night with improper blankets.

Sweat pores open when sweating, Wind brings in Cold to the superficial level of the body, they both deplete Yang Qi, and block the local meridians in the shoulder region resulting in pain.

The sun delivers Yang energy, while the moon delivers Yin energy. Yin energy makes people cold at night. The body's natural temperatures continue to drop through the evening and at night while sleeping. A nice blanket can help warm up the body to prevent the invasion of Wind-Cold. When the shoulder is not covered with a blanket, Wind-Cold invades the shoulder causing blockage to result in pain.

In Chapter 10 of "The Generation and Functions of Five Viscera五 藏 生 成 篇 of Su Wen (素 問 *Plain Questions*)", a part of The Yellow Emperor's Canon of Internal Medicine (黃 帝 內 經, BC475–221), the Yellow Emperor and his physician Qi Bo岐 伯 had a discussion as follows:

When one walks to the outdoor immediately after sleeping and is attacked by the blow of wind, Bi (blockage) occurs when Blood is congealed on the surface of the body臥 出 而 風 吹 之，血 凝 於 膚 者 為 痹. The blockage from Wind Cold Invasion is usually only affecting the body's superficial level.

Phlegm Damp Obstruction Shoulder Pain痰 濕 肩 痛

The shoulder pain is caused by the invasion of Cold, Damp, and Wind. Sleeping on a wet and cold place, or exposure to cold water while sweating, causes the invasion the pathogens of Cold and Damp; such pathogens usually go deeper than the pathogen of Wind, but the disease process takes longer time. The longer duration of the condition leads to deficiency of Qi and Blood. The invasion stays in the muscle and tendon level, resulting in pain.

There is an explanation of how the pathogens invading the body in Chapter 27 of "Bi Syndrome Occurring Whole Body 週 痹 of Ling Shu (靈 樞Spiritual Pivot)" a part of The Yellow Emperor's Canon of Internal Medicine (黃 帝 內 經, BC475–221). The Yellow Emperor's physician Qi Bo岐 伯 explained how Bi Syndrome occurred under invasion of external pathogens:

> After the invasion Wind, Cold and Damp from the Exterior toward the Interior, the pathogens stay in between muscles, they squeeze the Body Fluid between the muscles into the condensed fluid but then it becomes Phlegm when meeting the Cold pathogen. Since Phlegm is a tangible thing, the muscles are more broken up, fission results pain.

Blood Stasis Shoulder Pain 淤 血 肩 痛

This type of shoulder pain is caused by Blood stasis. It can be the result from a direct blow to the shoulder (e.g., falling down with the shoulder hitting the ground, shoulder joint dislocation that strains the soft tissues), sprain or strain (e.g., throwing a tennis ball for your dog, serving a wet tennis ball when raining, badminton practice while fatigued, or overstretching the shoulder joint tissues while swimming or reaching to overhead cabinets).

Another cause of Blood stasis in the shoulder is a recalcitrant shoulder pain under the Phlegm Damp Obstruction. Chronic injury of the condition precipitates Blood stasis.

Wind invasion often attacks the chronic traumatic regions. Wind usually doesn't cause blockage by itself; it always brings Cold, Damp, or Heat to block meridians, resulting in pain in the shoulder region.

Blood Deficiency Shoulder Pain血 虛 肩 痛

Due to old age or indulgence and deficiency of Liver, Kidney, and Spleen under a chronic condition or after a severe disease, both Qi and Blood are insufficient to support the needs of the shoulder functions.

Movement of Blood follows Qi movements, and Blood reaches the shoulder when there is enough Qi in the shoulder. The movements of the shoulder joint rely on both Qi and Blood nourishment. The deficiency of Liver, Kidney, and Spleen causes undernourishment of the shoulder, resulting in pain.

Physical Examination

- **Medical History**
  - Inquire about the general health conditions, such as systemic diseases that could be risk factors. Cervical degeneration may lead to shoulder pain. Shoulder pain may be a referred pain from liver, gallbladder, heart, and spleen.

- The patient's age. Frozen shoulder often is seen in patients who are 40–60 years old.
- Job occupation that requires the arm reaching high above the shoulder is also a risk factor to consider with frozen shoulder.
- Sports that need to bring the arm over the shoulder or to throw an object overhead such as swimming, throwing a baseball, and playing tennis or volleyball.
- Pain severity. The pain is mild in the beginning when the range of motion is still not limited; the severity increases by the degrees of limitation. The pain decreases when the muscles are atrophic. The pain is improved or worsened by shoulder and arm movements.
- Location of the pain, the pain begins in the anterior aspect of the shoulder then gradually moves to the whole joint. Radiating pain to the biceps or triceps.
- Time of the pain occurrence. The pain is greater at night, when it is raining, and in cold weather.
- Self-care or other treatments the patient has tried and whether heat or cold application improves the pain.
- **Inspection**
  - Observe both shoulders and both scapulae to note if there is a presence of asymmetry. Both scapulae need to be observed posteriorly, looking for winging (if the winging is static, it is a neuromuscular pathology, while dynamic is usually from shoulder pain). Check the musculature for signs of atrophy in the trapezius, supraspinatus, and deltoid muscles.
  - Observe the shoulder shape: is it a round shoulder or square shoulder. A round shape shoulder may indicate the hypertonia of the muscles around the shoulder joint, while square shoulder may indicate an atrophy condition of the muscles or a dislocation of the glenohumeral (GH) joint. A round shoulder is more noticeable by a lateral inspection while a square shoulder is easier to observe by anterior inspection.
  - Inspect the shoulder's skin color for bruising, redness, deformity, scars, and edema. They may be associated with acute shoulder trauma, such as fractures and acute bursitis, or other shoulder conditions, such as bone tuberculosis. Usually, edema of the shoulder can be easily visualized in the supraclavicular fossa, but the fullness of the supraclavicular fossa may be due to an enlarged lymph node.
- **Palpation**
  - Examine the position between the humerus bone and the scapula: misalignment can lead to frozen shoulder.
  - Palpate to locate the tender spots, and mark them to determine the relationship between the anatomic structures and the spots.
  - Palpate the skin temperature for warm or cold, and note if there is a presence of swelling.

- Palpate for crepitus with movement, which may indicate osteoarthritis, tendinopathy, and fracture.
- There are key palpable structures to not miss, including the acromioclavicular (AC) joint, sternoclavicular (SC) joint, clavicle, coracoid process, greater/lesser tuberosity of the humerus, bicipital groove, rotator cuff muscle insertions, long head biceps tendon, pectoralis major, deltoid muscle, trapezius muscle, biceps/triceps muscles, teres major muscle, teres minor muscle, supraspinatus muscle, and infraspinatus muscle.
- Palpate on sternoclavicular (SC) and acromioclavicular (AC) joints for any tenderness on the transverse ligament, prominence of the joint, floating sensation of the medial/lateral end of the clavicle, or crepitus when shrugging the shoulders.
- **Range of Motion**

Normal Range
- Begin by testing the active movements by asking the patient to stand upright, ask the patient to perform the shoulder flexion, extension, abduction, adduction, internal rotation, and external rotation. While testing, gently place a hand on the patient's acromioclavicular (AC) joint to stabilize the scapula in order to prevent the patient raising it.

Abduction
- The therapist stands behind to observe the rhythm of arm movement and the scapular movement.
- Ask the patient to raise the arm (elbow extended) sideway from touching the body to a full range. The arm is able to touch the ear with the elbow extended. Then slowly lower the arm back down to the side of the body.
- The normal range of abduction is 180° with elbow extended and 90° with elbow flexed 90°.
- In abduction, the scapula almost has no movement in the first 30° abduction. The ratio of the movement of the glenohumeral (GH) joint and the scapulothoracic joint (ST) joint is 2:1, the scapula movement is approximately 1/3 of the 180° abduction.
- The clavicle elevation of the sternoclavicular (SC) joint in abduction is approximately 40°, the movement of the scapula finishes in the first 90°.
- The acromioclavicular (AC) joint movement is approximately 20°, the first part of the movement is in the first 30° abduction, the second part is in the abduction above 135°.

Adduction
- Ask the patient to bring the arm across in front of the body, the arm should be able to cross the abdomen by about 45° or the elbow is on the midline with the elbow flexed 90°.

Flexion

- Ask the patient to raise the arm (elbow extended) forward from touching the body to a full range above head.
- The normal range of flexion is 170° to 180°.

Hyperextension

- Ask the patient to swing the arm (elbow extended) backward from touching the body to a full range behind the body.
- The normal range of extension is 45° to 60°.

Internal Rotation

- Ask the patient to clench a fist on the sacrum with the thumb tip pointing upward.
- Lift the fist upward along the spine.
- The normal range of internal rotation is that the fist can touch T7

External Rotation

- With the arm at the side, elbow flexed at 90°
- Ask the patient to rotate the arm outward
- The normal range of external rotation is 45°

Pathogenic Range

- The test should include two parts. Always initially begins the active test with scapular stabilization to ensure an accurate measurement of movement; if loss of motion is observed, then it is followed by the passive tests to differentiate the true loss of motion from pain-related guarding.
- Shoulder pain occurs in abduction in the range before 30° and after 135, it indicates a subluxation of AC joint.
- Severe shoulder pain occurs in abduction in a very small range, it indicates a shoulder dislocation or a fracture.
- Shoulder pain occurs in abduction, the higher the more pain, it indicates a frozen shoulder.
- Shoulder pain occurs before 90° abduction, the pain worsens near 90°, then the pain decreases when the abduction passes 90°, it indicates a subdeltoid bursitis.
- Shoulder pain occurs in abduction between 60° and 120°, it indicates a rotator cuff tear or shoulder impingement syndrome.
- **Orthopedic Tests**

Empty Can Test (Jobe's Test)

- The patient is seated, the arm is elevated to 90° abduction and 30° forward flexion in the scapular plane, the elbow is extended. The forearm is in pronation, and the hand is facing out, the thumb is pointing down.
- The therapist stabilizes the shoulder while applying a downwardly directed force to the arm, and the patient tries to resist the downward pressure.

- The test is positive if the patient experiences shoulder pain or weakness. The positive result indicates the supraspinatus tendon pathology.

Patte Test (Hornblower's Test)

- The patient is seated, the arm is elevated to 90° abduction and 30° forward flexion in the scapular plane, the elbow is flexed in 90° and is externally rotated.
- The therapist applies an internal rotation force to the tested forearm while the force is resisted by the patient's externally rotated force.
- The test is positive if the patient experiences shoulder pain or weakness. The positive result indicates teres minor and infraspinatus pathology.

Bear Hug Test

- The patient is seated and asked to place the palm of the hand onto the opposite shoulder with the elbow anterior to the body
- The therapist applies an external rotation force to the tested forearm while the force is resisted by the patient's internal rotated force to maintain placing the palm on the opposite shoulder.
- The test is positive if the patient experiences shoulder pain or weakness. The positive result indicates the subscapularis muscle pathology.

Apley's Scratch Test

- The patient is seated and places the hand on the sacrum.
- The patient is asked to raise the hand from the sacrum upward along the spine to reach as high as possible.
- The test is positive if the patient is unable to reach T7. The positive result indicates a pathology of the glenohumeral adduction, internal rotation and scapular retraction with downward rotation.

Hawkins Test

- The patient is seated, the arm is flexed to the shoulder height, the elbow is flexed in 90°.
- The therapist applies an internal rotation force to the tested forearm.
- The test is positive if the patient experiences shoulder pain. The positive result indicates subacromial impingement or rotator cuff tendonitis.

Speed's Test

- The patient can be seated or standing, the therapist stands at the tested site.
- The arm is forward flexed at 60° with the elbow extended and the hand supinated.
- The therapist places a hand on the shoulder to stabilize the shoulder, and places the other hand on the elbow to apply downward pressure.
- The patient tries to resist the downward force by forward flexion of the shoulder.

- The test is positive if the patient experiences shoulder pain in the bicipital groove. The positive result indicates long head biceps tendonitis.

Yergason Test

- The patient is seated, the elbow is flexed in 90° with the thumb tip pointing up.
- The therapist applies force to stabilize the wrist to resist the patient's movements, while the patient tries to actively supinate and flex the elbow.
- The test is positive if the patient experiences shoulder pain. The positive result indicates biceps tendonitis.

Cross-Arm Adduction Test (Scarf Test)

- The patient is seated, the arm is forward flexed at 90° with the elbow flexed at 90°. The hand is placed onto the opposite shoulder.
- The therapist pushes the arm into further horizontal adduction.
- The test is positive if the patient experiences shoulder pain over the AC joint. The positive result indicates AC joint pathology.

MANIFESTATIONS

Wind Cold Invasion Shoulder Pain风 寒 肩 痛

The pain usually is not severe, the duration is short, and the intensity is low. The pain quality is dull, numb, or tight. The shoulder range of motion is not limited. The area of the shoulder pain may occur only in the shoulder joint or may radiate to the scapula, or it may occur in the anterior aspect of the shoulder and radiate to the arm.

It is commonly accompanied by tightness in the neck, cold in the shoulder, and the pain is alleviated by warm application.

Tongue: white coating.

Pulse: superficial or normal pulse.

Phlegm Damp Obstruction Shoulder Pain痰 濕 肩 痛

The pain usually is severe and the duration long. The range of motion is usually not limited, but the patient may not want to move the arm because movement aggravates the shoulder pain.

Range of motion of the affected shoulder may become limited if the condition becomes chronic. It is commonly accompanied by feeling cold in the shoulder and aversion to cold. The pain intensity is alleviated when using a warm application, but it is usually just temporary: the pain returns to the prior intensity when the warm application is removed. The quality of sleep is affected due to the severe pain, which may also affect appetite and work. Symptoms include spontaneous sweating, shortness of breath, and fatigue. If the duration of pain is too long, Qi becomes deficient, and the patient could easily catch a common cold.

Tongue: pale body, thin white coating.

Pulse: wiry or wiry and thin.

Blood Stasis Shoulder Pain 淤 血 肩 痛

The pain usually is severe and lingering, the quality is stabbing with tender spots, and is more severe at night and less in the day. Treatments with warming and Damp draining may be able to relieve some degree of pain, but it doesn't stop the pain and the result doesn't last long.

Swelling may occur in acute trauma; it may not be visible in the chronic state, and the area of pain may be wide with no obvious tender spot.

Tongue: ecchymosis.

Pulse: thin and choppy.

Blood Deficiency Shoulder Pain 血 虚 肩 痛

The pain in the shoulder is deeper and feels like spasm. The pain is worse when fatigued, and improves after rest. It is accompanied by a pale face, shortness of breath, fatigue, dizziness, palpitation, and insomnia.

Tongue: pale body, thin white coating.

Pulse: deep, thin.

TREATMENTS

Wind Cold Invasion Shoulder Pain 风 寒 肩 痛

Treatment Principle

Expel Wind, disperse Cold, open the meridians, and stop pain.

Herbal Treatments

Herbal Formula

Juan Bi Tang 蠲 痹 湯

This formula is from "A Book of Formulas to Promote Well-Being 嚴 氏 濟 生 方", authored by Yan Yonghe 嚴 用 和 (1200~1268), in Song dynasty 南 宋 (1127–1239), and published in 1253.

Herbal Ingredients

Dang Gui當 歸9g, Chi Shao赤 芍9g, Jiang Huang薑 黃9g, Huang Qi黃 耆9g, Qiang Huo羌 活9g, Gan Cao甘 草3g, Sheng Jiang生 薑15g, Da Zao大 棗3 pieces

Ingredient Explanations

- Dang Gui當 歸 nourishes the Blood, benefits the Liver, and regulates menstruation.
- Chi Shao赤 芍 moves the Blood, disperses Bruising, and stops pain.
- Jiang Huang薑 黃 moves the Blood, opens the meridians, expels Wind, and reduces swelling.
- Huang Qi黃 耆 tonifies Qi, strengthens the Spleen, raises the Yang Qi of the Spleen and Stomach, tonifies Wei Qi, stabilizes the exterior, and tonifies the Blood.
- Qiang Huo羌 活 expels Wind-Cold-Dampness, unblocks painful obstruction, and alleviates pain.

- Gan Cao甘 草 harmonizes the herbs, tonifies Middle Jiao Qi, clears Heat, expectorates phlegm, and stops cough.
- Sheng Jiang生 薑 unblocks the pure Yang pathway and harmonizes rebellious Qi.
- Da Zao大 棗 tonifies the Spleen Qi, nourishes the Blood, moderates and harmonizes the harsh properties of the fragrant herbs.

Modifications
- For bringing the formula to the shoulder, add Gui Zhi桂 枝.
- For Cold symptoms, add Fu Zi附 子.
- For Damp symptoms, add Cang Zhu蒼 朮, Fang Ji防 己, and Yi Yi Ren薏 苡 仁.
- For Wind symptoms, add Fang Feng防 風.
- For Blood stasis, add Tao Ren桃 仁, Hong Hua紅 花, and Di Long 地 龍.

Formula Indication
- Shoulder pain due to Wind-Damp-Cold Bi Syndrome.

Formula Contraindication
- It is contraindicated for patients with Damp-Heat condition.

Phlegm Damp Obstruction Shoulder Pain痰 濕 肩 痛
Treatment Principle
Expel Wind and dissipate Cold, regulate Blood and stop pain.

Herbal Treatments
Herbal Formula
Qiang Huo Sheng Shi Tang 羌 活 勝 濕 湯
  This formula is from "Clarifying Doubts about Damage from Internal and External Causes 內 外 傷 辨 惑 論", authored by Li Dongyuan李 東 垣 (1180–1251), in Jin dynasty金 朝 (1115–1234), and published in 1232.

Herbal Ingredients
Qiang Huo羌 活3g, Du Huo獨 活3g, Gao Ben藁 本1.5g, Fang Feng防 風1.5g, Chuan Xiong川 芎1.5g, Man Jing Zi蔓 荊 子0.9g, Zhi Gan Cao甘 草 炙1.5g

Ingredient Explanations
- Qiang Huo羌 活 expels Wind-Cold-Dampness, unblocks painful obstruction, alleviates pain.
- Du Huo獨 活 expels Wind, Damp, and Cold in Lower Jiao and joints; removes chronic Bi stagnation in lower limb.
- Gao Ben藁 本 expels Wind and drains Damp in Tai Yang meridians.
- Fang Feng防 風 expels Wind and drains Damp.
- Chuan Xiong川 芎 moves the Blood and moves Qi.

- Man Jing Zi蔓 荊 子 expels Wind and stops pain.
- Zhi Gan Cao甘 草 炙 tonifies the Spleen, harmonizes the other herbs, and guides the herbs to all 12 meridians.

Modifications
- To make the formula stronger for draining Damp, add Er Chen Tang二 陳 湯.

Formula Indication
- Headache due to Damp, Phlegm, and Wind and heavy sensation of the head.

Contraindications
- Caution for patients who suffer from Yin deficiency and Heat in any condition, such as Wind Heat.

Blood Stasis Shoulder Pain 淤 血 肩 痛
Treatment Principle
Unblock and relax the meridians, and move the Blood.

Herbal Formula
Shu Jin Huo Xie Tang 舒 筋 活 血 湯
  This formula is from "Restoration of Health from the Myriad Diseases 萬 病 回 春", authored by Gong Tingxian 龔 廷 賢 (1522~1619), in Ming dynasty 明 朝 (1368–1644), and published in 1587.

Herbal Ingredients
Qiang Huo 羌 活6g, Fang Feng 防 風9g, Jing Jie 荊 芥6g, Du Huo 獨 活9g, Dang Gui 當 歸12g, Xu Duan 續 斷12g, Qing Pi 青 皮5g, Niu Xi 牛 膝9g, Wu Jia Pi 五 加 皮 9g, Du Zhong 杜 仲9g, Hong Hua 紅 花 6g, Zhi Ke 枳 殼6g

Ingredient Explanations
- Qiang Huo 羌 活 expels Wind-Cold-Dampness, unblocks painful obstruction, alleviates and stops pain, and guides Qi to the Tai Yang and Du channels.
- Fang Feng 防 風 releases the exterior, expels External Wind, and expels Wind-Dampness.
- Jing Jie 荊 芥 stops bleeding, dispels Wind, and relieves muscle spasm.
- Du Huo 獨 活 expels Wind Damp and stops pain.
- Dang Gui 當 歸 moves and tonifies the Blood, disperses Cold, and stops pain due to Blood Stasis.
- Xu Duan 續 斷 nourishes the Liver and Kidney, strengthens Bones and Tendons, and alleviates pain.
- Qing Pi 青 皮 soothes Liver Qi and breaks up stagnant Qi, dries Dampness, and transforms Phlegm.
- Niu Xi 牛 膝 expels Wind Dampness, tonifies the Liver and Kidneys, and directs herbs downward.

- Wu Jia Pi 五加皮 expels Wind and Damp, nourishes and warms the Liver and Kidney, and strengthens Bones and Tendons.
- Du Zhong 杜仲 tonifies the Kidneys and Liver, strengthens Bones and Tendons and lower blood pressure.
- Hong Hua 紅花 moves the Blood, releases Bruising, opens the meridians, and stops pain.
- Zhi Ke 枳殼 removes stagnant food, promotes Qi movement, and reduces distention and pressure.

Modifications

- For deficiency of Blood陽氣虛, add Dan Shen丹參 and E Jiao阿膠.
- For deficiency of Qi氣虛, add Ren Shen人參.
- For spasm, add Zhi Jing San 止痙散.

Formula Indication

- Bi Syndrome of Blood.

Formula Contraindication

- It is contraindicated for patients during pregnancy and patients with weak digestion.

Blood Deficiency Shoulder Pain血虛肩痛

Treatment Principle

Warm the meridian, benefit Qi, and open Bi blockage.

Herbal Treatments

Herbal Formula

Huang Qi Gui Zhi Wu Wu Tang 黃耆桂枝五物湯

This formula is from "Shang Han Lun 傷寒雜病論", authored by Zhang Zhong-Jing 張仲景 (150–219) in eastern Han dynasty 東漢 (25–220), and published in 200–210.

Formula Ingredients

Huang Qi黃耆9g, Bai Shao白芍9g, Gui Zhi桂枝9g, Sheng Jiang生薑18g, Da Zao大棗12 pieces

Ingredient Explanation

- Huang Qi黃耆 tonifies Qi, strengthens the Spleen, raises the Yang Qi of the Spleen and Stomach, tonifies Wei Qi, stabilizes the exterior, and tonifies the Blood.
- Bai Shao白芍 nourishes Blood, soothes Liver, and stops pain.
- Gui Zhi桂枝 warms the meridians and relieves pain.
- Sheng Jiang生薑 unblocks the pure Yang pathway, and harmonizes rebellious Qi.

- Da Zao大棗 tonifies the Spleen Qi, nourishes the Blood, and moderates and harmonizes the harsh properties of the fragrant herbs.

Modifications

- For excessive Cold, add Xi Xin細辛 and Gan Jiang乾薑.
- For deficiency of Yang陽氣虛, add Fu Zi附子.
- For deficiency of Qi氣虛, add Ren Shen人參.

Formula Indication

- Bi Syndrome of Blood.

Acupuncture Treatments

In treating shoulder pain, Excess or Deficiency or mix of both needs to be clearly differentiated. Applying reducing treatments for deficient condition worsens the shoulder pain, and tonifying the excessive condition slows down the treatment process.

The patient's constitution changes quickly according to shoulder activities, and deficiency can become excess if there is anger or a new shoulder injury or overusing the shoulder in overhead activities.

Acupuncture Point Formula

Basic Points

SJ14肩髎Jianliao, LI15肩髃Jianyu, SI10臑俞Naoshu, SJ5外關Waiguan, LI4合谷Hegu, Jianqian肩前

Modifications

- For more prominent in Wind, add UB12風門Fengmen and GB20風池Fengchi.
- For more prominent in Damp, add LI11曲池Quchi and SP9陰陵泉Yinlingquan.
- For more prominent in Blood stasis, add SI9肩貞Jianzhen, UB17膈俞Geshu, ST38條口Tiaokou, and SP10血海Xuehai.

Electro-Acupuncture Therapy

Jianqian肩前~ LI15肩髃Jianyu; 2/100 hz for 20 minutes daily

Moxibustion Therapy

Warming needling techniques on Jianqian肩前, LI15肩髃Jianyu, SJ14肩髎Jianliao, SI10臑俞Naoshu for 2–3 moxa corn (3 gram each corn)

Cupping Therapy

Jianqian肩前, LI15肩髃Jianyu, SJ14肩髎Jianliao for 10 minutes Figure 7.1.1 from shutterstock.com #1997808338

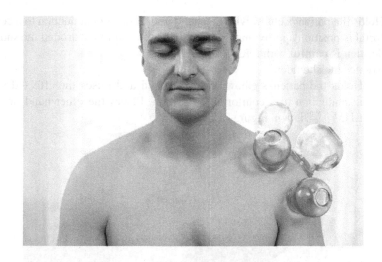

Bleeding Therapy

For Blood stasis condition, select Jianqian肩 前, LI15肩 髃Jianyu, and SJ14肩 髎Jianliao for bleeding before cupping Therapy.

Auricular Therapy

Shoulder, Neck, Endocrine, Liver, Shenmen, Kidney

Teding Diancibo Pu (TDP) Therapy

Jianqian肩 前, LI15肩 髃Jianyu, SJ14肩 髎Jianliao for 30 minutes

Tui Na Treatments

Tui Na Techniques

Pressing, kneading, grasping, pinching, plucking, rotating

Tui Na Procedures

Patient is in sitting position, the therapist stands near the affected shoulder.

1. Kneads with the thumb on the ipsilateral neck and shoulder Gallbladder meridian from GB20風 池Fengchi, GB21肩 井Jianjing for 3–5 times.
2. Grasps GB21肩 井Jianjing and the surrounding soft tissues for 3–5 times.
3. Holds the patient's elbow with one hand and raises the affected arm in abduction to a comfortable position approximately 45°. Pinches and kneads with the fingers on both sides of the deltoid muscle toward the elbow for 3–5 times.
4. Repeat step 3 with grasping and kneading techniques for 3–5 times.
5. Presses and kneads with thumb on SI11天 宗Tianzong, LI15肩 髃Jianyu, and LI11曲 池Quchi for 3–5 times on each acupuncture point.

6. Holds the patient's elbow with one hand and raises it in abduction to a comfortable position approximately 45°. Plucks the soft tissue around the shoulder that is painful with force for 3–5 times.
7. Rotates the shoulder
    a. Holds the patient's elbow with one hand and raises the affected arm in abduction to a comfortable position. Places the other hand on the LI15肩 髃Jianyu. Figure 7.1.2

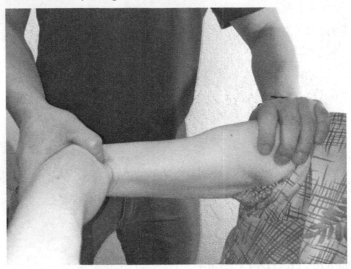

    b. Rotates the shoulder joint beginning with small circles and increases the range of circles gradually at a comfortable level until there is a slight pain. The rotation needs to be gentle and slow.
8. Abducting the shoulder
    a. Holds the patient's elbow with one hand and raises it in abduction to a position that begins to feel pain.
    b. Pinches the anterior and posterior aspects of deltoid muscle to relax and ease the pain.
    c. Continues to abduct the shoulder to another height that causes pain.
    d. Repeat step b.
    e. Increases the abducting angle approximately 5–10 degrees per treatment.
    f. Daily treatment is recommended.
    g. The abduction needs to be gentle and slow.
9. Flexing the shoulder
    a. Holds the patient's elbow with one hand and raises it in flexion to a position that begins to feel pain.
    b. Pinches the anterior and posterior aspects of deltoid muscle to relax and ease the pain.
    c. Continues to flex the shoulder to another height that causes pain.
    d. Repeat step b.

    e. Increases the flexing angle approximately 5–10 degrees per treatment.

    f. Daily treatment is recommended.

    g. The flexion needs to be gentle and slow.

10. Hyperextending the shoulder

    a. Holds the patient's elbow with one hand and raises it in hyperextension to a position that begins to feel pain.

    b. Pinches the anterior and posterior aspects of the deltoid muscle to relax and ease the pain.

    c. Continues to hyperextend the shoulder to another height that causes pain.

    d. Repeat step b.

    e. Increases the hyperextending angle approximately 5–10 degrees per treatment.

    f. Daily treatment is recommended.

    g. The hyperextension needs to be gentle and slow.

11. Holds the patient's elbow with one hand and raises it in abduction to a comfortable position approximately 45°. Applies rolling techniques in the anterior and posterior aspects of the shoulder for 1–3 minutes.

12. Extends the affected arm in abduction to a comfortable level, holds the wrist with one hand, and places the other hand over the acromion. Slightly shakes the arm for about a minute.

13. End the treatment by either one of two below:

    a. For deficiency, rub the arm with both palms.

    b. For excess, strike the shoulder joint with fingers. Figure 7.1.3

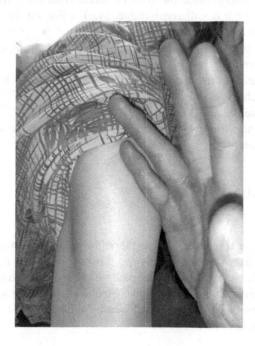

PATIENT ADVISORY

- Keep the shoulder warm, avoid wearing cold and/or wet clothes.
- Avoid drinking and eating cold, raw foods.
- Avoid sleeping on the side of the shoulder pain.
- Avoid overwork or any activity that lasts too long.
- Regular exercise is recommended, it helps to promote the circulation of the Qi and the Blood.
- If the range of motion of the shoudler is limited, practice the shoulder in flexion. I highly recommend an exercise called "climbing wall".
  1. Stand near a wall, face the wall for flexion climbing wall.
  2. Place the unaffected hand on the wall at the highest point it can reach to, mark a line where the fingertips touch on the wall. The line is the destination where the affected hand needs to reach to.
  3. Place the affected hand on the wall at the highest point it can reach to, and mark this point on the wall.
  4. Slowly walk the fingers to climb the wall to a higher point where the affected shoulder has more pain. Mark this point with a line and write the date and time on the line. Please do not give up the climbing due to the pain – keep the hand on the point for about 2–3 minutes.
  5. Repeat step 4 a few times; the goal is to repeat 20 times daily.
  6. Mark a new line whenever the affected hand climbs to a higher point.
- If the range of motion of the shoudler is limited, practice the shoulder in abduction.
  1. Stand at a door, place the affected forearm on the wall to the highest point where it is still comfortable. Mark a line on the wall where the fingertips can touch to.
  2. Lean the trunk forward and let the forearm on the wall stretch the chest.
  3. Keep stretching for a minute. Repeat 3–5 times.
  4. Place the affected forearm at a new higher point where it is still comfortable.
  5. Repeat steps 2 and 3.
  6. Mark a new line whenever the affected hand is able to reach a new, higher point.

FROZEN SHOULDER五十肩

INTRODUCTION

The medical term for frozen shoulder is "shoulder adhesive capsulitis" or "scapulo-humeral periarthritis" in western medicine. "Frozen" describes the condition that is painful and hard to move; the more the pain is felt, the less likely the patients will use the shoulder. Unfortunately, the less the shoulder is used, the harder it is to move, and it becomes "frozen" in its position. Frozen shoulder is most commonly seen in people between the ages of 40 and 60.

ANATOMY

- The shoulder consists of 3 bones
  - Humerus bone.
  - Scapula bone.
  - Clavicle bone.
- The shoulder comprises of three anatomic joints and one functional joint
  - Glenohumeral joint (GH, anatomic joint).
  - Acromioclavicular joint (AC, anatomic joint).
  - Sternoclavicular joint (SC, anatomic joint).
  - Scapulothoracic joint (ST, functional joint).

The Glenohumeral Joint

- The GH joint is between the scapula and the humerus. It is one of the most mobile joints in the human body.
- The glenohumeral joint is articulated at the head of the humerus with the glenoid cavity of the scapula. The joint is covered with hyaline cartilage.
- The joint capsule extends from the anatomical neck of the humerus to the rim of the glenoid fossa, and it encloses the structures of the joint.
- Ligaments: Superior glenohumeral, middle glenohumeral, inferior glenohumeral, coracohumeral, and transverse humeral.
- Nerve innervation: subscapular nerve, suprascapular nerve, axillary nerve, and lateral pectoral nerve.
- Blood supply: anterior and posterior circumflex humeral, circumflex scapular, and suprascapular arteries.
- Movements: flexion, extension, abduction, adduction, lateral rotation, and medial rotation. During the overhead movement of the arm abduction, the GH joint has approximately 120° elevation (Giangarra and Manske, 2018).

Acromioclavicular Joint

- The AC joint articulates the acromion of scapula and the acromial end of the clavicle.
- Ligaments: superior and inferior acromioclavicular ligament and coracoclavicular ligament.
- Nerve innervation: lateral pectoral nerve, suprascapular nerve.
- Blood supply: thoracoacromial artery and suprascapular artery.
- Movements: flexion, extension; abduction and adduction; medial and lateral rotation. During the overhead movement of the arm abduction, the AC joint has approximately 20° elevation.

Sternoclavicular Joint

- The SC joint articulates the sternum with the clavicle.
- Ligaments: anterior and posterior sternoclavicular ligaments; interclavicular and costoclavicular ligaments.

- Nerve innervation: medial supraclavicular nerve and nerve to subclavius.
- Blood supply: suprascapular artery and internal thoracic artery.
- Movements: protraction, retraction, elevation, depression, and axial rotation. During the overhead movement of the arm abduction, the SC joint has approximately 40° elevation (Kiel, Ponnarasu, and Kaiser 2021).

The Scapulothoracic Joint

Although the ST joint doesn't articulate with a bone, the articulation depends on the integrity of the anatomic AC and the SC joints. Any movement of the scapula must result in movement at the AC and SC joints. During the overhead movement of the arm abduction, the ST joint has approximately 60° upward rotation of the scapula.

The Synovial Bursae

The synovial bursae line the inner surface of the joint capsule to produce synovial fluid to the articular surfaces. In the shoulder region, there are six bursae – the first four are the major bursae in the shoulder.[1]

1) Subdeltoid bursa is inferior to the deltoid muscle and superior to the head of the humerus.
2) Subacromial bursa is superficial to the supraspinatus tendon, which separates the tendon from the acromion, the coracoid, and the coracoacromial ligament which lie above; its function is to reduce friction beneath the deltoid. The friction causes subacromial bursitis, which often causes shoulder impingement syndrome.[2]
3) Subscapular bursa is between the subscapularis tendon and the anterior surface of the scapula, where it reduces wear and tear on the tendon during movement at the shoulder joint. It communicates with the glenohumeral joint.
4) Subcoracoid bursa is superficial to the subscapularis tendon and deep to the conjoined tendons of the coracobrachialis and short biceps tendons.
5) Infraspinatus bursa is between the infraspinatus tendon and the capsule of the joint.
6) Subcutaneous acromial bursa is superior to the acromion and inferior to the skin.

The primary muscles involved in the action of arm abduction include the supraspinatus, deltoid, trapezius, and serratus anterior.

WESTERN MEDICINE ETIOLOGY AND PATHOLOGY

The exact cause in western medicine is not fully understood. It is an adhesive capsulitis, it is a disorder in which the shoulder capsule becomes inflamed and stiff, greatly restricting motion and causing chronic pain. It is believed that there are many possible risk factors.

Some risk factors may cause frozen shoulder including diabetes,[3] an autoimmune disease, stroke, cardiovascular disease, Parkinson's disease,[4] tuberculosis, reduced mobility of the shoulder or prolonged immobilization after shoulder joint dislocation, fractures, soft tissue injury (e.g., rotator cuff injury, bursitis), and shoulder surgery.

CHINESE MEDICINE ETIOLOGY & PATHOLOGY

1. **Qi and Blood Deficiency (氣血不足) Frozen Shoulder, Tendons 筋 (tendons, nerves, vessels, ligaments, fascia, bursae, joint capsules, and may even include cartilages and glenoid labrum) Lack of Nourishment from** Blood
   a. Due to old age or indulgence and deficiency of Liver, Kidney, and Spleen under a chronic condition or after a severe disease, both Qi and Blood are insufficient to support the needs of the shoulder functions.
   b. Movement of Blood follows Qi movements, Blood reaches the shoulder when there is enough Qi in the shoulder. The movements of the shoulder joint rely on both Qi and Blood nourishment. The deficiency of Liver, Kidney, and Spleen causes undernourishment of the shoulder, resulting in pain.

2. **Qi and Blood Stagnation (氣滯血瘀) Frozen Shoulder**
   The frozen shoulder is followed by the existing shoulder pain including Wind Cold Invasion shoulder pain, Phlegm Damp Obstruction shoulder pain, and Blood Stasis shoulder pain.

3. **Liver and Kidney Deficiency (肝腎虧虛) Frozen Shoulder**
   The shoulder lacks nourishment from Qi and Blood due to old age or a lingering disease leading to deficiency of Liver and Kidney. Insufficient Yin degrades the quality of Tendons 筋, insufficient Yang weakens the Blood supply to the shoulder joints. Liver manifests on the quality of Tendons 筋, Kidney manifests on the quality of bones and joint. Chinese medicine holds that Liver and Kidney are in one family 肝腎一家, they support each other, and the deficiency of either one will always affect the other.

MANIFESTATIONS

In Western Medicine
The Three Stages of Frozen Shoulder

1. *Pain Stage*
   Shoulder pain slowly increases, any movement of the shoulder causes pain, the pain can be dull or sharp, and the location can be around the shoulder joint. The pain may worsen at night especially when lying on the side of the painful shoulder. The shoulder range of motion in this stage is still not limited passively, but it is found that patients are afraid to move the shoulder.

2. *Adhesion Stage*
   The shoulder pain intensity is high, and the duration of pain is long; the pain is worse at night and can disturb sleep, especially when the weather is cold and rainy. The shoulder range of motion is limited.

3. *Atrophy Stage*

The shoulder range of motion is limited. The shoulder pain is less. Muscle atrophy occurs on deltoid muscle, trapezius muscle, supraspinatus muscle, and infraspinatus muscle.

In Chinese Medicine

1. *Qi and Blood Deficiency*氣血不足

The pain is dull or tight, sometimes it is like a spasm. The pain can be worsened by fatigue, it is improved by rest. The shoulder is still in full range of movement passively, but patients often are afraid to move it. It is accompanied by a pale face, fatigue, dizziness, shortness of breath, palpitation, and insomnia.

Tongue: pale body, thin white coating.

Pulse: deep, thin.

2. *Qi and Blood Stagnation*氣滯血瘀

Shoulder pain is a severe stabbing pain, with tenderness on the shoulder that refuses pressure. The pain is worse at night, with sleep disturbance. The shoulder range of movement is limited. The shoulder pain may radiate to the arm and neck. The pain is aggravated by emotional upset.

Tongue: ecchymosis.

Pulse: thin and choppy.

3. *Liver and Kidney Deficiency*肝腎虧損

Shoulder pain is sore. It is accompanied by a pale face, dizziness, tinnitus, sore and weak lower back and knee, poor appetite, loose stool, spontaneous sweating, aversion to wind, and nocturia.

Tongue: small body size, pink body, thin or peeled coating.

Pulse: thin, short or wiry, thin.

PHYSICAL EXAMINATION

- **Medical History**
  - General health conditions, such as systemic diseases that could be risk factors. Cervical degeneration may lead to shoulder pain. Shoulder pain may be a referred pain from liver, gallbladder, heart, and spleen.
  - The patient's age. Frozen shoulder often is seen in patients around 40–60 years old.
  - Job occupation that requires reaching high above the shoulder is also a risk factor to consider with frozen shoulder.
  - Sports that require bringing the arm over the shoulder or throwing an object overhead such as swimming, baseball, tennis, or volleyball.
  - Pain severity. The pain is mild in the beginning when the range of motion is still not limited; the severity increases by the degrees of limitation. The pain decreases when the muscles are atrophic. The pain is improved or worsened by shoulder and arm movement.

- Location of the pain: the pain begins in the anterior aspect of the shoulder then gradually moves to the whole joint. Radiating pain to biceps or triceps.
- Time of the pain occurrence: the pain is worse at night, especially during rainy and cold weather.
- Self-care or other treatments the patient has tried; whether heat or cold application improves the pain.
- **Inspection**
  - Patients with severe frozen shoulder may have lost the natural arm swing that occurs with walking (Ewald 2011).
  - Observe both shoulders and both scapulae to note if there is a presence of asymmetry. Both scapulae need to be observed posteriorly; look at musculature for signs of atrophy of the trapezius, supraspinatus, and deltoid muscles.
  - Observe the shoulder shapes: is it a round shoulder or square shoulder. A round shoulder may indicate the hypertonia of the muscles around the shoulder joint, while a square shoulder may indicate an atrophy condition of the muscles or a dislocation of the GH joint. A round shape shoulder is more noticeable by a lateral inspection, while a square shoulder is more noticeable by anterior inspection.
  - Inspect the shoulder's skin color for bruising, redness, deformity, scars, and edema, as they may be associated with acute shoulder trauma, such as fractures and acute bursitis, or other shoulder conditions, such as bone tuberculosis. Usually, edema of the shoulder can be easily visualized in the supraclavicular fossa, but the fullness of the supraclavicular fossa may be due to an enlarged lymph node.
- **Palpation**
  - Examine the position between the humerus bone and the scapula; misalignment can lead to frozen shoulder.
  - Palpate to locate the tender spots, diffuse tenderness over the anterior and posterior shoulder, and mark them to determine the relationship between the anatomic structures and the spots. They may suggest another shoulder diagnosis or concomitant pathology, such as rotator cuff or biceps tendinopathy (Ewald 2011).
  - Palpate the skin temperature for warm or cold; palpation can also detect if there is swelling. Frozen shoulder palpation usually presents a cooler temperature.
  - Palpation for crepitus with movement may indicate osteoarthritis, tendinopathy, and fracture.[5]
  - There are key palpable structures to not miss, including the AC and SC joints, clavicle, coracoid process, greater/lesser tuberosity of the humerus, bicipital groove, rotator cuff muscle insertions, long head biceps tendon, pectoralis major, deltoid muscle, trapezius muscle, biceps/triceps muscles, teres major muscle, teres minor muscle, supraspinatus muscle, and infraspinatus muscle.

- Palpate the SC and AC joints for tenderness on the transverse ligament, prominence of the joint, floating sensation of the medial/lateral end of the clavicle, or crepitus when shrugging the shoulders.[6]
- **Range of Motion**

Normal Range

- Test the active movements by asking the patient to stand upright, ask the patient to perform the shoulder flexion, extension, abduction, adduction, internal rotation, and external rotation. While testing, gently place a hand on the patient's AC joint to stabilize the scapula in order to prevent the patient raising it.

Abduction

- The therapist stands behind to observe the rhythm of arm movement and the scapular movement.
- Ask the patient to raise the arm, elbow extended sideways from touching the body to a full range. The arm is able to touch the ear with the elbow extended. Then slowly lower the arm back down to the side of the body.
- The normal range of abduction is 180° with elbow extended and 90° with elbow flexed 90°.
- In abduction, the scapula has almost no movement in the first 30° of abduction. The ratio of the movement of the GH joint and the ST joint is 2:1, the scapula movement is approximately one-third of the 180° abduction.
- The clavicle elevation of the SC joint in abduction is approximately 40°, the movement of the scapula finishes in the first 90°.
- The AC joint movement is approximately 20°, the first part of the movement is in the first 30° of abduction, the second part is in the abduction above 135°.

Adduction

- Ask the patient to bring the arm across in front of the body, the arm should be able to cross the abdomen by about 45°, or the elbow is on the midline with the elbow flexed 90°.

Flexion

- Ask the patient to raise the arm, elbow extended forward from touching the body to a full range above the head.
- The normal range of flexion is 170° to 180°.

Hyperextension

- Ask the patient to swing the arm (elbow extended) backward from touching the body to a full range behind the body.
- The normal range of extension is 45° to 60°.

Internal Rotation

- Ask the patient to clench a fist on the sacrum with the thumb tip pointing upward.
- Lift the fist upward along the spine.
- The normal range of internal rotation is that the fist can touch T7.

External Rotation

- With the arm at the side, elbow flexed at 90°.
- Ask the patient to rotate the arm outward.
- The normal range of external rotation is 45°.

Pathogenic Range

- The test should include two parts. Always begin the active test with scapular stabilization to ensure an accurate measurement of movement; if loss of motion is observed, then it is followed by the passive tests to differentiate the true loss of motion from pain-related guarding.
- Shoulder pain occurs in abduction, the higher the more pain, and it indicates a frozen shoulder.
- Loss of motion with flexion, abduction, and external and internal rotation can be common in frozen shoulder. Apley scratch test is used for the internal rotation movement, mark down the highest vertebral level the fist can reach to. Patients with second and third stage frozen shoulder often present flexion below 110°, abduction below 90°, and internal rotation with the fist touching only the sacrum.

- **Orthopedic Tests**

Testing should not be selected only for frozen shoulder; it should be performed to assess for other conditions for the purpose of differentiation.

Empty Can Test (Jobe's Test)

- The patient is seated, the arm is elevated to 90° abduction and 30° forward flexion in the scapular plane, the elbow is extended. The forearm is in pronation and the hand is facing out, the thumb is pointing down.
- The therapist stabilizes the shoulder while applying a downwardly directed force to the arm, and the patient tries to resist the downward pressure.
- The test is positive if the patient experiences shoulder pain or weakness. The positive result indicates the supraspinatus tendon pathology.

Patte Test (Hornblower's Test)

- The patient is seated, the arm is elevated to 90° abduction and 30° forward flexion in the scapular plane, and the elbow is flexed in 90° and is externally rotated.
- The therapist applies an internal rotation force to the tested forearm while the force is resisted by the patient's externally rotated force.
- The test is positive if the patient experiences shoulder pain or weakness. The positive result indicates teres minor and infraspinatus pathology.

Bear Hug Test

- The patient is seated and asked to place the palm of the hand onto the opposite shoulder with the elbow anterior to the body
- The therapist applies an external rotation force to the tested forearm while the force is resisted by the patient's internal rotated force to maintain placing the palm on the opposite shoulder.

- The test is positive if the patient experiences shoulder pain or weakness. The positive result indicates the subscapularis muscle pathology.

Apley's Scratch Test

- The patient is seated and places the hand on the sacrum.
- The patient is asked to raise the hand from the sacrum upward along the spine to reach as high as possible.
- The test is positive if the patient is unable to reach T7. The positive result indicates a pathology of the glenohumeral adduction, internal rotation, and scapular retraction with downward rotation.

Hawkins Test

- The patient is seated, the arm is flexed to shoulder height, the elbow is flexed in 90°.
- The therapist applies an internal rotation force to the tested forearm.
- The test is positive if the patient experiences shoulder pain. The positive result indicates subacromial impingement or rotator cuff tendonitis.

Neer's Test

- The patient is standing, the therapist stabilizes the patient's scapula with one hand and, meanwhile, passively flexes the arm while it is internally rotated.
- The test is positive if the patient experiences shoulder pain. The positive result indicates subacromial impingement.

Yergason Test

- The patient is seated, the elbow is flexed in 90° with the thumb tip pointing up.
- The therapist applies force to stabilize the wrist to resist the patient's movements while the patient tries to actively supinate and flex the elbow.
- The test is positive if the patient experiences shoulder pain. The positive result indicates biceps tendonitis.

Speed's Test

- The patient is seated, the arm is 60° to 90° forward flexion, the elbow is fully extended and the forearm is supinated.
- The therapist applies a downward pressure force to the tested forearm while the force is resisted by the patient's upward flexion force.
- The test is positive if the patient experiences shoulder pain in the bicipital groove. The positive result indicates the biceps tendon, subacromial impingement or a superior labral anterior to posterior (SLAP) tear pathology. Repeat the test with forearm pronation. The patient may still report pain, but the severity of the shoulder pain will be less and the pain in the bicipital groove is vague.

Cross-arm Adduction Test (Scarf Test)

- The patient is seated, the arm is forward flexed at 90° with the elbow flexed at 90°. The hand is placed onto the opposite shoulder.
- The therapist pushes the arm into further horizontal adduction.
- The test is positive if the patient experiences shoulder pain over the AC joint. The positive result indicates AC joint pathology.

Apprehension Test

- The patient is seated. The therapist stabilizes the scapula with one hand, and with the other hand, grasps the wrist and moves the patient's arm to 90° abduction and at 90° elbow external rotation.
- The test is positive if the patient doesn't experience shoulder pain but the patient has an apprehension over the face. The positive result indicates GH joint instability (e.g., GH joint dislocation).

DIFFERENTIAL DIAGNOSIS

Biceps Tendinopathy

Character of pain: Tenderness over long head of the biceps tendon.
   Test: positive Yergason test

Cervical Disk Degeneration

Character of pain: Shoulder pain radiating from neck with limited range of motion in neck.
   Test: positive Spurling test

Rotator Cuff Tendinopathy

Character of pain: Shoulder pain in painful arc (abduction 60°–120°)
   Test: positive Hawkins test

Subdeltoid Bursitis

Character of pain: Shoulder pain occurs before 90° abduction, the pain worsens near 90°, then the pain decreases when the abduction passes 90°
   Test: positive Speed's test

GH Joint Dislocation

Character of pain: Shoulder pain especially with movement and/or square shoulder
   Test: positive Apprehension Test

TREATMENTS

1. **Qi and Blood Deficiency氣血不足型**

Treatment Principle

Warm the meridian, benefit Qi, and open Bi blockage.

Herbal Treatments

Herbal Formula

Huang Qi Gui Zhi Wu Wu Tang 黃耆桂枝五物湯

This formula is from "Shang Han Lun 傷寒雜病論", authored by Zhang Zhong-Jing 張仲景 (150–219), in eastern Han dynasty 東漢 (25–220), and published in 200–210.

Herbal Ingredients

Huang Qi黃耆9g, Bai Shao白芍9g, Gui Zhi桂枝9g, Sheng Jiang生薑18g, Da Zao Da Zao大棗12 pieces

Ingredient Explanations

- Huang Qi黃耆 tonifies Qi, strengthens the Spleen, raises the Yang Qi of the Spleen and Stomach, tonifies Wei Qi, stabilizes the exterior, and tonifies the Blood.
- Bai Shao白芍 nourishes the Blood, soothes the Liver, and stops pain.
- Gui Zhi桂枝 warms the meridians and relieves pain.
- Sheng Jiang生薑 unblocks the pure Yang pathway and harmonizes rebellious Qi.
- Da Zao大棗 tonifies the Spleen Qi, nourishes the Blood, and moderates and harmonizes the harsh properties of the fragrant herbs.

Modifications

- For excessive Cold, add Xi Xin細辛 and Gan Jiang乾薑.
- For deficiency of Yang陽氣虛, add Fu Zi附子.
- For deficiency of Qi氣虛, add Ren Shen人參.

Formula Indication

Bi Syndrome of Blood.
   2. **Qi and Blood Stagnation氣滯血瘀型**

Treatment Principle

Unblock and relax the meridians and moves the Blood.

Herbal Formula

Shu Jin Huo Xie Tang 舒筋活血湯

This formula is from "Restoration of Health from the Myriad Diseases 萬病回春", authored by Gong Tingxian 龔廷賢 (1522～1619), in Ming dynasty 明朝 (1368–1644), and published in 1587.

Herbal Ingredients

Qiang Huo 羌活6g, Fang Feng 防風9g, Jing Jie 荊芥6g, Du Huo 獨活9g, Dang Gui 當歸12g, Xu Duan 續斷12g, Qing Pi 青皮5g, Niu Xi 牛膝9g, Wu Jia Pi 五加皮 9g, Du Zhong 杜仲9g, Hong Hua 紅花 6g, Zhi Ke 枳殼6g

Ingredient Explanations

- Qiang Huo 羌活 expels Wind-Cold-Dampness, unblocks painful obstruction, alleviates and stops pain, and guides Qi to the Tai Yang and Du channels.
- Fang Feng 防風 releases the exterior, expels External Wind, and expels Wind-Dampness.
- Jing Jie 荊芥 stops bleeding, dispels Wind, and relieves muscle spasm.
- Du Huo 獨活 expels Wind Damp and stops pain.
- Dang Gui 當歸 moves and tonifies the Blood, disperses Cold, and stops pain due to Blood Stasis.
- Xu Duan 續斷 nourishes the Liver and Kidney, strengthens Bones and Tendons, and alleviates pain.
- Qing Pi 青皮 soothes Liver Qi and breaks up stagnant Qi, dries Dampness, and transforms Phlegm.
- Niu Xi 牛膝 expels Wind Dampness, tonifies the Liver and Kidneys, and directs herbs downward.
- Wu Jia Pi 五加皮 expels Wind, Damp, nourishes and warms the Liver and Kidney, and strengthens Bones and Tendons.
- Du Zhong 杜仲 tonifies the Kidneys and Liver, strengthens Bones and Tendons, and lowers blood pressure.
- Hong Hua 紅花 moves the Blood, releases Bruising, opens the meridians, and stops pain.
- Zhi Ke 枳殼 removes stagnant food, promotes Qi movement, and reduces distention and pressure.

Modifications

- For deficiency of Blood陽氣虛, add Dan Shen丹參 and E Jiao阿膠.
- For deficiency of Qi氣虛, add Ren Shen人參.

Formula Indication

- Bi Syndrome of Blood.

Formula Contraindication

It is contraindicated for patients during pregnancy and patients with weak digestion.

3. **Liver and Kidney Deficiency** 肝腎虧損型

Treatment Principle

Warm the meridian and benefit Qi.

Herbal Treatments

Herbal Formula

Gui Fu Di Huang Tang 桂 附 地 黃 湯

This formula is from "Imperially Commissioned Golden Mirror of the Orthodox Lineage of Medicine醫 宗 金 鑑", authored by the Qianlong emperor 乾 隆 (1711–1799), in Qing dynasty 清 朝 (1636–1912), and published in 1742.

Herbal Ingredients

Shu Di Huang熟 地 黃12g, Shan Zhu Yu山 茱 萸6g, Shan Yao山 藥6g, Mu Dan Pi牡 丹 皮4.5g, Ze Xie澤 瀉4.5g, Fu Ling茯 苓4.5g, Zhi Fu Zi附 子(製) 3g, Ruo Gui 肉 桂3g

Ingredient Explanations

- Shu Di Huang熟 地 黃 tonifies the Liver and Kidney Yin and benefits Essence and the Blood.
- Shan Zhu Yu山 茱 萸 tonifies Liver and Kidney, strengthens Bones, benefits Essence and the Blood.
- Shan Yao山 藥 tonifies Kidneys and benefits the Yin and the Spleen Qi.
- Ze Xie澤 瀉 drains Damp, especially Damp Heat in Lower Jiao and clears Kidney deficient Heat.
- Fu Ling茯 苓 drains Damp and tonifies the Spleen.
- Mu Dan Pi牡 丹 皮 clears Heat, cools the Blood, and moves the Blood.
- Zhi Fu Zi附 子(製) disperses Cold and Dampness, warms the meridians, and stops pain.
- Ruo Gui 肉 桂 warms the meridians, expels Cold and opens the meridians to promote Blood circulation.

Modifications

- For stronger Qi tonic, add Bu Zhong Yi Qi Wan補 中 益 氣 丸.

Formula Indication

- Cold in joints and Wei syndrome.

Acupuncture Treatments

Acupuncture Point Formula

Basic Points

GB34陽 陵 泉Yanglingquan, ST38條 口Tiaokou, LI2二 間Erjian, LI4合 谷Hegu, LI11曲 池Quchi, SJ5外 關Waiguan, UB65束 骨Shugu

Jianqian肩 前 or Jianneiling肩 內 陵 (Midway between LI 15 and the anterior axillary crease), Jianhou肩 後 (1.5 Cun 寸 superior to the posterior axillary crease)

Special Points
Shoulder Three Needle 肩 三 針

- This is a set of 3 points, originally found by Jin Rui 靳 瑞. This set of points was a part of the points of the treatment for mentally retarded children. (Dr. Jin Rui directed the treatment in every workday morning, I studied and worked under his supervision at his experimental clinic in Guang Zhou in July 1993.)
- The first point is LI15肩 髃Jianyu. Traditionally, needling this point requires the shoulder abduction to locate a depression lateral to the acromion. Abduction is not needed when the point is a part of "Shoulder Three Needle".
- The second and the third points are 2 Cun 寸 to the first point.
- 1.5 Cun 寸 or 2 Cun 寸 needles are used to needle these 3 points. The needles are inserted at 1 Cun 寸 depth obliquely toward the deltoid muscle.
- Needle retention time is 30 minutes.
- Electro stimulation can be added to the needles.
- Moxibustion Therapy can be added to the needles for warming affects.

Modifications
- For shoulder pain near the acromion, add LI3三 間Sanjian.
- For shoulder pain in the posterior aspect and scapula, add SI3後 溪Houxi.
- For shoulder pain in the anterior aspect, add Yujian魚 肩 穴.

Yujian魚 肩 穴
- Yujian is a distal point for frozen shoulder, originally found by Gao Shuzhong高 樹 中.
- It is located 0.5 Cun 寸 distal to LU10魚 際Yuji.
- Before placing a needle, palpate the point to locate the painful knots for needling.
- Contralateral point selection is recommended.
- Ask the patient to exercise the affected shoulder while performing needle manipulation.
- Needle retention time is 2–5 minutes.

Auricular Therapy
Shoulder, Neck, Shenmen, Brain Point, Liver, Kidney, Endocrine

Cupping Therapy
- LI15肩 髃Jianyu
- Jianqian肩 前 or Jianneiling肩 內 陵 (Midway between LI 15 and the anterior axillary crease) Figure 7.2 from shutterstock.com #72466168

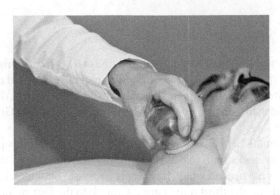

- Jianhou肩 後 (1.5 Cun 寸 superior to the posterior axillary crease).

Moxibustion Therapy

Warming needling techniques on

- LI15肩 髃Jianyu.
- Jianqian 肩 前 or Jianneiling肩 內 陵 (Midway between LI 15 and the anterior axillary crease).
- Jianhou肩 後 (1.5 cun superior to the posterior axillary crease).

Bleeding Therapy

- LI15肩 髃Jianyu

Electro-Acupuncture Therapy

- LI15肩 髃Jianyu ~ ST38條 口Tiaokou, 2/100 hz for 20 minutes.

Tui Na Treatments

Tui Na Techniques

Rubbing, rolling, kneading, grasping, rotating, shaking

Tui Na Procedures

The patient is seated, the therapist stands by the affected shoulder.

1. The therapist grasps the wrist with one hand and lifts the affected arm in abduction to the highest point that the patient can tolerate without pain.
2. Applies grasping technique on the trapezius muscle, deltoid muscle, biceps muscle, and triceps muscle for 3–5 minutes.

The patient is seated. The therapist stands by the affected shoulder, puts their foot on a stool, and places their knee in the affected armpit.

1. The therapist grasps LI4合 谷Hegu, SI3後 溪Houxi with one hand and, at the same time, with the other hand applies rolling techniques on the anterior, posterior and LI15肩 髃Jianyu for 3–5 times.
2. Applies grasping and kneading techniques on the shoulder and the arm.

The patient is seated, and the therapist stands by the affected shoulder.

1. The therapist places a hand on the shoulder joint while holding the affected elbow with the other hand.
2. Rotates the shoulder joint in slow speed, begin with small circles then gradually increases to the pain level the patient can tolerate for 3–5 circles in each direction.
3. If the affected hand is able to reach the top of the head, rotate the shoulder joint by circling the affected hand around the head.

The patient is seated, and the therapist stands by the affected shoulder.

1. The therapist grasps the shoulder joint with one hand and grasps the wrist with the other hand.
2. Shakes the shoulder.
   a. Draws the patient's hand toward the shoulder.
   b. Extends the hand slowly toward the side to the height and pain level the patient can tolerate.
   c. Shakes the arm at the end of the extending.
   d. Repeats the steps and shakes the arm posteriorly to the pain level the patient can tolerate.

The patient is seated, and the therapist stands in front of the affected shoulder.

1. The therapist overlaps the hands and places the patient's hand in between the hands.
2. Extends and flexes the affected arm to the front of the shoulder.
3. Rotates the affected shoulder for 3–5 times from small to large circles.
4. Grasps the patient's wrist on the affected side and pulls with one hand and, meanwhile, rubs the affected arm with the other hand from the wrist to the shoulder and back to the wrist.
5. Overlaps the hands and places the patient's hand in between.
6. Repeats step 3–5 for 3–5 times.

The patient is seated, and the therapist stands by the affected shoulder.

1. The patient's arm hangs down naturally.
2. The therapist rubs the affected arm from both biceps and triceps muscles to the wrist for 3–5 times.

PATIENT ADVISORY

- Keep the shoulder warm, avoid wearing cold or wet clothes.
- Avoid sleeping on the side of the shoulder pain.
- Avoid both physical and mental overwork or any activity that lasts too long.

- Use the shoulder appropriately to promote the circulation of Qi and Blood flow; avoid overhead activities for too long.
- If the range of motion of the shoudler is limited, practice the shoulder in flexion. I highly recommend an exercise called "climbing wall".
  1) Stand near a wall and face it for flexion climbing wall.
  2) Place the unaffected hand on the wall at the highest point it can reach to, and mark the point with a line. The line is the destination where the affected hand will try to reach.
  3) Place the affected hand on the wall at the highest point it can reach to, and mark it with a line.
  4) Slowly walk the fingers to climb the wall to a higher point where the affected shoulder has more pain. Mark a line and write the date and time on the line. Please do not give up the climbing due to the pain; keep the hand on the point for about 2–3 minutes.
  5) Repeat step 4 a few times; the goal is to repeat 20 times daily.
  6) Mark a new line whenever the affected hand climbs to a higher point.
- When the arms are always reaching high above the shoulder or in the flexion positions during work, such as typing or driving, the position is prone for frozen shoulder. The below exercise is helpful for preventing and improving the shoulder condition.
- If the range of motion of the shoudler is limited, practice the shoulder in abduction.
  1) Stand at a door and place the affected forearm on the wall to the highest where it is still comfortable. Mark a line on the wall where the finger-tips touch.
  2) Lean the trunk forward and let the forearm on the wall stretch the chest.
  3) Hold the stretch for a minute. Repeat 3–5 times.
  4) Place the affected forearm at a new higher point where it is still comfortable.
  5) Repeat steps 2 and 3.
  6) Mark a new line whenever the affected hand is able to reach a new, higher point.

SUPRASPINATUS TENDONITIS岡上肌肌腱炎

INTRODUCTION

Supraspinatus tendonitis is a common shoulder pain seen in laborers, athletes, housekeepers, and others whose work involves overhead shoulder activities. Due to its anatomic structure and location, it is one of four rotator cuff muscles, the pain from this tendonitis is so annoying, athletes may have to put the training on hold, it can quite disturb anyone's daily life from throwing a tennis ball for the dog to catch to put on a shirt or even just combing the hair, the pain affects near endless list of the daily arm movements.

ANATOMY

Four Rotator Cuff Muscles

The *supraspinatus muscle* arises from the supraspinous fossa superior portion of the scapula and is located deep in the trapezius muscle.

The supraspinatus tendon passes laterally beneath the cover of the acromion and inserts into the superior facet of the greater tubercle of the humerus. The tendons of the infraspinatus and teres minor also insert into the greater tubercle of the humerus respectively and inferiorly to the supraspinatus tendon. The subscapularis muscle attaches to the lesser tubercle of the humerus (https://en.wikipedia.org/wiki/Supraspinatus_muscle).

Nerve Supply

Both the supraspinatus muscle and the infraspinatus muscle are innervated by the suprascapular nerve, it comes from the upper trunk of the brachial plexus, which is a branch of C5-C6.

Blood Supply

The supraspinatus receives its blood supply from the suprascapular and the dorsal scapular arteries.

Function

The supraspinatus muscle is responsible for arm abduction; it stabilizes the shoulder by pulling the head of the humerus medially towards the glenoid cavity and preventing the head of the humerus from slipping inferiorly.

WESTERN MEDICINE ETIOLOGY AND PATHOLOGY

Although the causes of the supraspinatus can be many, the primary cause is impingement.

The impingement changes of the supraspinatus tendon of the rotator cuff commonly occur under the acromion as it passes beneath the acromion to attach the greater tubercle of the humerus head.

These are some of the extrinsic factors causing impingement:

- Increased subacromial loading from using poor form when lifting or pulling on something heavy leading to an acute tear.[7]
- Direct repetitive microtrauma.
- Bone spurs developed on the bones near the supraspinatus tendon can create friction to the tendon causing inflammation, tears, and pain.
- Aging is a factor; with age, poor blood circulation can occur.
- Glenohumeral instability.
- Long head biceps tendon instability or weakness.

If the tendonitis is not properly treated in the early stages, it can become worse and lead to supraspinatus tendinosis.

- Subacromial bursal thickening and fibrosis.
- Calcific tendinopathy.

CHINESE MEDICINE ETIOLOGY AND PATHOLOGY

Jin Bi Syndrome 筋 痹

The cause of supraspinatus tendonitis in Chinese medicine is a stagnation in the meridians by Wind-Cold invasion in the shoulder region. The stagnation is called "Bi Syndrome痹 證"; generally, its etiopathology is believed to be from the invasion of the external pathogen and the deficiency of the internal Zang-Fu臟 腑.

Externally, Wind, an external pathogenic factor of mysterious nature, is thought to carry the pathogenic Cold into the body, such as from exposure to a wet, windy, and cooler environment, and contribute to the variable nature of the shoulder pain.

Internally, there may be a deficiency of Liver and Kidney due to chronic repetitive supraspinatus tendon impingement. When a person works or sits outdoors in a rainy and cold environment and plays sports that require overhead arm movements, such as swimming, tennis, badminton, or does work that requires overhead arm-lifting, such as librarians, painters, or retail grocery store clerk. Such rainy, cold environment will affect the body by Wind bringing in the Damp and Cold into the meridians underlining the deficiency of Liver and Kidney due to the shoulder repetitive micro-trauma from overhead movement. When any soft tissue is injured or inflamed, Liver provides Blood and Yin to repair the trauma; Kidney provides Qi, Yin, and Yang, or even Essence, to assist in healing the joints. When the injury lesion takes too long to recover, and it becomes chronic, or the patient has been suffering from Liver and Kidney deficiency, or due to old age, both Liver and Kidney become deficient; hence, both Liver and Kidney are insufficient to assist the healing and repairing of the injured tendons and joints.

MANIFESTATIONS

In Western Medicine

- Shoulder pain or discomfort or feeling weakness when raising the arm above the shoulder, such as brushing hair, or putting the arm down from an overhead position, such as putting on a shirt or jacket.
- Shoulder pain occurs at the first 30° of abduction, and the painful arc ranges from 60° to 120° in abduction.[8]
- Initially the pain is felt only during arm activity. The progressive subdeltoid aching sensation is aggravated by abduction and overhead arm activity, and ultimately the pain can occur at rest.
- Shoulder pain may occur at night or even at rest during daytime.

- Clicking or grinding sensation in the shoulder acromion region with arm movements.
- The pain may radiate to the deltoid or the arm and may radiate to the neck.

In Chinese Medicine

Blood Stasis瘀 滯 型

Pain and swelling in the shoulder region; it is worse at night, the location of the pain is fixed, and pressure cannot be tolerated pressure. There is a clicking noise in the joint with shoulder movement.

Tongue: purple body with ecchymosis and a thin white or yellow coating.

Pulse: wiry or thin and choppy.

Wind-Cold-Damp Bi Syndrome 虚 寒 痹 型

Sore shoulder, fatigue, and exposure to cold worsen the pain; it is relieved by warm application. Pain occurs in both arms when lifting and dropping. Joint stiffness especially in abduction, and patients often shrug the shoulder in abduction.

Tongue: pale, thin, white coating.

Pulse: deep, thin, wiry, tight, and weak.

PHYSICAL EXAMINATION

- **Medical History**
  - General health condition, such as the injury history of the affected shoulder, arm activity at work, and any sports activity that require overhead movement.
  - It is also helpful to learn about the environment of the place where the patient resides in order to determine the patient's constitution and educate them about self-care during recovery. cold, dark, and moist conditions are factors that can cause the blockages.
  - A diet that is raw, cold, fatty, and sweet may lead to Cold accumulation in the meridians, resulting in blockages.
  - Going to bed later than midnight can worsen Liver and Kidney, leading to a slow recovery.
  - Sleep posture may be the cause or may delay the recovery. Sleeping with the affected shoulder near an open window, or sleeping on the affected shoulder should be avoided.
  - Emotional factors such as stress, anger, depression, and anxiety can directly affect Liver.
  - Other health conditions that can drain the Essence from Liver and Kidney will be an important medical history to obtain, as these medical conditions may be the cause and delay the recovery.
  - Obtain information regarding the pain location, the quality of pain, and the arm activity associated with the shoulder pain. Pain locations can indicate the associated muscles causing the shoulder pain; for instance,

the supraspinatus muscle is one of four muscles to cause rotator cuff pain, and the impingement of the supraspinatus often is the beginning of rotator cuff tendinosis. The quality of pain describes the condition of the supraspinatus muscle and tendon: sharp or stabbing can indicate a tear, while tenderness and a burning may indicate an inflammation of the tendon or muscle. The arm activity leading to the shoulder pain may indicate the type of shoulder pain; if the pain doesn't only occur in the painful arc but in higher abduction the pain is greater, then the shoulder pain is probably not supraspinatus tendonitis but rather a frozen shoulder.

- **Inspection**
  - The shoulders are inspected for symmetry, localized swelling, and muscle atrophy. Observe both shoulder joints and both scapulae, and compare the size of muscles and the position of the joints. Supraspinatus muscle atrophy may occur as a consequence of muscle disuse, tendon tear, or denervation. Reducing the use of the affected shoulder due to pain can lead to supraspinatus muscle atrophy. Denervation of the suprascapular nerve due to a history of traumatic shoulder dislocation may lead to supraspinatus muscle atrophy (Alomar, Powell, and Burman 2011).
- **Palpation**
  - Examine the shoulder joint. There may be tenderness below the acromion and over the greater tuberosity of the humerus. Internal rotation of the shoulder can facilitate palpation of the supraspinatus insertion on the greater tuberosity.
- **Range of Motion**
  - Shoulder pain occurs at the range of painful arc which is from 60° to 120° in abduction. As the condition progresses, the shoulder pain may occur at the first 30° of abduction (Yu, Cao and Wu 1985, 131).
- **Orthopedic Tests**
  - *Empty Can Test (Jobe's Test)*
  - The patient is seated, the arm is elevated to 90° abduction and 30° forward flexion in the scapular plane, the elbow is extended. The forearm is in pronation and the hand is facing out, the thumb is pointing down.
  - The therapist stabilizes the shoulder while applying a downwardly directed force to the arm, and the patient tries to resist the downward pressure.
  - The test is positive if the patient experiences shoulder pain or weakness. The positive result indicates the supraspinatus tendon pathology.
- **Special Tests**

X-Ray

Supraspinatus tendonitis cannot be diagnosed just by orthopedic tests and symptoms. If it is due to bone spurs or arthritis, the diagnosis can be challenging. X-rays, ultrasound, or an MRI can get a good look and tell the exact abnormal structure surrounding the tendons and the bones.

DIFFERENTIAL DIAGNOSIS

Frozen Shoulder

- The pain of frozen shoulder doesn't only occur in the painful arc, but it is the higher abduction that results in more pain.

Acromioclavicular Arthritis

- The shoulder pain occurs when crossing the arm over to the opposite shoulder.
- Tenderness is commonly limited to the acromioclavicular joint.

Suprascapular neuropathy and cervical radiculopathy

- These two conditions are common imitators of rotator cuff pain as the pain often radiates to biceps muscle and triceps muscle. Supraspinatus muscle is innervated by the suprascapular nerve, a branch of C5-C6. Meanwhile, the biceps muscle is also innervated by C5-C6. Triceps muscle is innervated by C7-C8. Patients with a pathology involving C5-C8 may present weakness of the rotator cuff muscles in during arm elevation and external rotation.[9]
- Performing the neurological examination can elicit the pathology of the cutaneous distribution of the nerve roots from C5 to T1.[10]
  1) C5 and C6 screen biceps reflex and elbow flexion.
  2) C7 and C8 screen triceps reflex and elbow extension.
  3) C5 screens abduction and external rotation.
  4) C6,7 and 8 screen adduction and internal rotation.
  5) C6 and C7 screen wrist extension and flexion.
  6) C7 and C8 screen finger flexion and extension.
  7) T1 screens finger abduction and adduction.
- The conditions affecting the nerve roots include the following[11]:
  1) Cervical spondylosis involving C5 and C6.
  2) Brachial plexopathy involving the suprascapular nerve.
  3) Suprascapular nerve entrapment at the suprascapular notch.
  4) Ganglion cyst at the spinoglenoid notch pressing the inferior branch of the suprascapular nerve.
  5) Traumatic severance in fractures.
  6) Iatrogenic injury.

TREATMENTS

Blood Stasis瘀滞型

Treatment Principle

Move the Blood, disperse Bruising, relax Tendons, open the meridians, move Qi, and stops pain.

Herbal Treatments

Fu Yuan Huo Xue Tang復元活血湯

This formula is from "Clarifying the Study of Medicine醫 學 發 明", authored by Li Dongyuan 李 東 垣, in Yuan dynasty元 朝 (1271–1368), and published in 1315.

Herbal Ingredients

Chai Hu 柴 胡9g Tian Hua Fen 天 花 粉9g, Dang Gui 當 歸9g, Hong Hua 紅 花6g Gan Cao 甘 草6g, Chuan San Jia 炮 穿 山 甲6g, Da Huang 大 黄12g, Tao Ren 桃 仁9g

Ingredient Explanations

- Chai Hu 柴 胡 harmonizes Shao Yang disorders and relieves stagnation.
- Tian Hua Fen 天 花 粉 clears Heat, expels pus, and reduces swelling.
- Dang Gui 當 歸 moves and tonifies the Blood, disperses Cold, and stops pain due to Blood Stasis.
- Hong Hua 紅 花 moves the Blood, releases Bruising, opens the meridians, and stops pain.
- Gan Cao 甘 草 releases spasms, alleviates pain, harmonizes the other herbs, and guides the herbs to all 12 meridians.
- Chuan San Jia 穿 山 甲 moves the Blood, reduces swelling, and expels Wind Damp.
- Da Huang 大 黄 moves the Blood and disperses Bruising.
- Tao Ren 桃 仁 breaks up stubborn Blood Stasis.

Formula action

Moves the Blood, disperses Bruising, soothes the Liver, and opens the meridians.

Formula Indication

Physical trauma, Blood stasis, severe pain.

Wind-Cold-Damp Bi Syndrome 虛 寒 痺 型

Treatment Principle

Move the Blood, open the channel, and tonify the Kidney and Liver.

Herbal Treatments

Huo Xue Shu Jin Tang活 血 舒 筋 湯

This formula is from "Handouts of Chinese Medicine Traumatology中 醫 傷 科 學 講 義", authored by Shanghai College of Traditional Chinese Medicine上 海 中 醫 學 院 in 1964.

Herbal Ingredients

Dang Gui Wei當 歸 尾12g, Chi Shao赤 芍12g, Jiang Huang姜 黄12 g, Shen Jin Cao伸 筋 草9 g, Song Jie松 節9g, Hai Tong Pi海 桐 皮9g, Lu Lu Tong路 路 通9g, Qiang Huo羌(獨)活9g, Fang Feng防 風9g, Xu Duan續 斷12g, Gan Cao甘 草5g

Ingredient Explanations

- Dang Gui Wei當 歸 尾 moves the Blood, disperses Bruising, unblocks stagnation, opens the meridians, and stops pain.
- Chi Shao赤 芍 moves the Blood, disperses Bruising, and stops pain.
- Jiang Huang薑 黃 moves the Blood, opens the meridians, expels Wind, and reduces swelling.
- Shen Jin Cao伸 筋 草 expels Wind, drains Damp, and relaxes the Tendons.
- Song Jie 松 節Dispels Wind, drains Damp, opens the meridians, and stops pain.
- Hai Tong Pi海 桐 皮 expels Wind, drains Damp, and reduces swelling.
- Lu Lu Tong路 路 通 moves the Blood, opens the meridians, expels Wind, and promotes water metabolism.
- Qiang Huo羌(獨)活 expels Wind-Cold-Dampness, unblocks painful obstruction, and alleviates pain.
- Fang Feng防 風 expels Wind and drains Damp.
- Xu Duan續 斷 tonifies Kidney and strengthens Bone.
- Gan Cao甘 草 harmonizes the herbs, tonifies Middle Jiao Qi, clears Heat, expectorates Phlegm, and stops cough.

Modifications

For severe pain, add Ru Xiang乳 香 and Mo Yao沒 藥.

Formula Actions

- Moves the Blood, disperses Bruising, soothes the Tendons, and opens the meridians.

Acupuncture Treatments

Acupuncture Point Formula
Basic Points
LI15肩 髃Jianyu, SJ14肩 髎Jianliao, SI9肩 貞Jianzhen, GB21肩 井Jianjing, SI11天 宗 Tianzong, Ashi acupoints 阿 是 穴, LI4合 谷Hegu

Modifications

UB23腎 俞Shenshu, UB20脾 俞Pishu, UBI8肝 俞Ganshu, UB17膈 俞Geshu

Auricular Therapy

Liver, Kidney, Shenmen, Shoulder, Endocrine

Cupping Therapy

Ashi acupoints 阿 是 穴, LI15肩 髃Jianyu, GB21肩 井Jianjing Figure 7.3.1 from Shutterstock.com #609056036

Moxibustion Therapy

Warming needling techniques on Ashi acupoints 阿是穴

Bleeding Therapy

Ashi acupoints 阿是穴

Electro-Acupuncture Therapy

Ashi acupoints 阿是穴 to GB21肩井Jianjing 2/100 hz for 20 minutes daily
Figure 7.3.2 from Shutterstock.com #1499365292

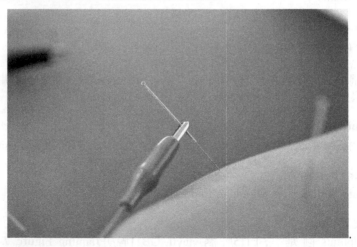

Tui Na Treatments

Tui Na Techniques
Pressing, kneading, rolling, grasping, pulling

Tui Na Procedures

The patient is in sitting position, the therapist stands behind the patient. Tui Na techniques should be with light force if the condition is in the acute stage. The force can be stronger in the chronic stage.

1. Applies rolling techniques on the shoulder for 2–3 minutes.
2. Applies grasping techniques on GB21肩 井Jianjing for 2–3 minutes.
3. Applies kneading techniques with thenar on the supraspinatus muscle for 2–3 minutes.
4. The patient bends the elbow. The therapist holds the patient's elbow and swings the shoulder forward and backward while pressing the greater tuberosity for 2–3 minutes.
5. The therapist stands beside the affected shoulder, both hands holding the patient's elbow, and the patient resists the force, while the therapist pulls the patient's arm downward for 2–3 minutes. Figure 7.3.3

PATIENT ADVISORY

1. Rest the shoulder in the acute stage, and avoid moving the shoulder in big range of movement especially when raising the arm overhead.
2. Keep the shoulder warm, avoid exposure to cold and rain.
3. Appropriate shoulder movements will be helpful in promoting circulation in the chronic stage.
4. Do not stay up late, go to bed before 11 p.m. Ensure eight hours of high quality sleep.

SHORT HEAD BICEPS INJURY 肱 二 頭 肌 短 頭 腱 損 傷

INTRODUCTION

The injury to the short head biceps tendon is not as common as the long head. But it is severe when the injury occurs, as the shoulder may lose a certain degree of movement. And the prognosis is worse than injury to the long head, as the short head injury may cause adhesion, causing permanent loss of some shoulder functions.

ANATOMY

The biceps brachii s a two-headed muscle. It is located anteriorly to the humerus.

The long head is located on the lateral side of the biceps brachii, while the short head is located on the medial side.

The long head biceps tendon originates from the supraglenoid tubercle of the scapula.

The short head biceps tendon originates from the coracoid process of the scapula. It passes anterior to the head and shaft of the humerus, converges with the long head around the middle of the humerus to the large muscle belly. The biceps brachii muscle inserts on the radius at the radial tuberosity and the fascia of the forearm via the bicipital aponeurosis.

Innervation: it is innervated by musculocutaneous nerve which is the terminal branch of the lateral cord of the brachial plexus (C5, C6, and C7).

Blood supply: it is supplied by the muscular branches of the brachial artery.

Function: the short head working together with the long head of the biceps brachii flexes and supinates the forearm at the elbow. At the shoulder joint, the short head adducts the humerus toward the body's midline and pulls the arm closer to the trunk It also acts as a fixator to stabilize the shoulder joint.

WESTERN MEDICINE ETIOLOGY AND PATHOLOGY

When the upper arm rotates, the lesser tuberosity of the humerus has rolling friction with the short head tendon. When the shoulder joint is abducted by 90°, the short head is stretched by 2 cm; when the shoulder joint is hyperextended by 40°, the short head is stretched by 3 cm. In such circumstances, there is no obvious change in tension of the long tendon. The long tendon of the biceps brachii is attached to the supraglenoid tubercle of the shoulder joint, and the short head is attached to the coracoid process. These two points and the fused point of the two heads of the biceps brachii around the middle of the humerus form a triangle. The short head becomes the long side of the triangle, and the drawing force is the greatest when the upper arm is abducted and hyperextended.

The shoulder can still function even after a complete tear of the long head because the short head takes parts of the functions of the long head; abduction can be lost and the shoulder movement can be affected if the short head is torn.

Short head biceps injury is from the excessive eccentric force as the arm is brought into abduction and hyperextension abruptly or forcefully from flexion leading to chronic or acute excessive traction on the tendon (Hsu, Anand, Mabrouk, and Chang 2022).

The long head biceps tendon is more likely to be injured. This is because it is vulnerable as it travels through the shoulder joint to its attachment point in the socket. Fortunately, the biceps has two attachments at the shoulder. The short head biceps tendon rarely tears. Because of this second attachment, most people can still use their biceps even after a complete tear of the long head.[12]

When you tear your biceps tendon, you can also damage other parts of your shoulder, such as the rotator cuff tendons.

CHINESE MEDICINE ETIOLOGY AND PATHOLOGY

1. Acute Stage: Qi and Blood Stagnation 氣滯血瘀

When the elbow is flexed, the biceps brachii muscle is in a tense state. If at this time, an external force excessively abducts or hyperextends the flexed upper limb, it could lead to laceration on the short head of the biceps attached to the coracoid process, resulting Qi and Blood stagnation.

2. Chronic Stage: Tendon Undernourishment 筋脈失養

Aging, long-standing illness, or long-term repetitive friction of the short head biceps tendon leads to Liver and Kidney deficiency and Qi and Blood insufficiency, resulting in Tendon undernourishment.

Repetitive exposure to Wind, Cold, and Damp leads to obstruction of Blood circulation, resulting in Tendon undernourishment.

MANIFESTATIONS

Stabbing pain in the anterior shoulder that is worse at night. Shoulder movement is restricted, especially when the upper arm is extended in the abduction and hyperextension position. In the severe condition, the affected hand is not able to reach the contralateral acromion or comb the hair.

PHYSICAL EXAMINATION

- **Medical History**
  - Establish general health conditions, such as a long-standing illness, diabetes, or sleep deprivation.
  - Inquire about the patient's occupation or sports activities that may require flexing the elbow and dragging it backward or arm-reaching, as these can be the risk factors that cause the short head biceps tendon to have chronic friction to the lessor tuberosity.
  - Obtain the pain sensation of the injured shoulder, when the shoulder pain is worse, and whether the pain location is in the anterior shoulder.

- **Inspection**
  - Observe the shoulder to note if there is bruised skin, redness, swelling, or atrophy.
- **Palpation**
  - Examine the shoulder, determine if the muscles in the anterior shoulder are stiff, and look for corresponding tender spots in the coracoid process of the shoulder joint.
- **Range of Motion**
  - Passive movement of the affected shoulder, especially during shoulder flexion, abduction, external rotation, and hyperextension, can aggravate the shoulder pain in the coracoid process.[13]The pain may decrease during shoulder adduction and inward rotation.
  - Inspect the shoulder range of motion when the affected hand is crossing over to the contralateral shoulder; the hand may not be able to reach the contralateral acromion in a severe condition.
- **Orthopedic Tests**
  - The shoulder resistance test can be positive in flexion and adduction, the elbow resistance flexion, and the forearm resistance pronation test.

DIFFERENTIAL DIAGNOSIS

- Pain in the short head biceps tendon needs to be differentiated from the long head biceps injury because they both have pain near each other in the anterior shoulder. The pain in long head biceps is located in the bicipital groove and it is positive in the Yergason test.
- Supraspinatus tendonitis has a painful arc, and the location of the pain is in the lateral and posterior aspect of the shoulder. The pain can radiate to the deltoid and triceps muscles.

TREATMENTS

Treatment Principle

Move Qi and Blood, drain Damp, reduce swelling, and stop pain.

Herbal Treatments

Acute Stage

Herbal Formula

Huo Xue Zhi Tong Tang 活 血 止 痛 湯

This formula is from "The Great Compendium of Chinese traumatology 傷 科 大 成", authored by Zhao Lian 趙 濂 (1798–1860), in Qing dynasty 清 朝 (1636–1912), and published in 1891.

Herbal Ingredients

Dang Gui 當 歸6g, Su Mu 蘇 木6g, Luo De Da 6g落 得 打6g, Chuan Xiong 川 芎2g, Hong Hua 紅 花1.5g, Ru Xiang 乳 香3g, Mo Yao 沒 藥3g, San Qi 三 七3g, Chi Shao 赤 芍3g, Chen Pi陳 皮 3g, Zi Jing Teng 紫 荆 藤9g, Di Bie Cong 地 鱉 蟲9g

Ingredient Explanations

- Dang Gui 當歸 moves and tonifies the Blood, disperses Cold, and stops pain due to Blood Stasis.
- Su Mu 蘇木 moves the Blood, reduces swelling, and stops bleeding.
- Luo De Da 6g落得打 reduces swelling, clears Heat, and drains Damp.
- Chuan Xiong 川芎 moves the Blood and Qi, expels Wind, and stops pain.
- Hong Hua 紅花 moves the Blood, releases Bruising, opens the meridians, and stops pain.
- Ru Xiang 乳香 moves the Blood and stops pain.
- Mo Yao 沒藥 moves the Blood, reduces swelling, and stops pain.
- San Qi 三七 moves the Blood, stops bleeding, reduces swelling, and stops pain.
- Chi Shao 赤芍 moves the Blood, reduces swelling, and stops pain.
- Chen Pi陳皮 dries Damp and transforms Phlegm.
- Zi Jing Pi 紫荊皮 moves the Blood, reduces swelling, and promotes urination.
- Di Bie Cong 地鱉蟲 breaks up stubborn Blood stasis.

Modifications

- For swelling, add Ze Lan澤蘭, Mu Tong木通, Che Qian Zi車前子, and Jiang Huang薑黃.
- For severe pain, add Yan Hu Suo延胡索, Niu Xi牛膝, and Mu Xiang木香.
- For swelling, redness, and hot sensation, add Mu Dan Pi牡丹皮, Zhi Zi栀子, and Huang Bai 黃柏.

Formula Indications

Blood Stagnation

Contraindications

Contraindicated during pregnancy.

Chronic Stage

Treatment Principle

Unblock and relax the meridians and move the Blood.

Herbal Formula

Shu Jin Huo Xie Tang 舒筋活血湯

This formula is from "Restoration of Health from the Myriad Diseases 萬病回春", authored by Gong Tingxian 龔廷賢 (1522~1619), in Ming dynasty 明朝 (1368–1644), and published in 1587.

Herbal Ingredients

Qiang Huo 羌活6g, Fang Feng 防風9g, Jing Jie 荊芥6g, Du Huo 獨活9g, Dang Gui 當歸12g, Xu Duan 續斷12g, Qing Pi 青皮5g, Niu Xi 牛膝9g, Wu Jia Pi 五加皮 9g, Du Zhong 杜仲9g, Hong Hua 紅花 6g, Zhi Ke 枳殼6g.

Ingredient Explanations

- Qiang Huo 羌活 expels Wind-Cold-Dampness, unblocks painful obstruction, alleviates and stops pain, and guides Qi to the Tai Yang and Du channels.
- Fang Feng 防風 releases the exterior, expels External Wind, and expels Wind-Dampness.
- Jing Jie 荊芥 stops bleeding, dispels Wind, and relieves muscle spasm.
- Du Huo 獨活 expels Wind Damp and stops pain.
- Dang Gui 當歸 moves and tonifies the Blood, disperses Cold, and stops pain due to Blood Stasis.
- Xu Duan 續斷 nourishes Liver and Kidney, strengthens Bones and Tendons, and alleviates pain.
- Qing Pi 青皮 soothes Liver Qi and breaks up stagnant Qi, dries Dampness, and transforms Phlegm.
- Niu Xi 牛膝 expels Wind Dampness, tonifies the Liver and Kidneys, and directs herbs downward.
- Wu Jia Pi 五加皮 expels Wind, Dampness, nourishes and warms the Liver and Kidney, and strengthens Bones and Tendons.
- Du Zhong 杜仲 tonifies the Kidneys and Liver, strengthens Bones and Tendons, and lowers blood pressure.
- Hong Hua 紅花 moves the Blood, releases Bruising, opens the meridians, and stops pain.
- Zhi Ke 枳殼 removes stagnant food, promotes Qi movement, and reduces distention and pressure.

Modifications

- For deficiency of Blood陽氣虛, add Dan Shen丹參 and E Jiao阿膠.
- For deficiency of Qi氣虛, add Ren Shen人參.
- For spasm, add Zhi Jing San 止痙散.

Formula Indication

- Bi Syndrome of Blood

Formula Contraindication

- It is contraindicated for patients during pregnancy and patients with weak digestion.

Acupuncture Treatments

Acupuncture Point Formula

Basic Points

LI15肩髃Jianyu,   Jianqian肩前,   LI11曲池Quchi,   SI11天宗Tianzong, GB34陽陵泉Yanglingquan, Ashi acupoints 阿是穴 (it is for chronic shoulder pain; the point should not be used if it is swelling and red in acute state)

Modifications
For chronic stage, add UB20脾 俞Pishu, UB23腎 俞Shenshu, UB24氣 海 俞Qihaishu, UBI8肝 俞Ganshu, and KD3太 溪Taixi.

Auricular Therapy
Shoulder, Shenmen, Liver, Kidney, Endocrine

Cupping Therapy
Jianqian肩 前

Moxibustion Therapy
Jianqian肩 前

Bleeding Therapy
Jianqian肩 前

Electro-acupuncture Therapy
Jianqian肩 前 ~LI11曲 池Quchi 2/100 hz for 20 minutes daily.

Tui Na Treatments
Tui Na Techniques
Rubbing, plugging, kneading, grasping,

Tui Na Procedures
Acute Stage
The force during Tui Na should be light for shoulder pain in acute stage.
   The patient is in sitting position, the therapist stands next to the affected shoulder.

1. The therapist grasps the patient's wrist with one hand and lifts the arm to the scapular plane (the arm is elevated to 90° abduction and 30° forward flexion), rubs the anterior shoulder with the other hand for 3–5 minutes. Figure 7.4

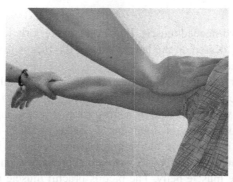

2. Applies kneading techniques with thumb pad on the Jianqian肩 前 for 3–5 minutes.
3. Applies grasping techniques on the biceps brachii muscle for 3–5 minutes.

Chronic Stage

Based on the procedures in acute stage, add the following steps.

1. While grasping the biceps brachii muscle, the therapist passively rotates the affected arm in abduction and hyperextension.
2. Applies plugging techniques on the tendons in the anterior shoulder.
3. Rubs the shoulder and the deltoid, biceps, and triceps muscles with both hands until warm.

PATIENT ADVISORY

1. The shoulder joint needs to rest in acute stage; for chronic stage, shoulder exercise is helpful in recovery.
2. Keep the shoulder warm.
3. Avoid exposure to wind and cold.

LONG HEAD BICEPS TENDONITIS 肱 二 頭 肌 長 頭 肌 腱 炎

INTRODUCTION

The tendonitis (or tendinitis) of long head biceps is an inflammation of the tendon that is commonly seen in arm activities involving overhead motion. Such repetitive motion causes impingement to the tendon. The risk factors are athletes whose sport requires overhead movements, such as swimming, tennis, and baseball. Many jobs and routine chores require overhead arm movement, like getting an object from the overhead kitchen cabinets or lifting an object above the shoulder or pulling objects overhead. These all can cause overuse damage to the tendon. Cigarette smokers are also at risk, as nicotine decreases the quality of the tendon making it prone to injury.

ANATOMY

Biceps brachii s a two-headed muscle. It is located anteriorly to the humerus.

The long head is located on the lateral side of the biceps brachii while the short head is located on the medial side.

Long head biceps tendon originates from the supraglenoid tubercle of the scapula. Prior to entering the intertubercular sulcus (bicipital groove) it arches through the rotator cuff interval where it is held by a sling formed by the superior glenohumeral ligament and the coracohumeral ligament. It makes a sharp turn at the humeral head and continues its course in the bicipital groove where it is held by the transverse humeral ligament.[14] It converges with the short head around the middle of the humerus to the large muscle belly. The biceps brachii muscle inserts on the radius at the radial tuberosity and the fascia of the forearm via the bicipital aponeurosis.

Innervation: it is innervated by the musculocutaneous nerve, which is the terminal branch of the lateral cord of the brachial plexus (C5, C6, and C7).

Blood supply: it is supplied by the muscular branches of the brachial artery.

Function: the functions of the long head itself are abduction and inward rotation. When it works with the short head, the functions are flexion and supination of the forearm at the elbow joint.

The long head's important task is to dynamically stabilize the shoulder joint in an extended downward position, especially in the initial 30° of abduction.[15]

WESTERN MEDICINE ETIOLOGY AND PATHOLOGY

Long head biceps tendonitis is inflammation of the tendon around the long head of the biceps muscle. The inflammation may be from several factors.

- Repetitive overhead motion of the arm precipitates the symptoms.
- Common causative factors are sports that involve shoulder direct contact, like swimming, gymnastics, and martial arts.
- Nicotine use can affect the quality of the tendon; tobacco smoking is an important risk factor that affects the nutrition of the tendon, resulting in long head biceps tendonitis (Zabrzyński et al. 2019).

CHINESE MEDICINE ETIOLOGY AND PATHOLOGY

1. Qi and Blood Stagnation 氣滯血瘀

The chronic repetitive microtrauma in friction, strain of the tendon from improper exertion of the biceps brachii muscle leads to edema and thickening of the long head biceps tendon. The microtrauma injures the meridians and impedes the circulation of Qi and Blood.

2. Liver and Kidney Insufficiency 肝腎虧虛

Aging or long-standing illness impairs Liver and Kidney and results in thickening of the bicipital groove of the humerus leading to friction of the tendon in the sheath.

MANIFESTATIONS

The shoulder pain is in the anterior shoulder; it is a deep, throbbing ache, and it can be exacerbated by overhead arm activities such as reaching an overhead kitchen cabinet, throwing a tennis ball, and swimming. The pain is usually localized to the bicipital groove but may radiate to the deltoid tuberosity and biceps brachii muscle. It can be worsened at night, especially when the sleeping on the affected side.

There may be biceps brachii muscle spasm in the acute onset; in the chronic condition, the pain is mild and can often be tolerated. However, if there is too much overhead activity or after minor trauma or exposure to cold, the symptoms can be aggravated, and in severe cases, shoulder joint activities may be limited.

PHYSICAL EXAMINATION

- Begin with obtaining medical history. Long head biceps tendonitis may occur with other shoulder conditions, so the interview should include questions to identify frozen shoulder, torn labrum, chronic shoulder joint instability, or trauma to the shoulder.
- **Inspection**

  Before palpation and testing the range of motion of the tendon, inspect the biceps at rest during contraction to detect any abnormalities. Flex the patient's elbow at 90°; if the muscle belly is more distal, a proximal rupture of the long head tendon should be suspected.
- **Palpation**
  - The patient's elbow is flexed at 90°, palpate the tendon of the long head within the bicipital groove between the greater and minor tubercles.
  - While palpating, the tendon should be detectable while moving the arm in medial and lateral rotation.
  - Note any pain with palpation over the bicipital groove; tenderness may be a sign of an inflammatory lesion or degeneration of the tendon.
  - A noticeable clicking sensation suggests a subluxation of the tendon.
- **Orthopedic Tests**

Speed's Test

- The patient can be seated or standing, the therapist stands at the tested site.
- The arm is forward flexed at 60° with the elbow extended, and the hand is supinated.
- The therapist places a hand on the shoulder to stabilize the shoulder, and places the other hand on the elbow to apply downward pressure.
- The patient tries to resist the downward force by forward flexion of the shoulder.
- The test is positive if the patient experiences shoulder pain in the bicipital groove. The positive result indicates long head biceps tendonitis.

Yergason Test

- The patient is seated, the elbow is flexed in 90° with the thumb tip pointing up.
- The therapist applies force to stabilize the wrist to resist the patient's movements while the patient tries to actively supinate and flex the elbow.
- The test is positive if the patient experiences shoulder pain. The positive result indicates biceps tendonitis.

DIFFERENTIAL DIAGNOSIS

- **Rupture of the Long Head Biceps Tendon**

  Sharp pain, an audible popping sound in the anterior shoulder, and bruising with a bulge like "Popeye" muscle visible near the shoulder.

- **Biceps Tendon Displacement/Subluxation/Dislocation**

  A snapping sound with arm rotating. Pain or discomfort in the anterior shoulder and the pain may be referred to the biceps muscle. The displacement often is caused by a deficiency of the ligaments near the long head tendon, including the superior glenohumeral ligament, the coracohumeral ligament, and the transverse humeral ligament.
- **Short Head Biceps Injury**

  Pain in the long head biceps tendon needs to be differentiated from short head biceps injury because they both have pain near each other in the anterior shoulder. The pain in short head biceps is located in medial aspect to the bicipital groove, and it is negative in the Speed's Test and the Yergason Test.
- **Supraspinatus Tendonitis**

  It has a painful arc, and the location of the pain is in the posterior aspect of the shoulder. The pain can radiate to the deltoid and triceps muscles.

TREATMENTS

Treatment Principle
Move Qi and Blood, drain Damp, reduce swelling, and stop pain.

Herbal Treatments
Acute Stage
Herbal Formula
Huo Xue Zhi Tong Tang 活血止痛湯

This formula is from "The Great Compendium of Chinese traumatology 傷科大成", authored by Zhao Lian 趙濂 (1798–1860), in Qing dynasty 清朝 (1636–1912), and published in 1891.

Herbal Ingredients
Dang Gui 當歸6g, Su Mu 蘇木6g, Luo De Da 6g落得打6g, Chuan Xiong 川芎2g, Hong Hua 紅花1.5g, Ru Xiang 乳香3g, Mo Yao 沒藥3g, San Qi 三七3g, Chi Shao 赤芍3g, Chen Pi陳皮3g, Zi Jing Teng 紫荆藤9g, Di Bie Cong 地鱉蟲9g

Ingredient Explanations
- Dang Gui 當歸 moves and tonifies the Blood, disperses Cold, and stops pain due to Blood Stasis.
- Su Mu 蘇木 moves the Blood, reduces swelling, and stops bleeding.
- Luo De Da 6g落得打 reduces swelling, clears Heat, and drains Damp.
- Chuan Xiong 川芎 moves the Blood and Qi, expels Wind, and stops pain.
- Hong Hua 紅花 moves the Blood, releases Bruising, opens the meridians, and stops pain.
- Ru Xiang 乳香 moves the Blood and stops pain.
- Mo Yao 沒藥 moves the Blood, reduces swelling, and stops pain.

- San Qi 三七 moves the Blood, stops bleeding, reduces swelling, and stops pain.
- Chi Shao 赤芍 moves the Blood, reduces swelling, and stops pain.
- Chen Pi陳皮 dries Damp and transforms Phlegm.
- Zi Jing Pi 紫荊皮 moves the Blood, reduces swelling, and promotes urination.
- Di Bie Cong 地鱉蟲 breaks up stubborn Blood stasis.

Modifications

- For swelling, add Ze Lan澤蘭, Mu Tong木通, Che Qian Zi車前子, and Jiang Huang薑黃.
- For severe pain, add Yan Hu Suo延胡索, Niu Xi 牛膝, and Mu Xiang木香.
- For swelling, redness, and hot sensation, add Mu Dan Pi牡丹皮, Zhi Zi栀子, and Huang Bai 黃柏.

Topical Application

Herbal Formula

Xiao Yu Gao 消瘀膏

This formula is from "Chinese Traumatology中醫傷科學", authored by Wang Heming王和鳴, and published by China Press of Traditional Chinese Medicine中國中醫藥出版社 in 2002.

Herbal Ingredients

Da Huang 大黃250g, Zhi Zi 栀子50g, Mu Gua 木瓜100g, Pu Gong Ying 蒲公英100g, Jiang Huang 薑黃100g, Huang Bai 黃柏150g

Ingredient Explanations

- Da Huang 大黃 moves the Blood and disperses Bruising.
- Zhi Zi 栀子 clears Heat, cools the Blood, resolves Damp, and reduces swelling.
- Mu Gua 木瓜 promotes Qi and Blood circulation, unblocks the channels, and releases spasm.
- Pu Gong Ying 蒲公英 reduces abscesses, clears Heat in Blood, and resolves Dampness.
- Jiang Huang 薑黃 moves the Blood, expels Wind, reduces swelling, and stops pain.
- Huang Bai 黃柏 clears deficient Kidney Yin Heat.

Application Procedures
Instructions

1. Grind the herbs to a fine powder.
2. Make a paste by mixing the powders with honey and water (2:1).

3. Apply the certain amount of the paste on the proximal fibular capitulum for about 2 mm, and the application region needs to be bigger than the affected area.
4. Cover the region with gauze.
5. Change the gauze and the paste daily.

Formula action: Cools the Blood and reduces inflammation and swelling.
Indications: contusion injury, joint dislocation.

Chronic Stage

Treatment Principle

Unblock and relax the meridians and move the Blood.

Herbal Formula

Shu Jin Huo Xie Tang 舒 筋 活 血 湯

This formula is from "Restoration of Health from the Myriad Diseases 萬 病 回 春", authored by Gong Tingxian 龔 廷 賢 (1522～1619), in Ming dynasty 明 朝 (1368–1644), and published in 1587.

Herbal Ingredients

Qiang Huo 羌 活6g, Fang Feng 防 風9g, Jing Jie 荊 芥6g, Du Huo 獨 活9g, Dang Gui 當 歸12g, Xu Duan 續 斷12g, Qing Pi 青 皮5g, Niu Xi 牛 膝9g, Wu Jia Pi 五 加 皮 9g, Du Zhong 杜 仲9g, Hong Hua 紅 花 6g, Zhi Ke 枳 殼6g

Ingredient Explanations

- Qiang Huo 羌 活 expels Wind-Cold-Dampness, unblocks painful obstruction, alleviates and stops pain, and guides Qi to the Tai Yang and Du channels.
- Fang Feng 防 風 releases the exterior, expels External Wind, expels Wind-Dampness.
- Jing Jie 荊 芥 stops bleeding, dispels Wind, and relieves muscle spasm.
- Du Huo 獨 活 expels Wind Damp and stops pain.
- Dang Gui 當 歸 moves and tonifies the Blood, disperses Cold, and stops pain due to Blood Stasis.
- Xu Duan 續 斷 nourishes the Liver and Kidney, strengthens Bones and Tendons, and alleviates pain.
- Qing Pi 青 皮 soothes Liver Qi and breaks up stagnant Qi, dries Dampness, and transforms Phlegm.
- Niu Xi 牛 膝 expels Wind Dampness, tonifies the Liver and Kidney, and directs herbs downward.
- Wu Jia Pi 五 加 皮 expels Wind and Damp, nourishes and warms the Liver and Kidney, and strengthens Bones and Tendons.
- Du Zhong 杜 仲 tonifies the Kidneys and Liver, strengthens Bones and Tendons, and lowers blood pressure.

- Hong Hua 紅 花 moves the Blood, releases Bruising, opens the meridians, and stops pain.
- Zhi Ke 枳 殼 removes stagnant food, promotes Qi movement, and reduces distention and pressure.

Modifications

- For deficiency of Blood陽 氣 虛, add Dan Shen丹 參 and E Jiao阿 膠.
- For deficiency of Qi氣 虛, add Ren Shen人 參.
- For spasm, add Zhi Jing San 止 痙 散.

Formula Indication

- Bi Syndrome of Blood.

Formula Contraindication

- It is contraindicated for patients during pregnancy and patients with weak digestion.

Acupuncture Treatments

Acupuncture Point Formula

Basic Points

LI15肩 髃Jianyu,   Jianqian肩 前,   LI11曲 池Quchi,   SI11天 宗Tianzong, GB34陽 陵 泉Yanglingquan, Ashi acupoints 阿 是 穴 (it is for chronic shoulder pain; the point should not be used if it is swelling and red in acute state).

Modifications

For chronic stage, add UB20脾 俞Pishu, UB23腎 俞Shenshu, UB24氣 海 俞Qihaishu, UBI8肝 俞Ganshu, and KD3太 溪Taixi.

Auricular Therapy

Shoulder, Shenmen, Liver, Kidney, Endocrine

Cupping Therapy

Jianqian肩 前

Moxibustion Therapy

Jianqian肩 前

Bleeding Therapy

Jianqian肩 前

Electro-Acupuncture Therapy

Jianqian肩 前 ～LI11曲 池Quchi 2/100 hz for 20 minutes daily.

Tui Na Treatments

Tui Na Techniques

Pressing, kneading, rolling, rotating, rubbing, plugging

Tui Na Procedures

The patient is in sitting position, the therapist stands next to the affected shoulder.

1. The therapist applies deep rolling techniques on the shoulder for 3–5 minutes. Figure 7.5.1

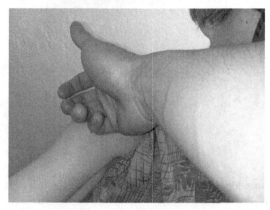

2. Applies pressing and kneading techniques along the direction of the deltoid muscle fibers for 3–5 minutes.
3. Applies rotating techniques on the shoulder in abduction and inward rotation. Rotates the shoulder joint in slow speed, beginning with small circles, then gradually increasing to the level of pain the patient can tolerate for 3–5 circles in each direction.
4. Applies grasping techniques on the deltoid brachii muscle and gradually shifts the grasping from the deltoid muscle to the elbow for 3–5 minutes. Figure 7.5.2

5. The therapist holds the ipsilateral wrist with one hand and passively swings the shoulder between abduction and adduction for 3–5 minutes. At the same time, the therapist plugs the long head tendon in the bicipital groove with the other hand. Figure 7.5.3

6. Rubs the shoulder both anterior and posterior aspects until warm. Figure 7.5.4

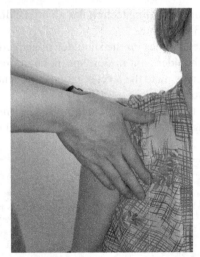

PATIENT ADVISORY

- The shoulder joint needs to rest in high pain intensity, and abduction activities should especially be avoided.
- Keep the shoulder warm.
- Avoid exposure to wind and cold.

SUBLUXATION OF THE LONG HEAD BICEPS TENDON 肱二頭肌長頭腱滑脫

INTRODUCTION

Medial subluxation of the long head biceps tendon can occur with repetitive mechanical wear, overuse, or acute trauma. It is a total loss of contact between the tendon and the bicipital groove. Certain sports that place the shoulder in abduction and rotating the arm such as swimming, volleyball, tennis, baseball, and some contact sports, such as martial arts, wrestling, and football, may be the cause of this injury.

ANATOMY

Biceps brachii s a two-headed muscle. It is located anteriorly to the humerus.

The long head is located on the lateral side of the biceps brachii, while the short head is located on the medial side.

Long head biceps tendon originates from the supraglenoid tubercle of the scapula. Prior to enter the intertubercular sulcus (bicipital groove), it arches through the rotator cuff interval where it is held by a sling formed by the superior glenohumeral ligament, the coracohumeral ligament, subscapularis tendon, and supraspinatus tendon. It makes a sharp turn at the humeral head and continues it course in the bicipital groove, where it is held by the transverse humeral ligament. It converges with the short head around the middle of the humerus to the large muscle belly. The biceps brachii muscle inserts on the radius at the radial tuberosity and the fascia of the forearm via the bicipital aponeurosis.

Innervation: it is innervated by the musculocutaneous nerve, which is the terminal branch of the lateral cord of the brachial plexus (C5, C6, and C7).

Blood supply: it is supplied by the muscular branches of brachial artery.

Function: the functions of the long head itself are abduction and inward rotation. When it works with the short head, the functions are flexion and supination of the forearm at the elbow joint.

The long head's important task is to dynamically stabilize the shoulder joint in an extended downward position especially in the initial 30° of abduction.[16]

WESTERN MEDICINE ETIOLOGY AND PATHOLOGY

- The transverse humeral ligament is a broad band bridging from the lesser to the greater tubercle of the humerus, it converts the bicipital groove into a canal. Its functions are to hold the long head of the biceps tendon within the bicipital groove. The long head tendon of the biceps brachii normally only slides longitudinally, it does not slide laterally. During shoulder abduction and external rotation, the long head tendon of the biceps has the greatest range of motion in the bicipital groove. When the shoulder is often used for excessive abduction and external rotation or sudden force for abduction and external rotation, for example, during a forcefully stopped overhead throwing motion, such as pitching in baseball, the thrower's shoulder is set into a position of maximum shoulder abduction and external rotation (Varacallo, Seaman, and Mair 2022). The strain increases in the biceps while the elbow is decelerated in extension, a maximal contraction of the long head biceps tendon is provoked,[17] and avulsion of the attachment of the pectoralis major or subscapularis occurs, causing the long head biceps tendon to slide on the medial edge of the bicipital groove.

- The subluxation of the long head tendon often is accompanied by other shoulder injury conditions, such as rotator cuff tears, including the supraspinatus and infraspinatus tendons.
- Due to congenital hypoplasia of the lesser tubercle of the humerus, shallow bicipital groove, when the shoulder joint is excessively abducted and externally rotated, the transverse humeral ligament is ruptured, and the long head of the biceps tendon is displaced in the anterior bicipital groove or medial to the lesser tubercle of humerus.

CHINESE MEDICINE ETIOLOGY AND PATHOLOGY

Liver and Kidney Deficiency 肝腎虧虛

Subluxation of the long head biceps tendon in Chinese medicine is considered a Slipped Tendon筋出槽. The main reason for Slipped Tendon is excessive fatigue of the tendon. The definition of Tendon 筋 in Chinese medicine includes muscles, tendons, ligaments, fascias, bursae, nerves, vessels, cartilages, and labrums.

Traditional Chinese medicine holds that chronic traumatic change to soft tissues causes Liver Blood and Yin deficiency; according to the energy flow of the Five Elements, when Liver is deficient, the demand of the energy from Kidney to Liver is high; when Kidney is exhausted, Liver becomes more deficient. When Kidney is insufficient, the bicipital groove becomes shallow. When physical trauma on tendons, bones, and joints occurs and becomes chronic, Liver and Kidney are affected and become insufficient.

Because Liver governs Tendons 筋, when both Liver and Kidney are insufficient, long head biceps tendon becomes fatigued and slips out of the bicipital groove when the shoulder is in excessive abduction and external rotation or there is a sudden force for abduction and external rotation, the long head tendon of the biceps has the greatest range of motion in the bicipital groove, the strain increases in the biceps brachii, and avulsion of the attachment of the pectoralis major or subscapularis occurs, causing the long biceps tendon to slide on the medial edge of the bicipital groove.

MANIFESTATIONS

The symptoms can be quite variable, some describe the sensation as a sharp pain, some patients have the sensation of a rope, string, or band flicking forward and back across the front of the shoulder. Some have a vague deep anterior shoulder pain in certain positions and under certain loads.[18]

Patients may present shoulder swelling, internal rotation of the upper arm, and flexion of the elbow, and the arm is supported by the other hand to relieve the pain in the shoulder. The pain is aggravated when extending the elbow and externally rotating the forearm.

Most patients have a feeling of a "clunk" which is the tendon slipping over the groove.

PHYSICAL EXAMINATION

1. **Medical History**
   - Establish the general health conditions, such as a long-standing illness, diabetes, or sleep deprivation, as these conditions may weaken Liver and Kidney.
   - Inquire about the patient's occupation or sports activities that require abducting and rotating the shoulder, as these can be risk factors that debilitate the long head biceps tendon.
   - Note the pain sensation of the injured shoulder: whether it is sharp, tight, or dull and whether the pain location is in the anterior shoulder.
2. **Inspection**
   - Observe the shoulder to detect edema and skin color changes, such as redness or bruising; if they present, the dislocation may be acute.
   - Differentiate the shoulder condition with long head biceps rupture, and inspect the biceps at rest during contraction to detect any abnormalities. Flex the patient's elbow at 90°; if the muscle belly is more distal, a proximal rupture of the long head tendon should be suspected.
3. **Palpation**
   - Examine the anterior shoulder, note any pain over the bicipital groove, there may be a tenderness along the bicipital sheath.
   - While palpating, the slipped tendon is not detectable in the bicipital groove; it is palpable medially to the groove.
4. **Orthopedic Tests**

Speed's Test
   a. The patient can be seated or standing, the therapist stands at the tested site.
   b. The patient flexes the arm at 60° with the elbow extended and the hand supinated.
   c. The therapist places a hand on the shoulder to stabilize the shoulder, and places the other hand on the elbow to apply downward pressure.
   d. The patient tries to resist the downward force by forward flexion of the shoulder.
   e. The test is positive if the patient experiences shoulder pain in the bicipital groove. The positive result indicates long head biceps pathology.

Shoulder Abduction 90° and External Rotation 90° Test
   a. The patient can be seated or standing, and the therapist stands at the tested site.
   b. The patient raises the arm to 90° abduction and 90° external rotation.

  c. The therapist asks the patient to move the arm to a "throw forward" motion while resisting the forward movement.
  d. The test is positive if the patient experiences shoulder pain in the bicipital groove. The positive result indicates long head biceps pathology.
     • **Special Tests**
       – MRI is useful in identifying the location of the long head biceps tendon; other abnormalities, including degenerative changes, will also be seen.
       – Ultrasound can visualize the movements of the unstable tendon in the bicipital groove.

DIFFERENTIAL DIAGNOSIS

• **Long Head Biceps Tendonitis**
   The shoulder pain lacks the feeling of a "clunk" which is the tendon slipping over the groove.
• **Rupture of the Long Head Biceps Tendon**
   Sharp pain, an audible popping sound in the anterior shoulder, and bruising with a bulge like "Popeye" muscle visible near the shoulder.
• **Short Head Biceps Injury**
   Pain in the long head biceps tendon needs to be differentiated from short head biceps injury because they both have pain near each other in the anterior shoulder. The pain in short head biceps is located in medial aspect to the bicipital groove and it is negative in the Speed's Test and the Yergason Test.
• **Supraspinatus tendonitis**
   It has a painful arc, and the location of the pain is in the posterior aspect of the shoulder. The pain can radiate to the deltoid and triceps muscles.

TREATMENTS

Treatment Principle
Move Qi and Blood, drain Damp, reduce swelling, and stop pain.

Herbal Treatments
Acute Stage
Herbal Formula
Huo Xue Zhi Tong Tang 活血止痛湯

   This formula is from "The Great Compendium of Chinese traumatology 傷科大成", authored by Zhao Lian 趙濂 (1798–1860), in Qing dynasty 清朝 (1636–1912), and published in 1891.

Herbal Ingredients
Dang Gui 當歸6g, Su Mu 蘇木6g, Luo De Da 6g落得打6g, Chuan Xiong 川芎2g, Hong Hua 紅花1.5g, Ru Xiang 乳香3g, Mo Yao 沒藥3g, San Qi 三七3g, Chi Shao 赤芍3g, Chen Pi陳皮 3g, Zi Jing Teng 紫荊藤9g, Di Bie Cong 地鱉蟲9g

Ingredient Explanations

- Dang Gui 當歸 moves and tonifies the Blood, disperses Cold, and stops pain due to Blood Stasis.
- Su Mu 蘇木 moves the Blood, reduces swelling, and stops bleeding.
- Luo De Da 6g落得打 reduces swelling, clears Heat, and drains Damp.
- Chuan Xiong 川芎 moves the Blood and Qi, expels Wind, and stops pain.
- Hong Hua 紅花 moves the Blood, releases Bruising, opens the meridians, and stops pain.
- Ru Xiang 乳香 moves the Blood and stops pain.
- Mo Yao 沒藥 moves the Blood, reduces swelling, and stops pain.
- San Qi 三七 moves the Blood, stops bleeding, reduces swelling, and stops pain.
- Chi Shao 赤芍 moves the Blood, reduces swelling, and stops pain.
- Chen Pi陳皮 dries Damp and transforms Phlegm.
- Zi Jing Pi 紫荊皮 moves the Blood, reduces swelling, and promotes urination.
- Di Bie Cong 地鱉蟲 breaks up stubborn Blood stasis.

Modifications

- For swelling, add Ze Lan澤蘭, Mu Tong木通, Che Qian Zi車前子, and Jiang Huang薑黃.
- For severe pain, add Yan Hu Suo延胡索, Niu Xi 牛膝, and Mu Xiang木香.
- For swelling, redness, and hot sensation, add Mu Dan Pi牡丹皮, Zhi Zi栀子, and Huang Bai 黃柏.

Formula Indications

Blood Stagnation

Contraindications

Contraindicated during pregnancy.

Chronic Stage

Treatment Principle

Unblock and relax the meridians and moves the Blood.

Herbal Formula

Shu Jin Huo Xie Tang 舒筋活血湯

This formula is from "Restoration of Health from the Myriad Diseases 萬病回春", authored by Gong Tingxian 龔廷賢 (1522~1619), in Ming dynasty 明朝 (1368–1644), and published in 1587.

Herbal Ingredients

Qiang Huo 羌活6g, Fang Feng 防風9g, Jing Jie 荊芥6g, Du Huo 獨活9g, Dang Gui 當歸12g, Xu Duan 續斷12g, Qing Pi 青皮5g, Niu Xi 牛膝9g, Wu Jia Pi 五加皮9g, Du Zhong 杜仲9g, Hong Hua 紅花6g, Zhi Ke 枳殼6g

Ingredient Explanations

- Qiang Huo 羌活 expels Wind-Cold-Dampness, unblocks painful obstruction, alleviates and stops pain, and guides Qi to the Tai Yang and Du channels.
- Fang Feng 防風 releases the exterior, expels External Wind, and expels Wind-Dampness.
- Jing Jie 荊芥 stops bleeding, dispels Wind, and relieves muscle spasm.
- Du Huo 獨活 expels Wind Damp and stops pain.
- Dang Gui 當歸 moves and tonifies the Blood, disperses Cold, and stops pain due to Blood Stasis.
- Xu Duan 續斷 nourishes the Liver and Kidney, strengthens Bones and Tendons, and alleviates pain.
- Qing Pi 青皮 soothes Liver Qi, breaks up stagnant Qi, dries Dampness, and transforms Phlegm.
- Niu Xi 牛膝 expels Wind Dampness, tonifies the Liver and Kidney, and directs herbs downward.
- Wu Jia Pi 五加皮 expels Wind and Damp, nourishes and warms the Liver and Kidney, and strengthens Bones and Tendons.
- Du Zhong 杜仲 tonifies the Kidneys and Liver, strengthens Bones and Tendons, and lowers blood pressure.
- Hong Hua 紅花 moves the Blood, releases Bruising, opens the meridians, and stops pain.
- Zhi Ke 枳殼 removes stagnant food, promotes Qi movement, and reduces distention and pressure.

Modifications

- For deficiency of Blood陽氣虛, add Dan Shen丹參 and E Jiao阿膠.
- For deficiency of Qi氣虛, add Ren Shen人參.
- For spasm, add Zhi Jing San 止痙散.

Formula Indication

- Bi Syndrome of Blood.

Formula Contraindication

- It is contraindicated for patients during pregnancy and patients with weak digestion.

Acupuncture Treatments

Acupuncture Point Formula

Basic Points

LI15肩髃Jianyu, Jianqian肩前, LI11曲池Quchi, LI4合谷Hegu, SP10血海Xuehai, GB34陽陵泉Yanglingquan, SJ5外關Waiguan, Ashi acupoints 阿是穴 (it is for chronic shoulder pain; the point should not be used if it is swelling and red in acute state).

Modifications

For chronic stage, add UB20脾 俞Pishu, UB23腎 俞Shenshu, UB24氣 海 俞Qihaishu, UBI8肝 俞Ganshu, and KD3太 溪Taixi.

Auricular Therapy

Shoulder, Shenmen, Liver, Kidney, Endocrine

Cupping Therapy

Jianqian肩 前

Moxibustion Therapy

Jianqian肩 前

Bleeding Therapy

Jianqian肩 前

Electro-Acupuncture Therapy

Jianqian肩 前 ~LI11曲 池Quchi 2/100 hz for 20 minutes daily

Tui Na Treatments

Tui Na Techniques

Pressing, kneading, swinging, grasping, rubbing, stretching, and relocating

Tui Na Procedures

Tui Na therapy is effective for both habitual and traumatic long head biceps tendon subluxation; habitual subluxation takes longer treatment periods, and traumatic subluxation has better treatment outcome.

Relax Muscles

1. The patient is in sitting position, the therapist stands next to the affected shoulder.
2. The therapist kneads and presses the anterior and the lateral shoulder while swinging the affected shoulder in abduction in small ranges for 3–5 minutes.
3. Grasps trapezius and deltoid muscles for 3–5 minutes.
4. Presses and kneads LI11曲 池Quchi, LU5尺 澤Chize, LI4合 谷Hegu, ST12缺 盆Quepen for 3–5 minutes.
5. Rubs the shoulder and upper arm to relax the muscles.

Relocate Tendon

1. The patient is in sitting position, the therapist stands in front of the affected shoulder.
2. The patient flexes the affected shoulder forward to the comfortable position, the elbow is extended, and the arm is pronated.

3. The therapist places a thumb on the lessor tuberosity while grasping the affected wrist with the other hand.
4. Stretches the affected arm, supinates the affected arm, and gradually abducts the affected shoulder to 60°. Figure 7.6.2; Figure 7.6.3

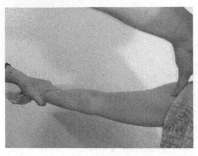
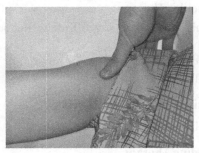

5. While stretching the affected shoulder, the therapist quickly pronates the arm; at the same time, the thumb placed at the lessor tuberosity plugs the tendon laterally and superiorly toward the greater tuberosity. This step is repeated 3–5 times. Figure 7.6.4

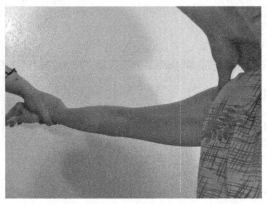

6. Kneads the anterior shoulder with the thumb or the palm 3–5 times, then rubs the whole shoulder joint with both palms 3–5 times to promote the circulation of Blood and relax the Tendons.
7. When the reduction is successful, set the shoulder in adduction and pronation positions for immobilization for 2–4 weeks.
8. If the dislocation of the long head biceps tendon is a secondary cause of a dislocation or fracture of the shoulder, the reduction of the dislocation or fracture of the shoulder is the primary treatment principle.

PATIENT ADVISORY

- Immolization of the shoulder after the reduction is crucial and nessessary; this is to provide a healing environment for the traumatic tendon.
- Keep the shoulder warm, and avoid exposure to wind and cold.

SUBACROMIAL BURSITIS 肩峰下滑囊炎

INTRODUCTION

Subacromial bursitis is a commonly seen shoulder condition that often relates to overuse of the shoulder in the overhead motion. It occurs in any aged adults, but it is more common in the older patient. The manifestation can be vague because the type and the location of pain depends on the severity of the pain, and it often leads to a wrong diagnosis.

ANATOMY

There are five main bursae around the shoulder, but the subacromial bursa is the most common one to have pathology.[19] The other four synovial cavities are the subscapular recess, subcoracoid bursa, coracoclavicular bursa, and supra-acromial bursa.

Subacromial bursa is located in the subacromial space inferior to the acromion process and superior to the greater tubercle of the humerus. The subdeltoid bursa is located between the deltoid muscle and the shoulder joint cavity.

The subacromial bursa and the subdeltoid bursa are often considered as a single bursa.[20] The subacromial bursa extends and communicates with the subdeltoid bursa to form[21] the subacromial-subdeltoid bursa.

The subacromial-subdeltoid bursa is located deep in the deltoid muscle and extends medially to the acromioclavicular joint, laterally beyond the humeral attachment of the rotator cuff, anteriorly to overlie the intertubercular groove, and posteriorly over the rotator cuff.

Function: it reduces friction in the space under the acromion. It protects the supraspinatus muscle from friction between the humeral head and the acromion (Faruqi and Rizvi 2021).

Blood supply: it is supplied by the thoracoacromial artery, the suprascapular artery, the axillary artery, and the anterior and posterior circumflex humeral arteries.

Innervation: suprascapular axillary nerves and lateral pectoral nerves.

WESTERN MEDICINE ETIOLOGY AND PATHOLOGY

It is the inflammation of the subacromial bursa. There are primary and secondary causes. The primary cause is rare, the inflammation may arise from gout, rheumatoid arthritis and infection.

The secondary cause is from subacromial impingement from direct trauma, repetitive overhead painting or overhead sports activities, such as badminton and swimming. Such activities irritate the bursa causing it to become inflamed.

During the overhead activity, abduction of the arm elevates the humerus, the space under the acromion is reduced, the structures in the space are closer (the abduction is at 90°), and irritation of the subacromial bursa and the supraspinatus muscle results in inflammation (Faruqi and Rizvi, 2021).

Chinese Medicine Etiology and Pathology

1. Qi and Blood Stagnation 氣 滯 血 瘀 型

The acute onset of the shoulder pain is commonly caused by the blockages in the meridians.

- The stagnation can be due to a direct blow to the shoulder (e.g., falling down with the shoulder hitting the ground or a direct attack during a contact sport such as boxing, wresting, or martial arts).
- The Blood stasis can be due to a sprain or strain to the anatomic structures in the subacromial space. The microtrauma of sprain or strain causes hypertrophy of the soft tissues which directly narrows the space leading to the impingement of the bursae. Multiplication of the synovial cells that leads to production of the fluid within the bursa blocks the Blood circulation, resulting in pain. Aggravating activities include repetitive overhead movements and reaching up and forward. Certain sports are prone to aggravate the pathology including pushups, throwing a tennis ball for your dog, serving a wet tennis ball when raining, badminton practice while fatigued, or swimming.
- The shoulder can be overused if the aggravating shoulder activities are under emotional upset. Liver Qi stagnation can easily degrade the quality of the tendons.

2. Liver and Kidney Deficiency 肝腎虧虛型

The shoulder lacks nourishment from Qi and Blood due to old age or a lingering disease, leading to deficiency of Liver and Kidney. Insufficient Yin degrades the quality of Tendons 筋, and insufficient Yang weakens the Blood supply to the shoulder joints. Liver manifests on the quality of Tendons 筋, and Kidney manifests on the quality of bones and joint. Chinese medicine holds that Liver and Kidney are in one family 肝 腎 一 家, they support each other, the deficiency of either one will always affect the other.

3. Wind Cold Invasion風寒侵淫型

Invasion of Wind while sweating or sleeping at night with improper blankets.

Sweat pores open when sweating, Wind brings in Cold to the superficial level of the body, and they both deplete Yang Qi and block the local meridians in the shoulder region, resulting in pain.

The sun delivers Yang energy, while the moon delivers Yin energy. Yin energy makes people cold at night. The body's natural temperature continue to drop through the evening and at night while sleeping. A nice blanket can help warm up the body to prevent the invasion of Wind-Cold. When the shoulder is pressed when sleeping on the affected side, or it is not covered with a blanket, Wind-Cold invades the shoulder causing blockage and resulting in pain.

MANIFESTATIONS

- Typically, symptoms include shoulder tenderness, swelling, reduced shoulder range of motion and weakness in shoulder movement, and overhead lifting or reaching activities become uncomfortable.
- The location of the pain is wide, it is commonly in the anterolateral aspect of the shoulder, the pain may radiate to the scapula, neck, elbow, and hand. The location of the chronic stage usually is found near the greater tubercle. The location of pain can shift with the rotation of the humerus.
- The pain can be worse at night, especially if sleeping on the affected shoulder, and may lead to interrupted sleep.
- The pain depends on the degree of inflammation in the shoulder, the acute onset usually is sudden, during which swelling, redness, and warm to touch may be seen. The pain, if it is severe inflammation, may start gradually, originating deep inside the shoulder and develops over a few weeks or months.[22]

PHYSICAL EXAMINATION

- Begin with obtaining medical history. Patients usually present with pain in the anterolateral aspect of the shoulder. There may be a traumatic shoulder history, such a fall with direct impact to the shoulder or a history of repetitive overhead shoulder activities, such as overhead sports (i.e., swimming, badminton, or throwing sports), and reaching for overhead kitchen cabinets.
- **Inspection**
  - The affected shoulder often is in adduction and inward rotation to alleviate the pressure to the bursa.
  - Erythema may be seen in the acute stage if the shoulder is not in abduction.
  - Atrophic muscles may be seen in the late stage.
- **Palpation**
  - There is local tenderness at the anterolateral aspect of the shoulder below the acromion.
  - There is also a palpable crepitus at the site.
  - In acute injury, the skin may be warm.
  - Swelling may be noted as there is puffiness at the site in acute shoulder injury.
- **Range of Motion**
  - The pain can be found in inferior acromion and greater tubercle, but the location of pain can be shifted by the rotation of the humerus.
  - The shoulder range of motion will be reduced when the bursa is thickened, fibrotic, and adhesive.
  - Abduction and outward rotation worsen the pain. Pain occurs when the patient actively abducts the arm and a painful arc occurs between 60° and 120°. Patients often experience a painful catch at midrange, approximately at 90°.

- **Orthopedic Tests**

Hawkins Test

- The patient is seated, the arm is flexed to the shoulder height, the elbow is flexed at 90°.
- The therapist applies an internal rotation force to the tested forearm.
- The test is positive if the patient experiences shoulder pain. The positive result indicates subacromial impingement or rotator cuff tendonitis.

Neer's Test

- The patient is standing, the therapist stabilizes the patient's scapula with one hand and, meanwhile, passively flexes the arm while it is internally rotated.
- The test is positive if the patient experiences shoulder pain. The positive result indicates subacromial impingement.

Speed's Test

- The patient is seated, the arm is 60° to 90° forward flexion, the elbow is fully extended, and the forearm is supinated.
- The therapist applies a downward pressure force to the tested forearm while the force is resisted by the patient's upward flexion force.
- The test is positive if the patient experiences shoulder pain in the anterior shoulder. The positive result indicates the biceps tendon, subacromial impingement, or a SLAP tear pathology.
- Repeat the test with forearm pronation. The patient may still report pain, but the severity of the shoulder pain will be less, and the pain in the bicipital groove is vague.
- **Special Test**

MRI may reveal an effusion or hemorrhage into the site of injury.

Differential Diagnosis

- Supraspinatus tendonitis: the location of the pain is near the greater tubercle of the humerus: Empty Can Test is positive.
- Long head biceps tendonitis: the pain is usually localized to the bicipital groove: Yergason Test is positive.
- Lung cancer: the cancer can affect the nerve to lead to shoulder pain and lung cancer coughs with blood.

Treatments

Treatment Principle
Nourish the Blood, move the Blood, disperse stasis, and stop pain.

Herbal Treatments

Herbal Formula

Zheng Gu Zi Jin Dan正骨紫金丹

This formula is from "Imperially Commissioned Golden Mirror of the Orthodox Lineage of Medicine醫宗金鑑", authored by the Qianlong emperor 乾隆 (1711–1799), in Qing dynasty 清朝 (1636–1912), and published in 1742.

Herbal Ingredients

Ding Xiang丁香2g, Mu Xiang木香2g, Xue Jie血竭2g, Er Cha兒茶2g, Da Huang大黃2g, Hong Hua紅花2g, Dang Gui當歸(頭) 4g, Lian Zi蓮子4g, Fu Ling茯苓4g, Bai Shao芍藥4g, Mu Dan Pi牡丹皮1g, Gan Cao甘草0.6g

Ingredient Explanations

- Ding Xiang丁香 warms Middle Jiao and relieves pain.
- Mu Xiang木香 promotes the movement of Qi and alleviates pain.
- Xue Jie血竭 moves the Blood, stops bleeding, and stops pain.
- Er Cha兒茶 clears Heat and stops bleeding.
- Da Huang大黃 moves the Blood and disperses Bruising.
- Hong Hua紅花 moves the Blood, disperses Bruising, and stops pain due to Blood stasis.
- Dang Gui當歸 moves the Blood, disperses Bruising, and stops pain due to Blood stasis.
- Lian Zi蓮子 tonifies the Spleen.
- Fu Ling茯苓 drains Damp and tonifies the Spleen.
- Bai Shao白芍 nourishes Blood, soothes Liver, and stops pain.
- Mu Dan Pi牡丹皮 clears Heat, cools the Blood, and moves the Blood.
- Gan Cao甘草 harmonizes the herbs, tonifies Middle Jiao Qi, and clears Heat.

Modifications

- For Wind-Damp Bi, add Ru Xiang乳香 and Mo Yao沒藥.
- For severe pain due to Blood stasis, add San Qi田七.
- For pain in the shoulder, add Sheng Ma升麻, Gui Zhi桂枝, and Jie Geng桔梗.

Formula Indications

- Pain, stiffness, swelling, and inflammation in the traumatic area.

Contraindications

- It is contraindicated for pregnancy.

Administrations

Mix the herbs with honey; form into 9-gram pills, and take 1 pill 2–3 times a day with warm wine.

Acupuncture Treatments

Acupuncture Point Formula

Basic Points

LI15肩 髃Jianyu,   SI11天 宗Tianzong,   SJ14肩 髎Jianliao,   LI16巨 骨Jugu, Jianqian肩 前, SI9肩 貞Jianzhen, LI4合 谷Hegu, Ashi acupoints 阿 是 穴 (it is for chronic shoulder pain; the point should not be used if it is swelling and red in acute state).

Modifications

For acute with Blood stasis, add LI11曲 池Quchi, SP10血 海Xuehai, and SJ5外 關Waiguan.

For chronic stage, add UB20脾 俞Pishu, GB34陽 陵 泉Yanglingquan, UB23腎 俞 Shenshu, UB24氣 海 俞Qihaishu, UBI8肝 俞Ganshu, and KD3太 溪Taixi.

Auricular Therapy

Shoulder, Shenmen, Liver, Kidney, Adrenal, Endocrine

Cupping Therapy

Jianqian肩 前, Ashi acupoints 阿 是 穴 (it is for chronic shoulder pain; the point should not be used if there is swelling and redness in the acute state).

Moxibustion Therapy

Warming needle techniques on Jianqian肩 前 , LI15肩 髃Jianyu, SJ14肩 髎Jianliao, Ashi acupoints 阿 是 穴 (it is for chronic shoulder pain; the point should not be used if there is swelling and redness in the acute state).

Bleeding Therapy

Jianqian肩 前 , LI15肩 髃Jianyu, SJ14肩 髎Jianliao, Ashi acupoints 阿 是 穴 (it is for chronic shoulder pain; the point should not be used if there is swelling and redness in the acute state).

Electro-Acupuncture Therapy

Jianqian肩 前 ～LI11曲 池Quchi 2/100 hz for 20 minutes daily.

Tui Na Treatments

Tui Na Techniques

Pressing, kneading, grasping, rubbing, rolling, chafing, swinging and plugging

Tui Na Procedures

Acute Stage

1. The patient is seated, the therapist applies pressing and kneading techniques on the deltoid muscle. The treatment focuses in the subacromial region for 3–5 minutes.

2. The therapist applies grasping techniques on the deltoid muscle and the trapezius muscle for 3–5 minutes. Figure 7.7.1

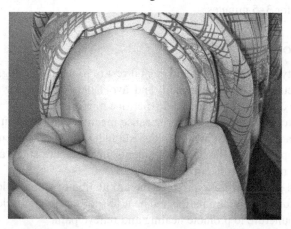

3. Applies rubbing techniques on the deltoid muscle and the trapezius muscle for 3–5 minutes. Figure 7.7.2

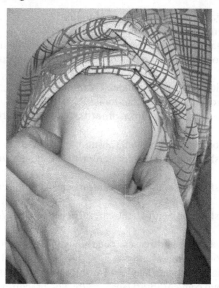

Chronic Stage

1. The patient is seated, the therapist applies rolling techniques on the deltoid muscle while passively rotating the affected shoulder in abduction, adduction, flexion, and extension for 3–5 minutes.
2. Applies grasping techniques on the deltoid muscle and the trapezius muscle for 3–5 minutes. The grasping force should be deep.

3. Sets the affected shoulder in 30° abduction, applies chafing, swinging, and plugging techniques on all the aspects of the deltoid and the supraspinatus muscles for 3–5 minutes.

PATIENT ADVISORY

- Get plenty of both physical and mental rest to promote healing.
- Avoid activities that cause pain, and avoid heavy lifting or movements. Vary the overhead tasks to avoid repetitive motion.
- Do not stop using your shoulder because inactivity can cause it to "freeze", causing a condition known as frozen shoulder. It is recommended to have certain appropriate shoulder movements in the acute stage and increase shoulder movements in chronic stage.
- Maintain good posture to promote the circulation of Qi and Blood.
- Use ice for acute inflammation and heat to sooth aching muscles.
- Get acupuncture to promote healing and relieve pain.

NOTES

1. https://www.physio-pedia.com/Shoulder_Bursitis.
2. https://anatomyzone.com/articles/subacromial-bursa/.
3. https://orthoinfo.aaos.org/en/diseases--conditions/frozen-shoulder/.
4. https://www.mayoclinic.org/diseases-conditions/frozen-shoulder/symptoms-causes/syc-20372684.
5. https://www.physio-pedia.com/Shoulder_Examination.
6. https://rheuminfo.com/docs/physician-tools/SHOULDER.pdf.
7. https://medicalwaveus.com/supraspinatus-tendon-and-rotator-cuff-tendonitis-treatment/.
8. https://www.18zhongyao.com/post/4037.html.
9. https://orthop.washington.edu/patient-care/articles/shoulder/rotator-cuff-differential-diagnosis.html.
10. https://orthop.washington.edu/patient-care/articles/shoulder/rotator-cuff-differential-diagnosis.html.
11. https://orthop.washington.edu/patient-care/articles/shoulder/rotator-cuff-differential-diagnosis.html.
12. https://orthoinfo.aaos.org/en/diseases--conditions/biceps-tendon-tear-at-the-shoulder/.
13. https://www.med66.com/new/40a188aa2009/20091112liuhon11176.shtml.
14. https://radiopaedia.org/articles/long-head-of-biceps-tendon-dislocation?lang=us.
15. https://www.kenhub.com/en/library/anatomy/biceps-brachii-muscle.
16. https://www.kenhub.com/en/library/anatomy/biceps-brachii-muscle.
17. https://www.sportsinjurybulletin.com/the-long-head-of-the-biceps-tendon-part-i.
18. https://www.perthsportsmedicine.com.au/biceps-tendon-subluxation-perth-claremont-cockburn-wa.html.
19. https://onlinelibrary.wiley.com/doi/10.1002/ca.22823.
20. https://www.physio-pedia.com/Shoulder_Bursitis.
21. https://anatomyzone.com/articles/subacromial-bursa.
22. https://www.physio-pedia.com/Shoulder_Bursitis#:~:text=The%20Neer's%20sign%20is%20performed,positive%20if%20they%20produce%20pain.

8 Elbow Pain

LATERAL EPICONDYLITIS 肱 骨 外 上 髁 炎

INTRODUCTION

Lateral epicondylitis is a commonly seen elbow pain. It is associated with the forearm and wrist twisting, such as using a screwdriver or playing tennis, so it is commonly called "tennis elbow". People in middle age present with it more often due to the overuse and degeneration of the elbow tendons.

ANATOMY

Bones: The elbow joint is comprised of three bones.

- The humerus widens distally forming lateral and medial epicondyles. (Mose 2022) The distal end of the humerus has two articulation areas (capitulum and trochlea).
- The ulna has two articulation areas (radial notch of ulnar and trochlear notch of ulnar).
- The radial head of the radius.

Articulations

- The trochlear notch of the ulna articulates with the trochlea of the humerus.[1] It allows flexion and extension of the elbow.
- The head of the radius articulates with the capitulum of the humerus. It allows flexion and extension and also allows supination and pronation of the forearm.
- The head of the radius articulates with the radial notch of the ulna.[2] It allows supination and pronation of the forearm.

Muscles and Tendons

- Triceps muscle attaches to the olecranon posteriorly.
- Distal biceps tendon from biceps brachii muscle attaches to the radial tuberosity.
- Brachialis originates from the distal anterior humerus and inserts onto the ulnar tuberosity.
- Brachioradialis muscle originates from the lateral supracondylar ridge of the humerus and attaches to the distal styloid process of the radius.
- Extensor-supinator muscle attaches to lateral epicondyle.

DOI: 10.1201/9781003203018-8

- Extensor carpi radialis brevis originates from the lateral epicondyle of humerus, attaches to the posterior aspect of the base of the third metacarpal bone, with a few fibers inserting into the medial dorsal surface of the second metacarpal bone.
- Anconeus muscle originates from the lateral epicondyle of humerus and attaches to the lateral surface of olecranon.

Bursae: these three are of clinical importance.

- **Intratendinous** – located within the tendon of the triceps brachii.
- **Subtendinous** – between the olecranon and the tendon of the triceps brachii, reducing friction between the two structures during extension and flexion of the arm.
- **Subcutaneous (olecranon) bursa** – between the olecranon and the overlying connective tissue.

Ligaments

- The radial collateral ligament extends from the lateral epicondyle and attaches to the annular ligament of the radius.
- The ulnar collateral ligament originates from the medial epicondyle and attaches to the coronoid process and olecranon of the ulna.
- The annular ligament of radius encircles the head of the radius and retains it in contact with the ulna anteriorly and posteriorly. It allows the head of the radius to rotate inward during supination and pronation while maintaining the stability of the radial ulnar joint.

Blood supplies: the **brachial** and **deep brachial** arteries.
Innervation:

- Musculocutaneous Nerve
  a. The inferior lateral cutaneous nerve of the arm and the posterior cutaneous nerve of the forearm supply the lateral aspect of the elbow.
  b. The medial cutaneous nerve of the forearm supplies the medial aspect of the elbow.
  c. The lateral cutaneous nerve of the forearm supplies the cubital fossa.
- Radial Nerve.
- Ulnar Nerve.

Functions:

- **Extension** – triceps brachii and anconeus.
- **Flexion** – brachialis, biceps brachii, brachioradialis.

Landmarks: when the elbow is flexed at 90°, these three bones form an equal triangle and form a straight line when the elbow is fully extended.

- Olecranon.
- Lateral epicondyle.
- Medial epicondyle.

WESTERN MEDICINE ETIOLOGY AND PATHOLOGY

Repetitive contraction of the extensor tendons can cause microscopic tearing in the tendons leading to formation of granulation tissue, fibrosis, and tendinosis. The tendon usually involved in lateral epicondylitis (tennis elbow) is called the extensor carpi radialis brevis.[3]

Such degeneration usually is the result of weak constitution in older patients who are involved in activities with wrist extension, pronation, or supination, such as screwdriver using, tennis playing, or other racquet sports. These activities are risk factors of repetitive overuse of the wrist, elbow, and forearm.

CHINESE MEDICINE ETIOLOGY AND PATHOLOGY

1. Qi and Blood Stagnation 氣滯血瘀
   - Direct blow to the tendons or overuse of the elbow injures the tendons leading to stagnation of both Qi and Blood.
   - Wind-Cold invasion to the meridians in the elbow when Wei Qi is weakened by sweating while fatigued. The invasion results in both Qi and Blood stagnation.
2. Liver Blood Deficiency肝 血 虛 虧

According to Chinese medicine, lateral epicondylitis is not an isolated condition, but rather a weak constitution due to Liver deficiency, where the tendons surrounding the elbow joint are prone to be predisposed for pain.

Tendon injury is considered "Jin Shang筋 伤" (or "traumatic injury on tendons"). In traditional Chinese medicine (TCM) theory, Liver governs Tendons 筋. Overuse of the tendons is consuming Liver energy during which mostly its Yin and Blood are consumed. As we get older, Liver is naturally deficient,

Liver Blood is not able to govern the Tendons 筋 when it is deficient; the tendons become degenerated if they lack of blood supply to repair when they are overused or injured. The injury to the tendons around the elbow underlying Blood deficiency is typically seen in middle-aged patients who participate in physical activities that involve wrist extension, pronation, or supination.

MANIFESTATIONS

In Western Medicine

Elbow pain is the main complaint, and the dominant arm is often affected, but it can occur on both arms. The location is at the anterior and distal from the lateral epicondyle of the elbow.[4] It is a burning pain that can radiate upward along the upper arm and downward along the backside of the forearm. Aggravating factors are forearm activity, such as holding a racquet, turning a wrench, holding a coffee mug, or shaking hands.

Patients report weakness of the hand grip or difficulty carrying a coffee mug in the hand, especially with the elbow extended.

There is usually no specific injury history associated with the start of symptoms.

In Chinese Medicine

  1. Qi and Blood Stagnation 氣滯血瘀

  a) Blood Stasis瘀 滯 型

There is usually a history of overuse or a direct blow to the elbow. It is stabbing pain, and the pain radiates to the forearm and the wrist. It is aggravated by holding or gripping objects, or the arm feels weak when holding an object such as a coffee mug.
Tongue: thin white coating.
Pulse: wiry and tight.

  b) Damp-Cold Invasion 寒 濕 外 侵

Dull pain in the elbow, the forearm is difficult to twist, and the patient is unable to hold objects. The pain is relieved when using warm application. It can be aggravated when fatigued.
Tongue: thin white coating.
Pulse: tight, floating, and slow.

  2. Liver Blood Deficiency肝 血 虛 虧

Elbow pain that is less severe in the daytime but worse at night. The arm is weak when holding objects. It is accompanied by dizziness, vertigo, tinnitus, weak knees, and sore lower back.
Tongue: pale tongue body with thin coating.
Pulse: thin and weak.

PHYSICAL EXAMINATION

- Begin with obtaining a medical history. Epicondylitis is commonly associated with a history of occupation or sport activity that requires repetitive forearm movements. Inquire about any elbow injury or overuse injury history. Inquire about activities/occupation relating to the possibility of elbow injury, such as a car mechanic, professional tennis player, or golf player. Obtain information relating to the occurrence of the elbow pain under repetitive forearm and wrist extension, pronation, or supination during hand labor, such as turning a doorknob or shaking hands with another person.
- **Inspection**
     Before palpation or testing the range of motion of the tendon, inspect the lateral epicondyle region to see if there are any abnormal findings, to

eliminate warmth, erythema, edema, or other findings of acute inflammation (Goldberg 2018).

- **Palpation**

  Elbow pain is produced by palpation on the extensor muscles, originating on the lateral epicondyle; the extensor carpi radialis brevis tendon is the most common one involving epicondylitis.

- **Range of motion**

  Reproducibility of pain with resisted wrist extension and supination (Goldberg 2018).

- **Orthopedic tests**

Mill's Test

- The patient is standing, the therapist stands near the tested elbow.
- The therapist palpates the patient's lateral epicondyle with one hand, while passively pronating the effected forearm.
- The therapist flexes the effected wrist and extends the tested elbow at the same time.
- The positive test is reproduction of lateral elbow pain. The positive result indicates lateral epicondylitis.

Cozen's Test

- The patient is seated with the tested arm comfortably placed on a table. The therapist sits next to the test arm.
- The therapist palpates the patient's lateral epicondyle by one hand, asks the patient to first, pronate the forearm, then radially deviated flex the wrist.
- The therapist holds the patient's fist with the other hand and gives resistance to the radially deviated flexion of the wrist.
- The positive test is a reproduction of lateral elbow pain and indicates lateral epicondylitis.

DIFFERENTIAL DIAGNOSIS

Cervical radiculopathy

It is often associated with neck pain. There is a sensorimotor deficit in the affected dermatome onto the arm and may radiate to fingers. C6 and C7 radiculopathy may cause referral of pain down the arm near the lateral epicondyle.

Elbow osteoarthritis

Elbow joint pain and stiffness, reduced range of motion made worse by use or toward the end of the day.

Radial nerve entrapment

Elbow pain is commonly at the proximal forearm near the supinator but can occur at any location within the course of the nerve distribution from C5 to C8, where patients may experience elbow pain, posterior arm pain, numbness, weakness, and reduced motor functions. (The radial nerve arises from C5 to C8, and it provides both motor function and sensory function to the posterior forearm.) (Buchanan; Maini and Varacallo 2022)

TREATMENTS

Herbal Treatments

1. Qi and Blood Stagnation氣滯血瘀型
a) Blood Stasis瘀滯型

Treatment Principle

Move Blood, disperse Bruising, relax the Tendons, open the meridians, move Qi, and stop pain.

Herbal Treatments

Fu Yuan Huo Xue Tang復元活血湯
This formula is from "Clarifying the Study of Medicine醫學發明", authored by Li Dongyuan李東垣, in Yuan dynasty元朝 (1271–1368), and published in 1315.

Herbal Ingredients

Chai Hu 柴胡9g Tian Hua Fen 天花粉9g, Dang Gui 當歸9g, Hong Hua 紅花6g Gan Cao 甘草6g, Chuan San Jia 炮穿山甲6g, Da Huang 大黃12g, Tao Ren 桃仁9g

Ingredient Explanations

- Chai Hu 柴胡 harmonizes Shao Yang disorders and relieves stagnation.
- Tian Hua Fen 天花粉 clears Heat, expels pus, and reduces swelling.
- Dang Gui 當歸 moves and tonifies the Blood, disperses Cold, and stops pain due to Blood Stasis.
- Hong Hua 紅花 moves the Blood, releases Bruising, opens the meridians, and stops pain.
- Gan Cao 甘草 releases spasms, alleviates pain, harmonizes the other herbs, and guides the herbs to all twelve meridians.
- Chuan San Jia 穿山甲 moves the Blood, reduces swelling, and expels Wind Damp.
- Da Huang 大黃 moves the Blood and disperses Bruising.
- Tao Ren 桃仁 breaks up stubborn Blood stasis.

Formula Action

Moves the Blood, disperses Bruising, soothes the Liver, and opens the meridians.

Formula Indication

Physical trauma, Blood stasis, and severe pain.

b) Damp-Cold Invasion 寒濕外侵

Treatment Principle

Expel Wind, disperse Cold, open the meridians, and stop pain.

Herbal Formula

Juan Bi Tang蠲痺湯

This formula is from "A Book of Formulas to Promote Well-Being 嚴氏濟生方", authored by Yan Yonghe 嚴用和 (1200~1268), in Song dynasty 南宋 (1127–1239), and published in 1253.

Herbal Ingredients

Dang Gui當歸9g, Chi Shao赤芍9g, Jiang Huang薑黃9g, Huang Qi黃耆9g, Qiang Huo羌活9g, Gan Cao甘草3g, Sheng Jiang生薑15g, Da Zao大棗3 pieces

Ingredient Explanations

- Dang Gui當歸 nourishes the Blood, benefits the Liver, and regulates menstruation.
- Chi Shao赤芍 moves the Blood, disperses Bruising, and stops pain.
- Jiang Huang薑黃 moves the Blood, opens the meridians, expels Wind, and reduces swelling.
- Huang Qi黃耆 tonifies Qi, strengthens the Spleen, raises the Yang Qi of the Spleen and Stomach, tonifies Wei Qi, stabilizes the exterior, and tonifies the Blood.
- Qiang Huo羌活 expels Wind-Cold-Dampness, unblocks painful obstruction, and alleviates pain.
- Gan Cao甘草 harmonizes the herbs, tonifies Middle Jiao Qi, clears Heat, expectorates Phlegm, and stops coughing.
- Sheng Jiang生薑 unblocks the pure Yang pathway and harmonizes rebellious Qi.
- Da Zao大棗 tonifies the Spleen Qi, nourishes the Blood, and moderates and harmonizes the harsh properties of the fragrant herbs.

Modifications

- For bringing the formula to the arm, add Gui Zhi桂枝.
- For Cold symptoms, add Fu Zi附子.
- For Damp symptoms, add Cang Zhu蒼朮, Fang Ji防己, and Yi Yi Ren薏苡仁.
- For Wind symptoms, add Fang Feng防風.
- For Blood stasis, add Tao Ren桃仁, Hong Hua紅花, and Di Long 地龍.

Formula Indication

- Shoulder pain due to Wind-Damp-Cold Bi Syndrome.

Formula Contraindication

It is contraindicated for patients with Damp-Heat condition.

2. Liver Blood Deficiency肝血虛虧

Treatment Principle

Tonify Qi, nourish the Blood, nourish the Heart, calm Shen.

Herbal Formula

Ren Shen Yang Rong Tang人參養榮湯

This formula is from "Prescriptions of the Bureau of Taiping People's Welfare Pharmacy太平惠民和劑局方", authored by Imperial Medical Bureau太醫局 in southern Song dynasty南宋 (1127–1239), and published in 1134.

Herbal Ingredients

Ren Shen人參8g, Bai Zhu白朮8g, Huang Qi黃耆8g, Gan Cao甘草8g, Chen Pi陳皮8g, Rou Gui肉桂8g, Dang Gui當歸8g, Shu Di Huang熟地黃6g, Wu Wei Zi五味子6g, Fu Ling茯苓6g, Yuan Zhi遠志4g, Bai Shao白芍15g, Da Zao大棗4 pieces, Shen Jiang生薑10g

Ingredient Explanations

- Ren Shen人參 tonifies Qi and Yang.
- Bai Zhu白朮 tonifies the Spleen Qi and dries Dampness.
- Huang Qi黃耆 tonifies Qi, strengthens the Spleen, raises the Yang Qi of the Spleen and Gan Cao甘草, harmonizes the herbs, tonifies Middle Jiao Qi, clears Heat, expectorates Phlegm, and stops coughing.
- Chen Pi陳皮 regulates Qi and removes blockage.
- Rou Gui肉桂 warms the meridians, expels Cold, and opens the meridians to promote Blood circulation.
- Dang Gui當歸 nourishes the Blood, benefits the Liver, and regulates menstruation.
- Shu Di Huang熟地黃 tonifies the Liver and Kidney, strengthens Bone, and benefits Essence and the Blood.
- Wu Wei Zi五味子 replenishes Qi, nourishes the Kidney, calms Shen, and stops diarrhea.
- Fu Ling茯苓 drains Damp and tonifies the Spleen.
- Yuan Zhi遠志 expels Phlegm, calms Shen, and reduces abscesses.
- Bai Shao白芍 nourishes the Blood, soothes the Liver, and stops pain.
- Da Zao大棗 tonifies the Spleen Qi, nourishes the Blood, and moderates and harmonizes the harsh properties of the fragrant herbs.
- Sheng Jiang生薑 unblocks the pure Yang pathway and harmonizes rebellious Qi.
- Da Zao大棗 tonifies the Spleen Qi, nourishes the Blood, and moderates and harmonizes the harsh properties of the fragrant herbs.
- Shen Jiang生薑 unblocks the pure Yang pathway and harmonizes rebellious Qi.

Modifications

- For bringing the formula to the arm, add Gui Zhi桂 枝.
- For Cold symptoms, add Fu Zi附 子.
- For Blood deficiency, add Zi He Che紫 河 車 and Lu Rong鹿 茸.
- For Wind symptoms, add Fang Feng防 風.
- For poor appetite, add Shan Zha山 查 and Mai Ya麥 芽.
- For Essence deficiency, add Lu Jiao Jiao鹿 角 膠, Gui Ban Jiao龜 板 膠, and Ba Ji Tian巴 戟 天.

Formula Indication

- Elbow pain due to Qi, Yang, and Blood deficiency.

Formula Contraindication

- It is contraindicated for patients with Excessive Heat syndrome or a common cold.

Acupuncture Treatments

Acupuncture Point Formula

Basic Points

LI11曲 池Quchi, LI10手 三 里Shousanli, SJ5外 關Waiguan, LI4合 谷Hegu, LI3三 間sanjian, SJ3中 渚ZhongZhu, SI3後 溪Houxi, Shouwuli手 五 里 (3 Cun 寸 proximal to LI11曲 池Quchi), Ashi acupoints 阿 是 穴

Modifications

a) For Blood stasis瘀 滯 型, add LU5尺 澤Chize and LI4合 谷Hegu.
b) For Damp-Cold Invasion 寒 濕 外 侵, add LI12肘 髎Zhouliao, LI10手 三 里Shousanli, and LU5尺 澤Chize.
c) For Liver Blood Deficiency肝 血 虛 虧, add UB23腎 俞Shenshu, LI10手 三 里Shousanli, UBI8肝 俞Ganshu, ST36足 三 里Zusanli, and SP6三 陰 交Sanyinjiao.

Auricular Therapy

Elbow, Adrenal, Shenmen

Moxibustion Therapy

LI12肘 髎Zhouliao, LI11曲 池Quchi for Liver Blood Deficiency肝 血 虛 虧.

Bleeding Therapy

LI11曲 池Quchi, LI10手 三 里Shousanli. This is for Blood stasis瘀 滯 型.

Cupping Therapy

LI11曲 池Quchi, LI10手 三 里Shousanli after the bleeding Therapy.

Tui Na Treatments

Tui Na Techniques

Grasping, kneading, pushing, pressing, rubbing, plugging, and stretching.

Tui Na Procedures

Patient is seated, the therapist stands beside the affected arm. The therapist performs the following procedure:

1. Grasps the affected wrist with one hand and the other hand applies grasping and kneading techniques on the arm from the shoulder to the wrist for 3–5 minutes.
2. Pushes with the palm on the arm on the Large Intestine Meridian from the shoulder to the wrist for 3–5 minutes.
3. Presses and kneads with the tip of the thumb on LI11曲 池Quchi for 1–3 minutes with light pressure Figure 8.1

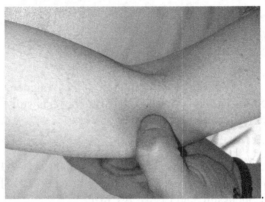

4. Presses and kneads with the thumb pad SJ10天 井Tianjing for 1–3 minutes with light pressure.
5. Presses and kneads with the thumb pad LI10手 三 里Shousanli for 1–3 minutes with light pressure.
6. Presses and kneads with the thumb pad SJ5外 關Waiguan for 1–3 minutes with light pressure.
7. Presses and kneads with the thumb pad LI4合 谷Hegu for 1–3 minutes with light pressure.
8. Rubs with the thumb pad on LI11曲 池Quchi until warm.
9. Plugs the tendons around LI11曲 池Quchi for 1–3 minutes.
10. Stretches the elbow joint.

Patient Advisory

1. Avoid activities or movements that trigger the elbow pain.
2. The elbow pain may reoccur easily, so avoid movements with the arm in hyperextension, especially dragging or pulling objects with force.

3. Always keep the elbow warm and avoid exposure to Cold.
4. Wearing an elbow brace will prevent further overuse of the muscle and tendons while laboring.
5. Patients are recommended to do the following exercises of the elbow joint in a seated or standing position:
    a. Clench a loose fist with the palm facing downward and straighten the affected arm in front of the shoulder.
    b. Flex the elbow, draw the fist towards the front of the shoulder with the palm facing the shoulder and the wrist palmar flexed.
    c. Pronate the forearm to face the palm forward.
    d. Straighten the elbow by punching the fist forward to the front of the shoulder.
    e. The punching should be slow but with force and as far as possible.

MEDIAL EPICONDYLITIS 肱骨內上髁炎

INTRODUCTION

Medial epicondylitis is commonly seen in students and golfers, so medial epicondylitis is also named Student's Elbow and Golfer's Elbow; but it is also seen in badminton players, table tennis players, chefs, and housewives. The patients who suffer from this condition are the people whose jobs or sport activities involve repeatedly and forcefully bending the elbow with the wrist bending toward the palm side. It is most common in middle-age people due to the overuse and degeneration of the elbow tendons.

ANATOMY

The medial epicondyle is located on the distal end of the humerus. It is inferior to the medial supracondylar ridge and proximal to the olecranon fossa.

Attachments: The ulnar collateral ligament, and the flexor tendons of the flexor carpi ulnaris, the flexor digitorum superficialis, pronator teres, the flexor carpi radialis, and the palmaris longus muscles attach to the medial epicondyle.

Functions: the medial epicondyle protects the ulnar nerve which runs in a groove on the posterior aspect of the epicondyle.

Bones: the elbow joint is articulated of 3 bones.
- The humerus widens distally, forming lateral and medial epicondyles (Mose 2022). The distal end of the humerus has two articulation areas, the capitulum and the trochlea.
    - The ulna has two articulation areas, the radial notch of ulnar and the trochlear notch of ulnar.
    - The radius has a head of radius.

Articulations

- The trochlear notch of the ulna articulates with the trochlea of the humerus.[5] It allows flexion and extension of the elbow.

- The head of the radius articulates with the capitulum of the humerus.[6] It allows flexion, extension, and supination and pronation of the forearm.
- The head of the radius articulates with the radial notch of ulna.[7] It allows supination and pronation of the forearm.

Muscles and Tendons

- The flexor carpi ulnaris muscle originates from the medial epicondyle and the medial margin on olecranon of ulna, and it inserts onto the pisiform at the base of the fifth metacarpal bone in the volar aspect. It is supplied by the ulnar artery and innervated by the ulnar nerves from C8 to T1. Its functions are the flexion and adduction of the wrist.
- The flexor digitorum superficialis muscle originates from the medial epicondyle of the humerus and inserts onto the anterior margins on the base of the middle phalanges of the four fingers. It is supplied by the ulnar artery and innervated by the median nerve, and its function is the flexion of the fingers.
- The flexor carpi radialis muscle originates from the medial epicondyle of the humerus and inserts onto the bases of the second and the third metacarpal bones. It is supplied by the ulnar artery and innervated by the median nerve. Its functions are the flexion and abduction of the wrist.
- The pronator teres muscle has two origins: the first origin is from the medial supracondylar ridge of the humerus slightly above the medial epicondyle of humerus; the second origin is from the coronoid process of ulna head and inserts onto the middle of the lateral surface of the body of the radius. It is supplied by the ulnar artery and radial artery and innervated by the median nerve. Its functions are the pronation of the forearm and flexion of the elbow.
- The palmaris longus muscle originates from the medial epicondyle of the humerus and inserts onto the palmar aponeurosis and flexor retinaculum of the hand. It is supplied by the ulnar artery and innervated by the median nerve. Its function is the flexion of the wrist.

Bursae: these three are of clinical importance.

- Intratendinous – located within the tendon of the triceps brachii.
- Subtendinous – between the olecranon and the tendon of the triceps brachii, reducing friction between the two structures during extension and flexion of the arm.
- Subcutaneous (olecranon) bursa – between the olecranon and the overlying connective tissue.

Ligaments

- The ulnar collateral ligament is at the medial aspect of the elbow. It originates from the medial epicondyle and attaches to the coronoid process and olecranon of the ulna.

Landmarks: when the elbow is flexed at 90°, these three bones form an equal triangle.

- Olecranon.
- Lateral epicondyle.
- Medial epicondyle.

Western Medicine Etiology and Pathology

Medial epicondylitis is caused by a repetitive activity that places extreme stress on the flexor muscles of the elbow. This can be from placing an excessive valgus force onto the elbow which involves forcefully flexing the volar forearm muscles while the wrist is in fully palmar flexion. The causative activities include poor golf swing technique, throwing a javelin, carrying a heavy suitcase, chopping wood with an ax, hammering, typing, or serving a tennis ball – particularly if the tennis ball is wet, the racket is too heavy, the strings are too tight, or has an undersized grip; it is also often caused by a top spin serve (Liebert 2022).

Manifestations

Pain is felt in the flexor pronator tendons on the medial aspect of the elbow and volar forearm. It is worse when the wrist is flexed or pronated against resistance.

The elbow may be stiff, and making a fist can aggravate the pain. The onset is usually slow. The elbow may be hot when there is a tear on the tendon. The arm may feel weak when lifting a heavy object, such as using a hammer. The pain sometimes may extend to the forearm and may radiate into one or more fingers and it is more common in the ring and little fingers.

Physical Examination

- Begin with obtaining medical history. The interview should include the type of occupation of the patient and any activities involving the flexion of the arm and wrist that might have precipitated the elbow injury.
- **Inspection**
  Medial epicondylitis usually is not warmth, erythema, edema.
- **Palpation**
  Pain is felt in the flexor pronator tendons on the medial aspect of the elbow and volar forearm.
- **Range of Motion**
  The elbow may be stiff but the range of motion usually is not affected. The pain is reproduced or worse when the wrist is flexed or pronated against resistance.
- **Orthopedic Tests**

Moving Valgus Stress Test

1. The patient is seated, and the therapist stands behind the affected side.
2. The therapist holds the patient's wrist and bring the arm to 90° abduction.

3. Fully flexes the patient's elbow to 120° and applies the maximal external rotation of the shoulder while stabilizing the distal humerus.
4. Asks the patient to palmar flex the wrist while applying a moderate valgus force to extend the elbow to 30°.
5. The test is positive if pain is reproduced in the medial aspect of the elbow in between 120° and 70°. The positive result indicates the laxity or instability of the elbow medial collateral ligament.

DIFFERENTIAL DIAGNOSIS

Cervical Radiculopathy

Cervical radiculopathy is often associated with neck pain. There is a sensorimotor deficit in the affected dermatome into the arm, and it may radiate to the fingers.

Elbow Osteoarthritis

Elbow joint pain, stiffness, and reduced range of motion that is worsened by use or toward the end of the day.

Anterior Interosseous Nerve Entrapment

Pain in the forearm and cubital fossa, with difficulty bringing the thumb and index finger together or making a fist.

TREATMENTS

Herbal Treatments

1. Qi and Blood Stagnation氣滯血瘀型
a) Blood Stasis瘀滯型

Treatment Principle

Move the Blood, disperse Bruising, relax Tendons, open the meridians, move Qi, and stop pain.

Herbal Treatments

Fu Yuan Huo Xue Tang復元活血湯

This formula is from "Clarifying the Study of Medicine醫學發明", authored by Li Dongyuan 李東垣, in Yuan dynasty元朝 (1271–1368), and published in 1315.

Herbal Ingredients

Chai Hu 柴胡9g Tian Hua Fen 天花粉9g, Dang Gui 當歸9g, Hong Hua 紅花6g Gan Cao 甘草6g, Chuan San Jia 炮穿山甲6g, Da Huang 大黃12g, Tao Ren 桃仁9g

Ingredient Explanations

- Chai Hu 柴胡 harmonizes Shao Yang disorders and relieves stagnation.
- Tian Hua Fen 天花粉 clears Heat, expels pus, and reduces swelling.

- Dang Gui 當歸 moves and tonifies the Blood, disperses Cold, and stops pain due to Blood stasis.
- Hong Hua 紅花 moves the Blood, releases Bruising, opens the meridians, and stops pain.
- Gan Cao 甘草 releases spasms, alleviates pain, harmonizes the other herbs, and guides the herbs to all twelve meridians.
- Chuan San Jia 穿山甲 moves the Blood, reduces swelling, and expels Wind Damp.
- Da Huang 大黃 moves the Blood and disperses Bruising.
- Tao Ren 桃仁 breaks up stubborn Blood stasis.

Formula Action

Moves the Blood, disperses Bruising, soothes the Liver, and opens the meridians.

Formula Indication

Physical trauma, Blood stasis, severe pain.

b) Damp-Cold Invasion 寒濕外侵

Treatment Principle

Expel Wind, disperse Cold, open the meridians, and stop pain.

Herbal Formula

Juan Bi Tang蠲痹湯

This formula is from "A Book of Formulas to Promote Well-Being 嚴氏濟生方", authored by Yan Yonghe 嚴用和 (1200~1268), in Song dynasty 南宋 (1127–1239), and published in 1253.

Herbal Ingredients

Dang Gui當歸9g, Chi Shao赤芍9g, Jiang Huang薑黃9g, Huang Qi黃耆9g, Qiang Huo羌活9g, Gan Cao甘草3g, Sheng Jiang生薑15g, Da Zao大棗3 pieces

Ingredient Explanations

- Dang Gui當歸 nourishes the Blood, benefits the Liver, and regulates menstruation.
- Chi Shao赤芍 moves the Blood, disperses Bruising, and stops pain.
- Jiang Huang薑黃 moves the Blood, opens the meridians, expels Wind, and reduces swelling.
- Huang Qi黃耆 tonifies Qi, strengthens the Spleen, raises the Yang Qi of the Spleen and Stomach, tonifies Wei Qi, stabilizes the exterior, and tonifies the Blood.
- Qiang Huo羌活 expels Wind-Cold-Dampness, unblocks painful obstruction, and alleviates pain.
- Gan Cao甘草 harmonizes the herbs, tonifies Middle Jiao Qi, clears Heat, expectorates Phlegm, and stops coughing.

- Sheng Jiang生 薑 unblocks the pure Yang pathway, and harmonizes rebellious Qi.
- Da Zao大 棗 tonifies the Spleen Qi, nourishes the Blood, and moderates and harmonizes the harsh properties of the fragrant herbs.

Modifications
- For bringing the formula to the arm, add Gui Zhi桂 枝.
- For Cold symptoms, add Fu Zi附 子.
- For Damp symptoms, add Cang Zhu蒼 朮, Fang Ji防 己, and Yi Yi Ren薏 苡 仁.
- For Wind symptoms, add Fang Feng防 風.
- For Blood stasis, add Tao Ren桃 仁, Hong Hua紅 花, and Di Long 地 龍.

Formula Indication
- Shoulder pain due to Wind-Damp-Cold Bi Syndrome.

Formula Contraindication
- It is contraindicated for patients with Damp-Heat condition.

2. Liver Blood Deficiency肝 血 虛 虧

Treatment Principle
Tonify Qi, nourish the Blood, nourish the Heart, calm Shen.

Herbal Formula
Ren Shen Yang Rong Tang人 參 養 榮 湯
This formula is from "Prescriptions of the Bureau of Taiping People's Welfare Pharmacy 太 平 惠 民 和 劑 局 方", authored by Imperial Medical Bureau 太 醫 局 in southern Song dynasty南 宋 (1127–1239), and published in 1134.

Herbal Ingredients
Ren Shen人 參8g, Bai Zhu白 朮8g, Huang Qi黃 耆8g, Gan Cao甘 草8g, Chen Pi陳 皮8g, Rou Gui肉 桂8g, Dang Gui當 歸8g, Shu Di Huang熟 地 黃6g, Wu Wei Zi五 味 子6g, Fu Ling茯 苓6g, Yuan Zhi遠 志4g, Bai Shao白 芍15g, Da Zao 大 棗4 pieces, Shen Jiang生 薑10g

Ingredient Explanations
- Ren Shen人 參 tonifies Qi and Yang.
- Bai Zhu白 術 tonifies the Spleen Qi and dries Dampness.
- Huang Qi黃 耆 tonifies Qi, strengthens the Spleen, raises the Yang Qi of the Spleen and Gan Cao甘 草, harmonizes the herbs, tonifies Middle Jiao Qi, clears Heat, expectorates Phlegm, and stops coughing.
- Chen Pi陳 皮 regulates Qi and removes blockage.
- Rou Gui肉 桂 warms the meridians, expels Cold, and opens the meridians to promote Blood circulation.

- Dang Gui當 歸 nourishes the Blood, benefits the Liver, and regulates menstruation.
- Shu Di Huang熟 地 黄 tonifies the Liver and Kidney, strengthens Bone, and benefits Essence and the Blood.
- Wu Wei Zi五 味 子 replenishes Qi, nourishes the Kidney, calms Shen, and stops diarrhea.
- Fu Ling茯 苓 drains Damp and tonifies the Spleen.
- Yuan Zhi遠 志 expels Phlegm, calms Shen, and reduces abscesses.
- Bai Shao白 芍 nourishes the Blood, soothes the Liver, and stops pain.
- Da Zao 大 棗 tonifies the Spleen Qi, nourishes the Blood, and moderates and harmonizes the harsh properties of the fragrant herbs.
- Sheng Jiang生 薑 unblocks the pure Yang pathway and harmonizes rebellious Qi.
- Da Zao大 棗 tonifies the Spleen Qi, nourishes the Blood, and moderates and harmonizes the harsh properties of the fragrant herbs.
- Shen Jiang生 薑 unblocks the pure Yang pathway and harmonizes rebellious Qi.

Modifications

- For bringing the formula to the arm, add Gui Zhi桂 枝.
- For Cold symptoms, add Fu Zi附 子.
- For Blood deficiency, add Zi He Che紫 河 車 and Lu Rong鹿 茸.
- For Wind symptoms, add Fang Feng防 風.
- For poor appetite, add Shan Zha山 查 and Mai Ya麥 芽.
- For Essence deficiency, add Lu Jiao Jiao鹿 角 膠, Gui Ban Jiao龜 板 膠, and Ba Ji Tian巴 戟 天.

Formula Indication

- Elbow pain due to Qi, Yang, and Blood deficiency.

Formula Contraindication

- It is contraindicated for patients with Excessive Heat syndrome and during a common cold.

Acupuncture Treatments

Acupuncture Point Formula

Basic Points

HT3少 海Shaohai, SI8小 海Xiaohai, SJ5外 關Waiguan, LI4合 谷Hegu, LI3三 間 sanjian, SJ3中 渚ZhongZhu, SI3後 溪Houxi, Ashi acupoints 阿 是 穴

Modifications

a) For Blood stasis瘀 滯 型, add LU5尺 澤Chize and LI4合 谷Hegu.
b) For Damp-Cold invasion 寒 濕 外 侵, add HT3少 海Shaohai, SI8小 海 Xiaohai, and LU5尺 澤Chize.

   c) For Liver Blood deficiency肝血虛虧, add UB23腎俞Shenshu, KD3太溪Taixi, UBI8肝俞Ganshu, ST36足三里Zusanli, and SP6三陰交Sanyinjiao.

Auricular Therapy

Elbow, Adrenal, Shenmen

Moxibustion Therapy

SI8小 海Xiaohai, KD3太 溪Taixi for Liver Blood deficiency肝 血 虛 虧.

Bleeding Therapy

SI8小 海Xiaohai, Ashi acupoints 阿 是 穴. It is for Blood stasis瘀 滯 型.

Cupping Therapy

SI8小 海Xiaohai, Ashi acupoints 阿 是 穴after the bleeding Therapy.

Tui Na Treatments

Tui Na Techniques

Rubbing, chafing, rotating, grasping, plugging

Tui Na Procedures

The patient is seated, and the therapist stands near the affected elbow. The patient raises the affected arm in forward flexion and places the arm on a desk with a pillow under the arm to prevent further irritation to the elbow.

   1. While the patient rotates the wrist joint, the therapist applies grasping Figure 8.2 and kneading techniques on the medial side of the forearm from the elbow to the wrist for 3–5 minutes. The pressure should be mild. This step is to relax the flexor muscles.

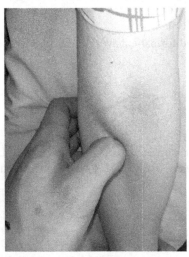

2. The therapist applies plugging techniques on the Ashi acupoints 阿是穴 for 1–2 minutes.
3. Applies grasping techniques with light force on the pronator teres muscle, the flexor carpi radialis muscle, the palmaris longus muscle, the flexor carpi ulnaris muscle, and the flexor digitorum superficialis muscle for 3–5 minutes.
4. Applies rubbing and chafing techniques on the medial aspect of the elbow and the forearm until hot.

Patient Advisory

1. Avoid movements that trigger the elbow pain.
2. Recommend the patient to apply a heating pad on the elbow after each treatment. Always keep the elbow warm and avoid exposure to Cold.
3. Avoid overusing the elbow, forearm, or wrist.
4. Wearing an elbow brace will help to prevent further overuse of the muscle and tendons while laboring.
5. The elbow pain may reoccur easily, so avoid movements of the arm and wrist that involve repetitive forceful bending.

OLECRANON BURSITIS 尺骨鷹嘴滑囊炎

Introduction

Olecranon bursitis occurs when, during a fall, the tip of elbow directly and forcefully hits the floor. Overuse of the elbow joint during work, such as digging or mining; hitting the tip of the elbow while placing it on the desk; or using the elbow to push a heavy object can also injure the bursa. The elbow pain is often called "minor's elbow" or "student's elbow"; when the bursa swells like a water balloon, it is also called "Popeye's elbow" or "water elbow".

Anatomy

The olecranon is the proximal end of the ulna; it articulates posteriorly with the olecranon fossa of the humerus. It receives Blood supply from ulnar artery and is innervated by the radial and ulnar nerves. It is attached by the triceps brachii muscle superiorly, the flexor carpi ulnaris muscle and the ulnar collateral ligament medially, and the anconeus muscle laterally. The ulnar nerve passes in between the olecranon and the medial epicondyle of the humerus, and the elbow joint capsule is attached to the olecranon surface.

Bursae in the elbow are categorized into two groups according to their locations: the olecranon and the cubital fossa (Mercadante and Marappa-Ganeshan 2021).

Olecranon Bursae

- The superficial olecranon bursa is located between the subcutaneous tissue and where the triceps attaches to the olecranon process of the proximal ulna. It is a fibrous sac that is adherent to the olecranon process and the skin or the subcutaneous tissue.

- The subtendinous olecranon bursa is located between the tip of the olecranon, the posterior elbow ligament, and the triceps brachii tendon.
- The intratendinous olecranon bursa is located within the triceps brachii tendon near its insertion onto the olecranon.

Cubital Bursae

- The bicipitoradial bursa is located between the distal biceps brachii tendon and the tuberosity of the radius.
- The interosseous bursa is located alongside the brachialis muscle and adjacent to the biceps brachii tendon in the medial antecubital fossa.

Synovial fluid is an oily substance also called hyaluronic acid. The fluid allows the olecranon process to slide dependently from the skin and prevents the process from protruding through the skin (O'Shea and Tadi 2021).

In the normal state, the olecranon bursa does not communicate with the elbow joint cavity, and no detectable fluid is present (Villaseñor-Ovies 2012); it promotes the secretion of synovial fluid when it senses the irritation.

The internal pressure in the olecranon bursa when it is in bursal effusion increases with elbow flexion; the pressure is low in mid flexion, and increases with further flexion, but not in extension (Villaseñor-Ovies 2012).

Because the olecranon bursa lies so close to the skin, traumatic bursitis from repetitive direct blows, and septic bursitis from infection caused by skin bacteria, can easily occur.

WESTERN MEDICINE ETIOLOGY AND PATHOLOGY

Traumatic changes to the olecranon bursa can easily occur because it is located at the tip of the olecranon and just under the skin; it is not protected by muscle or other soft tissues.

- Traumatic bursitis usually occurs by a direct blow or direct fall on a flexed elbow.
- Chronic bursitis usually occurs by repetitive mechanisms, such as friction and pressure on the bursa, or the repetitive flexion and extension of the elbow joint.
- Septic bursitis usually occurs by bacterial infection of the skin.
- Olecranon bursitis can also be caused by inflammatory disorders, such as gout, rheumatoid arthritis, bone spur, and pseudogout. Bone spurs on the tip of the elbow bone could repeatedly cause elbow bursitis.

CHINESE MEDICINE ETIOLOGY AND PATHOLOGY

Qi and Blood Stagnation 氣 滯 血 瘀 型

The acute onset of the elbow pain is commonly caused by the blockages in the meridians.

- The stagnation can be due to a direct blow to the olecranon, for example, falling down onto the shoulder or direct contact during sports, such as boxing, wrestling, or martial arts.
- The Blood stasis can be due to repetitive mechanisms, such as friction and pressure, on the bursa. The increase of bloody fluid within the bursa blocks the Blood circulation, resulting pain. Aggravating activities include placing the olecranon on a desk, while reading a book or a tablet, or repetitive flexion and extension of the elbow during activities, such as mining or digging with a pickaxe.

MANIFESTATIONS

Swelling, not pain, at the tip of the elbow is usually the first sign. In some cases, the swelling flares up quickly right after the injury. The swollen bursa may look like a golf ball at the tip of the elbow, and there is a wave like sensation.

Elbow pain usually comes later when the swollen bursa gets too big, especially during elbow flexion; there is usually no pain when the elbow is extended. The affected arm becomes weak. It is stiff or limited in elbow movement. The mass becomes hard when the condition becomes chronic.

In some cases, if the injured elbow is red and warm, the bursa may be infected.

PHYSICAL EXAMINATION

- Begin with obtaining a medical history. There may be an injury history or an incident of the elbow overuse, such as digging or pushing an object with the elbow.
- **Inspection**

Patients usually prefer bending the elbow in a semi-flexion position with a hand supporting the injured arm. Skin inspection may reveal contusions or abrasions if the injury is recent.

- **Palpation**

In the acute stage, the affected bursa may be warm or hot, tender, and soft; and it may produce wave like sensation when it is touched.

- **Range of Motion**

The elbow range of motion is usually not affected but may be limited at the end of flexion due to pain. The pain increases with elbow flexion, the pain is less in flexion, and may totally subside with extension. Unusual restriction of active or passive movements with a history of trauma raises the suspicion of fracture of the olecranon process.[8] The landmark between olecranon, lateral epicondyle, and the medial epicondyle changes if there is a fracture of one of them. These three bones form an equal triangle when the elbow is flexed at 90°, and form a straight line when the elbow is fully extended.

- **Special Test**

X-ray can be used to look for bone fracture, bone spur, and bursa calcification. A Blood test or a sample of the synovial fluid may be taken if an infected bursa is suspected.[9]Fever and chill present if there is advanced infection.

DIFFERENTIAL DIAGNOSIS

Fracture of the Olecranon Process

The landmark between olecranon, lateral epicondyle, and the medial epicondyle changes if there is a fracture of the olecranon process.

Distal Partial Ruptures of Triceps Brachii Tendon

Limited range of motion and difficulty extending the elbow.

TREATMENTS

Treatment Principle

Nourish the Blood, move the Blood, disperse stasis, and stop pain.

Herbal Treatments

Herbal Formula
    Zheng Gu Zi Jin Dan正骨紫金丹
    This formula is from "Imperially Commissioned Golden Mirror of the Orthodox Lineage of Medicine醫宗金鑑", authored by the Qianlong emperor 乾隆 (1711–1799), in Qing dynasty 清朝 (1636–1912), and published in 1742.

Herbal Ingredients

Ding Xiang丁香2g, Mu Xiang木香2g, Xue Jie血竭2g, Er Cha兒茶2g, Da Huang大黃2g, Hong Hua紅花2g, Dang Gui當歸（頭）4g, Lian Zi蓮子4g, Fu Ling茯苓4g, Bai Shao芍藥4g, Mu Dan Pi牡丹皮1g, Gan Cao甘草0.6g

Ingredient Explanations

- Ding Xiang丁香 warms Middle Jiao and relieves pain.
- Mu Xiang木香 promotes the movement of Qi and alleviates pain.
- Xue Jie血竭 moves the Blood, stops bleeding and stops pain.
- Er Cha兒茶 clears Heat and stops bleeding.
- Da Huang大黃 moves the Blood and disperses Bruising.
- Hong Hua紅花 moves the Blood, disperses Bruising, and stops pain due to Blood stasis.
- Dang Gui當歸 moves the Blood, disperses Bruising, and stops pain due to Blood stasis.
- Lian Zi蓮子 tonifies the Spleen.
- Fu Ling茯苓 drains Damp and tonifies the Spleen.

- Bai Shao白 芍 nourishes the Blood, soothes the Liver, and stops pain.
- Mu Dan Pi牡 丹 皮 clears Heat, cools the Blood, and moves the Blood.
- Gan Cao甘 草 harmonizes the herbs, tonifies Middle Jiao Qi, and clears Heat.

Modifications

- For Wind-Damp Bi, add Ru Xiang乳 香 and Mo Yao沒 藥.
- For severe pain due to Blood stasis, add San Qi 三 七.
- For pain in the shoulder, add Sheng Ma升 麻, Gui Zhi桂 枝, and Jie Geng桔 梗.

Formula Indications

- Pain, stiffness, swelling, and inflammation in the traumatized area.

Contraindications

- It is contraindicated for pregnancy.

Administrations

Mix the herbs with honey, form into 9-gram pills, and take 1 pill 2–3 times a day with warm wine.

Topical Herbal Formula

Yunnan Baiyao Aerosol (topical application) 雲 南 白 藥 氣 霧 劑

Formula Ingredients

The main ingredient in Yun Nan Bai Yao雲 南 白 藥 is San Qi三 七. San Qi 三 七is used in the treatment of internal and external bleeding, and it also relieves any type of pain from injury.

The topical spray helps to promote Blood circulation, stop pain, and reduce swelling. The spray is easier to apply to closed wounds than the Yun Nan Bai Yao powder form.

Dose

The product comes in two bottles: a cream-colored Yunnan Baiyao Aerosol 雲 南 白 藥 氣 霧 劑and a red-colored Yunnan Baiyao Aerosol Baoxianye雲 南 白 藥 氣 霧 劑 保 險 液.

Contraindication

This product is contraindicated for pregnancy.

Suggested Usage

Shake the bottle several times before use. Hold the bottle vertically with the spray nozzle 5–10 cm from the affected skin and spray for about 3–5 seconds.

Red-Colored Spray – Yunnan Baiyao Aerosol
Baoxianye雲南白藥氣霧劑保險液

- For serious trauma, apply Aerosol Baoxianye 氣霧劑保險液to the injured part first; reapply it 1–2 minutes later if the pain doesn't dissipate.
- Aerosol Baoxianye 氣霧劑保險液should not be used more than three times a day. If pain continues, apply Yunnan Baiyao Aerosol雲南白藥氣霧劑.
- Aerosol Baoxianye氣霧劑保險液 should not be used on open wounds, cuts, or grazes.

Cream-Colored Spray – Yunnan Baiyao Aerosol雲南白藥氣霧劑

Externally used to spray on wounds 3–5 times daily.

Acupuncture Treatments

Acupuncture Formula

Basic Points

LI11曲池Quchi, SJ10天井Tianjing, SJ9四瀆Sidu, SJ11清冷淵Qinglengyuan

Modifications

- For severe pain, add LI4合谷Hegu and Ashi acupoints 阿是穴.
- For infection, add LU5尺澤Chize, LI11曲池Quchi, and ST36足三里Zusanli.

Auricular Therapy

Elbow, Adrenal, Shenmen

Bleeding Therapy

Procedures:

- Prior to the procedure, the tools should be prepared and ready to be used in the working area.
- Set up a clean field with paper towels.
- Place the tools on the clean field along with dry sterile cotton balls, medical alcohol wipes or cotton swabs soaked in alcohol, sterile gloves, a biohazard sharps container, a biohazard trash container, goggles and face mask, and paper towels. A medical lancet (a traditional three-edge needle) is selected for bleeding olecranon in this case.
- The room should be well lit.
- The patient lies supine, the shoulder is at 45° abduction, the elbow is fully flexed.
- The therapist washes and dries their hands thoroughly and puts on sterile gloves.
- The therapist sits facing the olecranon.
- Wipe the olecranon with a medical alcohol wipe or a cotton swab soaked in alcohol. Move from the center of the acupuncture point to the periphery.

- Place the medical lancet at an angle of 90 degrees to the skin.
- In 3–4 quick moves, prick the needle into the olecranon about 0.1 Cun 寸 deep, eliciting a few drops of fluid.
- If more drops of blood are needed, absorb the blood with a sterile cotton ball.
- Lastly, the olecranon is pressed with a sterile cotton ball until the bleeding ceases.
- Apply a bandage to the prick wound.
- Dispose of the tools in the proper containers.
- Wash the hands thoroughly.

Cupping Therapy
SJ10天 井Tianjing after the bleeding Therapy.

Tui Na Treatments
Tui Na Techniques
Pressing, kneading, pushing, grasping

Tui Na Procedures
Patient is seated, the therapist sits near the affected elbow and performs the following procedure:

1. Applies grasping techniques on the triceps brachii muscle and its tendon for 3–5 minutes.
2. Applies kneading and pressing on the triceps brachii tendon for 3–5 minutes.
3. Applies pushing techniques on the triceps brachii muscle, the pushing direction is from the distal tendon towards its proximal tendon for 3–5 minutes or until warm.

PATIENT ADVISORY

- Heating should be applied after treatments.
- Advise the patient to keep the elbow warm.
- Wearing a brace is helpful to protect the affected elbow from any direct blows.
- Educate the patient to avoid any activities or movements that may irritate the bursa, including resting the elbow on a desk, using the elbow to push a heavy object, and repetitive flexion and extension of the elbow.

NOTES

1. https://boneandspine.com/anatomy-of-elbow-joint.
2. https://boneandspine.com/anatomy-of-elbow-joint.
3. https://www.physio-pedia.com/Lateral_Epicondylitis.

4. https://www.physio-pedia.com/Lateral_Epicondylitis.
5. https://boneandspine.com/anatomy-of-elbow-joint.
6. https://boneandspine.com/anatomy-of-elbow-joint.
7. https://boneandspine.com/anatomy-of-elbow-joint.
8. https://patient.info/doctor/olecranon-bursitis#:~:text=The%20differential%20diagno-ses%20can%20include,sepsis%20is%20present%20or%20absent.
9. https://www.webmd.com/arthritis/olecranon-bursitis.

9 Wrist Pain

CARPAL TUNNEL SYNDROME 腕管綜合症

INTRODUCTION

Carpal tunnel syndrome is a common condition that causes wrist pain, numbness, and tingling in the forearm, wrist, hand, and fingers. It is prevalent in people who extensively use vibrating power tools, spend long hours typing on a keyboard/using a mouse with poor wrist position; and it is often seen in workplaces where the environment is cold, windy, and rainy. Patients often notice weak finger gripping force and may drop a coffee cup when trying to grip it.

ANATOMY

The carpal tunnel is a canal in the volar wrist. The boundaries are the carpal bones forming the floor, and the roof is the transverse carpal ligament (flexor retinaculum).

The inside of the carpal tunnel contains one nerve and nine tendons.

The Nerves

- The median nerve arises from the lateral and medial branches of the brachial plexus which originates in C5, C6, and C7 in the lateral root and C8, T1 in the medial root.[1] It enters the arm at the axilla, then travels with the brachial artery between the biceps brachii and triceps brachii muscles to the anterior forearm.
- The median nerve provides motor and sensory innervation to parts of the forearm and hand. It is associated with carpal tunnel syndrome directly via its sensory innervation through two branches to the skin over the thenar eminence palmar aspect of the thumb, index, middle finger, and the radial half of the ring finger.
  1. Prior to entering the carpal tunnel, the radial side of the median nerve gives off its palmar cutaneous branch; this branch is spared in carpal tunnel syndrome.[2] It travels in relation to the tendons of the palmaris longus and flexor carpi radialis muscles then courses superficially through fascial planes to reach the surface of the palm (Smith and Ebraheim 2019).
  2. The palmar digital cutaneous branch arises in the hand, and it innervates the palmar surface and fingertips at the dorsal nail beds of the thumb, index, and middle fingers and the radial half of the ring finger.

The Tendons

The tendons comprise one flexor pollicis longus, four flexor digitorum superficialis, and four flexor digitorum profundus.

The Boundary Forming the Tunnel

The hook of the hamate and the pisiform form the boundaries medially, while the scaphoid and trapezium tubercles form the boundaries laterally.

The transverse carpal ligament bridges the carpal tunnel between the medial and the lateral aspects from the hook of the hamate and pisiform in the medial aspect to the scaphoid and trapezium in the lateral aspect Figure 9.1.1- Item ID: 88897531.

The Carpal Tunnel

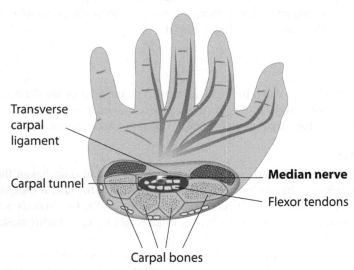

Transverse carpal ligament

Carpal tunnel

Median nerve

Flexor tendons

Carpal bones

WESTERN MEDICINE ETIOLOGY AND PATHOLOGY

Carpal tunnel syndrome is caused by pressure on the median nerve. Anything that irritates the median nerve in the carpal tunnel may lead to carpal tunnel syndrome.

Risk factors include wrist fracture or dislocation, workplace factors (e.g., working with vibrating tools, especially if the workplace is in a cold, windy, and rainy environment); repetitive wrist, hand, and finger motions (e.g., typing or mouse using); especially if the hands are lower than the wrists[3]; arthritis that deforms the bones in the wrist; some medical conditions that increase the risk of nerve damage (e.g., diabetes, thyroid disorders, kidney failure, lymphedema, and menopause); rheumatoid arthritis (inflammatory component affects the tendon lining in the carpal tunnel and increases internal pressure within the tunnel, resulting in irritation to the median nerve); and obesity and fluid retention within the carpal tunnel (common during pregnancy and menopause).

Chinese Medicine Etiology and Pathology

Qi and Blood stagnation 氣滯血瘀

- According to Chinese medicine, carpal tunnel syndrome is not an isolated condition. Overuse of the tendons causes Qi and Blood to become stagnant when the tendons become weak, which manifests as swelling of the tendons, then results in increasing pressure in the tunnel.
- The invasion of Cold and Damp are followed by the weakness of the tendons. It is more prevalent in a workplace that is located in a cold, windy, and rainy environment.

Manifestations

Tingling, numbness, or an electric shock in the thumb, index, middle fingers, and the radial half of the ring finger. The sensation may begin from the wrist and go to the hand. Aggravating factors include holding a steering wheel, a phone, or a newspaper. The sensation may occur at night, when patients experience the fingers getting numb at night while sleeping. Shaking the hand and finger may relieve the symptoms. When carpal tunnel syndrome gets worse, patients may experience weakness in the hand and start dropping objects, as the muscles in the hand shrink and cramp.

Physical Examination

- Begin with obtaining a medical history. Usually, carpal tunnel syndrome is not associated with a traumatic injury. Rather, patients may have a history of activities and workplace environments, as previously outlined, that make them vulnerable to this syndrome.
- Inspection
  Examine the neck, arms, wrists, and hands to inspect any abnormality that may be the cause of median nerve damage. Also compare the strength and appearance of the affected side to the contralateral side. Check the grip strength of the thumb and other 3 fingers.
- Palpation
  Patients may report that the fingers seem swollen, but they are not visibly swollen (LeBlanc and Cestia 2011).
- Range of motion
  Patients report loss of strength when gripping or performing certain tasks, but the wrist range of motion is normal.
- Orthopedic tests
  Phalen's test
  - The patient is seated. Ask the patient to maximally flex the wrist in front of the chest with fingertips pointing downward, place the dorsum of hands together with pressure, flex elbows, forearms in horizontal position.

- The patient holds the posture for 60 seconds.
- The test is positive if the tingling or numbness is felt in the thumb, index finger, middle finger, and the lateral half of the ring finger, as this is the distribution of the median nerve. The positive result indicates carpal tunnel syndrome Figure 9.1.2 – Item ID: 1126688009.

Durkan's carpal compression test

- With the patient seated, ask them to supinate the arm and place the hand on a flat surface.
- The therapist places even pressure with both thumbs overlapped directly over the patient's median nerve in the carpal tunnel for 30 seconds.
- The test is positive if tingling or numbness is felt in the thumb, index finger, middle finger, and the lateral half of the ring finger, as this is the distribution of the median nerve. The positive result indicates carpal tunnel syndrome Figure 9.1.3- Item ID: 580423459.

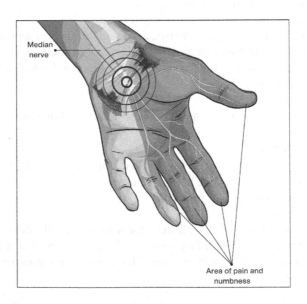

Hoffman-Tinel sign

- The patient is seated, the elbow is extended and the palm is supinated.
- The therapist taps the carpal tunnel at the wrist.
- The test is positive if tingling or numbness is felt in the thumb, index finger, middle finger, and the lateral half of the ring finger, as this is the distribution of the median nerve. The positive result indicates carpal tunnel syndrome Figure 9.1.4-1125937310.

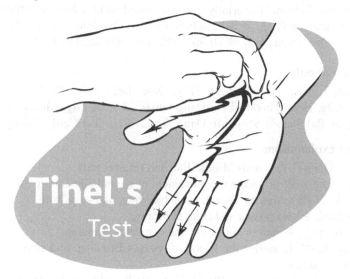

DIFFERENTIAL DIAGNOSIS

Cervical Radiculopathy

Cervical radiculopathy is the most common disorder that mimics carpal tunnel syndrome. C6 and C7 disorders present the symptoms in arm pain and paresthesia similar to carpal tunnel syndrome. Symptoms that favor cervical radiculopathy include neck pain that radiates to the shoulder and arm, exacerbation of symptoms with neck movement, sensory loss of the muscles in the forearm, and weakness of proximal arm muscles involving elbow flexion and extension.

Median Neuropathy in the Forearm

The neuropathy proximal to the carpal tunnel may occur in the forearm where the nerve passes through the pronator teres muscle. The condition can look remarkably similar to carpal tunnel syndrome. In pronator syndrome, symptoms that favor proximal median neuropathy include forearm pain with activity; patients often complain of discomfort in the forearm with activity, such as twisting to dry a towel or operating a manual screwdriver, where the elbow is extended with repetitive pronation. Other symptoms include numbness and tingling of the thumb and the first two digits; there is often loss of sensation over the thenar eminence.

TREATMENTS

Treatment Principle

Nourish the Blood, move the Blood, disperse stasis, and stop pain.

Herbal Treatments

Herbal Formula

Zheng Gu Zi Jin Dan正骨紫金丹

This formula is from "Imperially Commissioned Golden Mirror of the Orthodox Lineage of Medicine醫宗金鑑", authored by the Qianlong emperor乾隆 (1711–1799), in the Qing dynasty清朝 (1636–1912), and published in 1742.

Herbal Ingredients

Ding Xiang丁香2g, Mu Xiang木香2g, Xue Jie血竭2g, Er Cha兒茶2g, Da Huang大黃2g, Hong Hua紅花2g, Dang Gui當歸（頭）4g, Lian Zi蓮子4g, Fu Ling茯苓4g, Bai Shao芍藥4g, Mu Dan Pi牡丹皮1g, Gan Cao甘草0.6g

Ingredient Explanations

- Ding Xiang丁香 warms Middle Jiao and relieves pain.
- Mu Xiang木香 promotes the movement of Qi and alleviates pain.
- Xue Jie血竭 moves the Blood, stops bleeding, and stops pain.
- Er Cha兒茶 clears Heat and stops bleeding.
- Da Huang大黃 moves the Blood and disperses Bruising.
- Hong Hua紅花 moves the Blood, disperses Bruising, and stops pain due to Blood stasis.
- Dang Gui當歸 moves the Blood, disperses Bruising, and stops pain due to Blood stasis.
- Lian Zi蓮子 tonifies the Spleen.
- Fu Ling茯苓 drains Damp, tonifies the Spleen.
- Bai Shao白芍 nourishes the Blood, soothes Liver, and stops pain.
- Mu Dan Pi牡丹皮 clears Heat, cools the Blood, and moves the Blood.
- Gan Cao甘草 harmonizes the herbs, tonifies Middle Jiao Qi, and clears Heat.

Modifications

- For Wind-Damp Bi, add Ru Xiang乳香 and Mo Yao沒藥.
- For severe pain due to Blood stasis, add San Qi三七.

Formula Indications

- Pain, stiffness, swelling, and inflammation in the traumatized area.

Contraindications

- It is contraindicated for pregnancy.

Administrations

Mix the herbs with honey; form into 9-gram pills, and take 1 pill 2–3 times a day with warm wine.

Acupuncture Treatments

Acupuncture Point Formula

Basic Points

PC7大 陵Daling, SJ4陽 池Yangchi, SI5陽 谷Yanggu, LU8經 渠 Jingqu, SJ5外 關Waiguan through PC6內 關Neiguan

Moxibustion Therapy

Warming needling techniques on PC7大 陵Daling

Tui Na Treatments

Tui Na Techniques

Kneading, grasping, rolling, plugging, rubbing

Tui Na Procedures

Patient is seated, ask patient to place the forearm, wrist, and hand on a flat surface, with the hand supinated. Place a thin cushion under the wrist. The therapist performs the following procedures:

1. Applies rolling techniques on the forearm toward the flexor tendons for 3–5 minutes Figure 9.1.5.

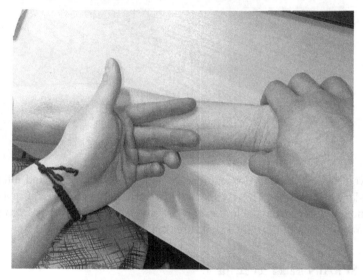

2. Applies grasping techniques on the affected wrist for 3–5 minutes.
3. Asks the patient to rotate the wrist while kneading on the carpal tunnel for 3–5 minutes.
4. Applies plugging techniques on the tendons that pass through the carpal tunnel for 1–3 minutes.
5. Applies rubbing techniques on the volar forearm from the wrist to the elbow until warm.

1. Immobilization is recommended in acute stage; the forearm should be set at semi-supination.
2. Avoid forearm supination and pronation rotation forcefully in the chronic stage.
3. Avoid the exposing the affected wrist to cold.
4. Keep the hands free from menthol, wintergreen, or alcohol products as they bring Cold, Damp, and Wind into the meridians. If you cannot avoid using them, such as massage therapists using these products during massage, wear groves before using them or wash the hands afterward with warm water. Avoid washing hands with cold water.
5. Keep the hands warm. If the workplace is cold and humid, warm up the hands as soon as there is opportunity.
6. Change the way you use the computer mouse and keyboard. Hit the keys softly and keep the wrists straight. Avoid bending the wrists while working for long periods of time Figure 9.1.6- 417113656.

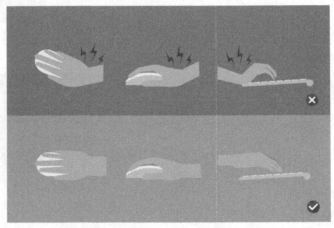

7. Avoid using vibrating power tools for long periods of time and alternate tasks when possible. This is especially important if the tools require a great amount of force from the forearm, wrist, and hand. Take a break from using the tools, even if it is only a short break; it makes a difference.

WRIST SPRAIN 腕 關 節 扭 傷

INTRODUCTION

Wrist sprain is a very commonly seen wrist pain and can be acute or chronic. It is a condition where there is traumatic change to the wrist ligaments, the ligaments have been stretched beyond their limits, or have been torn. This occurs when the wrist is bent or twisted forcefully, repetitively, or abruptly such as when using the hands to break a fall.

ANATOMY

The wrist joint:

- The wrist joint is articulated by seven carpal bones (trapezium, trapezoid, capitate, hamate, triquetrum, lunate, and scaphoid); although pisiform is considered one of the eight carpal bones, it is not part of the wrist joint movement, unlike all the other carpal bones. It is situated where the ulna and the wrist meet but articulates with the triquetrum only.

The joint by these seven carpal bones is known as the midcarpal joint. This is the intersection of the proximal row and the distal row of the carpal bones. There is not much movement in this joint; rather, they allow for smooth motion and assist and boost the radiocarpal joint's range of motion.

- The wrist is articulated by the distal end of the radius bone known as a radiocarpal joint; it is the only true articulating joint between the forearm and the carpal bones. It is called a "synovial joint", as there is a capsule containing lubricating synovial fluid within the joint. The radiocarpal joint allows the hand and the wrist to perform palmar flexion, extension of dorsiflexion, and also allows side to side gliding.

The outer layer of the synovial membrane is a fibrous layer that attaches to the radius, ulna, and the proximal row of the carpal bones.

- The ulna bone articulates with the radius bone also known as the distal radioulnar joint. The joint is not just a synovial joint, it is also a pivot joint that allows the forearm to pronate and supinate. The ulna bone does not articulate with the carpal bone directly; rather, it is prevented from articulating with the carpal bones by a fibrocartilaginous ligament, an articular disc known as the Triangular Fibrocartilage Complex (TFCC). Located in the superior surface of the ulna, the disc serves as a meniscus (like the knee joint) and can be compressed, torn, or damaged when there is weight and twisting on the wrist. Thus, it is commonly injured in weight-bearing activities.[4]
- Wrist sprain occurs in the wrist ligaments.

Ligaments Associated with the Radiocarpal Joint
- The palmar radiocarpal ligaments are located in the volar wrist. They pass from the radius bone to the scaphoid, lunate, triquetrum, and capitate of the carpal bones. There are four distinct parts:
  1) The radioscaphocapitate ligament: from the styloid process of radius bone to the distal aspect of scaphoid bone, with a small number of fibers continuing, and proximal aspect of the capitate bone.
  2) The long radiolunate ligament: from the distal radius to the lunate bone.

3) The radioscapholunate ligament: arises ulnar to the long radiolunate ligament and merges with the scapholunate interosseous ligament.

4) The short radiolunate ligament: from the distal radius to the palmar margin of the lunate bone. Its function is to assist the hand to follow the forearm supination and to increase wrist stability.

- The dorsal radiocarpal ligament, located in the dorsal wrist, passes from the radius bone to the lunate and triquetral carpal bones, and passes the scaphoid carpal bone without direct attachment (Wozasek and Laske 1991). Its function is to assist the hand to follow the forearm pronation and to increase wrist stability.

- Palmar ulnocarpal ligament, located in the ulnar wrist, passes from the styloid process of the ulna to the lunate, capitate, and triquetral carpal bones. Its components include ulnolunate, ulnocapitate, and ulnotriquetal ligaments. The ulnotriquetal ligament is also known as the ulnar collateral ligament.

- The radial collateral ligament, located in the radial wrist, passes from the radial styloid process to scaphoid and trapezium carpal bones. Its function is to prevent excessive ulnar deviation of the hand.

- The blood supply to the wrist joint is from the dorsal and palmar carpal arches which are derived from the ulnar and radial arteries.

- The nerve innervation to the wrist is from median, radial and ulnar nerves.

WESTERN MEDICINE ETIOLOGY AND PATHOLOGY

- Acute wrist spraining may be caused by a forceful impact on the wrist from bending or twisting, from a falling on an outstretched arm or hand when slipping on ice, snow, or wet surfaces; or trauma from a direct blow to the wrist such as in motor vehicle accidents.

- Sports involving repetitive use of the wrist, such as push-ups or weight lifting, are common risk factors for wrist sprains.[5]

- Chronic repeated trauma to the wrist ligaments may also cause sprains.

- Instability makes the wrist joint more prone to sprains and usually may take longer to heal.

- The injury causes the tearing of the ligaments.

CHINESE MEDICINE ETIOLOGY AND PATHOLOGY

Qi and Blood Stagnation 氣滯血瘀

Normally both Qi and Blood should circulate freely in the meridians of the wrist to support wrist functions. Sprain from a fall, overstain from push-ups, physical trauma from a direct blow on the wrist may obstruct the circulation of Blood. Then Qi flow is impeded, leading to the stagnation of Qi and Blood in the meridians, and pain occurs, as the saying goes, "If there is no free flow, there is pain 不通則痛".

MANIFESTATIONS

Wrist pain usually occurs immediately following the injury. The symptoms may include swelling, bruising, a warm sensation around the wrist, and a popping sensation inside the wrist.

The symptoms may not be obvious in chronic or mild wrist sprain; patients usually notice more symptoms at rest after a busy day. As the severity increases, the wrist pain appears more frequent, the pain is felt at work and at rest, the patient may even wake up in pain while sleeping. Common activities aggravate the wrist pain, including holding a baby, doing laundry and ironing, twisting a hand towel, brushing teeth, cooking, washing dishes, and wiping the table.

PHYSICAL EXAMINATION

- Begin with obtaining medical history. During the interview, the patient should be questioned about the onset of trauma or injury, the onset of pain and the type of pain, the location of the hand numbness, and how soon after injury the pain occurred. Inquire about the types of activities such as sports, what kinds of sports, and if it involves repetitive wrist movements (especially important for weight-bearing wrist movements). Inquire about the cause of trauma: if it was a fall, it is important to know the hand position when the fall happened (an outstretched hand is more commonly seen in wrist sprain); whether the fall was from a high place, such as from the roof or a ladder; and whether it is chronic sprain, where the patient may have a history of repeated trauma to the wrist's ligaments.

Inspection

- Palpation
  Tenderness may be found by gently pressing the wrist joint. It is also important to carefully palpate the wrist and arm to check for fracture.
- Range of motion
  All wrist range of motion should be tested to eliminate the pathology, the complete range of motion include extension and flexion, and radial and ulnar deviation. The flexibility and the range of motion of the affected wrist is decreased. Wrist joint movement causes pain. The strength and sensation of the wrist is deficient.
  - Injury of the dorsal radiocarpal ligaments: the pain is located on the dorsal wrist and is aggravated by palmar flexion.
  - Injury of the palmar radiocarpal ligaments: the pain is located on the volar wrist and is aggravated by dorsal extension.
  - Injury of the palmar ulnocarpal ligaments: the pain is located on the ulnar wrist and is aggravated by radial deviation.

- Injury of the radial collateral ligaments: the pain is located on the radial wrist and is aggravated by ulnar deviation.
- If pain occurs during all ranges of motion and the wrist movement is limited, it suggests that both ligaments and tendons have been injured.
- **Orthopedic Tests**
  - Lunotriquetral Ballottement Test
  1) The patient is seated, the affected hand in pronated on a desk.
  2) The therapist grasps the triquetral carpal bone between the thumb and the second finger of one hand and the lunate carpal bone with the thumb and the second finger of the other hand.
  3) The therapist mobilizes the lunate carpal bone both palmarly and dorsally in feeling the excessive mobility of the lunotriquetral joint, checking the pain, and hearing the grinding or crepitation. The positive result indicates a distal radioulnar joint instability.

- Special Tests
- X-rays will not show the ligament, but they can suggest a ligament injury if there is a fracture or an abnormal position of the carpal bones.

DIFFERENTIAL DIAGNOSIS

De Quervain's Tendinosis

The location of the wrist pain is felt over the radius side of the wrist, the pain can radiate upward toward the forearm. The pain is aggravated when using the hand and the thumb, such as twisting the wrist or grasping objects. The Finkelstein test is positive.

Ganglion Cyst

Ganglion cysts are typically round or oval and are filled with a jellylike fluid. Ganglion cyst may not have symptoms, as it usually does not interfere with the wrist joint range of motion. But it can cause pain when it presses on a nearby nerve.

Scaphoid Fracture of the Wrist

Scaphoid fractures usually cause pain and swelling in the anatomic snuffbox. X-rays can reveal the fracture.

Distal Radius Fracture

There can be a visible deformity in the wrist, it is known as a "dinner fork" deformity where the broken bone angles upward at the back of the hand.

TREATMENTS

Treatment Principle

Nourish the Blood, move the Blood, disperse stasis, stop pain 養血活血、祛瘀止痛.

Herbal Treatments

Herbal Formula

Zheng Gu Zi Jin Dan正 骨 紫 金 丹

This formula is from "Imperially Commissioned Golden Mirror of the Orthodox Lineage of Medicine醫 宗 金 鑑", authored by the Qianlong emperor 乾 隆 (1711– 1799), in Qing dynasty 清 朝 (1636–1912), and published in 1742.

Herbal Ingredients

Ding Xiang丁 香2g, Mu Xiang木 香2g, Xue Jie血 竭2g, Er Cha兒 茶2g, Da Huang大 黃2g, Hong Hua紅 花2g, Dang Gui當 歸（頭）4g, Lian Zi蓮 子4g, Fu Ling茯 苓4g, Bai Shao芍 藥4g, Mu Dan Pi牡 丹 皮1g, Gan Cao甘 草0.6g

Ingredient Explanations

- Ding Xiang丁 香 warms Middle Jiao and relieves pain.
- Mu Xiang木 香 promotes the movement of Qi and alleviates pain.
- Xue Jie血 竭 moves the Blood, stops bleeding, and stops pain.
- Er Cha兒 茶 clears Heat and stops bleeding.
- Da Huang大 黃 moves the Blood and disperses Bruising.
- Hong Hua紅 花 moves the Blood, disperses Bruising, and stops pain due to Blood stasis.
- Dang Gui當 歸 moves the Blood, disperses Bruising, and stops pain due to Blood stasis.
- Lian Zi蓮 子 tonifies the Spleen.
- Fu Ling茯 苓 drains Damp and tonifies Spleen.
- Bai Shao白 芍 nourishes the Blood, soothes the Liver, and stops pain.
- Mu Dan Pi牡 丹 皮 clears Heat, cools the Blood, moves the Blood.
- Gan Cao甘 草 harmonizes the herbs, tonifies Middle Jiao Qi, and clears Heat.

Modifications

- For Wind-Damp Bi, add Ru Xiang乳 香 and Mo Yao沒 藥.
- For severe pain due to Blood stasis, add San Qi 三 七.

Formula Indications

- Pain, stiffness, swelling, and inflammation in the traumatized area

Contraindications

- It is contraindicated for pregnancy.

Administrations

- Mix the herbs with honey, form into 9-gram pills, and take 1 pill 2–3 times a day with warm wine.

External Wash Formula
Herbal Formula for Late Stage
Shu Jin Huo Luo Bathing 舒筋活絡澡

Herbal Ingredients
Shen Jin Cao 伸筋草9g, Hai Tong Pi 海桐皮9g, Qin Jiao 秦艽9g, Du Huo 獨活9g, Dang Gui 當歸9g, Gou Teng 鉤藤9g, Ru Xiang 乳香6g, Mo Yao 沒藥6g, Hong Hua 紅花6g

Ingredient Explanations
- Shen Jin Cao 伸筋草 expels Wind, drains Damp, and relaxes Tendons.
- Hai Tong Pi 海桐皮 expels Wind, drains Damp, and reduces swelling.
- Qin Jiao 秦艽 expels Wind Damp.
- Du Huo 獨活 expels Wind Damp and stops pain.
- Dang Gui 當歸 nourishes the Blood, reduces edema, and stops pain.
- Gou Teng 鉤藤 expels Wind and releases spasm.
- Ru Xiang 乳香 moves the Blood and stops pain.
- Mo Yao 沒藥 moves the Blood and stops pain.
- Hong Hua 紅花 moves the Blood, releases bruising, opens the meridians, and stops pain.

Modification
To improve Blood movement, add Tao Ren桃仁 and Chuan Xiong川芎.

Direction
Wash the affected area with this herbal soup.

Topical Herbal Formula
Yunnan Baiyao Aerosol (topical application) 雲南白藥氣霧劑

Formula Ingredients
The main ingredient in Yun Nan Bai Yao雲南白藥 is San Qi三七. San Qi 三七is used in the treatment for internal and external bleeding, and it also relieves any type of pain from injury.

The topical spray helps to promote Blood circulation, stop pain, and reduce swelling. The spray is easier to apply to closed wounds than the Yun Nan Bai Yao powder form.

DOSE

The product comes in two bottles: a cream-colored Yunnan Baiyao Aerosol 雲南白藥氣霧劑 and a red-colored Yunnan Baiyao Aerosol Baoxianye雲南白藥氣霧劑保險液.

Contraindication

This formula is contraindicated for pregnancy.

Suggested Usage

Shake the bottle several times before use. Hold the bottle vertically with the spray nozzle 5–10 cm from the affected skin. Spray for about 3–5 seconds.

Red-Colored Spray – Yunnan Baiyao Aerosol Baoxianye雲南白藥氣霧劑保險液

- For serious trauma, apply Aerosol Baoxianye 氣霧劑保險液to the injured part first. Reapply it 1–2 minutes later if the pain doesn't dissipate.
- Aerosol Baoxianye氣霧劑保險液should not be use more than three times a day. If pain continues, apply Yunnan Baiyao Aerosol雲南白藥氣霧劑.
- Aerosol Baoxianye氣霧劑保險液 should not be used on open wounds, cuts, or grazes.

Cream-Colored Spray – Yunnan Baiyao Aerosol雲南白藥氣霧劑

Externally used to spray on wounds, 3–5 times daily.

Acupuncture Treatments

Acupuncture Point Formula
Basic Points
HT7神門henmen, PC6內關Neiguan, LI4合谷Hegu, PC7大陵Daling, LI11曲池Quchi, SI5陽谷Yanggu, LU7列缺Lieque

Modification

For Blood stasis, add SP10血海Xuehai.

Auricular Therapy

Wrist, Shenmen, Sympathetic, Liver, Kidney

Moxibustion Therapy

Warming needling techniques on PC7大陵Daling for chronic wrist sprain.

Bleeding Therapy

PC7大陵Daling for acute wrist sprain.

Tui Na Treatments

Tui Na Techniques
Pressing, kneading, stretching, rotating

Tui Na Procedures

The patient is seated, the therapist is sitting in front of the affected wrist. The therapist performs the following procedures:

1. Presses and kneads the origin and the attachment of the affected ligaments, while gently rotating the affected wrist for 3–5 minutes.
2. Holds the affected hand with one hand while holding the affected distal forearm with the other hand Figure 9.2.1, then applies stretching techniques, starting with mild intensity and increasing to the patient's tolerance for 3–5 minutes. The stretching direction depends on the injured ligaments.

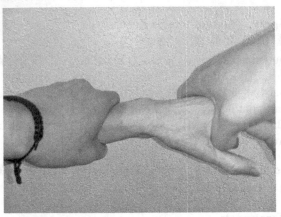

   a.   Stretch the wrist joint by volar flexion if the injury is on the dorsal side.
   b.   Stretch the wrist joint by dorsal flexion if the injury is on the palmar side Figure 9.2.2.

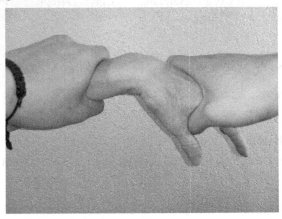

   c.   Stretch the wrist joint by ulnar flexion if the injury in on the radius side.
   d.   Stretch the wrist joint by radial flexion if the injury is on the ulnar side.
3. Applies rubbing technique on the affected wrist until warm.

PATIENT ADVISORY

1. Avoid exposure to cold.
2. Keep the wrist, hand, and arm warm.

3. Wear a wrist brace during the treatment period.
4. Rest is highly recommended if the injury is acute.
5. Apply compression on the affected wrist if the wrist is swelling.

DISLOCATION OF DISTAL RADIOULNAR JOINT 遠端橈尺關節脫位

INTRODUCTION

Dislocation of the distal radioulnar joint can be either dorsal or volar. Dorsal dislocation is more common than volar dislocation. It is an injury that can be missed initially due to subtle signs.

ANATOMY

Articulation

The distal ulna bone articulates with the distal radius bone and is known as the distal radioulnar joint. The articular surfaces are the distal head of ulna to the ulnar notch of radius. The joint is not just a synovial joint; it is also a pivot joint that allows the forearm to pronate and supinate.

Tendons

The tendons of extensor carpi ulnaris, pronator quadratus and the interosseous membrane of forearm are the extrinsic stabilizers.

Ligaments

The ulnar collateral ligament, dorsal and palmar radioulnar ligaments, base of extensor carpi ulnaris sheath, ulnolunate and ulnotriquetral ligaments form the triangular fibrocartilage complex (TFCC), which acts as intrinsic stabilizers, transmitting and distributing the load from the head to the ulna. The TFCC lies in close association with the base of the distal ulna and ulnar border of the radius; such location makes it susceptible to damage with volar dislocation of the distal radioulnar joint (Muralikuttan 2008).

Innervation

The anterior interosseous nerve is a branch of the median nerve, and the posterior interosseous nerve is a branch of the radial nerve.

Blood Supply

Anterior interosseous, posterior interosseous, and ulnar arteries.

Movements

Pronation: the muscles that pronate the forearm at the distal radioulnar joint are the pronator quadratus (without resistance) and the pronator teres (with resistance).

Supination: the muscles that supinate the forearm at the distal radioulnar joint are the supinator muscle (without resistance when the forearm is extended) and the biceps brachii muscle (with resistance when the forearm is flexed).

WESTERN MEDICINE ETIOLOGY AND PATHOLOGY

The main function of the distal radioulnar joint is for pronation and supination. Centered on the distal head of ulna during rotation, the ulnar notch of the radius rotates around the articular surface of the ulnar head. However, when the hand is fixed and the forearm is rotated, the rotational stress acts on the radius using the hand as a lever (the center of the radius is no longer the ulna), which can cause abnormal movement of the distal radioulnar joint and result in joint dislocation.

Volar Dislocation

The injury is caused by an excessive hypersupination of the forearm on a fixed hand with a strengthened pronator quadratus muscle, a pronation injury to the hand in which the forearm is fixed, or by a direct dorsally applied force to the distal ulna. The injury often is associated with sports such as gymnastics, weight lifting, and rugby.[6]

Dorsal Dislocation

Dorsal dislocation can be caused by three mechanisms: the first is by an excessive hyperpronation of the forearm on a fixed hand with the strengthened extensor carpi ulnaris and ulnar carpal ligaments; the second is by a violent direct force on the ulna driving it dorsally with fixed radius and carpal bones, and the ulnar head breaks through the dorsal capsule; and the third is by a direct force on the radius driving it to slide palmarly with the ulna held in a fixed position (Muralikuttan 2008).

CHINESE MEDICINE ETIOLOGY AND PATHOLOGY

Qi and Blood Stagnation 氣滯血瘀

Over-stretching of the forearm in supination or pronation or a direct blow to the wrist joint lead to injury to the ligaments. Soft tissues trauma causes blockages to the flow of the local Qi and Blood in the meridians. The trauma leads to both Qi stagnation and Blood stasis simultaneously. As a result of Qi and Blood stagnation in the meridians, pain occurs; according to the classic quote: "If there is no free flow, there is pain不通則痛".

MANIFESTATIONS

There is usually a history of wrist trauma. The affected wrist is swelling and hemorrhaging.

In the case of a volar dislocation, the pain, swelling, and hemorrhaging are particularly on the volar aspect. The forearm is locked in full supination, and in the absence of the ulna styloid prominence, the ulnar head may be palpable on the volar surface but may also show no evidence of deformity that it is usually not visibly prominent on the volar wrist. Instead, there can be a hollow dorsally where the ulnar head is usually visible, and the wrist can appear narrow in the transverse dimension.

In the case of a dorsal dislocation, the pain, swelling, and hemorrhaging are particularly on the dorsal aspect. The forearm is locked in pronation with limited supination, and the protrusion of the head of the ulna dorsally is palpable and visible.

PHYSICAL EXAMINATION

- Begin with obtaining medical history. During the interview, inquire about how the wrist injury occurred, whether it was a direct force to either the head of the radius or the ulna, the position of the hand and forearm when the accident occurred, and the activities leading to the injury. Early diagnosis and treatment usually result in a more favorable outcome. The interview can provide the details of the traumatic change to the wrist joint, and it may detect whether this is a simple dislocation or there is a fracture and ligamentous damage.
- **Inspection**

  Inspect the wrist for swelling, hemorrhaging, and any evidence of deformity. The ulna styloid prominence may not be visible if it is a volar ulna dislocation. The forearm will be locked in supination if it is a volar ulna dislocation, or if it is locked in pronation, it is a dorsal ulna dislocation.
- **Palpation**

  Palpation on the distal radioulnar joint may reveal the absence of the ulna styloid prominence, and the ulnar head may be palpable on the volar or the dorsal aspect.
- **Range of Motion**

  If the forearm is locked in either supination or pronation, it will not be able to rotate toward the opposite.
- **Special Test**

  X-rays will not show the ligament, but they can suggest a ligament injury if there is a fracture or an abnormal position of the radius and ulnar bones.

DIFFERENTIAL DIAGNOSIS

Distal Radius Fracture

- The most common type of distal radius fracture is a Colles fracture. This type of fracture moves the radius bone slightly upward but the wrist joint remains unaffected,[7]which produces a very distinctive condition known as the "dinner fork deformity". Viewed from the side, the wrist has the appearance of an overturned fork[8] due to a distinct bump in the wrist similar to the neck of the fork.[9]
- The distal part of the radius bone shows an obvious deformity by a protrusion of the broken end of the distal radius bone shifting upward toward the dorsal aspect of the hand.
- A tingling sensation in the fingertips or there may be limited finger movement.

TREATMENTS

Treatment Principle

Nourish the Blood, move the Blood, disperse stasis, and stop pain.

Herbal Treatments

Herbal Formula

Zheng Gu Zi Jin Dan正 骨 紫 金 丹

This formula is from "Imperially Commissioned Golden Mirror of the Orthodox Lineage of Medicine醫 宗 金 鑑", authored by the Qianlong emperor 乾 隆 (1711–1799), in Qing dynasty 清 朝 (1636–1912), and published in 1742.

Herbal Ingredients

Ding Xiang丁 香2g, Mu Xiang木 香2g, Xue Jie血 竭2g, Er Cha兒 茶2g, Da Huang大 黃2g, Hong Hua紅 花2g, Dang Gui當 歸（頭）4g, Lian Zi蓮 子4g, Fu Ling茯 苓4g, Bai Shao芍 藥4g, Mu Dan Pi牡 丹 皮1g, Gan Cao甘 草0.6g

Ingredient Explanations

- Ding Xiang丁 香 warms Middle Jiao and relieves pain.
- Mu Xiang木 香 promotes the movement of Qi and alleviates pain.
- Xue Jie血 竭 moves the Blood, stops bleeding, and stops pain.
- Er Cha兒 茶 clears Heat, and stops bleeding.
- Da Huang大 黃 moves the Blood and disperses Bruising.
- Hong Hua紅 花 moves the Blood, disperses Bruising, and stops pain due to Blood stasis.
- Dang Gui當 歸 moves the Blood, disperses Bruising, and stops pain due to Blood stasis.
- Lian Zi蓮 子 tonifies the Spleen.
- Fu Ling茯 苓 drains Damp and tonifies the Spleen.
- Bai Shao白 芍 nourishes the Blood, soothes Liver, and stops pain.
- Mu Dan Pi牡 丹 皮 clears Heat, cools the Blood, and moves the Blood.
- Gan Cao甘 草 harmonizes the herbs, tonifies Middle Jiao Qi, and clears Heat.

Modifications

- For Wind-Damp Bi, add Ru Xiang乳 香 and Mo Yao沒 藥.
- For severe pain due to Blood stasis, add San Qi 三 七.

Formula Indications

- Pain, stiffness, swelling, and inflammation in the traumatized area.

Contraindications

- It is contraindicated for pregnancy.

Administrations

Mix the herbs with honey, form into 9-gram pills, and take 1 pill 2–3 times a day with warm wine.

External Wash Formula

Herbal Formula for Late Stage

Shu Jin Huo Luo Bathing 舒 筋 活 絡 澡

Herbal Ingredients

Shen Jin Cao 伸 筋 草9g, Hai Tong Pi 海 桐 皮9g, Qin Jiao 秦 艽9g, Du Huo 獨 活9g, Dang Gui 當 歸9g, Gou Teng 鈎 藤9g, Ru Xiang 乳 香6g, Mo Yao 沒 藥6g, Hong Hua 紅 花6g

Ingredient Explanations

- Shen Jin Cao 伸 筋 草 expels Wind, drains Damp, and relaxes Tendons.
- Hai Tong Pi 海 桐 皮 expels Wind, drains Damp, and reduces swelling.
- Qin Jiao 秦 艽 expels Wind Damp.
- Du Huo 獨 活 expels Wind Damp and stops pain.
- Dang Gui 當 歸 nourishes the Blood, reduces edema, and stops pain.
- Gou Teng 鈎 藤 expels Wind and releases spasm.
- Ru Xiang 乳 香 moves the Blood and stops pain.
- Mo Yao 沒 藥 moves the Blood and stops pain.
- Hong Hua 紅 花 moves the Blood, releases Bruising, opens the meridians, and stops pain.

Modification

To improve Blood movement, add Tao Ren桃 仁 and Chuan Xiong川 芎.

Procedure

Wash the affected area with this herbal soup.

Topical Herbal Formula

Yunnan Baiyao Aerosol (topical application) 雲 南 白 藥 氣 霧 劑

Formula Ingredients

The main ingredient in Yun Nan Bai Yao雲 南 白 藥 is San Qi三 七. San Qi 三 七is used in the treatment for internal and external bleeding. It also relieves any type of pain from injury.

The topical spray helps to promote Blood circulation, stop pain, and reduce swelling. The spray is easier to apply on closed wounds than the Yun Nan Bai Yao powder form.

DOSE

The product comes in two bottles: a cream-colored Yunnan Baiyao Aerosol 雲 南 白 藥 氣 霧 劑and a red-colored Yunnan Baiyao Aerosol Baoxianye雲 南 白 藥 氣 霧 劑 保 險 液 。

Contraindication

This formula is contraindicated for pregnancy.

Suggested Usage

Shake the bottle several times before use. Hold the bottle vertically with the spray nozzle 5–10 cm from the affected skin and spray for about 3–5 seconds.

Red-Colored Spray-Yunnan Baiyao Aerosol Baoxianye雲南白藥氣霧劑保險液

- For serious trauma, apply Aerosol Baoxianye 氣霧劑保險液to the injured part first; reapply it 1–2 minutes later, if the pain doesn't dissipate.
- Aerosol Baoxianye 氣霧劑保險液should not be use more than three times a day. If pain continues, apply Yunnan Baiyao Aerosol雲南白藥氣霧劑.
- Aerosol Baoxianye氣霧劑保險液 should not be used on open wounds, cuts, or grazes.

Cream-Colored Spray-Yunnan Baiyao Aerosol雲南白藥氣霧劑

Externally used to spray on wounds 3–5 times daily.

Acupuncture Treatments

Acupuncture Point Formula

Basic Points

SJ5外關Waiguan,　　PC6內關Neiguan,　　SI5陽谷Yanggu,　　SJ4陽池Yangchi, SI4腕骨Wangu, SI6養老Yanglao, HT7神門Shenmen

Modifications

For Blood stasis, add SP10血海Xuehai.

Moxibustion Therapy

Warming needling techniques on PC7大陵Daling for chronic wrist sprain.

Bleeding Therapy

PC7大陵Daling for acute wrist sprain.

Auricular Therapy

Wrist, Shenmen, Sympathetic, Liver, Kidney

Tui Na Treatments

Tui Na Techniques

Rotating, pushing

Tui Na Procedures

Volar Ulna Dislocation

The patient is seated, the affected forearm is in a pronated position. The therapist stands in front of the affected wrist. These procedures are for the right distal radioulnar joint dislocation as an example. The therapist performs the following procedure:

1. Places the thumb and the index fingers of the right hand on the dorsal and volar aspects of the distal radius, respectively; and the remaining fingers support the thenar. The index finger of the therapist's left hand is semi-flexed, the distal radial aspect of the left index finger presses against the volar surface of the ulna, and the thumb supports the dorsal aspect of the ulnar head.
2. Asks the patient to relax the wrist. The therapist rotates the affected wrist in clockwise rounds. When the affected wrist rotates to the radius side, the left index finger pushes the ulnar head upwards.
3. When the affected wrist rotates to the neutral position, the thumb and the index fingers of both hands move both radius and ulna bones toward each other.
4. This procedure can be repeated several times if the reduction is not initially successful.

Dorsal Ulna Dislocation

The patient is seated, the affected forearm is in pronation position. The therapist stands in front of the affected wrist. These procedures are for the right distal radioulnar joint dislocation as an example. The therapist performs the following procedure:

1. Places the thumb and the index fingers of the right hand on the dorsal and volar aspects of the distal radius, respectively; the remaining fingers support the thenar. The index finger of the therapist's left hand is semi-flexed, the distal radial aspect of the left index finger presses against the volar surface of the ulna, and the thumb supports the dorsal aspect of the ulnar head.
2. Asks the patient to relax the wrist. The therapist rotates the affected wrist in counterclockwise rounds. When the affected wrist is rotating, the left thumb finger pushes the ulnar head downwards, then the thumb and the index fingers of both hands move both radius and ulna bones toward each other.
3. This procedure can be repeated several times if the reduction is not initially successful.

PATIENT ADVISORY

- After reduction, place the wrist in a short arm plaster for 4–6 weeks. This provides sufficient immobilization for dorsal ulna dislocation in forearm supination and for volar ulna dislocation in forearm pronation Figure 9.3-528015841.

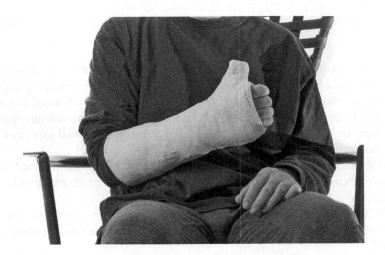

- Avoid forearm supination and pronation during the immobilization for four weeks.
- Management for swelling of the forearm, hand, and fingers: raise the arm above head and exercise the fingers in repetitive fist-clenching movements. Use gentle movements and avoid causing pain.
- Exercise after the immobilization, beginning with finger movements in the first 1-2 weeks after immobilization.
- After 2–4 weeks of immobilization, exercise palmar flexion and dorsal flexion for 10 times each set, 2 sets per day. Hold for 10 seconds in each flexion.
- After 4–6 weeks of immobilization, exercise ulnar flextion and radial flexion for 10 times each set, 2 sets per day. Hold for 10 seconds in each flexion.
- After 6–8 weeks of immobilization, flex the elbow, place the forearm in front of the abdomen, exercise supination and pronation for 10 times each set, 2 sets per day. Hold for 10 seconds in each flexion.
- After 9–12 weeks of immobilization, exercise the forearm and wrist while bearing weight.
- Early diagnosis and proper treatments will give a favorable prognosis.

DISLOCATION OF THE WRIST JOINT 腕 關 節 脫 位

INTRODUCTION

Dislocation of the wrist joint is a common hand injury. It has an obvious injury history and onset is usually abrupt. The wrist looks weird, and some wrist bones may be visibly shifted. Tears may happen in the tendons and ligaments, bones may fracture, and nerves may be also damaged. Early diagnosis and treatment are crucial for successful recovery. Chronic injury may lead to the loss of hand muscles and functions.

Anatomy

Wrist Joint

- is articulated by the distal end of the radius bone known as a radiocarpal joint, it is an only true articulating joint between the forearm and the carpal bones. It is a synovial joint, there is a joint capsule containing lubricating synovial fluid within the capsule. The joint allows the hand and the wrist to perform palmar flexion and extension of dorsiflexion, and also allows a side to side gliding.[10]

  The outer layer of the synovial membrane is a fibrous layer that attaches to the radius, ulna, and the proximal row of the carpal bones.
- **Ligaments Associated with the Radiocarpal Joint**
  - The palmar radiocarpal ligaments are located in the volar wrist, passing from the radius bone to the scaphoid, lunate, triquetrum, and capitate of the carpal bones. There are four distinct parts:
    1) The radioscaphocapitate ligament: from the styloid process of radius bone to the distal aspect of scaphoid bone, with a small number of fibers continuing and proximal aspect of the capitate bone.
    2) The long radiolunate ligament: from the distal radius to the lunate bone.
    3) The radioscapholunate ligament: arises ulnar to the long radiolunate ligament and merges with the scapholunate interosseous ligament.
    4) The short radiolunate ligament: from the distal radius to the palmar margin of the lunate bone.

    Its function is to assist the hand to follow the forearm supination and to increase the wrist stability.
  - The dorsal radiocarpal ligament, located in the dorsal wrist, passes from the radius bone to the lunate and triquetral carpal bones, and it passes the scaphoid carpal bone without direct attachment (Wozasek and Laske 1991). Its function is to assist the hand in following the forearm pronation and to increase wrist stability.
  - The palmar ulnocarpal ligament is located in the ulnar wrist and passes from the styloid process of the ulna to the lunate, capitate, and triquetral carpal bones. Its components include ulnolunate, ulnocapitate, and ulnotriquetal ligaments. The ulnotriquetal ligament is also known as the ulnar collateral ligament.
  - The radial collateral ligament is located in the radial wrist, and it passes from the radial styloid process to scaphoid and trapezium carpal bones. Its function is to prevent excessive ulnar deviation of the hand.
- Blood supply: the blood supply to the wrist joint is from the dorsal and palmar carpal arches, which are derived from the ulnar and radial arteries.
- Nerve innervation: the nerve innervation to the wrist is from the median, radial, and ulnar nerves.

WESTERN MEDICINE ETIOLOGY AND PATHOLOGY

The traumatic changes are usually in ligaments. They hold the bones in place and allow them to move properly. A tear in any of the ligaments can result in two or more of the carpal bones being shifted out of the usual position with the other carpal bones or to the distal end of the radius.

While it is possible for any of the eight carpal bones to become dislocated, the lunate and scaphoid are most commonly seen in clinical dislocation. These two bones are directly associated with the radius and ulna and form a bridge between them.

Perilunate Wrist Dislocation

The carpal dislocations are typically caused by an acute hyperextension in the ulnar deviation and intercarpal supination simultaneously. A fall on the thenar eminence is an example, where the body weight falls onto the thenar eminence to the fixed point of the ground, leading ligamentous failure beginning on the radial side of the wrist and resulting in four stages of joint instability.

Stage 1: It is scaphoid rotation, and the affected joint is the scapholunate joint. The affected ligaments are the scapholunate interosseous, palmar radioscaphoid, and radiocapitate ligaments.

Stage 2: It is capitate dislocation, and the affected joints are the scapholunate and capitolunate joints. The affected ligaments are radioscaphoid, scapholunate interosseous, radiocapitate, and radial collateral ligaments.

Stage 3: It is triquetral dislocation, and the affected joint is the triquetrolunate joint. The affected ligaments are radioscaphoid, scapholunate interosseous, radiocapitate, palmar radiotriquetral, ulnotriquetral, and radial collateral ligaments.

Stage 4: It is lunate dislocation, and the affected joints are the scapholunate, capitolunate, triquetral lunate, and radiolunate joints. The affected ligaments are radioscaphoid, scapholunate interosseous, radiocapitate, radial collateral ligaments palmar radiotriquetral, and ulnotriquetral and dorsal radiocarpal ligaments.

CHINESE MEDICINE ETIOLOGY AND PATHOLOGY

Qi and Blood Stagnation 氣滯血瘀

It can be caused by the impact of external force on the radial wrist during a fall. Serious falls, such as in motorcycle or skiing accidents, can cause direct blows to the carpal bones and may also entrap the median nerve and break the vessels. This acute traumatic injury leads to Blood stasis, causing blockage in the meridians and resulting in wrist pain.

MANIFESTATIONS

1. Patients may have an obvious injury history; the pain is sudden onset with a history of trauma. The distal radius of the affected side is bulged and swelling, with dorsal tenderness.

2. Wrist joint functionality may be lost, grip weakened, range of motion reduced.
3. If the median nerve is damaged, its distribution area is tingling, numb, or there may be a sensation of electric shock in the thumb, index, middle fingers, and the radial half of the ring finger. The sensation may begin in the wrist and travel to the hand. Aggravating factors include push-ups and holding a steering wheel.
4. Even if the anterior interosseous branch of the median nerve is damaged, since it does not have a cutaneous branch, there are no sensory deficits. The neuropathy presents with muscle weakness, patients are unable to approximate the thumb and index finger, the thumb pad cannot draw to touch the other fingertips, and patients are unable to make the OK sign with the thumb and the index finger (Dydyk et al. 2022).

PHYSICAL EXAMINATION

- Begin with obtaining medical history.
  - Obtain a detailed traumatic history. The injury happens when the wrist is hyperdoriflexed, the hand is deviated to the ulna side, the arm is supinated, and the thenar hand lands with great force on the ground.
  - Ask the patient about symptoms, including hand grip force and the wrist range of motion. Damage to the median nerve is the most commonly associated injury in lunate and perilunate dislocations of the wrist, so inquire about any abnormal sensation on the fingers, especially the first four fingers.
  - Ask about aggravating factors such as doing push-ups.
- **Inspection**

  The wrist has marked swelling. Look for deformities, like dorsal perilunate dislocation or a volarly prominent radius. If it is a pure lunate dislocation, the lunate is prominent volarly. Observe the hand muscles, looking for muscle wasting in the thenar eminence, which may lead to atrophy. In severe cases, the hand may have "Ape hand deformity".[11]
- **Palpation**

  Palpation on hands and wrists detects swelling, skin temperature, and tenderness.

  Palpation on the wrist includes radial styloid, ulnar styloid, scaphoid, trapezium/first metacarpal joint, lunate, TFCC, triquetrum, pisiform, and hook of hamate.

  Palpation on hand includes metacarpals and phalanges.
- **Range of motion**

  Test the patient's range of motions on fingers, hands, and wrists. Test the wrist for flexion and extension, radial and ulnar deviation.

  Test the extension, abduction, and opposition of the first carpometacarpal (CMC) joint. In "Ape hand deformity", the thumb may lack these movements. This is easily tested by trying to make an OK sign, in which the index fingertip and thumb tip meet.

- **Special Test**
    Although X-rays will not show the ligaments, they can suggest a ligament injury if there is a fracture or an abnormal position of the radius, ulnar, or the carpal bones.

DIFFERENTIAL DIAGNOSIS

Distal Radius Fracture

- The most common type of distal radius fracture is a Colles fracture, this type of fracture moves the radius bone slightly upward, but the wrist joint remains unaffected which produces a very distinctive condition known as the "dinner fork deformity". Viewed from the side, the wrist has the appearance of an overturned fork due to a distinct bump in the wrist similar to the neck of the fork.
- The distal part of the radius bone shows an obvious deformity by a protrusion of the broken end of the distal radius bone shifting upward toward the dorsal aspect of the hand.
- A tingling sensation in the fingertips or there may be limited finger movement.

TREATMENTS

Treatment Principle

Nourishes the Blood, moves the Blood, disperses stasis, stops pain

Herbal Treatments

Herbal Formula

Zheng Gu Zi Jin Dan正 骨 紫 金 丹

    This formula is from "Imperially Commissioned Golden Mirror of the Orthodox Lineage of Medicine醫 宗 金 鑑", authored by the Qianlong emperor 乾 隆 (1711–1799), in Qing dynasty 清 朝 (1636–1912), and published in 1742.

Herbal Ingredients

Ding Xiang丁 香2g, Mu Xiang木 香2g, Xue Jie血 竭2g, Er Cha兒 茶2g, Da Huang大 黃2g, Hong Hua紅 花2g, Dang Gui當 歸（ 頭 ）4g, Lian Zi蓮 子4g, Fu Ling茯 苓4g, Bai Shao芍 藥4g, Mu Dan Pi牡 丹 皮1g, Gan Cao甘 草0.6g

Ingredient Explanations

- Ding Xiang丁 香 warms Middle Jiao and relieves pain.
- Mu Xiang木 香 promotes the movement of Qi and alleviates pain.
- Xue Jie血 竭 moves the Blood, stops bleeding, and stops pain.
- Er Cha兒 茶 clears Heat and stops bleeding.
- Da Huang大 黃 moves the Blood and disperses Bruising.
- Hong Hua紅 花 moves the Blood, disperses Bruising, and stops pain due to Blood stasis.

- Dang Gui當歸 moves the Blood, disperses Bruising, and stops pain due to Blood stasis.
- Lian Zi蓮子 tonifies the Spleen.
- Fu Ling茯苓 drains Damp and tonifies the Spleen.
- Bai Shao白芍 nourishes the Blood, soothes the Liver, and stops pain.
- Mu Dan Pi牡丹皮 clears Heat, cools the Blood, and moves the Blood.
- Gan Cao甘草 harmonizes the herbs, tonifies Middle Jiao Qi, and clears Heat.

Modifications
- For Wind-Damp Bi, add Ru Xiang乳香 and Mo Yao沒藥.
- For severe pain due to Blood stasis, add San Qi 三七.

Formula Indications
- Pain, stiffness, swelling, and inflammation in the traumatized area.

Contraindications
- It is contraindicated for pregnancy.

Administrations
Mix the herbs with honey, for into 9-gram pills, and take 1 pill 2–3 times a day with warm wine.

External Wash Formula
Herbal Formula for Late Stage
Shu Jin Huo Luo Bathing 舒筋活絡澡

Herbal Ingredients
Shen Jin Cao 伸筋草9g, Hai Tong Pi海桐皮9g, Qin Jiao秦艽9g, Du Huo獨活9g, Dang Gui 當歸9g, Gou Teng 鉤藤9g, Ru Xiang 乳香6g, Mo Yao 沒藥6g, Hong Hua 紅花6g

Ingredient Explanations
- Shen Jin Cao 伸筋草 expels Wind, drains Damp, and relaxes Tendons.
- Hai Tong Pi 海桐皮 expels Wind, drains Damp, and reduces swelling.
- Qin Jiao 秦艽 expels Wind Damp.
- Du Huo 獨活 expels Wind Damp and stops pain.
- Dang Gui 當歸 nourishes the Blood, reduces edema, and stops pain.
- Gou Teng 鉤藤 expels Wind and releases spasm.
- Ru Xiang 乳香 moves the Blood and stops pain.
- Mo Yao 沒藥 moves the Blood and stops pain.
- Hong Hua 紅花 moves the Blood, releases Bruising, opens the meridians, and stops pain.

Modification
For moving the Blood, add Tao Ren桃 仁 and Chuan Xiong川 芎.

Direction
Wash the affected area with this herbal soup.

Topical Herbal Formula
Yunnan Baiyao Aerosol (topical application) 雲 南 白 藥 氣 霧 劑

Formula Ingredients
The main ingredient in Yun Nan Bai Yao雲 南 白 藥 is San Qi三 七. San Qi 三 七is used as a treatment for internal and external bleeding, and it also relieves any type of pain from injury.

The topical spray helps to promote Blood circulation, stop pain, and reduce swelling. The spray is easier to apply to closed wounds than the Yun Nan Bai Yao powder form.

Dose

The product comes in two bottles: a cream-colored Yunnan Baiyao Aerosol 雲 南 白 藥 氣 霧 劑and a red-colored Yunnan Baiyao Aerosol Baoxianye雲 南 白 藥 氣 霧 劑 保 險 液.

Contraindication
This product is contraindicated for pregnancy.

Suggested Usage
Shake the bottle several times before use. Hold the bottle vertically with the spray nozzle 5–10 cm from the affected skin and spray for about 3–5 seconds.

Red-Colored Spray – Yunnan Baiyao Aerosol Baoxianye雲 南 白 藥 氣 霧 劑 保 險 液
- For serious trauma, apply Aerosol Baoxianye 氣 霧 劑 保 險 液to the injured part first, then reapply it 1–2 minutes later if the pain doesn't dissipate.
- Aerosol Baoxianye 氣 霧 劑 保 險 液should not be used more than three times a day. If pain continues, apply Yunnan Baiyao Aerosol雲 南 白 藥 氣 霧 劑.
- Aerosol Baoxianye氣 霧 劑 保 險 液 should not be used on open wounds, cuts, or grazes.

Cream-Colored Spray – Yunnan Baiyao Aerosol雲 南 白 藥 氣 霧 劑
Externally used to spray on wounds 3–5 times daily.

Acupuncture Treatments

Acupuncture Point Formula

Basic Points

SJ5外 關Waiguan, PC6內 關Neiguan, SI5陽 谷Yanggu, SJ4陽 池Yangchi, SI4腕 骨Wangu, SI6養 老Yanglao, HT7神 門Shenmen

Modifications

For Blood stasis, add SP10血 海Xuehai.

For severe pain, add LI4合 谷Hegu.

Cupping Therapy

SJ4陽 池Yangchi, PC7大 陵Daling after bleeding Therapy, this is for acute wrist dislocation with edema.

Moxibustion Therapy

Warming needling techniques on SJ4陽 池Yangchi, PC7大 陵Daling for habitual wrist dislocation.

Bleeding Therapy

SJ4陽 池Yangchi, PC7大 陵Daling for acute wrist dislocation with edema.

Auricular Therapy

Wrist, Shenmen, Sympathetic, Liver, Kidney

Tui Na Treatments

Tui Na Techniques

Stretching, pressing, relocating.

Tui Na Procedures

The patient is seated, the affected forearm is fully supinated. Assistant 1 stands near the ipsilateral shoulder and holds the affected forearm, assistant 2 stands near the ipsilateral hand facing the patient and holds the affected hand. The therapist stands at the ipsilateral side near the affected wrist, places both thumbs on the radius aspect of the carpal bones and places the rest of fingers on the ulnar aspect of the affected wrist.

1. The therapist asks the patient to relax, and asks the assistants to stretch the affected wrist by consistently pulling the forearm and the hand away from the wrist.
2. The therapist's thumbs apply force to push the shifted carpal bones toward the volar ulnar aspect of the wrist.
3. After the successful reduction, immobilize the wrist in a 30–45° palmar flexion for the first week. Set the wrist in a neutral position for another 2

weeks. Advise the patient to attempt mild finger movements in the first three weeks of immobilization, and then attempt wrist movement after immobilization.

PATIENT ADVISORY

Treatment principles after wrist joint reduction in Chinese medicine:

- In the first two weeks, it is to move Blood, disperse Bruising, move Qi, and stop pain.
- In weeks 2–3, it is to harmonize with the new tissues, connect the Tendons, and promote healing of the Tendons.
- After week 3, it is to tonify Qi, nourish the Blood, nourish the Liver and Kidney in order to strengthen Tendons, Bones, and joints.

Advice for patients during recovery:

- Avoid foods that are hot, spicy, sour, fatty, deep-fried, and the foods that are listed as Hot temperature in Chinese dietary, such as lamb and goose.
- Keep the wrist warm and avoid exposure to cold water for three weeks.
- If the prognosis is good, full recovery is expected within 12 weeks. Early diagnosis and the proper treatments give a favorable outcome.

NOTES

1. https://radiopaedia.org/articles/median-nerve-2?lang=us.
2. https://teachmeanatomy.info/upper-limb/nerves/median-nerve.
3. https://www.webmd.com/pain-management/carpal-tunnel/carpal-tunnel-syndrome.
4. https://www.acropt.com/blog/2017/5/28/anatomy-of-the-wrist-mfkkp#:~:text=The%20Wrist%3A,the%20carpals%20in%20the%20hand.
5. https://www.sports-health.com/sports-injuries/hand-and-wrist-injuries/risk-factors-and-causes-wrist-sprain.
6. https://litfl.com/distal-radioulnar-injury/.
7. https://www.webmd.com/first-aid/what-to-know-about-distal-radius-fracture.
8. https://www.sports-health.com/sports-injuries/hand-and-wrist-injuries/symptoms-distal-radius-fracture.
9. https://www.hopkinsmedicine.org/health/conditions-and-diseases/distal-radius-fracture-wrist-fracture.
10. https://www.acropt.com/blog/2017/5/28/anatomy-of-the-wrist-mfkkp#:~:text=The%20Wrist%3A,the%20carpals%20in%20the%20hand.
11. https://www.physio-pedia.com/Wrist_and_Hand_Examination.

10 Hand and Finger Pain

TRIGGER FINGER 板機指

INTRODUCTION

If when you try to straighten a finger, it snaps or locks in a halfway position, and you have to use another hand to straighten it, you may already have a trigger finger. It is treatable and preventable. The finger condition is very common in jobs that overuse the tendons of the hand Figure 10.1.1-1640524138.

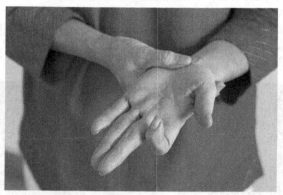

ANATOMY

Flexor Digitorum Superficialis Tendons

The origin of the flexor digitorum superficialis muscle has two heads – the humer-oulnar and the radial. In the distal forearm, there are four flexor digitorum superficialis tendons that extend from this muscle and travel through the carpal tunnel formed by the flexor retinaculum, with one flexor pollicis longus and four flexor digitorum profundus. The flexor digitorum superficialis tendons attach to the anterior margins on the bases of the intermediate phalanges of the four fingers; however, it is relatively common for the flexor digitorum superficialis tendon to be missing from the fifth digit bilaterally and unilaterally.[1]

The primary function of the flexor digitorum superficialis muscle and tendon is flexion of the middle phalanges of the four fingers (excluding the thumb) at the proximal interphalangeal joints.

A1 Pulley

- Each of the flexor tendons passes through a tunnel in the palm, the tunnel is called the tendon sheath. It is also called the synovial lining or the fibrous sheath.

DOI: 10.1201/9781003203018-10

- The tendon sheath is quite thin, but it is composed of a few layers of connective tissue (1) Stratum fibrosum, (2) Mesotendinum, (3) external layer of Stratum synoviale, (4) Synovia, (5) internal layer of Stratum synoviale, and (6) tendon. These layers are composed of fibrous and synovial layers. The functions of the fibrous layer are to support and protect, while the synovial layer produces synovial fluid that allows the tendons to move smoothly.
- Pulleys are the annular part of the fibrous sheaths within the tendon sheaths that hold the flexor digitorum superficialis tendons closely to the finger bones. In the human body, the annular part of the fibrous sheaths of the fingers, the annular ligaments, are often called "A pulleys". There are four or five pulleys on each finger, numbered in order from the proximal to the distal or from the metacarpophalangeal joint (MCP joint), proximal interphalangeal joint (PIP) to the distal interphalangeal joint (DIP), they are A1, A2, A3, A4, and A5. The A1, A3, and A5 pulleys are the joint pulleys located on the MCP, PIP, and DIP joints, respectively. Thus, the A1 pulley is the pulley at the base of the finger near the head of the MCP joint[2] Figure 10.1.2-2034874742.

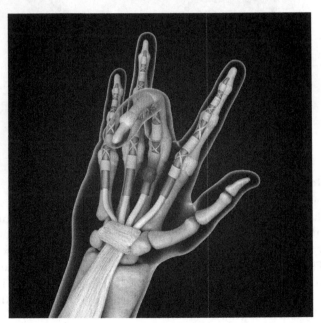

WESTERN MEDICINE ETIOLOGY AND PATHOLOGY

Trigger finger is caused by inflammation and thickening of the flexor digitorum superficialis tendons. The thickening of the flexor digitorum superficialis makes it difficult to glide through the A1 pulley.

Inflammation of the tendon is usually the result of repetitive motion of the digits or compression by the head of the metacarpal bone. Inflammation thickens the tendon sheath and develops nodules on the tendon surface, the nodules catch or pop in the tendon sheath when they strike the A1 pulley making the finger difficult to straighten. In a severe case of trigger finger, the nodule cannot pass through the A1 pulley at all, and the digit becomes stuck in the flexed position; often the patient has to pull the distal finger to strike through the tunnel.

Risk factors to develop trigger finger include repetitive overuse of the finger, such as pulling the cord of a gas lawn mower and holding the safety bar while it is running or extensive typing and computer mouse using while the wrist and the elbow are flexed. Certain medical conditions may also be prone to the injury, including diabetes and rheumatoid arthritis.[3]

CHINESE MEDICINE ETIOLOGY & PATHOLOGY

Qi and Blood Stagnation 氣滯血瘀

From the point of view of traditional Chinese medicine, trigger finger refers to Tendon Bi Syndrome 筋痹. Wind-Damp-Cold attack the Tendon due to deficiency of Righteous Qi正氣. The factors of Wing-Damp-Cold block the meridians and obstruct the flow of Qi and Blood, resulting in stiffness and painful Tendons 筋 (Tendons 筋 in Chinese medicine has a boarder meaning, it includes tendons, nerves, vessels, ligaments, fascia, bursae, joint capsules, and may even include cartilages and glenoid labrum when Kidney is deficient).

MANIFESTATIONS

- After the finger flexion, the finger joint is locked or catching during finger extension. In severe cases, passive manipulation may be needed to extend the digit.
- The digit is stiff, especially in the morning in the cold season.
- There is pain or tenderness in the palm, often seen in the A1 pulley, and it may radiate to a corresponding painful digit.
- A palpable nodule is observed in the line of the flexor digitorum superficialis tendon, just distal to the MCP joint in the palm.
- Swelling of the affected digit and trigger finger happening in flexion is sometimes seen, but it is rare.
- Trigger finger can happen on children as well, often reported on the thumb's interphalangeal joint (IP).

PHYSICAL EXAMINATION

- Begin with obtaining medical history.

    Patients with trigger finger may have a history of repetitive trauma to the finger. This includes machine operation, holding a steering wheel for long periods, pulling the cord on a lawn mower repeatedly and holding the safety bar while it is running, and extensive typing and clicking the computer mouse.

- Inspection
    For some, the affected digit may lock in flexion but may still be able to pass through the locked point to extend the digit. In severe cases, however, the digit is unable to move beyond the restriction; thus, triggering or snapping does not occur.[4]
- Palpation[5]
    - When the digit is flexed, a tender nodule can be palpated overlying the MCP joint.
    - During digit extension from a full flexion, a palpable snapping sensation can be detected approximately 5 mm distal from the proximal palmar crease of the index, middle, ring, and the little fingers. In the later stage, passive extension of the locked digit is very painful for the patient.
    - Some patients may have tenderness over the proximal tendon of flexor digitorum superficialis distal to the medial epicondyle.
- Range of motion
    Other than the stiffness, snapping, and locking, the finger movement is usually in the normal range. Pain usually occurs while extending the digit, although some patients have discomfort while flexing.

DIFFERENTIAL DIAGNOSIS

Metacarpophalangeal Joint Sprain

With MCP joint sprain, there is an obvious injury history. The joint is swollen and painful, but there is no trigger motion.

Metacarpophalangeal Joint Arthritis

In MCP joint arthritis, the joint is painful and swollen. The symptoms get worse when gripping or grasping, such as opening a jar or turning a key; patients may tend to drop objects due to sudden pain. Again, the MCP joint doesn't lock, and there is no trigger motion.

TREATMENTS

Treatment Principle

Nourish the Blood, move the Blood, disperse stasis, and stop pain.

Herbal Treatments

Herbal Formula

Zheng Gu Zi Jin Dan正骨紫金丹

This formula is from "Imperially Commissioned Golden Mirror of the Orthodox Lineage of Medicine醫宗金鑑", authored by the Qianlong emperor 乾隆 (1711–1799), in Qing dynasty 清朝 (1636–1912), and published in 1742.

Herbal Ingredients

Ding Xiang丁香2g, Mu Xiang木香2g, Xue Jie血竭2g, Er Cha兒茶2g, Da Huang大黃2g, Hong Hua紅花2g, Dang Gui當歸（頭）4g, Lian Zi蓮子4g, Fu Ling茯苓4g, Bai Shao芍藥4g, Mu Dan Pi牡丹皮1g, Gan Cao甘草0.6g

Ingredient Explanations

- Ding Xiang丁香 warms the Middle Jiao and relieves pain.
- Mu Xiang木香 promotes the movement of Qi and alleviates pain.
- Xue Jie血竭 moves the Blood, stops bleeding and stops pain.
- Er Cha兒茶 clears Heat and stops bleeding.
- Da Huang大黃 moves the Blood and disperses Bruising.
- Hong Hua紅花 moves the Blood, disperses Bruising and stops pain due to Blood stasis
- Dang Gui當歸 moves the Blood, disperses Bruising, and stops pain due to Blood stasis.
- Lian Zi蓮子 tonifies the Spleen.
- Fu Ling茯苓 drains Damp and tonifies the Spleen.
- Bai Shao白芍 nourishes the Blood, soothes the Liver, and stops pain.
- Mu Dan Pi牡丹皮 clears Heat, cools the Blood, moves the Blood.
- Gan Cao甘草 harmonizes the herbs, tonifies the Middle Jiao Qi, and clears Heat.

Modifications

- For Wind-Damp Bi, add Ru Xiang乳香 and Mo Yao沒藥.
- For severe pain due to Blood stasis, add San Qi三七.

Formula Indications

- Pain, stiffness, and inflammation in the traumatic area.

Contraindications

- It is contraindicated for pregnancy.

Administrations

Mix the herbs with honey; form into 9-gram pills, and take 1 pill, 2–3 times a day with warm wine.

Acupuncture Treatments

Acupuncture Point Formula

Basic Points

PC6內關Neiguan, PC7大陵Daling, A1 pulley

Modifications

For acute trigger finger, add SP10血海Xuehai and UB17膈俞Geshu.

For chronic trigger finger, add UBI8肝俞Ganshu, UB20脾俞Pishu, UB23腎俞Shenshu, and ST36足三里Zusanli.

Extra Points

Shouyili 手 一 里 (1 Cun 寸 distal to SI8小 海Xiaohai, elbow 90° flexion, 0.3–0.5 Cun 寸 insertion, while moving wrist in palmar and dorsal flexion), Shouerli手 二 里 (2 Cun 寸 distal to SI8小 海Xiaohai, elbow 90° flexion, 0.3–0.5 Cun 寸 insertion, while moving wrist in palmar and dorsal flexion), Shousili手 四 里 (4 Cun 寸 distal to PC3曲 澤Quze, 1–1.5 Cun 寸 oblique insertion along Pericardium meridian toward the wrist, while moving wrist in palmar and dorsal flexion), Ashi acupoints 阿 是 穴.

Auricular Therapy

Finger, wrist, Shenmen, Subcortex, Endocrine.

Moxibustion Therapy

Warming needling techniques on Shouyili 手 一 里, Shouerli手 二 里, Shousili手 四 里

Tui Na Treatments

Tui Na Techniques

Pressing, kneading

Tui Na Procedures

Patient is seated, the affected elbow is in 90° flexion, the forearm is fully supine. The therapist should perform the following procedures:

- Grasp the affected finger and stretches it away from the wrist, while pressing and pushing the affected A1 pulley toward the wrist. Hold for 1 minute each time and repeat 3–5 times.
- Press the A1 pulley with fingernail Figure 10.1.3. While pressing, stretch the finger and move it in a palmar and dorsal flexion. Hold for 1 minute each time and repeat 2–3 times. Do not press forcefully and avoid breaking the skin.

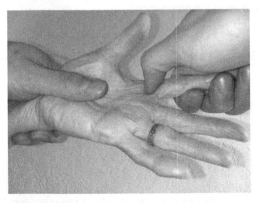

- Apply rolling and chafing techniques on the volar forearm with Hong Hua You 紅 花 油for 3–5 minutes.

Patient is seated, the affected elbow is in 90° flexion, and the forearm is in neutral position.

- Press Shouyili 手一里, Shouerli手二里, and Shousili手四里with fingertips.
- Rotates the patient's wrist in palmar and dorsal flexion.
- Hold the procedure for 1–3 minutes.

PATIENT ADVISORY

- Rest and avoid jobs that are mechanical factors, correct the positions of the elbow and wrist while using a computer.
- More locking in the morning may be due to flexion the elbow or pulling the comforter with the affected hand while sleeping. Recommend that the patient sleeps on the back to avoid flexion of the arm. Splinting the affected digit at night helps to keep the finger or thumb in a straight position while sleeping.
- Keep the arm, hand, and finger warm; application of a heating pad is useful. Wash hands with warm water.
- Avoid exposure to cold temperatures, wear long sleeves, avoid washing hands and arms with cold water, especially when menthol products are used on the hand.

GANGLION CYST IN WRIST 腱鞘囊腫

INTRODUCTION

A ganglion cyst is a small sac of fluid over the tendon or joint, it occurs commonly in the wrist but may also occur in the ankles and feet. It is one of the hand's benign tumors.[6] Ganglion cysts can be painful if they press on a nearby nerve; sometimes they may interfere with joint movement Figure 10.2.1-shutterstock.com #2038198556

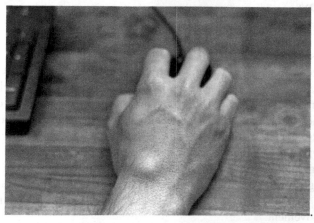

ANATOMY

Dorsal wrist ganglions are the most common type of ganglion cyst, usually originating from the dorsal portion of the scapholunate ligament. On the superficial aspect, they are located between the extensor pollicus longus and extensor digitorum communis.

The jelly-like fluid in the ganglion cyst (synovial fluid)[7] is a gelatinous mucoid material (Gregush and Habusta 2021) similar to that found in joints or around tendons[8]; thus, the ganglion cyst is also called a "synovial cyst". The fluid is from the joint or tendon tunnel surrounding the cyst and is swollen beneath the skin.

WESTERN MEDICINE ETIOLOGY AND PATHOLOGY

The exact etiology of the development of ganglion cysts is unknown, even though there are several suggestions. The most common and accepted theory is that the fluid in the cyst leaks from the surrounding tendon sheath or joint capsule. The cause of the leak is unknown, but it is believed to arise from repetitive microtrauma resulting in mucinous degeneration of connective tissue. It also has been suggested that the chronic damage from an antecedent trauma causes the mesenchymal cells or fibroblasts to produce mucin.[9]

CHINESE MEDICINE ETIOLOGY & PATHOLOGY

Phlegm-Damp Accumulation 痰濕內蘊

According to traditional Chinese medicine, acute ganglion cyst is due to trauma to the fascia leading to accumulation of Damp and Blood, causing fluid retention in joints and resulting in obstruction of the meridians. Chronic tendon or joint damage from an antecedent trauma ganglion leads to deficiency of Blood. Tendons' lack of Blood nourishment leads to the degeneration of connective tissue. Fluid leaks from the surrounding tendons or joint capsule forming the cysts.

MANIFESTATIONS

Symptoms of a ganglion cyst in the wrist include:

- The cysts are typically round or oval.
- The sizes can be from 1 to 2.5 centimeters in diameter. The size of a cyst can fluctuate; they can get larger when the joint is used repetitively.[10] Some cysts are deep in the wrist; they are not visible in the skin and cannot be felt. These are called hidden or occult ganglions, and they can be a source of unexplained wrist pain and disproportionate tenderness.
- It is a visible soft bump in the skin, but it does not move.
- They can quickly appear, disappear, and reappear.
- Although many ganglions manifest without other symptoms, some cysts can be painful if pressed by a nearby nerve; they can also interfere with joint movement.

PHYSICAL EXAMINATION

- Begin with obtaining medical history.

Although some patients with ganglions may not have an obvious injury history, prior injury can still be the cause of it. The interview should include the previous injury history, jobs, and sports that may associate with the ganglions. Inquire about whether the cysts change sizes or disappear/reappear and the pain quality, such as tingling and numbness on the first four digits.

- Inspection
  Shining a penlight on the cyst to observe transillumination will aid in diagnosis, as the ganglion is filled with fluid.[11] The size of the cysts may fluctuate with time and activity.
- Palpation
  Palpation on a cyst can determine if the cyst is a solid mass or filled with fluid (ganglion cysts are filled with fluid).
- Range of motion
  Other than the stiffness, the range of motion of the wrist, hand, and fingers should still be in the normal range. Some patients with ganglion cyst compression on the median nerve may demonstrate weak hand gripping force.
- Special Test
  Imaging tests, including x-ray, ultrasound, and MRI, can rule out other conditions, such as arthritis or a malignant tumor.

DIFFERENTIAL DIAGNOSIS

Lipoma
A freely movable, soft, fluctuant mass that does not transilluminate.

TREATMENTS

Treatment Principle
Nourishes the Blood, soothes the Liver, and regulates Tendons.

Herbal Treatments
Herbal Formula:
*Si Wu Tang*四 物 湯
This formula is from "Tai Ping Hui Min He Ji Ju Fang太 平 惠 民 和 劑 局 方", authored by Imperial Medical Bureau 太 醫 局 (1134AD), in southern Song dynasty南 宋 1127–1239AD.

Herbal Ingredients
Dang Gui當 歸10g, Chuan Xiong川 芎10g, Shu Di Huang熟 地 黃10g, Bai Shao白 芍10g

Ingredient Explanations

- Dang Gui當歸 nourishes the Blood, benefits the Liver, and regulates menstruation.
- Chuan Xiong川 芎 moves the Blood and moves Qi.
- Shu Di Huang熟 地 黃 tonifies the Liver and Kidney Yin; benefits Essence and the Blood.
- Bai Shao白 芍 nourishes the Blood, soothes the Liver, and stops pain.

Modifications

- For Damp, add Can Sha蚕 砂.
- For tenderness and stiffness, add Jiang Huang薑 黃, Mu Gua木 瓜, Wu Jia Pi五 加 皮, and Hai Tong Pi海 桐 皮.

Formula indication

- Liver Blood deficiency causes Tendon malnourishment.

Contraindications

- Caution for patients who are pregnant.
- Caution for patients with constitutional Spleen Yang deficiency.

Acupuncture Treatments

Bleeding Therapy

Ganglion cyst drainage:

- Location: four edges of the cyst.
- Remove a few drops with a cotton wool swab.
- Treatment frequency is once every 2–3 days, for 3–5 times.
- Tool: medical lancet.

Procedures

- Prior to the procedure, the tools should be prepared and ready to be used in the working area.
- Set up a clean field with paper towels.
- Place the lancets in the clean field, along with dry sterile cotton balls, medical alcohol wipes or cotton swabs soaked in alcohol, sterile gloves, a biohazard sharps container, a biohazard trash container, goggles and face mask, and paper towels.
- The room should be well lit.
- The patient is seated with forearm pronated for dorsal ganglion cyst drainage.
- The therapist washes and dries their hands thoroughly and puts on sterile gloves.

- The therapist wears a mask and sits facing the cyst.
- Wipe the cyst with a medical alcohol wipe or a cotton swab soaked in alcohol, moving from the center of the cyst to the periphery.
- Place the medical lancet at an angle of 90 degrees to the skin.
- In one quick move, prick the needle into the cyst about 0.1 Cun 寸 deep, eliciting a few drops of fluid.
- If more drops of fluid are needed, absorb the fluid with a sterile cotton ball.
- Lastly, the cyst is pressed with a sterile cotton ball until the bleeding ceases.
- Apply a bandage to the prick wound.
- Dispose of the tools in the proper containers.
- Wash hands thoroughly.

Tui Na Treatments

Tui Na Techniques

Pressing, kneading, pushing, stretching

Tui Na Procedures

The patient is seated.

- The therapist applies pressing, kneading, and rubbing techniques to the joints around the ganglion cyst for 3–5 minutes.
- Holding the affected wrist with one hand and the fingers with the other hand, the therapist then applies traction force to stretch the affected joint and tendon while a thumb presses and pushes the cyst along the tendon Figure 10.2.2.

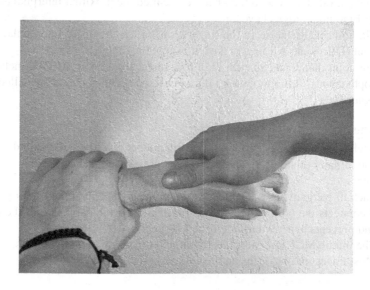

PATIENT ADVISORY

- Advise the patient to use only gentle force when using the affected joint.
- Keep the drainage site clean to avoid infection.
- Get plenty of mental and physical rest.

FINGER SPRAIN 手 指 扭 傷

INTRODUCTION

Finger sprains are common, they tend to be caused by sports injuries or other external factors that may overstretch finger ligaments, such as a fall, car accident, and work-related or sports-related injury. Finger injury can be very painful, and usually, the pain happens immediately. Improper diagnosis and treatment of finger sprain can cause deformity and dysfunction over time.

ANATOMY

Bones

The human hand has five digits, the thumb is the first digit and the rest four fingers the second to the fifth digits. The thumb has two phalanges, and each of the other fingers has three phalanges.

Joints

- The phalanges are separated by interphalangeal joints. The thumb has one interphalangeal joint, and there are 2 in each finger.
- The joint that is close to the knuckle is called the proximal interphalangeal (PIP) joint; all digits have one.
- The joint that is close to the fingertip is called the distal interphalangeal joint (DIP) joint. All fingers have it except the thumb.
- The joint that connects each digit to a metacarpal bone is called metacarpophalangeal (MCP) joint. All digits have one. This joint is also called the "knuckle".

Ligaments

- Each finger joint has two collateral ligaments on either side called radial and ulnar collateral ligaments. They prevent abnormal flexion of the joints.
- Each finger joint has a ligament in the palmar side called the "volar plate". It connects the proximal and middle phalanx on the palm side of the joint and prevents hyperextension of the PIP joint.
- The thumb MCP joint also has radial and ulnar collateral ligaments.
- Most finger sprains occur on the collateral ligaments.[12]

Muscles

- Each finger has six muscles: three extrinsic originating in the wrist and hand and three intrinsic muscles originating in the forearm or elbow.
- The second and fifth digits each have an extra extrinsic extensor.

Nerves

- The ulnar nerve crosses the wrist through Guyon's canal and branches to provide sensation to the fifth digit and the ulnar half of the fourth digit.
- The median nerve crosses the wrist through the carpal tunnel and branches to provide sensation to the palm, the first, second, and third digits; and the radial half of the fourth digit.
- The radial nerve runs down the radial aspect of the forearm and provides sensation to the back of the hand from the first to the third digit.

Blood Supplies

- Ulnar and radial arteries supply blood to the volar hand, fingers, and thumb.

WESTERN MEDICINE ETIOLOGY AND PATHOLOGY

The traumatic changes to the fingers are usually the result of sprain, strain, or fall. The external force impacts the ligaments on the digits. Ligamentous trauma may involve a sprain, lateral blow to the fingers, extreme finger dorsiflexion, palmar flexion, or twisting of the knuckles. The injury causes damage to the collateral ligaments, joint capsules, or articular cartilage of the knuckles. Commonly, finger sprains stretch the ligaments but do not tear them; other, more severe sprains can partly or completely tear the ligaments. This commonly occurs when a finger is hit by a ball during playing sports, such as basketball or volleyball or when a person falls and the fingers outstretch their normal range.

CHINESE MEDICINE ETIOLOGY AND PATHOLOGY

Qi and Blood Stagnation 氣滯血瘀

The acute onset of finger sprain causes blockages in the local hand meridians. The blockage or stagnation can be due to a direct blow to the hand and finger. The blockage impedes the flow of Qi and Blood to the local meridians, and pain occurs, as the saying goes, "if there is no free flow, there is pain 不通則痛".

MANIFESTATIONS

- Patients may have an obvious injury history.
- The pain occurs suddenly with a history of trauma.
- Swelling in the finger joint.
- Bruising in the affected finger.

- Tenderness in the finger joint when touching or pressing.
- Stiffness or pain in the finger.

PHYSICAL EXAMINATION

- Begin with obtaining medical history.

Finger sprains usually have an obvious injury history, and the patient may hear a "pop" sound as it happens. Inquire about the onset of injury, as even a side blow may cause the collateral ligament damage. In the case if chronic trauma, inquire about the symptoms while it happens, including how soon the finger pain, bruising, and swelling occur, the sooner the more severe.

- Inspection

Examine the affected finger for edema and bruising. Inspection can also detect a deformity of the finger, which may suggest fractures.

- Palpation

Palpation on the affected finger to sense the swelling and tenderness, Palpation of the joint over four planes (dorsal, volar, medial, and lateral) allow assessment of tenderness over ligamentous origins and insertions. The location of pain may suggest the traumatic site, and it is also able to detect fracture.[13]

- Range of motion

Active finger movements in extension and flexion can demonstrate the stiffness and weakness. In case the pain intensity increases suddenly to an extreme severe pain during finger movement, it may suggest a finger fracture. The finger injury may reduce the wrist joint range of motion.

- Special Test

X-ray is able to detect fractures.

DIFFERENTIAL DIAGNOSIS

Phalangeal Fracture

Whether the injury is in the proximal interphalangeal joint (PIP joint), distal phalanx, middle phalanx, or proximal phalanx, if fracture of them is suspected, radiographs should be performed before more forceful testing. Symptoms include inability to grasp or weakness of grasping between the finger and the thumb. Tenderness to the touch, blue or black discoloration of the skin over the injury site, high intensity of pain is produced with movement in any or all directions.[14]

TREATMENTS

Treatment Principle
Nourish the Blood, move the Blood, disperse stasis, and stop pain.

Herbal Treatments
Herbal Formula
Zheng Gu Zi Jin Dan正骨紫金丹
   This formula is from "Imperially Commissioned Golden Mirror of the Orthodox Lineage of Medicine醫宗金鑑", authored by the Qianlong emperor 乾隆 (1711–1799), in Qing dynasty 清朝 (1636–1912), and published in 1742.

Herbal Ingredients
Ding Xiang丁香2g, Mu Xiang木香2g, Xue Jie血竭2g, Er Cha兒茶2g, Da Huang大黃2g, Hong Hua紅花2g, Dang Gui當歸（頭）4g, Lian Zi蓮子4g, Fu Ling茯苓4g, Bai Shao芍藥4g, Mu Dan Pi牡丹皮1g, Gan Cao甘草0.6g

Ingredient Explanations
- Ding Xiang丁香 warms the Middle Jiao and relieves pain.
- Mu Xiang木香 promotes the movement of Qi and alleviates pain.
- Xue Jie血竭 moves the Blood, stops bleeding, and stops pain.
- Er Cha兒茶 clears Heat and stops bleeding.
- Da Huang大黃 moves the Blood and disperses Bruising.
- Hong Hua紅花 moves the Blood, disperses Bruising, and stops pain due to Blood stasis.
- Dang Gui當歸 moves the Blood, disperses Bruising, and stops pain due to Blood stasis.
- Lian Zi蓮子 tonifies the Spleen.
- Fu Ling茯苓 drains Damp and tonifies the Spleen.
- Bai Shao白芍 nourishes the Blood, soothes the Liver and stops pain.
- Mu Dan Pi牡丹皮 clears Heat, cools the Blood, and moves the Blood.
- Gan Cao甘草 harmonizes the herbs, tonifies the Middle Jiao Qi, and clears Heat.

Modifications
- For Wind-Damp Bi, add Ru Xiang乳香 and Mo Yao沒藥.
- For severe pain due to Blood stasis, add San Qi三七.

Formula Indications
- Pain, stiffness, and inflammation in the traumatized area.

Contraindications
- It is contraindicated for pregnancy.

Administrations

Mix the herbs with honey, form 9-gram pills, and take 1 pill 2–3 times a day with warm wine.

External Wash Formula:

Herbal Formula
  Shu Jin Huo Luo Bathing 舒筋活絡澡

Herbal Ingredients

Shen Jin Cao 伸筋草9g, Hai Tong Pi 海桐皮9g, Qin Jiao 秦艽9g, Du Huo 獨活9g, Dang Gui 當歸9g, Gou Teng 鉤藤9g, Ru Xiang 乳香6g, Mo Yao 沒藥6g, Hong Hua 紅花6g

Ingredient Explanations

- Shen Jin Cao 伸筋草 expels Wind, drains Damp, and relaxes Tendons.
- Hai Tong Pi 海桐皮 expels Wind, drains Damp, and reduces swelling.
- Qin Jiao 秦艽 expels Wind Damp.
- Du Huo 獨活 expels Wind Damp and stops pain.
- Dang Gui 當歸 nourishes the Blood, reduces edema, and stops pain.
- Gou Teng 鉤藤 expels Wind and releases spasm.
- Ru Xiang 乳香 moves the Blood and stops pain.
- Mo Yao 沒藥 moves the Blood and stops pain.
- Hong Hua 紅花 moves the Blood, releases Bruising, opens the meridians and stops pain.

Modification

To move the Blood, add Tao Ren桃仁 and Chuan Xiong川芎.

Procedures

Wash the affected area with this herbal soup.

Topical Herbal Formula

Yunnan Baiyao Aerosol (topical application) 雲南白藥氣霧劑

Formula Ingredients

The main ingredient in Yun Nan Bai Yao雲南白藥 is San Qi三七. San Qi三七is used in the treatment of internal and external bleeding and also relieves any type of pain from injury.

The topical spray helps to promote Blood circulation, stop pain, and reduce swelling. The spray is easier to apply for closed wounds than the Yun Nan Bai Yao powder form.

DOSE

The product comes in two bottles: a cream-colored Yunnan Baiyao Aerosol 雲南白藥氣霧劑and a red-colored Yunnan Baiyao Aerosol Baoxianye雲南白藥氣霧劑保險液.

Contraindication

This product is contraindicated for pregnancy.

Suggested Usage

Shake the bottle several times before use. Hold the bottle vertically with the spray nozzle 5 to 10 cm from the affected skin, and spray for about 3–5 seconds.

Red Colored Spray-Yunnan Baiyao Aerosol Baoxianye雲南白藥氣霧劑保險液

- For serious trauma, apply Aerosol Baoxianye 氣霧劑保險液to the injured part first, then reapply it 1–2 minutes later if the pain doesn't dissipate.
- Aerosol Baoxianye氣霧劑保險液should not be use more than three times a day. If pain continues, apply Yunnan Baiyao Aerosol雲南白藥氣霧劑.
- Aerosol Baoxianye氣霧劑保險液 should not be used on open wounds, cuts, or grazes.

Cream Colored Spray-Yunnan Baiyao Aerosol雲南白藥氣霧劑

Used externally to spray on wounds, 3–5 times daily.

Acupuncture Treatments

Acupuncture Point Formula
Basic Points
LI11曲池Quchi,   SI5陽谷Yanggu,   LU7列缺Lieque,   LI4合谷Hegu, HT7神門henmen, Neiguan, PC6內關, PC7大陵Daling

Modifications

- For Blood stasis瘀滯型, add SP10血海Xuehai and LU5尺澤Chize.
- For Damp-Cold Invasion 寒濕外侵, add LI12肘髎zhouliao, LI10手三里Shousanli, and LU5尺澤Chize.
- For Liver Blood Deficiency肝血虛虧, add UB23腎俞Shenshu, LI10手三里Shousanli, UBI8肝俞Ganshu, ST36足三里Zusanli, and SP6三陰交Sanyinjiao.

Auricular Therapy

Wrist, Finger, Shenmen, Sympathetic, Liver, Kidney

Bleeding Therapy

Ashi acupoints 阿是穴

Tui Na Treatments

Tui Na Techniques

Pressing, pushing, rotating

Tui Na Procedures

Tui Na therapy is selected only after phalangeal fracture has been resolved.
   The patient is seated, and the therapist is sitting in front of the affected wrist.

1. Acute Sprain: The therapist applies pressing and pushing techniques in very
   light force to any stripped like tissue under the skin on the digit. The direction of the pushing technique is along the tissues until the sensation of the
   strips disappear Figure 10.3.1.

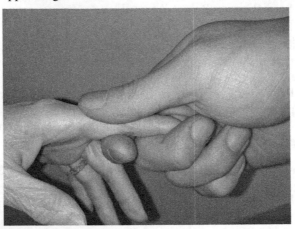

2. Chronic Sprain: The therapist holds the distal phalanx of the affected finger and applies rotating techniques; the center of rotation is in the affected
   joint. In the acute stage, the Tui Na procedures should not include rotating
   the finger.

PATIENT ADVISORY

- Stop the finger activity that caused the injury. If the injury occurs in the
  middle of a sport game, pull yourself out of it and rest.
- Apply ice on the affected finger and hand for 10–20 minutes at a time every
  1 to 2 hours for the first three days. Please note that this is done only when
  the patient is awake. Do not place ice over the skin directly; instead, use a
  thin cloth between the ice pack and the skin.
- Wrap an elastic compression bandage around the injured joint.
- Keep the hand elevated above the heart.

You can often treat a mild sprain at home. Nonsteroidal anti-inflammatory drugs
can help relieve pain. You can also use RICE therapy (Rest, Ice, Compression, and
Elevation):

- **Rest:** Let your finger rest, if possible. Stop doing the activity that injured it (such as playing basketball) while your finger heals.
- **Ice:** Ice the injured area for 15–20 minutes at a time, especially within the first 24 hours. Put ice in a towel or plastic bag rather than directly on your skin. Icing the injury can help relieve pain and swelling.
- **Compression:** Wrap an elastic compression bandage around the injured joint to support it and reduce swelling. Be sure the wrap is comfortable and not too tight.
- **Elevation:** Elevate your injured hand to minimize swelling, especially within the first 24–72 hours. Keep it elevated overnight, if possible.

NOTES

1. https://en.wikipedia.org/wiki/Flexor_digitorum_superficialis_muscle
2. https://en.wikipedia.org/wiki/Flexor_digitorum_superficialis_muscle
3. https://orthoinfo.aaos.org/en/diseases--conditions/trigger-finger
4. https://emedicine.medscape.com/article/1244693-clinical#b1
5. https://emedicine.medscape.com/article/1244693-clinical#b1
6. https://emedicine.medscape.com/article/1243454-overview
7. https://www.nhs.uk/conditions/ganglion/#:~:text=Ganglion%20cysts%20look%20and%20feel,and%20cushion%20them%20during%20movement
8. https://www.mayoclinic.org/diseases-conditions/ganglion-cyst/symptoms-causes/syc-20351156
9. https://emedicine.medscape.com/article/1243454-overview#a12.
10. https://www.mayoclinic.org/diseases-conditions/ganglion-cyst/symptoms-causes/syc-20351156
11. https://orthoinfo.aaos.org/en/diseases--conditions/ganglion-cyst-of-the-wrist-and-hand/
12. https://www.ypo.education/orthopaedics/hand-wrist/hand-anatomy-t191/video/
13. https://emedicine.medscape.com/article/98322-clinical
14. https://www.physio-pedia.com/Skier%27s_thumb.

Bibliography

"Bell's Palsy Fact Sheet." National Institute of Neurological Disorders and Stroke. U.S. Department of Health and Human Services. Accessed February 27, 2022. https://www.ninds.nih.gov/Disorders/Patient-Caregiver-Education/Fact-Sheets/Bells-Palsy-Fact-Sheet.

"Cancer Diagnosis." Cancer Diagnosis | SEER Training. Accessed February 27, 2022. https://training.seer.cancer.gov/disease/diagnosis/.

Agarwal, Nitin, Rut Thakkar, and Khoi Than. "Concussion." AANS. Accessed February 26, 2022. https://www.aans.org/en/Patients/Neurosurgical-Conditions-and-Treatments/Concussion.

Alomar, Abdulaziz Z., Tom Powell, and Mark L. Burman. "Isolated Supraspinatus Muscle Atrophy and Fatty Infiltration Associated with Recurrent Anterior Shoulder Instability: A Case Report and Review of the Literature." *International Journal of Shoulder Surgery.* Medknow Publications Pvt Ltd, July 2011. https://www.ncbi.nlm.nih.gov/pmc/articles/PMC3205528/.

Apichai, Benjamin. *Chinese Medicine for Lower Body Pain.* Boca Raton, FL: CRC Press, 2020, p.311, 332.

Ataullah, A. H. M. "Cerebellar Dysfunction." *StatPearls* [Internet]. U.S. National Library of Medicine, August 30, 2021. https://www.ncbi.nlm.nih.gov/books/NBK562317/#:~:text=Cerebellar%20dysfunction%20causes%20balance%20problems,part%20of%20the%20vestibulocerebellar%20system.

Azar, Frederick M., James H. Beaty, Kay Daugherty, Linda Jones, and Willis C. Campbell. *Campbell's Operative Orthopaedics.* 14th ed. Vol. 8. Philadelphia, PA: Elsevier, 2021.

Barón-Esquivias, Gonzalo, and Antoni Martínez-Rubio. "Tilt Table Test: State of the Art." *Indian Pacing and Electrophysiology Journal.* Indian Pacing and Electrophysiology Group. Accessed October 1, 2003. https://www.ncbi.nlm.nih.gov/pmc/articles/PMC1513525/.

Biundo, Joseph J. "Fibromyalgia - Musculoskeletal and Connective Tissue Disorders." Merck Manuals Professional Edition. Merck Manuals, February 22, 2022. https://www.merckmanuals.com/professional/musculoskeletal-and-connective-tissue-disorders/bursa-muscle-and-tendon-disorders/fibromyalgia.

Bondy, Robert K. *The Merck Manual.* 16th ed. Rahway, NJ: Merck, 1992, p.1369.

Bucak, Abdulkadir, Sahin Ulu, Abdullah Aycicek, Emre Kacar, and Murat Cem Miman. "Grisel's Syndrome: A Rare Complication Following Adenotonsillectomy." Case Reports in Otolaryngology. Hindawi Publishing Corporation, March 24, 2014. https://www.ncbi.nlm.nih.gov/pmc/articles/PMC3982280/.

Buchanan, Benjamin K. "Radial Nerve Entrapment." *StatPearls* [Internet]. U.S. National Library of Medicine, February 12, 2022. https://www.ncbi.nlm.nih.gov/books/NBK431097/.

Carreiro, Jane E. *An Osteopathic Approach to Children: The Perfect Companion to Pediatric Manual Medicine.* 2nd ed. Edinburgh: Churchill Livingstone/Elsevier, 2009.

Coco, Andrew S. "Primary Dysmenorrhea." *American Family Physician,* August 1, 1999. https://www.aafp.org/afp/1999/0801/p489.html#afp19990801p489-t1.

Cohen, Helen S., Jasmine Stitz, Haleh Sangi-Haghpeykar, Susan P. Williams, Ajitkumar P. Mulavara, Brian T. Peters, and Jacob J. Bloomberg. "Tandem Walking as a Quick Screening Test for Vestibular Disorders." *The Laryngoscope*. U.S. National Library of Medicine, July 1, 2019. https://www.ncbi.nlm.nih.gov/pmc/articles/PMC5995610/.

Corner, E. M. "Rotary Dislocations of the Atlas." *Annals of Surgery*. U.S. National Library of Medicine, January, 1907. https://www.ncbi.nlm.nih.gov/pmc/articles/PMC1414264/.

Dvorak, J., E. Schneider, P. Saldinger, and B. Rahn. "Biomechanics of the Craniocervical Region: The Alar and Transverse Ligaments." *Journal of Orthopaedic Research: Official Publication of the Orthopaedic Research Society*. U.S. National Library of Medicine. Accessed February 27, 2022. https://pubmed.ncbi.nlm.nih.gov/3357093.

Dydyk, Alexander M. "Median Nerve Injury." *StatPearls* [Internet]. U.S. National Library of Medicine, February 5, 2022. https://www.ncbi.nlm.nih.gov/books/NBK553109/.

Ewald, Anthony. "Adhesive Capsulitis: A Review." *American Family Physician*, February 15, 2011. https://www.aafp.org/afp/2011/0215/p417.html.

Farrell, Connor. "Anatomy, Back, Rhomboid Muscles." *StatPearls* [Internet]. U.S. National Library of Medicine, July 26, 2021. https://www.ncbi.nlm.nih.gov/books/NBK534856/.

Faruqi, Taha. "Subacromial Bursitis." *StatPearls* [Internet]. U.S. National Library of Medicine, June 29, 2021. https://www.ncbi.nlm.nih.gov/books/NBK541096/.

Fazekas, Dalton. "Intercostal Neuralgia." *StatPearls* [Internet]. U.S. National Library of Medicine, August 29, 2021. https://www.ncbi.nlm.nih.gov/books/NBK560865/.

Galetta, K. M., J. Barrett, M. Allen, F. Madda, D. Delicata, A. T. Tennant, C. C. Branas, et al. "The King-Devick Test as a Determinant of Head Trauma and Concussion in Boxers and MMA Fighters." *Neurology*. Lippincott Williams & Wilkins, April 26, 2011. https://www.ncbi.nlm.nih.gov/pmc/articles/PMC3087467/.

Gerdle, Björn, Bijar Ghafouri, Malin Ernberg, and Britt Larsson. "Chronic Musculoskeletal Pain: Review of Mechanisms and Biochemical Biomarkers as Assessed by the Microdialysis Technique." *Journal of Pain Research*. Dove Medical Press, June 12, 2014. https://www.ncbi.nlm.nih.gov/pmc/articles/PMC4062547/.

Giangarra, Charles E., and Robert C. Manske. *Clinical Orthopaedic Rehabilitation: A Team Approach*. Philadelphia, PA: Elsevier, 2018.

Goldberg, Charlie. "Exam of the Abdomen." UC San Diego's Practical Guide to Clinical Medicine, 2018a. https://meded.ucsd.edu/clinicalmed/abdomen.html.

Goldberg, Charlie. "UC San Diego's Practical Guide to Clinical Medicine", 2018b. https://meded.ucsd.edu/clinicalmed/joints4.html.

Government of Canada, Canadian Centre for Occupational Health and Safety. "Thoracic Outlet Syndrome: Osh Answers." Canadian Centre for Occupational Health and Safety, February 25, 2022. https://www.ccohs.ca/oshanswers/diseases/thoracic.html#:~:text=What%20are%20the%20occupational%20factors,the%20shoulders%20and%20upper%20arms.

Gregush, R. E., and S. F. Habusta. "Ganglion Cyst." 2021. Accessed July 20, 2022. https://www.ncbi.nlm.nih.gov/books/NBK470168/.

Grujičić, Roberto. "Atlantooccipital Joint." Kenhub, September 30, 2021. https://www.kenhub.com/en/library/anatomy/atlanto-occipital-joint.

Henson, Brandi. "Anatomy, Head and Neck, Nose Sinuses." *StatPearls* [Internet]. U.S. National Library of Medicine, July 31, 2021. https://www.ncbi.nlm.nih.gov/books/NBK513272/.

Hillam, Jeffery. "Mandible Dislocation." *StatPearls* [Internet]. U.S. National Library of Medicine, July 28, 2021. https://www.ncbi.nlm.nih.gov/books/NBK549809/.

Hsu, David. "Biceps Tendon Rupture." *StatPearls* [Internet]. U.S. National Library of Medicine, February 7, 2022. https://www.ncbi.nlm.nih.gov/books/NBK513235/.

"黃帝內經·素問:脈要精微論篇第十七-中國哲學書電子化計劃." n.d. Ctext .org. Accessed January 20, 2023. https://ctext.org/wiki.pl?if=gb&chapter=782710.

"黃帝內經:靈樞經-中國哲學書電子化計劃." n.d. Ctext.org. Accessed January 20, 2023. https://ctext.org/huangdi-neijing/ling-shu-jing/zh.

"黃帝內經·素問-中國哲學書電子化計劃." n.d. Ctext.org. Accessed January 20, 2023. https://ctext.org/wiki.pl?if=gb&res=69882.

https://www.aafp.org/afp/1999/0801/p489.html#afp19990801p489-t1.

Jahan, Firdous, Kashmira Nanji, Waris Qidwai, and Rizwan Qasim. "Fibromyalgia Syndrome: An Overview of Pathophysiology, Diagnosis and Management." *Oman Medical Journal* 27 (3): 192–95, 2012. https://doi.org/10.5001/omj.2012.44.

Jensen, Laura. "TIPS for Performing a Physical Examination of the Neck in Whiplash-Associated Disorders." *BCMJ* 51 (2): 60–61, March 2009. https://bcmj.org/icbc/tips -performing-physical-examination-neck-whiplash-associated-disorders.

Kaufmann, A. M., and M. Patel. "Your Complete Guide to Trigeminal Neuralgia." *CCND Winnipeg*, 2001. https://www.umanitoba.ca/cranial_nerves/trigeminal_neuralgia/man-uscript/anatomy.html.

Kiel, John. "Sternoclavicular Joint Injury." *StatPearls* [Internet]. U.S. National Library of Medicine, September 2, 2021. https://www.ncbi.nlm.nih.gov/books/NBK507894/.

Kim, Sung Deuk, and Marios Loukas. "Anatomy and Variations of Digastric Muscle." *Anatomy and Cell Biology*. Korean Association of Anatomists, March 2019. https:// www.ncbi.nlm.nih.gov/pmc/articles/PMC6449592/.

Konbaz, Faisal S., Sami I. Aleissa, Fahad H. Alhelal, Majed S. Abalkhail, Asim J. Alamri, Abdullah I. Saeed, Asim F. Mohabbat, et al. "Upper Cervical Spine Instability Due to Pyogenic Infection Successfully Treated by One Stage Posterior Debridement and Stabilization: Case Report and Literature Review." *Journal of Musculoskeletal Surgery and Research*. Scientific Scholar, November 13, 2021. https://journalmsr.com/upper -cervical-spine-instability-due-to-pyogenic-infection-successfully-treated-by-one -stage-posterior-debridement-and-stabilization-case-report-and-literature-review/.

Lacy, Jordan. "Atlantoaxial Instability." *StatPearls* [Internet]. U.S. National Library of Medicine, July 25, 2021. https://www.ncbi.nlm.nih.gov/books/NBK519563/.

Leas, Daniel P. "Atlantoaxial Instability Workup: Laboratory Studies, Plain Radiography, CT and MRI." Atlantoaxial Instability Workup: Laboratory Studies, Plain Radiography, CT and MRI. Medscape, June 14, 2021. https://emedicine.medscape.com/article /1265682-workup#c7.

LeBlanc, Kim Edward, and Wayne Cestia. "Carpal Tunnel Syndrome." *American Family Physician*, April 15, 2011. https://www.aafp.org/afp/2011/0415/p952.html.

Liebert, Paul L. "Hamstring Injury - Injuries and Poisoning." Merck Manuals Consumer Version. Merck Manuals, December 2021. https://www.merckmanuals.com/home/ injuries-and-poisoning/sports-injuries/hamstring-injury.

Liu, Chongyun, Angela Tseng, and Sue Yang. *Chinese Herbal Medicine Modern Applications of Traditional Formulas*. Boca Raton, FL: CRC Press, 2005.

Liu, Fang, Wei Wei, Jianhua Fang, Yuanyuan Wu, Qifei Zhang, Cunxin Wang, Lihong Ye, and Min Liu. "Effects of Laser Needle-Knife Therapy on Vertebroarterial Morphology and Protein Expression of PI-3K, Akt and VEGF in the Carotid Artery in a Rabbit Model of Cervical Spondylotic Arteriopathy." *Saudi Journal of Biological Sciences*. Elsevier, February 2019. https://www.ncbi.nlm.nih.gov/pmc/articles/PMC6717135/.

Macaluso, Christopher R., and Robert M. McNamara. "Evaluation and Management of Acute Abdominal Pain in the Emergency Department." *International Journal of General Medicine*. Dove Medical Press, September 26, 2012. https://www.ncbi.nlm.nih.gov/ pmc/articles/PMC3468117/.

Martin, Ryan M., and David E. Fish. "Scapular Winging: Anatomical Review, Diagnosis, and Treatments." *Current Reviews in Musculoskeletal Medicine*. Humana Press Inc, March 2008. https://www.ncbi.nlm.nih.gov/pmc/articles/PMC2684151/.

Matuszak, Jason M., Jennifer McVige, Jacob McPherson, Barry Willer, and John Leddy. "A Practical Concussion Physical Examination Toolbox." *Sports Health*. SAGE Publications, March 28, 2016. https://www.ncbi.nlm.nih.gov/pmc/articles/PMC4981071/.

McMurray, D., and K. Muralikuttan. "Volar Dislocation of the Distal Radio–Ulnar Joint without Fracture: A Case Report and Literature Review." *Injury Extra*. Elsevier, August 6, 2008. https://www.sciencedirect.com/science/article/pii/S1572346108000731.

"醫砭》難經本義》第二十二難是動、所生病." n.d. Yibian.hopto.org. Accessed January 21, 2023. http://yibian.hopto.org/shu/?sid=4575.

Mercadante, Juliette R., and Raghavendra Marappa-Ganeshan. "Anatomy, Skin Bursa." U.S. National Library of Medicine. *StatPearls* [Internet], July 27, 2021. https://www.ncbi.nlm.nih.gov/books/NBK554438/.

Meyer, Roger A. "The Temporomandibular Joint Examination." *Clinical Methods: The History, Physical, and Laboratory Examinations*. 3rd edition. U.S. National Library of Medicine, January 1, 1990. https://www.ncbi.nlm.nih.gov/books/NBK271/.

Miller, Jimmy D., Stephanie Pruitt, and Thomas J. McDonald. "Acute Brachial Plexus Neuritis: An Uncommon Cause of Shoulder Pain." *American Family Physician*, November 1, 2000. https://www.aafp.org/pubs/afp/issues/2000/1101/p2067.html.

Mose, Scott. "Elbow Anatomy." Elbow Anatomy, June 30, 2022. https://fpnotebook.com/mobile/ortho/anatomy/elbwantmy.htm

"Neurological Diagnostic Tests and Procedures Fact Sheet." National Institute of Neurological Disorders and Stroke. U.S. Department of Health and Human Services, April 10, 2019. https://www.ninds.nih.gov/Disorders/Patient-Caregiver-Education/Fact-Sheets/Neurological-Diagnostic-Tests-and-Procedures-Fact.

Nturibi, Eric. "Anatomy, Head and Neck, Greater Petrosal Nerve." *StatPearls* [Internet]. U.S. National Library of Medicine, November 5, 2021. https://www.ncbi.nlm.nih.gov/books/NBK553121/.

O'Shea, Noreen E. "Olecranon Bursa Aspiration." *StatPearls* [Internet]. U.S. National Library of Medicine, September 29, 2021. https://www.ncbi.nlm.nih.gov/books/NBK554617/

Ourieff, Jared. "Anatomy, Back, Trapezius." *StatPearls* [Internet]. U.S. National Library of Medicine, July 26, 2021. https://www.ncbi.nlm.nih.gov/books/NBK518994/.

Permenter, Cara M. "Postconcussive Syndrome." *StatPearls* [Internet]. U.S. National Library of Medicine, January 2, 2022. https://www.ncbi.nlm.nih.gov/books/NBK534786/.

Pickrell, Brent B., Arman T. Serebrakian, and Renata S. Maricevich. "Mandible Fractures." *Seminars in Plastic Surgery*. Thieme Medical Publishers, May 2017. https://www.ncbi.nlm.nih.gov/pmc/articles/PMC5423793/.

Pothiawala, Sohil, and Fatimah Lateef. "Bilateral Facial Nerve Palsy: A Diagnostic Dilemma." *Case Reports in Emergency Medicine*. Hindawi Publishing Corporation, January 23, 2012. https://www.ncbi.nlm.nih.gov/pmc/articles/PMC3542940/.

"QuickStats: Percentage* of Adults Aged ≥18 Years Who Reported Having a Severe Headache or Migraine in the Past 3 Months,† by Sex and Age Group — National Health Interview Survey,§ United States, 2015." . *MMWR. Morbidity and Mortality Weekly Report* 66(24): 654. 2017. https://doi.org/10.15585/mmwr.mm6624a8.

Ruoppi, Pirkko. "Isolated Sphenoid Sinus Diseases." *Archives of Otolaryngology–Head & Neck Surgery*. JAMA Network, June 1, 2000. https://jamanetwork.com/journals/jamaotolaryngology/fullarticle/404830.

Saikrishna, D., S. Shyam Sundar, and K. S. Mamata. " Superolateral Dislocation of Intact Mandibular Condyle: A Case Report and Review of Literature." Europe PMC. December 15, 2015. https://europepmc.org/article/med/27408459.

Saito, Eliza Tiemi, Paula Marie Hanai Akashi, and Isabel de Camargo Neves Sacco. "Global Body Posture Evaluation in Patients with Temporomandibular Joint Disorder." Clinics (Sao Paulo, Brazil). Hospital das Clinicas da Faculdade de Medicina da Universidade de São Paulo, January 2009. https://www.ncbi.nlm.nih.gov/pmc/articles/PMC2671968/.

Smith, Jennifer L., and Nabil A. Ebraheim. "Anatomy of the Palmar Cutaneous Branch of the Median Nerve: A Review." *Journal of Orthopaedics*. Elsevier, June 5, 2019. https://www.ncbi.nlm.nih.gov/pmc/articles/PMC6807303/.

Sweeney, C. J., and D. H. Gilden. "Ramsay Hunt Syndrome." *Journal of Neurology, Neurosurgery & Psychiatry*. BMJ Publishing Group Ltd, August 1, 2001. https://jnnp.bmj.com/content/71/2/149.

"The Merck Manual 16th Edition." *The Merck Manual*. 1992. Merck.

"Trigeminal Neuropathy with Loss of the Corneal Reflex." J. Willard Marriott Digital Library. Accessed March 2, 2022. https://collections.lib.utah.edu/ark:/87278/s6dz4654.

Traynelis, Vincent C. "Congenital Upper Cervical Disorders." *SpineUniverse*, June 25, 2019. https://www.spineuniverse.com/conditions/upper-neck-disorders/congenital-upper-cervical-disorders.

"傷寒論 - 中國哲學書電子化計劃." n.d. Ctext.org. Accessed January 20, 2023. https://ctext.org/shang-han-lun/zh.

Varacallo, Matthew, Travis J. Seaman, and Scott D. Mair. *Biceps Tendon Dislocation and Instability*. Treasure Island, FL: StatPearls Publishing, 2022. https://www.ncbi.nlm.nih.gov/books/NBK534102/.

Villaseñor-Ovies, Pablo, Angélica Vargas, Karla Chiapas-Gasca, Juan J. Canoso, Cristina Hernández-Díaz, Miguel Ángel Saavedra, José Eduardo Navarro-Zarza, and Robert A. Kalish. "Clinical Anatomy of the Elbow and Shoulder: Reumatología Clínica." *Reumatología Clínica* (English Edition). Elsevier, December 1, 2012. https://www.reumatologiaclinica.org/en-clinical-anatomy-elbow-shoulder-articulo-S1699258X12002471.

Voorhies, R. M. "Cervical Spondylosis: Recognition, Differential Diagnosis, and Management." *The Ochsner Journal*. Ochsner Clinic, L.L.C. and Alton Ochsner Medical Foundation, April 2001. https://www.ncbi.nlm.nih.gov/pmc/articles/PMC3116771/.

Wang, Xiao-Rong, Timothy C. Y. Kwok, James F. Griffith, Blanche Wai Man Yu, Jason C. S. Leung, and Yì Xiáng J. Wáng. "Prevalence of Cervical Spine Degenerative Changes in Elderly Population and Its Weak Association with Aging, Neck Pain, and Osteoporosis." *Annals of Translational Medicine*. AME Publishing Company, September 2019. https://www.ncbi.nlm.nih.gov/pmc/articles/PMC6803181/.

Whitney, Eric. "Hoffmann Sign." *StatPearls* [Internet]. U.S. National Library of Medicine, October 30, 2021. https://www.ncbi.nlm.nih.gov/books/NBK545156/.

Windsor, Robert E. "Cervical Spine Anatomy." Overview: Gross Anatomy. Medscape, June 11, 2020. https://emedicine.medscape.com/article/1948797-overview.

Wozasek, G. E., and H. Laske. "[The Ligaments of the Scaphoid Bone]." Handchirurgie, Mikrochirurgie, plastische Chirurgie: Organ der Deutschsprachigen Arbeitsgemeinschaft fur Handchirurgie: Organ der Deutschsprachigen Arbeitsgemeinschaft fur Mikrochirurgie der Peripheren Nerven und Gefasse: Organ der V... U.S. National Library of Medicine, 1991. https://pubmed.ncbi.nlm.nih.gov/2032630/.

Yu, D., R. Cao, and J. Wu. *Zhong Yi Tui na Xue*. Ren min wei sheng chu ban she, 1985.

Zabrzyński, J., J. Szukalski, Ł. Paczesny, D. Szwedowski, D. Grzanka, Jacenty Szukalski Jan, Łukasz Paczesny, Dawid Szwedowski, and Dariusz Grzanka. "Cigarette Smoking Intensifies Tendinopathy of the LHBT. A Microscopic Study after Arthroscopic Treatment." *Polish Journal of Pathology: Official Journal of the Polish Society of Pathologists*. U.S. National Library of Medicine, September 4, 2019. https://pubmed.ncbi.nlm.nih.gov/31556564/.

Index

A

A1 pulley, 535–536
Abdominal pain
  acupuncture treatments, 288
  anatomy, 279
  auricular therapy, 288
  Chinese medicine etiology and
    pathology, 280
  classifications and treatments,
    282–289
  dysmenorrhea, 361–377
  epigastric pain, 290–311
  hypochondriac pain, 311–324
  IBS, 324–339
  lower, 339–360
  manifestations, 281–282
  patient advisory, 289
  Tui Na treatments, 289
  Yang energy, 279
Abnormal vaginal bleeding, 343
Acromioclavicular arthritis, 441
Acromioclavicular (AC) joint, 421
Acupuncture treatments
  abdominal pain, 288
  acute cervical fibrositis/stiff neck, 215
  Bell's palsy, 206
  Bi Syndrome, 20
  carpal tunnel syndrome, 509
  cervical spondylosis, 237
  chest pain, 275–276
  distal radioulnar joint, 524
  dysmenorrhea, 374–375
  epigastric pain, 307–308
  fibromyalgia syndrome, 39–40
  finger sprain, 551
  frozen shoulder, 432–433
  ganglion cysts, 544–545
  headache, 92–93
  hypochondriac pain, 323
  IBS, 336–337
  intercostal neuralgia, 265–266
  lateral epicondylitis, 485
  long head biceps tendonitis, 458
  lower abdominal pain, 358–359
  medial epicondylitis, 493–494
  migraine headache, 120
  olecranon bursitis, 500
  post-concussion syndrome, 141
  rhomboid muscle pain, 393–394
  short head biceps injury, 450–451
  shoulder pain, 416
  sinusitis, 178–179
  subacromial bursitis, 474
  subluxation of long head biceps tendon,
    466–467
  supraspinatus tendonitis, 443
  temporomandibular joint dislocation, 166
  thoracic outlet syndrome, 30
  TMJ disorders, 158
  trapezius muscle, 384
  trigeminal neuralgia, 190–191
  trigger finger, 539
  upper cervical subluxation, 254
  wrist joint, 533
  wrist sprain, 517
Acute appendicitis, 349
Acute cervical fibrositis, 209–219, 227, 249; see
    also Stiff neck
Acute cervical muscle sprain, 227
Acute traumatic injury, 24
Adenomyosis, 364
Adenosine triphosphate (ATP), 125
Adenotonsillectomy, 245
Adhesive capsulitis, 228
Alar ligament test, 248
Allergic rhinitis, 171
Angina pectoris, 228, 273, 297
Anterior articulation, 257
Anterior interosseous nerve entrapment, 490
Anterior scalene muscle, 22
Aortic aneurysm, 348
Aortic rupture, 297
Ape hand deformity, 529
Apley's scratch test, 410, 428
Apprehension test, 429
A pulleys, 536
Atlantoaxial joint, 240, 242–243
Atlantoaxial joint disorder syndrome, 243
Atlantodental interval (ADI), 248
Atlantooccipital joint, 241
Atlas subluxation complex, 243
Atypical ribs, 257
Autonomic dysregulation, 128
Autonomic nervous system, 224, 225
Axial joint pain, 225

B

Babinski's sign, 225
Balance/coordination examination, 129–130
Bear hug test, 410, 427–428
Bell's palsy
    acupuncture treatments, 206
    anatomy, 193
    Chinese medicine etiology and pathology,
        195–196
    differential diagnosis, 199
    extracranial, 194
    facial paralysis, 192, 193
    herbal treatments, 199
    intracranial, 193
    manifestations
        Chinese medicine, 196–197
        Western medicine, 196
    patient advisory, 207
    physical examination, 197–198
    treatments, 199–207
    Tui Na treatments, 207
    Western medicine etiology and pathology,
        194–195
Biceps brachii, 446, 452, 461
Biceps tendinopathy, 429
Bicipital aponeurosis, 446, 452, 461
Bi Syndrome
    acupuncture treatments, 20, 40
    causative factors, 388
    Chinese medicine etiology and
        pathology, 1–4
    classification, 4–6
    external causes, 4
    herbal treatments, 38–39
    internal causes, 1–4
    manifestations, diagnosis and
        treatments, 6–21
    meridians, 1
    modifications, 20–21
    patient advisory, 21
    shoulder pain, 403–405
    supraspinatus tendonitis, 438
Bleeding therapy
    chest pain, 276
    costovertebral joint sprain, 399
    distal radioulnar joint, 524
    finger sprain, 551
    frozen shoulder, 434
    headache, 93–94
    lateral epicondylitis, 485
    long head biceps tendonitis, 458
    medial epicondylitis, 494
    migraine headache, 121
    olecranon bursitis, 500–501
    rhomboid muscle pain, 394

    short head biceps injury, 451
    shoulder pain, 417
    stiff neck, 216
    subacromial bursitis, 474
    subluxation of long head biceps tendon, 467
    supraspinatus tendonitis, 444
    trapezius muscle, 384
    wrist joint, 533
    wrist sprain, 517
Blood deficiency shoulder pain
    Chinese medicine etiology and pathology,
        406
    manifestations, 412
    treatments, 415–416
Blood stagnation in joints, 19–20
Blood stasis Bi Syndrome, 5–6
Blood stasis shoulder pain
    Chinese medicine etiology and
        pathology, 406
    manifestations, 412
    treatments, 414–415
Bone Bi Syndrome, 6
Boundary forming the tunnel, 504
Brachial plexus neuritis, 227
Breast tenderness, 316
Bruits, 347

C

Caffeine headache, 100
Cardio exercise, 123
Cardiovascular diseases, 261, 317
Carpal tunnel syndrome, 27, 227
    acupuncture treatments, 509
    anatomy, 503–504
    Chinese medicine etiology and
        pathology, 505
    differential diagnosis, 507
    herbal treatments, 508
    manifestations, 505
    moxibustion therapy, 509
    patient advisory, 510
    physical examination, 505–507
    Tui Na treatments, 509
    vibrating power tools, 503
    Western medicine etiology and
        pathology, 504
Central facial palsy, 194
Cerebral hemorrhage, 109
Cerebrospinal fluid, 124
Cervical disk degeneration, 429
Cervical myelopathy, 223–225
Cervical radiculopathy, 223, 224, 441, 481–482,
    490, 507
Cervical spinal nerves, 222, 243
Cervical spine, 220, 240

Cervical spondylosis
  acupuncture treatments, 237
  age-related neck pain, 220
  anatomy, 220–222
  Chinese medicine
    etiology and pathology, 228–229
    manifestations, 229
  differential diagnosis, 227–228
  herbal treatments, 229
  patient advisory, 239–240
  physical examination, 225–226
  spinal cord and nerve roots, 220
  treatments, 229–239
  Tui Na treatments, 237–239
  Western medicine
    etiology and pathology, 222–223
    manifestations, 223–225
Cervical spondylotic arteriopathy, 223, 225
Cervical strain, 249
Cervical trauma, 211
Cervix, 362–363
Charcot-Marie-Tooth disease (CMT), 243
Chest pain
  acupuncture treatments, 275–276
  anatomy, 268–270
  bleeding and cupping therapy, 276
  Chinese medicine etiology and
    pathology, 271
  differential diagnosis, 273–274
  heart attack/angina pectoris, 268
  herbal treatments, 274
  intercostal neuralgia, 257–268
  manifestations, 271–272
  patient advisory, 277–278
  physical examination, 272–273
  sprain of hypochondrium, 268–278
  strained/pulled muscle, 268
  treatments, 274–277
  Tui Na treatments, 276–277
  Western medicine etiology and pathology,
    270–271
Coffee ground emesis, 345
Cognitive ability, 128
Cold Bi Syndrome, 5, 7–11
Colorectal cancer, 329
Condition of pain, 282
Congenital-Dejerine-Sottas disease (DS), 243
Congenital muscular torticollis, 211
Coracoid process, 446, 447
Corticospinal tract dysfunction, 225
Costotransverse joint, 258
Costovertebral joint, 258
Costovertebral joint sprain
  acupuncture treatments, 399
  anatomy, 395
  bleeding therapy, 399

Chinese medicine etiology and
  pathology, 396
connective tissue, 395
cupping therapy, 399
differential diagnosis, 397
electro-acupuncture therapy, 400
herbal treatments, 397–399
manifestations, 396
moxibustion therapy, 399
patient advisory, 402
physical examination, 396–397
procedures, 399–400
Tui Na treatments, 400–402
Western medicine etiology and
  pathology, 396
Cozen's test, 481
Craniovertebral junction (CVJ), 244
Crohn's disease, 345
Cross-arm adduction test (Scarf test), 411, 429
Cruciform ligament, 242
Cubital bursae, 496
Cupping therapy
  chest pain, 276
  costovertebral joint sprain, 399
  epigastric pain, 308
  frozen shoulder, 433–434
  intercostal neuralgia, 266
  lateral epicondylitis, 485
  long head biceps tendonitis, 458
  medial epicondylitis, 494
  olecranon bursitis, 501
  rhomboid muscle pain, 394
  short head biceps injury, 451
  shoulder pain, 416
  stiff neck, 215–216
  subacromial bursitis, 474
  subluxation of long head biceps tendon, 467
  supraspinatus tendonitis, 443
  trapezius muscle, 384
  wrist joint, 533

D

Damp-Heat invasion, 365
Damp Heat type, 15–16
Dampness Bi Syndrome, 5, 11–13
Degenerative intervertebral disc disease, 249
De Quervain's tendinosis, 514
Dinner fork deformity, 514, 521, 530
Disk degeneration, 222
Distal interphalangeal joint (DIP joint), 536, 546
Distal radioulnar joint, 511
  acupuncture treatments, 524
  anatomy, 519
  auricular therapy, 524
  bleeding therapy, 524

Chinese medicine etiology and
pathology, 520
differential diagnosis, 521
dorsal/volar dislocation, 519
herbal treatments, 522–524
manifestations, 520–521
moxibustion therapy, 524
patient advisory, 525–526
physical examination, 521
Tui Na treatments, 524–525
Western medicine etiology and
pathology, 520
Distal radius fracture, 514, 521, 530
Dormant Evil, 48
Dorsal ulna dislocation, 525
Down syndrome, 244, 248
Duration of pain, 281
Durkan's carpal compression test, 506
Dysmenorrhea, 316
acupuncture treatments, 374–375
anatomy, 361–363
auricular therapy, 375
Chinese medicine etiology and pathology,
364–365
differential diagnosis, 369
herbal treatments, 370
manifestations, 365–367
moxibustion therapy, 375
painful menstrual periods, 361
patient advisory, 376–377
physical examination, 367–369
treatments, 369–376
Tui Na treatments, 375–376
Western medicine etiology and pathology,
363–364

E

Ectopic pregnancy, 349, 364, 366
Ehlers-Danlos syndrome, 244
Elbow osteoarthritis, 481, 490
Elbow pain
lateral epicondylitis, 477–487
medial epicondylitis, 487–495
olecranon bursitis, 495–501
Electric shock-like headache, 97–98
Electro-acupuncture therapy
costovertebral joint sprain, 400
frozen shoulder, 434
headache, 94
hypochondriac pain, 323
long head biceps tendonitis, 458
lower abdominal pain, 359
short head biceps injury, 451
stiff neck, 217
subacromial bursitis, 474

subluxation of long head biceps tendon, 467
supraspinatus tendonitis, 444
trapezius muscle, 385
Emotional disorder, 103, 104
Emotional stress, 387, 388
Emptiness headache, 97
Empty can test (Jobe's test), 409–410, 427, 440
Endometriosis, 364, 366, 369
Enteric nervous system, 325
Epicondylitis, 480
Epidural hematoma (EDH), 131–132
Epigastric pain
acupuncture treatments, 307–308
anatomy, 290
auricular therapy, 308
Chinese medicine etiology and pathology,
291–294
cupping therapy, 308
differential diagnosis, 297
digestive system disorders, 290
herbal treatments, 297–307
manifestations, 294–296
patient advisory, 311
physical examination, 296–297
Tui Na treatments, 308–310
Western medicine etiology and pathology,
290–291
Ethmoidal sinuses, 168, 170
Etiopathogenesis, 245
Extensor carpi radialis brevis, 479
External rotation 90° test, 463–464
Eyes, 66–69

F

Facet joints, 221
Fallopian (uterine) tubes, 362
Fatigue headache, 99–100
Female reproductive system, 361
Fibroids, 364
Fibromyalgia, 228
Fibromyalgia syndrome
acupuncture treatments, 39–40
anatomy, 32
Chinese medicine etiology and pathology, 32
differential diagnosis, 34–35
herbal treatments, 35
inflammation, 31
manifestations
in Chinese medicine, 33
in Western medicine, 32–33
mood disorders, 31
patient advisory, 43
physical examination, 33–34
treatments, 35–39
Tui Na treatments, 40–42

Western medicine etiology and pathology, 32
Fibrositis, 31
Fibrous sheath, 535
Finger sprain
    acupuncture treatments, 551
    anatomy, 546–547
    bleeding therapy, 551
    Chinese medicine etiology and
        pathology, 547
    differential diagnosis, 548
    dosage, 551
    external factors, 546
    herbal treatments, 549–550
    manifestations, 547–548
    patient advisory, 552–553
    physical examination, 548
    Tui Na treatments, 552
    Western medicine etiology and
        pathology, 547
Fixed Bi Syndrome, see Dampness
        Bi Syndrome
Flexor digitorum superficialis tendons, 535
Floating ribs, 257
Follicle-stimulating hormone, 363
Food triggered headache, 101
Frontal headache, 59–61
Frontal sinuses, 168, 169
Frozen shoulder, 228
    acupuncture treatments, 432–433
    adhesion stage, 423
    anatomy, 421–422
    atrophy stage, 424
    auricular therapy, 433
    bleeding therapy, 434
    Chinese medicine etiology and
        pathology, 423
    cupping therapy, 433–434
    description, 420
    differential diagnosis, 429, 441
    electro-acupuncture therapy, 434
    herbal treatments, 429, 441
    manifestations, 423–424
    moxibustion therapy, 434
    pain stage, 423
    patient advisory, 435–436
    physical examination, 424–429
    treatments, 429–435
    Tui Na treatments, 434–435
    Western medicine etiology and pathology,
        422–423

G

Ganglion cysts, 514
    acupuncture treatments, 544–545
    anatomy, 542

Chinese medicine etiology and
        pathology, 542
    differential diagnosis, 543
    hand's benign tumors, 541
    herbal treatments, 543–544
    manifestations, 542
    patient advisory, 546
    physical examination, 543
    Tui Na treatments, 545
    Western medicine etiology and
        pathology, 542
GH joint dislocation, 429
Glenohumeral (GH) joint, 421
Golfer's elbow, see Medial epicondylitis
Grief, 105
Grisel's syndrome, 245, 248

H

Half-exterior-half-interior syndrome, 52
Hangover headache, 100
Hawkins test, 410, 428, 472
Hay fever, 171
Headache
    acupuncture treatments, 92–93
    behind the eyes, 66–69
    bleeding therapy, 93–94
    Blood stasis, 84–86
    Chinese medicine etiology and pathology, 45
    chronic condition, 76–78
    circumstances, 98–101
    electro-acupuncture and moxibustion
        therapy, 94
    exterior factors, 46–55
    frontal, 59–61
    Liver Fire, 72–74
    Liver Qi stagnation, 68–70
    Liver Yang rising, 70–72, 83–84
    manifestations, 55–56
    menstrual cycle, 88–92
    migraine, 101–124
    occipital, 55–58
    pain types, 96–98
    patient advisory, 95–96
    Phlegm and Dampness accumulation, 78–80
    Phlegm-Dampness obstruction, 86–88
    source of pain, 45
    substance dependence, 45
    temporal, 61–64
    Tui Na treatments, 94–95
    vertex, 64–66
    whole head, 74–76
    Wind Damp, 81–83
    Wind Heat, 80–81
Head/cervicothoracic examination, 129
Head-up tilt test (HUTT), 129

Heart attack, 274
Heart disease, 35
Heat Bi Syndrome, 5, 13–14
Heaviness headache, 97
Herbal treatments
    Bi Syndrome, 38–39
    carpal tunnel syndrome, 508
    costovertebral joint sprain, 397–399
    distal radioulnar joint, 522–524
    epigastric pain, 297–307
    finger sprain, 549–550
    ganglion cysts, 543–544
    lateral epicondylitis, 482–485
    long head biceps tendonitis, 455–458
    medial epicondylitis, 490–493
    olecranon bursitis, 498–500
    rhomboid muscle pain, 392, 393
    short head biceps injury, 448–450
    subacromial bursitis, 473
    subluxation of long head biceps tendon,
        464–466
    trigger finger, 538–539
    wrist joint, 530–532
    wrist sprain, 515–517
Hoffmann's sign, 225, 226
Hoffman-Tinel sign, 507
Hormone replacement therapy, 343
Hornblower's test, 410, 427
Hyaluronic acid, 496
Hyperextension motion, 226
Hypochondriac pain
    acupuncture treatments, 323
    anatomy, 311–312
    auricular therapy, 323
    Chinese medicine etiology and pathology,
        313–315
    differential diagnosis, 317
    electro-acupuncture therapy, 323
    herbal treatments, 317
    manifestations, 316–317
    patient advisory, 324
    physical examination, 317
    stress, 311
    treatments, 317–324
    Tui Na treatments, 323–324
    Western medicine etiology and pathology,
        312–313
Hypochondrium, 268; see also Chest pain
Hypogastric region, 340

I

Idiopathic peripheral facial paralysis, 195
Indulgence, 103
Inferior thoracic aperture, 21
Inflammatory bowel disease (IBD), 329

Inflammatory rheumatoid arthritis (RA), 244
Inguinal hernia, 342
Insubstantial phlegm, 126
Intercostal muscle, 269–270
Intercostal nerves, 258
Intercostal neuralgia, 296, 317
    acupuncture treatments, 265–266
    anatomy, 257–258
    Chinese medicine etiology and pathology,
        258–259
    cupping therapy, 266
    cutaneous acupuncture therapy, 266
    differential diagnosis, 261
    herbal treatments, 261
    manifestations, 259–260
    nerve irritation, 257
    patient advisory, 268
    physical examination, 260–261
    treatments, 261–267
    Tui Na treatments, 267
    Western medicine etiology and
        pathology, 258
Interscalene triangle, 22
Intervertebral discs, 221
Intracellular calcium, 125
Intracranial hematoma, 132
Irregular menstruation, 316
Irritable bowel syndrome (IBS)
    acupuncture treatments, 336–337
    anatomy, 325
    Chinese medicine etiology and pathology,
        326–327
    differential diagnosis, 329
    gastrointestinal disorder, 324
    herbal treatments, 330
    manifestations, 327–328
    moxibustion therapy, 337
    patient advisory, 339
    physical examination, 328–329
    treatments, 330–339
    Tui Na treatments, 337–339
    Western medicine etiology and pathology,
        325–326

J

Jobe's test, 409–410, 427, 440
Jue Yin, 52–55
Jueyin headache, 64–66

K

Kidney Essence deficiency, 135–137
Kidney system, 126
King–Devick test, 130
Knuckle, 546, 547

L

Lateral cervical spine radiograph, 248
Lateral epicondylitis, 27
  acupuncture treatments, 485
  anatomy, 477–479
  auricular therapy, 485
  bleeding therapy, 485
  Chinese medicine etiology and pathology, 479
  cupping therapy, 485
  differential diagnosis, 481–482
  forearm and wrist twisting, 477
  herbal treatments, 482–485
  manifestations, 479–480
  moxibustion therapy, 485
  patient advisory, 486–487
  physical examination, 480–481
  Tui Na treatments, 486
  Western medicine etiology and
      pathology, 479
Left iliac region, 340, 341, 344
Left upper quadrant, 312, 313
Ligamentous trauma, 547
Lipoma, 543
Liver Blood deficiency, 26, 28–29
Liver Kidney deficiency type, 16–18
Location of pain, 281
Long head biceps tendonitis
  acupuncture treatments, 458
  anatomy, 452–453
  auricular therapy, 458
  bleeding therapy, 458
  Chinese medicine etiology and pathology, 453
  cupping therapy, 458
  differential diagnosis, 454–455
  electro-acupuncture therapy, 458
  herbal treatments, 455–458
  inflammation, 452
  manifestations, 453
  moxibustion therapy, 458
  patient advisory, 460
  physical examination, 454
  risk factors, 452
  Tui Na treatments, 459–460
  Western medicine etiology and
      pathology, 453
Lower abdomen, 341
Lower abdominal pain
  acupuncture treatments, 358–359
  anatomy, 340
  auricular therapy, 360
  Chinese medicine etiology and
      pathology, 342
  differential diagnosis, 348–349
  electro-acupuncture therapy, 359
  herbal treatments, 350

  manifestations, 342–344
  patient advisory, 360
  physical examination, 344–348
  symptoms, 339
  treatments, 350–360
  Tui Na treatments, 359–360
  Western medicine etiology and pathology,
      341–342
Lunotriquetral ballottement test, 514
Lyme disease, 199

M

McBurney's sign, 348
Magnesium, 125
Maxillary sinuses, 168, 169
Medial epicondylitis, 27
  acupuncture treatments, 493–494
  anatomy, 487–489
  auricular therapy, 494
  bleeding therapy, 494
  Chinese medicine etiology and pathology 490
  cupping therapy, 494
  differential diagnosis, 490
  herbal treatments, 490–493
  manifestations, 489
  moxibustion therapy, 494
  patient advisory, 495
  physical examination, 489–490
  Tui Na treatments, 494–495
  Western medicine etiology and
      pathology, 489
  wrist bending, 487
Median neuropathy, 507
Memory function check, 128
Meningitis, 109
Menstrual cycle, headache
  Blood deficiency, 91–92
  Blood stasis, 89–91
  liver fire, 88–89
Mental stress, 146
Metabolic acidosis, 346
Metacarpophalangeal joint (MCP joint), 536,
      538, 546
Microtrauma, 453
Midcarpal joint, 511
Middle scalene muscle, 22
Migraine headache
  acupuncture treatments, 120
  anatomy, 102–103
  aquapuncture treatments, 121
  auricular therapy, 121
  bleeding therapy, 121
  Chinese medicine etiology and pathology,
      103–105
  differential diagnosis, 109

family history, 101
herbal treatments, 110
manifestations
    Chinese medicine, 106–108
    Western medicine, 105–106
neurological condition, 101
patient advisory
    dietary recommendations, 123
    lifestyle, 123–124
physical examination, 108–109
procedures, 121
treatments, 110–122
Tui Na treatments, 121–122
Western medicine etiology and
    pathology, 103
Mild traumatic brain injury (mTBI), 125
Mill's test, 481
Minor's elbow, see Olecranon bursitis
Monosodium glutamate (MSG), 123
Morning headache, 98–99
Motor nucleus, 102
Movement, exercise, analgesia and treatment
    (MEAT) protocol, 277–278
Moving valgus stress test, 489–490
Moxibustion therapy
    carpal tunnel syndrome, 509
    costovertebral joint sprain, 399
    distal radioulnar joint, 524
    dysmenorrhea, 375
    frozen shoulder, 434
    headache, 94
    IBS, 337
    lateral epicondylitis, 485
    long head biceps tendonitis, 458
    medial epicondylitis, 494
    rhomboid muscle pain, 394
    short head biceps injury, 451
    shoulder pain, 416
    stiff neck, 216
    subacromial bursitis, 474
    subluxation of long head biceps tendon, 467
    supraspinatus tendonitis, 444
    trapezius muscle, 384
    trigger finger, 540
    wrist joint, 533
    wrist sprain, 517
Muscle Bi Syndrome, 6
Muscular atrophy, 224
Myofascial pain, 35

N

Neck pain
    acute cervical fibrositis/stiff neck, 209–219
    cervical spondylosis, 220–240
    upper cervical subluxation, 240–256

Neer's test, 428, 472
Nerve innervation, 269, 270
Nerve roots, 222
Neuroforamen, 221
Neurogenic abnormalities, 211
Neurological examinations, 129, 224
Neurologic bowel disorder, 324
Neurovascular pain syndrome, 103
Nighttime headache, 99
N-methyl-D-aspartate (NMDA) receptor, 125
Nonsteroidal anti-inflammatory drugs, 553

O

Occipital headache, 55–58, 249
Ocular/ophthalmologic examination, 130
Olecranon bursae, 495–496
Olecranon bursitis
    acupuncture treatments, 500
    anatomy, 495–496
    auricular therapy, 500
    bleeding therapy, 500–501
    Chinese medicine etiology and pathology,
        496–497
    cupping therapy, 501
    differential diagnosis, 498
    elbow joint, 495
    herbal treatments, 498–500
    manifestations, 497
    patient advisory, 501
    physical examination, 497–498
    Tui Na treatments, 501
    Western medicine etiology and
        pathology, 496
Olecranon process, fracture of, 498
Open mouth radiograph, 248–249
Orgasm headache, 100
Orthostatic hypotension, 128–129
Ovarian torsion, 349
Ovulation, 363

P

Painful Bi Syndrome, see Cold Bi Syndrome
Patte test (Hornblower's test), 410, 427
Pelvic inflammatory disease (PID), 364
Perilunate wrist dislocation, 528
Peripheral facial palsy, 194
Peripheral neuropathy, 224
Phalangeal fracture, 548
Phalen's test, 505
Phlegm, 126
Phlegm-Damp accumulation, 542
Phlegm-dampness obstruction, 86–88
Phlegm Damp obstruction shoulder pain
    Chinese medicine etiology and pathology,
        405–406

manifestations, 411
treatments, 413–414
Phlegm obstruction in joints, 18–19
Pinched nerve, 223
Pivot joint injury, 243
Pleurisy, 261
Pleuritis, 273
Pneumonia, 273
Popeye muscle, 454
Popeye's elbow, 495
Post-concussion syndrome
    acupuncture treatments, 141
    anatomy, 124–125
    Chinese medicine etiology and pathology,
        125–127
    differential diagnosis, 130–132
    herbal treatments 132
    manifestations, 127–128
    needling technique, 141
    patient advisory, 143
    physical activities, 124
    physical examination, 128–130
    traumatic brain injury, 124
    treatments, 132–143
    Tui Na treatments, 141–143
    Western medicine etiology and
        pathology, 125
Posterior articulation, 257–258
Post-traumatic upper cervical subluxation, 243
Postural torticollis, 211
Primary dysmenorrhea, 363–366
Pronator syndrome, 507
Prostaglandin, 363
Proximal interphalangeal joint (PIP joint), 536,
    546, 548
Pulmonary diseases, 261
Pulmonary embolism, 273

Q

Quality of pain, 281

R

Radial nerve entrapment, 481–482
Radial/ulnar collateral ligaments, 546
Radiocarpal joint, 511–512, 527
Ramsay Hunt syndrome, 199
Raynaud's disease, 27
Referred pain, 341
Rest, ice, compression and elevation (RICE),
    277, 553
Retrodiscal tissue, 144
Rheumatoid arthritis (RA), 34–35, 245, 248–249
Rhomboid muscle pain, 397
    acupuncture treatments, 393–394

anatomy, 387
auricular therapy, 394
bleeding therapy, 394
Chinese medicine etiology and pathology,
    387–388
cupping therapy, 394
differential diagnosis, 390–391
herbal treatments, 392, 393
manifestations, 388–389
moxibustion therapy, 394
patient advisory, 395
physical examination, 389–390
scapula, 387
treatments, 391–395
Tui Na treatments, 394–395
Western medicine etiology and
    pathology, 387
Right iliac region, 340, 341, 344
Right upper quadrant, 311–313
Rotator cuff tear, 228
Rotator cuff tendinopathy, 429

S

Scaphoid fractures, 514
Scapulohumeral periarthritis, see Frozen
    shoulder
Scapulothoracic (ST) joint, 422
Scarf test, 411, 429
Schepelmann's sign, 261
Secondary dysmenorrhea, 364, 366
Sensory disturbance, 227
Sensory nuclei, 102
Sexually transmitted diseases (STDs), 368
Shao Yang headache, 51–52, 61–64, 104
Shingle/chicken pox, 258
Shingles-related facial palsy, 199
Short head biceps injury
    acupuncture treatments, 450–451
    anatomy, 446
    auricular therapy, 451
    bleeding therapy, 451
    Chinese medicine etiology and
        pathology, 447
    cupping therapy, 451
    differential diagnosis, 448
    electro-acupuncture therapy, 451
    herbal treatments, 448–450
    manifestations, 447
    moxibustion therapy, 451
    patient advisory, 452
    physical examination, 447–448
    prognosis, 446
    Tui Na treatments, 451–452
    Western medicine etiology and pathology,
        446–447

Shoulder abduction 90° test, 463–464
Shoulder adhesive capsulitis, *see* Frozen shoulder
Shoulder Bi Syndrome, 403
Shoulder impingement syndrome, 228, 422
Shoulder pain
    acupuncture treatments, 416
    auricular therapy, 417
    bleeding therapy, 417
    causes, 403
    Chinese medicine etiology and pathology,
        405–406
    Chinese medicine mechanism
        Bi Syndrome, 403–405
        Wei syndrome, 405
    cupping therapy, 416
    frozen, 420–436
    long head biceps tendonitis, 452–460
    manifestations, 411–412
    moxibustion therapy, 416
    patient advisory, 420
    physical examination, 406–411
    short head biceps injury, 446–452
    subacromial bursitis, 469–476
    subluxation of long head biceps tendon,
        461–468
    supraspinatus tendonitis, 436–445
    Teding Diancibo Pu therapy, 417
    treatments, 412–419
    Tui Na treatments, 417–419
Shoulder withering syndrome, 403
Silent killer, 243
Sinus inflammation/infection, 168
Sinusitis
    acupuncture treatments, 178–179
    anatomy, 168
    auricular therapy, 179
    Chinese medicine etiology and pathology,
        168–169
    differential diagnosis, 171–172
    facial trauma, 167
    herbal treatments, 172
    inflammation, 167
    manifestations
        Chinese medicine, 170–171
        Western medicine, 169–170
    patient advisory, 180
    physical examination, 171
    treatments, 172–180
    Tui Na treatments, 179–180
    Western medicine etiology and
        pathology, 168
Six Meridian Headache, 48–55
Six Qi, 46
Six Yin
    characteristics
        Cold, 47
        Damp, 47–48

Dryness, 48
Fire, 48
Heat, 47
Wind, 46–47
        high-temperature environment, 46
Skin abnormalities, 346
Speed's test, 410–411, 428, 454, 463, 472
Sphenoidal sinuses, 168, 170
Sphenomandibular ligament, 144
Spinal cord tumors, 227
Spurling's sign, 226
Stabbing headache, 97
Stabbing pain, 447
Sternoclavicular (SC) joint, 421–422
Stiff neck
    acupuncture treatments, 215
    anatomy, 209
    bleeding therapy, 216
    cervical spondylosis, 220–240
    Chinese medicine etiology and pathology,
        209–210
    cupping therapy, 215–216
    differential diagnosis, 211
    electro-acupuncture therapy, 217
    herbal treatments, 212
    manifestations, 210–211
    moxibustion therapy, 216
    patient advisory, 219
    physical examination, 211
    treatments, 211–218
    Tui Na treatments, 217–218
    Western medicine etiology and
        pathology, 209
Stress factors, 145
Stroke-related facial palsy, 199
Student's elbow, *see* Medial epicondylitis;
        Olecranon bursitis
Stylomandibular ligament, 144
Subacromial bursitis
    acupuncture treatments, 474
    anatomy, 469
    auricular therapy, 474
    bleeding therapy, 474
    Chinese medicine etiology and
        pathology, 470
    cupping therapy, 474
    differential diagnosis, 472
    electro-acupuncture therapy, 474
    herbal treatments, 473
    manifestations, 471
    moxibustion therapy, 474
    patient advisory, 476
    physical examination, 471–472
    shoulder condition, 469
    Tui Na treatments, 474–476
    Western medicine etiology and
        pathology, 469

Subarachnoid hemorrhage, 109
Subcostal nerve, 258
Subdeltoid bursitis, 429
Subdural hematoma, 132
Subluxation of long head biceps tendon
    acupuncture treatments, 466–467
    anatomy, 461
    auricular therapy, 467
    bicipital groove, 461
    bleeding therapy, 467
    Chinese medicine etiology and
        pathology, 462
    cupping therapy, 467
    differential diagnosis, 464
    electro-acupuncture therapy, 467
    herbal treatments, 464–466
    manifestations, 462–463
    moxibustion therapy, 467
    patient advisory, 468
    physical examination, 463–464
    Tui Na treatments, 467–468
    Western medicine etiology and pathology,
        461–462
Substantial phlegm, 126
Superior thoracic aperture, 21–22
Suprapubic region, 340, 341, 343–344, 350–352
Suprascapular neuropathy, 441
Supraspinatus muscle, 437
Supraspinatus tendonitis
    acupuncture treatments, 443
    anatomy, 437
    auricular therapy, 443
    bleeding therapy, 444
    Chinese medicine etiology and
        pathology,438
    cupping therapy, 443
    differential diagnosis, 441
    electro-acupuncture therapy, 444
    four rotator cuff muscles, 436
    herbal treatments, 441
    manifestations, 438–439
    moxibustion therapy, 444
    patient advisory, 445
    physical examination, 439–440
    treatments, 441–445
    Tui Na treatments, 444–445
    Western medicine etiology and pathology,
        437–438
Sympathetic cervical spondylosis, 223, 225
Symptom severity scale (SS), 34
Synovial bursae, 422
Synovial cavities, 469
Synovial cyst, 542
Synovial fluid, 496
Synovial joint, 511
Synovial lining, 535

T

Tai Yang headache, 49–50, 55–58
Temporal headache, 61–64, 104
Temporomandibular joint dislocation
    acupuncture treatments, 166
    anatomy, 159
    Chinese medicine etiology and pathology, 161
    differential diagnosis, 163
    herbal treatments, 163
    ligamentous laxity, 159
    manifestations, 161
    patient advisory, 167
    physical examination, 161–162
    treatments, 163–167
    Tui Na treatments, 166–167
    Western medicine etiology and pathology,
        159–161
Temporomandibular joint (TMJ) disorders
    acupuncture treatments, 158
    anatomy, 143–144
    Chinese medicine etiology and pathology,
        145–147
    differential diagnosis, 150–151
    herbal treatments, 151
    lifestyle limitation, 143
    manifestations, 147–149
    patient advisory, 159
    physical examination, 149–150
    treatments, 151–159
    Tui Na treatments, 158–159
    Western medicine etiology and pathology, 145
Temporomandibular ligament, 144
Tendon Bi Syndrome, 537
Tendon lotion (linament), 10–11
Tendon sheath, 535, 536
Tennis elbow, 477
Tension headache, 98
Testicular torsion, 349
Thoracic aperture syndrome, *see* Thoracic outlet
        syndrome
Thoracic inlet syndrome, *see* Thoracic outlet
        syndrome
Thoracic outlet syndrome
    acupuncture treatments, 30
    anatomy, 21–23
    causes, 21
    Chinese medicine etiology and pathology,
        24–25
    differential diagnosis, 27
    herbal treatments, 27
    manifestations
        in Chinese medicine, 25–26
        in Western medicine, 25
    patient advisory, 31
    physical examination, 26
    tissues, 21

treatments, 27–28
Tui Na treatments, 30–31
Western medicine etiology and
    pathology, 23–24
Throbbing headache, 96
Thunderclap headache, 98
Traditional Chinese medicine (TCM),
    25, 45, 479
Transverse foramina, 221
Trapezius muscle
    acupuncture treatments, 384
    anatomy, 379
    bleeding therapy, 384
    Chinese medicine etiology and
        pathology, 380
    cupping therapy, 384
    differential diagnosis, 381–382
    electro-acupuncture therapy, 385
    herbal treatments, 382
    manifestations, 380
    moxibustion therapy, 384
    neck and shoulder pain, 379
    patient advisory, 386
    physical examination, 380–381
    procedures, 384–385
    treatment, 382–386
    Tui Na treatments, 385–386
    Western medicine etiology and pathology,
        379–380
Traumatic atlantoaxial rotatory dislocation, 243
Traumatic brain injury (TBI), 124, 125
Traumatic factors, 145
Traumatic injury, 243
Traumatic injury on tendons, 25, 479
Triangular fibrocartilage complex (TFCC),
    511, 519
Triceps brachii tendon, 498
Trigeminal ganglion, 103
Trigeminal neuralgia
    acupuncture treatments, 190–191
    anatomy, 181–182
    auricular therapy, 191
    Chinese medicine etiology and pathology,
        182–183
    differential diagnosis, 185
    facial pain, 181
    herbal treatments, 185
    manifestations, 183–184
    patient advisory, 192
    physical examination, 184–185
    treatments, 185–192
    Tui Na treatments, 191–192
    Western medicine etiology and
        pathology, 182
Trigger finger
    acupuncture treatments, 539
    anatomy, 535–536

auricular therapy, 540
Chinese medicine etiology & pathology, 537
differential diagnosis, 538
herbal treatments, 538–539
manifestations, 537
moxibustion therapy, 540
overuse, tendons, 535
patient advisory, 541
physical examination, 537–538
Tui Na treatments, 540–541
Western medicine etiology and pathology,
    536–537
Tui Na treatments
    abdominal pain, 289
    acute cervical fibrositis/stiff neck, 217–218
    Bell's palsy, 207
    carpal tunnel syndrome, 509
    cervical spondylosis, 237–239
    chest pain, 276–277
    costovertebral joint sprain, 400–402
    distal radioulnar joint, 524–525
    dysmenorrhea, 375–376
    epigastric pain, 308–310
    fibromyalgia syndrome, 40–42
    finger sprain, 552
    frozen shoulder, 434–435
    ganglion cysts, 545
    headache, 94–95
    hypochondriac pain, 323–324
    IBS, 337–339
    intercostal neuralgia, 267
    lateral epicondylitis, 486
    long head biceps tendonitis, 459–460
    lower abdominal pain, 359–360
    medial epicondylitis, 494–495
    migraine headache, 121–122
    olecranon bursitis, 501
    post-concussion syndrome, 141–143
    rhomboid muscle pain, 394–395
    short head biceps injury, 451–452
    shoulder pain, 417–419
    sinusitis, 179–180
    subacromial bursitis, 474–476
    subluxation of long head biceps tendon,
        467–468
    supraspinatus tendonitis, 444–445
    temporomandibular joint dislocation,
        166–167
    thoracic outlet syndrome, 30–31
    TMJ disorders, 158–159
    trapezius muscle, 385–386
    trigeminal neuralgia, 191–192
    trigger finger, 540–541
    upper cervical subluxation, 254–255
    wrist joint, 533–534
    wrist sprain, 517–518
Typical ribs, 257

U

Ulnar collateral ligament, 512
Uniaxial mechanical testing, 243
Upper back pain
  costovertebral joint sprain, 395–402
  inflammation of trapezius muscle, 379–386
  rhomboid muscle pain, 387–395
Upper cervical disharmony, *see* Upper cervical subluxation
Upper cervical disorder, 243
Upper cervical instability, 243
Upper cervical subluxation
  acupuncture treatments, 254
  anatomy, 240–243
  Chinese medicine etiology and pathology, 245–246
  differential diagnosis, 249
  herbal treatments, 249
  manifestations, 246–247
  neck bones, 240
  patient advisory, 255–256
  physical examination, 247–249
  treatments, 249–255
  Tui Na treatments, 254–255
  Western medicine etiology and pathology, 243–245
Upper motor neuron, 225
Upper quadrants, 290
Uterine contractions, 363
Uterus, 361–362

V

Vertebrae, 395
Vertebral subluxation (VS), 243
Vertebral subluxation complex (VSC), 243
Vertex headache, 64–66
Vessel Bi Syndrome, 6
Vestibulo-ocular examination, 130
Vision difficulties/dysfunction, 130
Volar plate, 546

W

Wandering Bi Syndrome, *see* Wind Bi Syndrome
Warming soak, 10
Water elbow, 495
Weather triggered headache, 101
Weekend headache, 100–101
Weight-bearing wrist movements, 513
Wei syndrome, 405
Widespread pain index (WPI), 34
Wind Bi Syndrome, 4–7
Wind-Cold-Damp type, 14–15
Wind damp headache, 81–83

Wind Heat headache, 80–81
Wind invading tendons, 26, 29–30
World Health Organization (WHO), 101
Wrist joint, 511
  acupuncture treatments, 533
  anatomy, 527
  auricular therapy, 533
  bleeding therapy, 533
  Chinese medicine etiology and pathology, 528
  chronic injury, 526
  cupping therapy, 533
  differential diagnosis, 530
  hand injury, 526
  herbal treatments, 530–532
  manifestations, 528–529
  moxibustion therapy, 533
  patient advisory, 534
  physical examination, 529–530
  Tui Na treatments, 533–534
  Western medicine etiology and pathology, 528
Wrist pain
  carpal tunnel syndrome, 503–510
  distal radioulnar joint, 519–526
  joint, 526–534
  sprain, 510–519
Wrist sprain
  acupuncture treatments, 517
  anatomy, 511–512
  auricular therapy, 517
  bleeding therapy, 517
  Chinese medicine etiology and pathology, 512
  differential diagnosis, 514
  herbal treatments, 515–517
  ligaments, 510
  manifestations, 513
  moxibustion therapy, 517
  patient advisory, 518–519
  physical examination, 513–514
  Tui Na treatments, 517–518
  Western medicine etiology and pathology, 512

X

Xiphoid process, 258

Y

Yang Ming headache, 50–51, 59–61
Yergason test, 411, 428, 454

Z

Zoom fatigue syndrome, 2

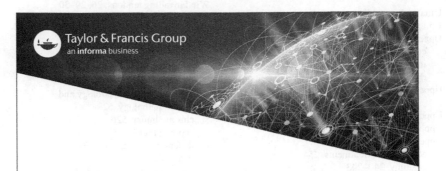

Printed in the United States
by Baker & Taylor Publisher Services